Graduate Texts in Mathematics 168

Springer
New York
Berlin
Heidelberg
Barcelona
Budapest
Hong Kong
London
Milan
Paris
Santa Clara
Singapore
Tokyo

Graduate Texts in Mathematics

1 TAKEUTI/ZARING. Introduction to Axiomatic Set Theory. 2nd ed.
2 OXTOBY. Measure and Category. 2nd ed.
3 SCHAEFER. Topological Vector Spaces.
4 HILTON/STAMMBACH. A Course in Homological Algebra.
5 MAC LANE. Categories for the Working Mathematician.
6 HUGHES/PIPER. Projective Planes.
7 SERRE. A Course in Arithmetic.
8 TAKEUTI/ZARING. Axiomatic Set Theory.
9 HUMPHREYS. Introduction to Lie Algebras and Representation Theory.
10 COHEN. A Course in Simple Homotopy Theory.
11 CONWAY. Functions of One Complex Variable I. 2nd ed.
12 BEALS. Advanced Mathematical Analysis.
13 ANDERSON/FULLER. Rings and Categories of Modules. 2nd ed.
14 GOLUBITSKY/GUILLEMIN. Stable Mappings and Their Singularities.
15 BERBERIAN. Lectures in Functional Analysis and Operator Theory.
16 WINTER. The Structure of Fields.
17 ROSENBLATT. Random Processes. 2nd ed.
18 HALMOS. Measure Theory.
19 HALMOS. A Hilbert Space Problem Book. 2nd ed.
20 HUSEMOLLER. Fibre Bundles. 3rd ed.
21 HUMPHREYS. Linear Algebraic Groups.
22 BARNES/MACK. An Algebraic Introduction to Mathematical Logic.
23 GREUB. Linear Algebra. 4th ed.
24 HOLMES. Geometric Functional Analysis and Its Applications.
25 HEWITT/STROMBERG. Real and Abstract Analysis.
26 MANES. Algebraic Theories.
27 KELLEY. General Topology.
28 ZARISKI/SAMUEL. Commutative Algebra. Vol.I.
29 ZARISKI/SAMUEL. Commutative Algebra. Vol.II.
30 JACOBSON. Lectures in Abstract Algebra I. Basic Concepts.
31 JACOBSON. Lectures in Abstract Algebra II. Linear Algebra.
32 JACOBSON. Lectures in Abstract Algebra III. Theory of Fields and Galois Theory.
33 HIRSCH. Differential Topology.
34 SPITZER. Principles of Random Walk. 2nd ed.
35 WERMER. Banach Algebras and Several Complex Variables. 2nd ed.
36 KELLEY/NAMIOKA et al. Linear Topological Spaces.
37 MONK. Mathematical Logic.
38 GRAUERT/FRITZSCHE. Several Complex Variables.
39 ARVESON. An Invitation to C^*-Algebras.
40 KEMENY/SNELL/KNAPP. Denumerable Markov Chains. 2nd ed.
41 APOSTOL. Modular Functions and Dirichlet Series in Number Theory. 2nd ed.
42 SERRE. Linear Representations of Finite Groups.
43 GILLMAN/JERISON. Rings of Continuous Functions.
44 KENDIG. Elementary Algebraic Geometry.
45 LOÈVE. Probability Theory I. 4th ed.
46 LOÈVE. Probability Theory II. 4th ed.
47 MOISE. Geometric Topology in Dimensions 2 and 3.
48 SACHS/WU. General Relativity for Mathematicians.
49 GRUENBERG/WEIR. Linear Geometry. 2nd ed.
50 EDWARDS. Fermat's Last Theorem.
51 KLINGENBERG. A Course in Differential Geometry.
52 HARTSHORNE. Algebraic Geometry.
53 MANIN. A Course in Mathematical Logic.
54 GRAVER/WATKINS. Combinatorics with Emphasis on the Theory of Graphs.
55 BROWN/PEARCY. Introduction to Operator Theory I: Elements of Functional Analysis.
56 MASSEY. Algebraic Topology: An Introduction.
57 CROWELL/FOX. Introduction to Knot Theory.
58 KOBLITZ. p-adic Numbers, p-adic Analysis, and Zeta-Functions. 2nd ed.
59 LANG. Cyclotomic Fields.
60 ARNOLD. Mathematical Methods in Classical Mechanics. 2nd ed.

continued after index

Günter Ewald

Combinatorial Convexity and Algebraic Geometry

With 130 Illustrations

Günter Ewald
Fakultät für Mathematik
Ruhr-Universität Bochum
Universitätsstrasse 150
D-44780 Bochum
Germany

Mathematics Subject Classification (1991): 52-01, 14-01

Ewald, Günter, 1929–
Combinatorial convexity and algebraic geometry / Günter Ewald.
p. cm.—(Graduate texts in mathematics; 168)
Includes bibliographical references and index.
ISBN 0-387-94755-8 (hard: alk. paper)
1. Combinatorial geometry. 2. Toric varieties. 3. Geometry.
Algebraic. I. Title. II. Series.
QA639.5.E93 1996
516′.08—dc20 96-11792

Printed on acid-free paper.

Production managed by Lesley Poliner; manufacturing supervised by Johanna Tschebull.
Photocomposed pages prepared from the author's TEX files.
Printed and bound by R.R. Donnelley and Sons, Harrisonburg, VA.
Printed in the United States of America.

9 8 7 6 5 4 3 2 1

ISBN 0-387-94755-8 Springer-Verlag New York Berlin Heidelberg SPIN 10424913

To Hanna
and our children
Daniel, Sarah, Anna, Esther, David

Preface

The aim of this book is to provide an introduction for students and nonspecialists to a fascinating relation between combinatorial geometry and algebraic geometry, as it has developed during the last two decades. This relation is known as the theory of toric varieties or sometimes as torus embeddings.

Chapters I–IV provide a self-contained introduction to the theory of convex polytopes and polyhedral sets and can be used independently of any applications to algebraic geometry. Chapter V forms a link between the first and second part of the book. Though its material belongs to combinatorial convexity, its definitions and theorems are motivated by toric varieties. Often they simply translate algebraic geometric facts into combinatorial language. Chapters VI–VIII introduce toric varieties in an elementary way, but one which may not, for specialists, be the most elegant.

In considering toric varieties, many of the general notions of algebraic geometry occur and they can be dealt with in a concrete way. Therefore, Part 2 of the book may also serve as an introduction to algebraic geometry and preparation for farther reaching texts about this field.

The prerequisites for both parts of the book are standard facts in linear algebra (including some facts on rings and fields) and calculus. Assuming those, all proofs in Chapters I–VII are complete with one exception (IV, Theorem 5.1). In Chapter VIII we use a few additional prerequisites with references from appropriate texts.

The book covers material for a one year graduate course. For shorter courses with emphasis on algebraic geometry, it is possible to start with Part 2 and use Part 1 as references for combinatorial geometry.

For each section of Chapters I–VIII, there is an addendum in the appendix of the book. In order to avoid interruptions and to minimize frustration for the beginner, comments, historical notes, suggestions for further reading, additional exercises, and, in some cases, research problems are collected in the Appendix.

Acknowledgments

This text is based on lectures I gave several times at Bochum University. Many collegues and students have contributed to it in one way or another.

There are seven people to whom I owe special thanks. Jerzy Jurkiewicz (Warsaw) gave me much advice and help in an early stage of the writing. Gottfried Barthel and Ludger Kaup (Konstanz) thoroughly analyzed and corrected large parts of the first six chapters, and even rewrote some of the sections. In a later stage, Jaroslav Włodarczyk (Warsaw) worked out strong improvements of Chapters VI and VII. Robert J. Koelman prepared the illustrations by computer. Finally, Bernd Kind suggested many changes and in addition has supervised the production of the text and patiently solved all arising technical problems.

Also Markus Eikelberg, Rolf Gärtner, Ralph Lehmann, and Uwe Wessels made important contributions. Michel Brion, Dimitrios Dais, Bernard Teissier, Günter Ziegler added remarks, and Hassan Azad, Katalin Bencsath, Peter Braß, Sharon Castillo, Reinhold Matmann, David Morgan, and Heinke Wagner made corrections to the text. Elke Lau and Elfriede Rahn did the word processing of the computer text.

I thank all who helped me, in particular, those who are not mentioned by name.

Günter Ewald

Contents

Introduction

Studying the complex zeros of a polynomial in several variables reveals that there are properties which depend not on the specific values of the coefficients but only on their being nonzero. They depend on the exponent vectors showing up in the polynomial or, more precisely, on the lattice polytope which is the convex hull of such vectors. This had already been discovered by Newton and was taken into consideration by Minding and some other mathematicians in the nineteenth century. However, it had practically been forgotten until its rediscovery around 1970, when Demazure, Oda, Mumford, and others developed the theory of toric varieties.

The starting point lay in algebraic groups. Properties of zeros of polynomials that depend only on the exponent vectors do not change if each coordinate of any solution is multiplied by a nonvanishing constant. Such transformations are effected by diagonal matrices with nonzero determinants. They form a group which can be represented by $\mathbb{C}^{*n}$ where $\mathbb{C}^* := \mathbb{C} \setminus \{0\}$ is the multiplicative group of complex numbers. $\mathbb{C}^{*n}$ (for $n = 2$ having, topologically, an ordinary torus as a retract) is called an algebraic torus. Demazure succeeded in combinatorially characterizing those regular algebraic varieties on which a torus operates with an open orbit. Oda, Mumford, and others extended this to the nonregular case and termed the introduced varieties torus embeddings or toric varieties.

Once the combinatorial characterization had been achieved, it gave way to defining toric varieties without starting from algebraic groups by use of combinatorial concepts like lattice cones and the algebras defined by monoids of all lattice points in cones. This is the path we follow in the present book.

Toric varieties—being a class of relatively concrete algebraic varieties—may appear to relate combinatorics to old-fashioned, say, up to 1950, algebraic geometry. This is not the case. Actually, the more recent way of thought provides the tools for building a wide bridge between combinatorial and algebraic geometry. Notions like sheaves, blowups, or the use of homology in algebraic geometry are such tools.

In the first part of the book, we have naturally limited the topics to those which are needed in the second part. However, there was not much to be omitted. Coming from combinatorial convexity, it is quite a surprise how many of the traditional notions like support function or mixed volume now appear in a new light.

In our attempt to present a compact introduction to the theory of convex polytopes, we have sought short proofs. Also, a coordinate-free approach to Gale transforms seemed to fit particularly well into the needs of later applications. Similarly, in Part 2 we spent much energy on simplifications. Our definition of intersection numbers and a discussion of the Hodge inequality working without the tools of algebraic topology are some of the consequences.

A natural question concerning the relationship between combinatorial and algebraic geometry is "Does the algebraic geometric side benefit more from the combinatorial side than the combinatorial side does from the algebraic geometric one?" In this text the former is true. We prove algebraic geometric theorems from combinatorial geometric facts, "turning around" the methods often applied in the literature. There is only one exception in the very last section of the book. We quote a toric version of the Riemann–Roch–Hirzebruch theorem without proof and draw combinatorial conclusions from it. A purely combinatorial version of the theorem due to Morelli [1993a] would require more work on so-called polytope algebra.

Many related topics have been omitted, for example, matroid theory or the theory of Stanley–Reisner rings and their powerful combinatorial implications. The reader familiar with such topics may recognize their links to those covered here and detect the common spirit of mathematical development in all of them.

Part 1

Combinatorial Convexity

I

Convex Bodies

1. Convex sets

Most of the sets considered in the first part of the book are subsets of Euclidean n-space. Many definitions and theorems could be stated in an affinely invariant manner. We do not, however, stress this point. If we use the symbol $\mathbb{R}^n$, it should be clear from the context whether we mean real vector space, real affine space, or Euclidean space. In the latter case, we assume the ordinary scalar product

$$\langle x, y\rangle = \xi_1\eta_1 + \cdots + \xi_n\eta_n \qquad \text{for } x = (\xi_1, \ldots, \xi_n), \quad y = (\eta_1, \ldots, \eta_n)$$

so that the square of Euclidean distance between points x and y equals

$$\|x - y\|^2 = \langle x - y, x - y\rangle.$$

Recall that an open ball with center x and radius r is the set $\{y \mid \|x - y\| < r\}$. By $\langle K, y\rangle \geq 0$, we mean $\langle x, y\rangle \geq 0$ for every $x \in K$. We assume the reader to be somewhat familiar with n-dimensional affine and Euclidean geometry.

1.1 Definition. A set $C \subset \mathbb{R}^n$ is called *convex* if, for all $x, y \in C$, $x \neq y$, the line segment

$$[x, y] := \{\lambda x + (1 - \lambda)y \mid 0 \leq \lambda \leq 1\}$$

is contained in C (Figure 1).

Examples of convex sets are a point, a line, a circular disc in $\mathbb{R}^2$, the platonic solids (see Figure 10 in section 6) in $\mathbb{R}^3$. Also $\emptyset$ and $\mathbb{R}^n$ are convex.

If B is an open circular disc in $\mathbb{R}^2$ and M is any subset of the boundary circle ∂B of B, then $B \cup M$ is also convex. So, a convex set need be neither open nor closed. In general we shall restrict ourselves to closed convex sets.

There is a simple way to construct new convex sets from given ones:

1.2 Lemma. *The intersection of an arbitrary collection of convex sets is convex.*

PROOF. If a line segment is contained in every set of the collection, it is also contained in their intersection. □

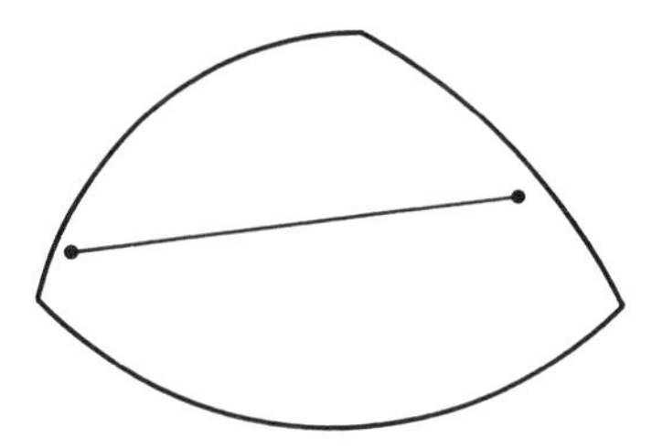

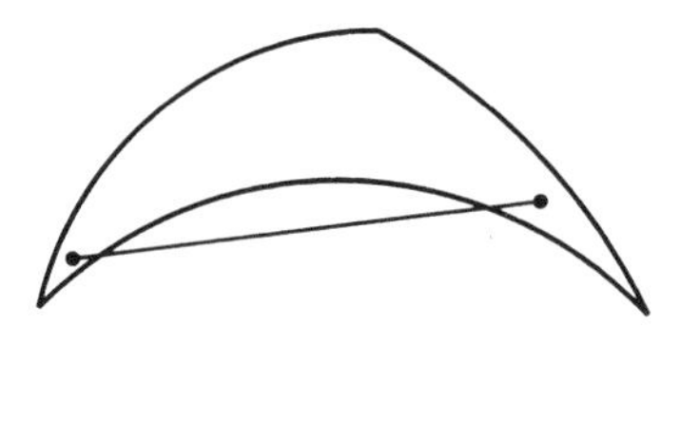

FIGURE 1. Left: convex. Right: nonconvex.

1.3 Definition. We say x is a *convex combination* of $x_1, \ldots, x_r \in \mathbb{R}^n$ if there exist $\lambda_1, \ldots, \lambda_r \in \mathbb{R}$ such that

$$x = \lambda_1 x_1 + \cdots + \lambda_r x_r, \tag{1}$$

$$\lambda_1 + \cdots + \lambda_r = 1, \tag{2}$$

$$\lambda_1 \geq 0, \ldots, \lambda_r \geq 0. \tag{3}$$

If condition (3) is dropped, we have an *affine combination* of $x_1, \ldots, x_r$, and $x, x_1, \ldots, x_r$ are called *affinely dependent*. If $x, x_1, \ldots, x_r$ are not affinely dependent, we say they are *affinely independent*.

So, convex combinations are special affine combinations (Figure 2).

If $x_1, \ldots, x_r$ are affinely independent, the numbers $\lambda_1, \ldots, \lambda_r$ are sometimes called *barycentric coordinates* of x (with respect to the *affine basis* $x_1, \ldots, x_r$).

1.4 Definition. The set of all convex combinations of elements of a set $M \subset \mathbb{R}^n$ is called the *convex hull*

$$\operatorname{conv} M$$

of M; in particular, $\operatorname{conv} \emptyset = \emptyset$. Analogously, the set of all affine combinations of elements of M is called the *affine hull*

$$\operatorname{aff} M$$

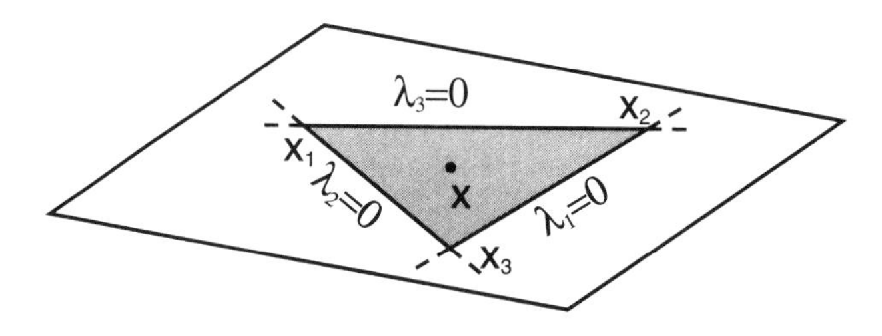

FIGURE 2.

of M. We will denote by lin M (*linear hull*) the linear space generated by M. It is the "smallest" linear space containing M.

If $M = \{x_1, \ldots, x_r\}$ is a finite set, we say $P := \text{conv}\, M$ is a *convex polytope*, or simply a *polytope*.

If $x_1, \ldots, x_r$ are affinely independent, we say

$$T_{r-1} := \text{conv}\{x_1, \ldots, x_r\}$$

is an $(r-1)$-*simplex* or, briefly, a *simplex*. aff T_{r-1} and T_{r-1} are said to have *dimension* $r-1$.

Remarks.
(1) Clearly, $M \subset \text{conv}\, M \subset \text{aff}\, M$.
(2) Every polytope is compact (that is, bounded and closed).

1.5 Theorem.
(a) *A set $M \subset \mathbb{R}^n$ is convex if and only if it contains all its convex combinations, that is, if and only if*

$$M = \text{conv}\, M.$$

(b) *The convex hull of $M \subset \mathbb{R}^n$ is the smallest convex set that contains M; this means $M \subset M'$ and M' convex imply* conv $M \subset M'$.

PROOF. First, we will show that conv M is convex.

If $x, y \in \text{conv}\, M$, there exist $x_1, \ldots, x_r, y_1, \ldots, y_s \in M$ and real numbers $\lambda_1, \ldots, \lambda_r, \mu_1, \ldots, \mu_s$ such that

$$x = \lambda_1 x_1 + \cdots + \lambda_r x_r, \quad \lambda_1 + \cdots + \lambda_r = 1, \quad \lambda_1 \geq 0, \ldots, \lambda_r \geq 0$$

and

$$y = \mu_1 y_1 + \cdots + \mu_s y_s, \quad \mu_1 + \cdots + \mu_s = 1, \quad \mu_1 \geq 0, \ldots, \mu_s \geq 0.$$

Employing 0 coefficients, if necessary, we may assume $r = s$ and $y_j = x_j$, $j = 1, \ldots, r$. For arbitrary $0 \leq \lambda \leq 1$,

$$\begin{aligned}\lambda x + (1-\lambda)y &= \lambda(\lambda_1 x_1 + \cdots + \lambda_r x_r) + (1-\lambda)(\mu_1 x_1 + \ldots + \mu_r x_r) \\ &= [\lambda\lambda_1 + (1-\lambda)\mu_1]x_1 + \cdots + [\lambda\lambda_r + (1-\lambda)\mu_r]x_r.\end{aligned}$$

Since all coefficients are nonnegative, and since

$$\lambda\lambda_1 + (1-\lambda)\mu_1 + \cdots + \lambda\lambda_r + (1-\lambda)\mu_r = \lambda + 1 - \lambda = 1,$$

$\lambda x + (1-\lambda)y$ is a convex combination of $x_1, \ldots, x_r$. So, conv M is convex and, in view of Remark 1, we obtain (a).

Now, to see (b), suppose M' is a convex set, $M' \supset M$, and that $x \in \text{conv}\, M$. Then there exist $x_1, \ldots x_r \in M$ such that $x = \lambda_1 x_1 + \cdots + \lambda_r x_r$, $\lambda_1 + \cdots + \lambda_r =$

1, and $\lambda_1, \dots, \lambda_r$ all > 0. Since $x_1, \dots, x_r \in M'$ as well, we find successively

$$\begin{aligned} y_1 &:= \lambda_1(\lambda_1+\lambda_2)^{-1}x_1 + \lambda_2(\lambda_1+\lambda_2)^{-1}x_2 \\ y_2 &:= (\lambda_1+\lambda_2)(\lambda_1+\lambda_2+\lambda_3)^{-1}y_1 + \lambda_3(\lambda_1+\lambda_2+\lambda_3)^{-1}x_3 \\ &\ \vdots \\ x &= (\lambda_1+\cdots+\lambda_{r-1})(\lambda_1+\cdots+\lambda_r)^{-1}y_{r-2} + \lambda_r(\lambda_1+\cdots+\lambda_r)^{-1}x_r \end{aligned}$$

which are all in M', hence, conv $M \subset M'$. □

1.6 Definition. If C is a convex set, we call

$$\dim C := \dim(\operatorname{aff} C)$$

the *dimension* of C. By convention, $\dim \emptyset = -1$.

1.7 Definition. A compact convex set C is called a *convex body*.

For example, note that points and line segments are convex bodies in $\mathbb{R}^n$, $n \geq 1$, so that a convex body in $\mathbb{R}^n$ need not have dimension n.

1.8 Definition. We say $x \in M \subset \mathbb{R}^n$ is in the *relative interior* of M, $x \in \operatorname{relint} M$, if x is in the interior of M relative to aff M (that is, there exists an open ball B in aff M such that $x \in B \subset M$). If aff $M = \mathbb{R}^n$, then relint $M =:$ int M (note that relint $\mathbb{R}^0 = \operatorname{int} \mathbb{R}^0 = \{0\}$).

Our main emphasis will be on convex polytopes and an unbounded counterpart of polytopes, called polyhedral cones:

1.9 Definition. If $M \subset \mathbb{R}^n$, the set of all nonnegative linear combinations

$$x = \lambda_1 y_1 + \cdots + \lambda_k y_k, \qquad y_1, \dots, y_k \in M, \quad \lambda_1 \geq 0, \dots, \lambda_k \geq 0$$

of elements of M is called the *positive hull*

$$\sigma := \operatorname{pos} M$$

of M or the *cone* determined by M. By convention, pos $\emptyset := \{0\}$.

For fixed $u \in \mathbb{R}^n$, $u \neq 0$, and $\alpha \in \mathbb{R}$, the set $H := \{x \mid \langle x, u\rangle = \alpha\}$ is a hyperplane. $H^+ := \{x \mid \langle x, u\rangle \geq \alpha\}$ and $H^- := \{x \mid \langle x, u\rangle \leq \alpha\}$ are called the *half-spaces bounded by* H. If $\sigma \subset H^+$ and $\alpha = 0$, we say σ has an *apex*, namely 0. (We use the symbol 0 for the number 0, the zero vector, and the origin).

If $M = \{x_1, \dots, x_r\}$ is finite, we call

$$\sigma = \operatorname{pos}\{x_1, \dots, x_r\}$$

a *polyhedral cone*. Unless otherwise stated, by a *cone* we always mean a polyhedral cone. Sometimes we write

$$\sigma = \mathbb{R}_{\geq 0}\, x_1 + \cdots + \mathbb{R}_{\geq 0}\, x_r,$$

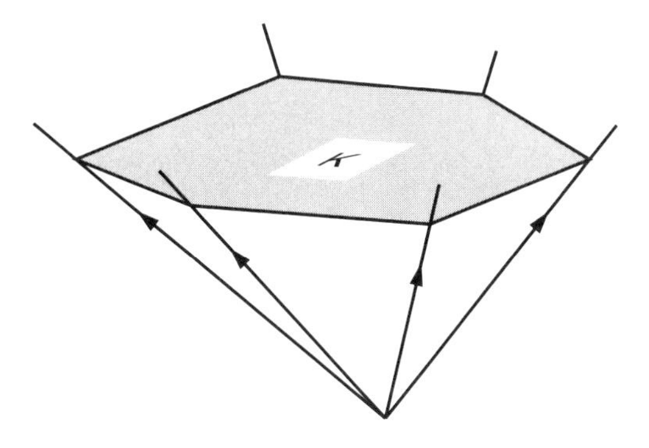

FIGURE 3.

$\mathbb{R}_{\geq 0}$ denoting the set of nonnegative real numbers.

Example. A quadrant in $\mathbb{R}^2$ and an octant in $\mathbb{R}^3$ are cones with an apex, whereas a closed half-space or the intersection of two closed half-spaces H_1^+, H_2^+ with $0 \in H_1, 0 \in H_2$ in $\mathbb{R}^3$, are cones without apex.

Since convex combinations are, by definition, nonnegative linear combinations, we have

1.10 Lemma. *The positive hull of any set M is convex.*

Figure 3 illustrates a polyhedral cone of dimension three which is the positive hull of a two-dimensional polytope K. Though pos M might generally be called a cone, we reserve this term for polyhedral cones.

Exercises

1. The convex hull of any compact (closed and bounded) set is again compact.
2. Find an example of a closed set M such that conv M is not closed.
3. Determine all convex subsets C of $\mathbb{R}^3$, for which $\mathbb{R}^3 \setminus C$ is also convex. (Except $\emptyset$, $\mathbb{R}^3$ there are, up to three such sets of affine transformations, that is, translations combined with linear maps.
4. Call a set M ϵ-convex if, for a given $\epsilon > 0$, each ball with radius ϵ and center in M intersects M in a convex set. Furthermore, call a set M connected if any two of its points can be joined by a rectifiable arc (as is defined in calculus) contained in M. Prove: (a) Any ϵ-convex closed connected set M in $\mathbb{R}^2$ is convex. (b) Statement (a) is false without the assumption of M being connected.

2. Theorems of Radon and Carathéodory

The following theorem is helpful when handling convex combinations.

2.1 Theorem (Radon's Theorem). *Let* $M = \{x_1, \dots, x_r\} \subset \mathbb{R}^n$ *be an arbitrary finite set, and let* M_1, M_2 *be a partition of* M*, that is,* $M = M_1 \cup M_2$, $M_1 \cap M_2 = \emptyset$, $M_1 \neq \emptyset$, $M_2 \neq \emptyset$.

(a) *If* $r \geq n + 2$ *then the partition can be chosen such that*

$$\operatorname{conv} M_1 \cap \operatorname{conv} M_2 \neq \emptyset.$$

(b) *If* $r \geq n + 1$ *and* 0 *is an apex of* pos M*, yet* $0 \notin M$ *or* $r \geq n + 2$*, then the partition can be chosen such that*

$$\operatorname{pos} M_1 \cap \operatorname{pos} M_2 \neq \{0\}.$$

(c) *The partition is unique if and only if, in case* (a), $r = n + 2$ *and any* $n + 1$ *points of* M *are affinely independent, in case* (b), $r = n + 1$ *and any* n *points of* M *are linearly independent.*

2.2 Definition. We call M_1, M_2 in Theorem 2.1 a *Radon partition* of M.

PROOF OF THEOREM 2.1.

(a) From $r \geq n + 2$, it follows that $x_1, \dots, x_r$ are affinely dependent. Hence

$$\lambda_1 x_1 + \cdots + \lambda_r x_r = 0 \text{ can hold with } \lambda_1 + \cdots + \lambda_r = 0, \text{ not all } \lambda_i = 0.$$

We may assume that, for a particular $j, 0 < j < r$,

$$\lambda_1 > 0, \dots, \lambda_j > 0; \quad \lambda_{j+1} \leq 0, \dots, \lambda_r \leq 0$$

We set

$$\lambda := \lambda_1 + \cdots + \lambda_j = -\lambda_{j+1} - \cdots - \lambda_r > 0 \qquad \text{and}$$
$$x := \lambda^{-1}(\lambda_1 x_1 + \cdots + \lambda_j x_j) = -\lambda^{-1}(\lambda_{j+1} x_{j+1} + \cdots + \lambda_r x_r).$$

Then, $x \in \operatorname{conv} M_1 \cap \operatorname{conv} M_2$ for

$$M_1 := \{x_1, \dots, x_j\}, \quad M_2 := \{x_{j+1}, \dots, x_r\}.$$

(b) By definition of an apex, there exists a hyperplane H such that $H \cap \operatorname{pos} M = \{0\}$ and $\operatorname{pos} M \subset H^+$. Let $H' \neq H$ be parallel to H and $H' \cap M \neq \emptyset$. Then, for any $x_j \in M$, the ray $\operatorname{pos}\{x_j\}$ intersects H' in a point x'_j. We apply (a) to $M' := \{x'_1, \dots, x'_r\}$ relative to the $(n-1)$-space H' and find a partition of M' into $M'_1 := \{x'_1, \dots, x'_j\}$, $M'_2 = \{x'_{j+1}, \dots, x'_r\}$ such that $\operatorname{conv} M'_1 \cap \operatorname{conv} M'_2 \neq \emptyset$. Now for $M_1 := \{x_1, \dots, x_j\}$, $M_2 := \{x_{j+1}, \dots, x_r\}$, we find

$$\operatorname{pos} M_1 \cap \operatorname{pos} M_2 \neq \{0\}.$$

(c) We prove the uniqueness only in case (a); case (b) is proved similarly.

First, assume $r = n+2$ and no $n+1$ points are affinely dependent. Suppose that

$$\tilde{M}_1 = \{x_{i_1}, \ldots, x_{i_k}\}, \quad \tilde{M}_2 = \{x_{i_{k+1}}, \ldots, x_{i_{n+2}}\}$$

is a second Radon partition of M and

$$y \in \operatorname{conv} \tilde{M}_1 \cap \operatorname{conv} \tilde{M}_2.$$

Then,

$$y = \mu^{-1}(\mu_1 x_{i_1} + \cdots + \mu_k x_{i_k}) = -\mu^{-1}(\mu_{k+1} x_{i_{k+1}} + \cdots + \mu_{n+2} x_{i_{n+2}})$$

where $\mu_1 > 0, \ldots, \mu_k > 0$; $\mu_{k+1} \leq 0, \ldots, \mu_{n+2} \leq 0$; $k \geq 1$, and $\mu = \mu_1 + \cdots + \mu_k = -\mu_{k+1} - \cdots - \mu_{n+2}$. We may assume

$$x_{i_1} = x_{j+1} \quad (\in M_2).$$

We choose $0 < \alpha < 1$ such that

$$\alpha\lambda^{-1}\lambda_{j+1} + (1-\alpha)\mu^{-1}\mu_1 = 0.$$

Then,

$$\begin{aligned}&\alpha\lambda^{-1}(\lambda_1 x_1 + \cdots + \lambda_{n+2} x_{n+2}) \\ &\quad + (1-\alpha)\mu^{-1}(\mu_1 x_{i_1} + \cdots + \mu_{n+2} x_{i_{n+2}}) = 0 + 0 = 0\end{aligned}$$

and

$$\alpha\lambda^{-1}(\lambda_1 + \cdots + \lambda_{n+2}) + (1-\alpha)\mu^{-1}(\mu_1 + \cdots + \mu_{n+2}) = 0$$

expresses an affine relation between $n+1$ of the points of M (x_{i_1} and x_{j+1} cancel out), unless all coefficients vanish. Therefore, $\lambda_\varrho = -\alpha^{-1}(1-\alpha)\lambda\mu^{-1}\mu_{i_\varrho}$, $\varrho = 1, \ldots, n+2$, and there is a map $\varrho \mapsto \varrho'$, $\varrho \in \{1, \ldots, j, j+2, \ldots, n+2\}$, $\varrho' \in \{i_2, \ldots, n+2\}$ such that $\lambda_\varrho = -\alpha^{-1}(1-\alpha)\lambda\mu_{\varrho'}$. Since $\alpha^{-1} > 0$, $1-\alpha > 0$, and $\lambda > 0$, the set of those ϱ' for which $\mu_{\varrho'} < 0$ is the same as the set of those ϱ for which $\lambda_\varrho > 0$. Therefore $M_1 = \{x_1, \ldots, x_j\} = \{x_{i_{k+1}}, \ldots, x_{i_{n+2}}\} = \tilde{M}_2$ and consequently $M_2 = \tilde{M}_1$, too.

To prove the converse, we distinguish two cases.

(I) $r = n+2$, and $x_1, \ldots, x_{n+1}$ are affinely dependent, $\overset{\circ}{M} := \{x_1, \ldots, x_{n+1}\}$.
(II) $r > n+2$.

In case I, $\overset{\circ}{M}$ is contained in a hyperplane so that, by (a), we find a partition of $\overset{\circ}{M}$ into $\overset{\circ}{M}_1, \overset{\circ}{M}_2$ with $\operatorname{conv} \overset{\circ}{M}_1 \cap \operatorname{conv} \overset{\circ}{M}_2 \neq \emptyset$. Then, $\overset{\circ}{M}_1 \cup \{x_{n+2}\}, \overset{\circ}{M}_2$ and $\overset{\circ}{M}_1, \overset{\circ}{M}_2 \cup \{x_{n+2}\}$ are two different Radon partitions of M.

In case II, consider a proper subset $\tilde{M}$ of M which has at least $n+2$ points. Let $\tilde{M}_1, \tilde{M}_2$ be a Radon partition of $\tilde{M}$. Then, $\tilde{M}_1 \cup (M \setminus \tilde{M}), \tilde{M}_2$ and $\tilde{M}_1, \tilde{M}_2 \cup (M \setminus \tilde{M})$ are different Radon partitions of M. □

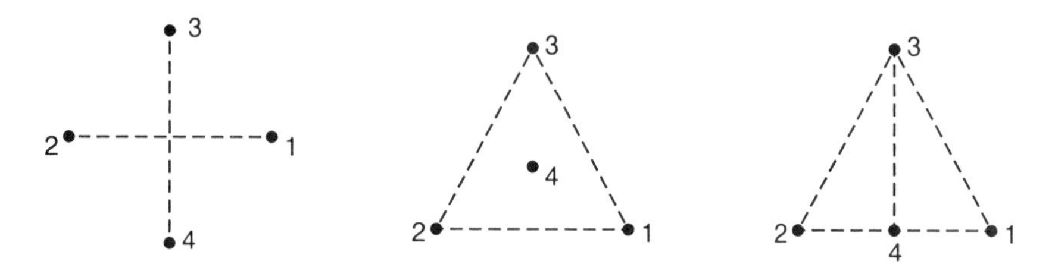

FIGURE 4. Left: $M_1 = \{1, 2\}$; $M_2 = \{3, 4\}$. Middle: $M_1 = \{1, 2, 3\}$; $M_2 = \{4\}$. Right: $M_1 = \{1, 2\}$; $M_2 = \{3, 4\}$, $M_1 = \{1, 2, 3\}$; $M_2 = \{4\}$

Examples for $M = \{1, 2, 3, 4\}$

2.3 Theorem (Carathéodory's theorem).

(a) *The convex hull* conv M *of a set* $M \subset \mathbb{R}^n$ *is the union of all convex hulls of subsets of* M *containing at most* $n + 1$ *elements.*

(b) *The positive hull* pos M *of a set* $M \subset \mathbb{R}^n$ *is the union of all positive hulls of subsets of* M *containing at most* n *elements of* M.

PROOF.

(a) Let

(1)
$$x = \lambda_1 x_1 + \cdots + \lambda_r x_r \in \operatorname{conv} M,$$

and let r be the smallest number of elements of M of which x is a convex combination. Contrary to the claim, $r \geq n + 2$ implies that there exists an affine relation

(2)
$$\mu_1 x_1 + \cdots + \mu_r x_r = 0, \quad \text{with } \mu_1 + \cdots + \mu_r = 0, \text{ but not all } \mu_j = 0.$$

For $\mu_j \neq 0$, we obtain from (1) and (2)

(3)
$$x = \lambda_1 x_1 + \cdots + \lambda_r x_r = \left(\lambda_1 - \frac{\lambda_j}{\mu_j}\mu_1\right) x_1 + \cdots + \left(\lambda_r - \frac{\lambda_j}{\mu_j}\mu_r\right) x_r.$$

We may assume $\mu_j > 0$, and, for all $\mu_k > 0$, $k = 1, \ldots, r$,

$$\frac{\lambda_j}{\mu_j} \leq \frac{\lambda_k}{\mu_k}.$$

Then,

$$\lambda_i - \frac{\lambda_j}{\mu_j}\mu_i \geq 0 \qquad \text{for } i = 1, \ldots, r.$$

Since $\lambda_j - \frac{\lambda_j}{\mu_j}\mu_j = 0$, equation (3) expresses x as a convex combination of less than r elements of M, a contradiction of the initial assumption.

(b) Replace in the proof of (a) "convex combination" by "positive linear combination" and "affine dependence of $n+1$ elements" by "linear dependence of n elements" to obtain a proof of (b).

□

Exercises

1. In analogy to the above examples in Figure 4, find all types of Radon partitions of $n+2$ points in $\mathbb{R}^n$ whose affine hull is $\mathbb{R}^n$.
2. If aff $M = \mathbb{R}^n$, then, conv M is the union of n-simplices with vertices in M.
3. Every n-dimensional convex polytope is the union of finitely many simplices, no two of which have an interior point in common.
4. Helly's Theorem. Suppose every $n+1$ of the convex sets $K_1, \ldots, K_m$ in $\mathbb{R}^n$ has a nonempty intersection, $m \geq n+1$. Then, $\bigcap_{i=1}^m K_i \neq \emptyset$. (Hint: For $m = n+1$ there is nothing to prove. Apply induction on m and use Radon's Theorem).

3. Nearest point map and supporting hyperplanes

Quite a few properties of a closed convex set K can be studied by using the map that assigns to each point in $\mathbb{R}^n$ its nearest point on K. First, we show that this map is well defined.

3.1 Lemma. *Let K be a closed convex set in $\mathbb{R}^n$. To each $x \in \mathbb{R}^n$ there exists a unique $x' \in K$ such that*

$$(*) \qquad \|x - x'\| = \inf_{y \in K} \|x - y\|.$$

Proof. The existence of an x' satisfying $(*)$ follows from K being closed. Suppose that, for $x'' \in K$, $x'' \neq x'$,

$$\|x - x''\| = \inf_{y \in K} \|x - y\|.$$

Consider the isosceles triangle with vertices x, x', x''. The midpoint $m = \frac{1}{2}(x' + x'')$ of the line segment between x' and x'' is, by convexity, also in K, but satisfies

$$\|x - m\| < \inf_{y \in K} \|x - y\|,$$

a contradiction. □

3.2 Definition. The map

$$\begin{aligned} p_K : \mathbb{R}^n &\longrightarrow K \\ x &\longmapsto p_K(x) = x' \end{aligned}$$

of lemma 3.1 is called the *nearest point map* relative to K.

Clearly,

3.3 Lemma.
(a) $p_K(x) = x$ *if and only if* $x \in K$*;*
(b) p_K *is surjective.*

Generalizing the concept of a *tangent hyperplane* is the following.

3.4 Definition. A hyperplane H is called a *supporting hyperplane* of a closed convex set $K \subset \mathbb{R}^n$ if $K \cap H \neq \emptyset$ and $K \subset H^-$ or $K \subset H^+$.

We call H^- (or H^+, respectively,) a *supporting half-space* of K (possibly $K \subset H$).

If u is a normal vector of H pointing into H^+ (or H^-, respectively), we say that u is an *outer normal* of K (Figure 5), and $-u$ an *inner normal* of K.

3.5 Lemma. *Let* $\emptyset \neq K \subset \mathbb{R}^n$ *be closed and convex. For every* $x \in \mathbb{R}^n \setminus K$ *the hyperplane* H *containing* $x' := p_K(x)$ *and perpendicular to the line joining* x *and* x' *is a supporting hyperplane of* K *described by* $H = \{y \mid \langle y, u \rangle = 1\}$*, for* $u := \frac{x-x'}{<x', x-x'>}$*, unless* H *contains* 0.

PROOF. The hyperplane $H := \{y \mid \langle y, u \rangle = 1\}$ (u as before) is perpendicular to $x - x'$ and satisfies $x' \in H$. Moreover, $\langle x - x', x - x' \rangle > 0$ implies $\langle x, x - x' \rangle > \langle x', x - x' \rangle$ and, thus, $x \in H^+$. Suppose H is not a supporting hyperplane of K. Then there exists some $y \in K \cap (H^+ \setminus H)$, $y \neq x$. By elementary geometry applied to the plane E spanned by x, x', and y, the line segment $[y, x']$ contains a point z interior to the circle in E about x with radius $\|x - x'\|$. Then, $\|x - z\| < \|x - x'\|$, a contradiction. □

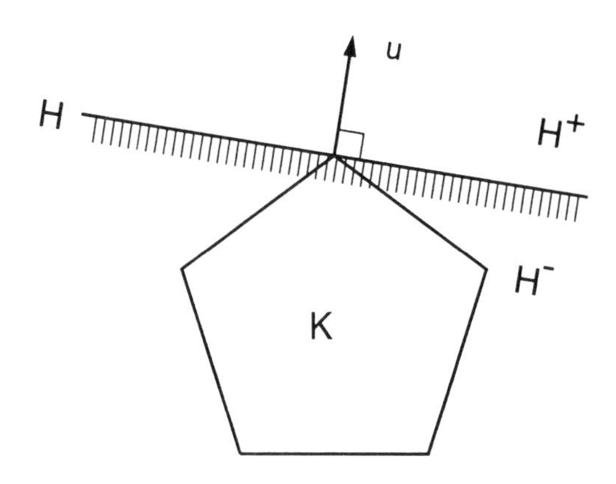

FIGURE 5.

3.6 Lemma. *Let $K \subset \mathbb{R}^n$ be closed and convex, and let $x \in \mathbb{R}^n \setminus K$. Suppose y lies on the ray emanating from x' and containing x. Then, $x' = y'$.*

PROOF. First, assume $y \in [x, x']$. Then, in the case $x' \neq y'$,

$$\|x - x'\| = \|y - x'\| + \|x - y\| > \|y - y'\| + \|x - y\| \geq \|x - y'\|,$$

a contradiction.

If $x \in [y, x']$, $x' \neq y'$, then, the line parallel to $[y, y']$ through x meets $[x', y']$ in a point $x_0 \neq x'$. From $\|x - x_0\| = \|x - x'\| \frac{\|y-y'\|}{\|y-x'\|}$ (similar triangles) and $\|y - y'\| < \|y - x'\|$ (Lemma 3.1), we obtain $\|x - x_0\| < \|x - x'\|$, a contradiction. □

3.7 Lemma (Busemann and Feller's lemma). *p_K does not increase distances, and, hence, is Lipschitz with Lipschitz constant* 1. *In particular, p_K is uniformly continuous.*

PROOF. Let $x, y \in \mathbb{R}^n \setminus K$. For $p_K(x) = p_K(y)$, the lemma is trivial; so, suppose $p_K(x) \neq p_K(y)$, and let g be the line through $x' := p_K(x)$ and $y' := p_K(y)$. We denote by H_1, H_2 the hyperplanes perpendicular to g in x', y', respectively.

Neither of x and y lies in the open stripe S bounded by H_1 and H_2, for if, say, x does, the foot x_0 (orthogonal projection) of x on g lies in K, and then

$$\|x - x_0\| < \|x - x'\|,$$

a contradiction. Also the points x, y cannot lie on the same side of H_1 or H_2 opposite to S since $[x, x'] \cap (S \setminus K) \neq \emptyset$ or $[y, y'] \cap (S \setminus K) \neq \emptyset$ would contradict what we just have shown and Lemma 3.6. □

3.8 Theorem. *A closed convex proper subset of $\mathbb{R}^n$ is the intersection of its supporting half-spaces.*

PROOF. By Lemma 3.5, there exists a supporting half-space of K. Let $K' := \bigcap H^+$ for all supporting half-spaces H^+ of K. Clearly, $K \subset K'$.

Suppose $x \in K' \setminus K$. Then, $p_K(x) \neq x$ and, hence, by Lemma 3.5, the hyperplane perpendicular in $p_K(x)$ to the line joining x and $p_K(x)$ separates x and K, so that $x \notin K'$, a contradiction. □

Remark. In general, not all supporting half-spaces of K are needed to represent K as their intersection. A triangle in $\mathbb{R}^2$, for example, has infinitely many supporting half-planes, but three half-planes already suffice to represent the triangle as their intersection.

3.9 Theorem. *Any closed convex set K possesses a supporting hyperplane at each of its boundary points.*

PROOF. Suppose $x_0 \in \partial K$ is a boundary point of K, that is, any open disc U_δ with center x_0 and radius $\delta > 0$ contains points from $\mathbb{R}^n \setminus K$. Then, x_0 is the limit point of a sequence $\{x_j\} \to x_0$ with $x_j \in \partial K$, such that there exist supporting hyperplanes H_i of K at x_i according to Lemma 3.5. Let s_i be the ray of outer normals of H_i in x_i, $i = 1, 2, \ldots$, and let S be a sphere with center x_0.

For sufficiently large i, $s_i \cap S$ is a point y_i, and $x_i = p_K(y_i)$, by Lemma 3.6. $\{y_i\}$ has a cluster point $y_0 \neq x_0$. Since p_K is continuous (Lemma 3.7), $p_K(y_0) = x_0$ and $y_0 \notin K$ otherwise $p_K(y_0) = y_0 = x_0$ would follow. Therefore, Lemma 3.5 applies, and the theorem follows. □

Exercises

1. Let $K \subset \mathbb{R}^n$ be closed and convex. Then, $\dim K = k$ if and only if, for any $x \in \operatorname{relint} K$, the set $p_K^{-1}(x)$ is an $(n-k)$-dimensional affine space, $0 \leq k \leq n$.
2. Every closed convex set is the intersection of countably many of its supporting half-spaces.
3. Let $M \subset \mathbb{R}^n$ be compact. pos M has an apex if $0 \notin \operatorname{conv} M$.
4. A closed set $K \subset \mathbb{R}^n$ that possesses a well-defined nearest point map is convex. (Hint: Reduce the problem to $n = 2$. Use increasing sequences $B_1 \subset B_2 \subset \cdots$ of circular discs $B_j \subset \mathbb{R}^2 \setminus K$, $j = 1, 2, \ldots$).

4. Faces and normal cones

Although faces and normal cones will mainly be used in the special case of polytopes, we introduce them for closed convex sets. This lets us see which are properties specific to polytopes.

4.1 Definition. If H is a supporting hyperplane of the closed convex set K, we call $F := K \cap H$ a *face* of K. By convention, $\emptyset$ and K are called *improper faces* of K.

If we speak about faces, it should be clear from the context whether we include $\emptyset$ or K or not.

By Lemma 1.2,

4.2 Lemma. *Every face of a closed convex set K is again a closed convex set.*

So we can speak about the dimension of a face. Recall the convention $\dim \emptyset = -1$.

4.3 Definition. By a *k-face* F of K, we mean a face of dimension k. We call F

(a) a *vertex* of K, if $k = 0$,
(b) an *edge* of K, if $k = 1$,
(c) a *facet* of K, if $k = \dim K - 1$.

We denote the set of vertices of K by vert K.

4.4 Lemma. *Let F_0 and F_1 be faces of a closed convex set K such that $F_0 \subset F_1$. Then, F_0 is a (possibly improper) face of F_1.*

PROOF. Let $F_0 = K \cap H_0$, where H_0 is a supporting hyperplane of K and, hence, also of F_1. Then,

$$F_1 \cap H_0 \subset K \cap H_0 = F_0 \subset F_1 \cap H_0,$$

hence, $F_0 = F_1 \cap H_0$ which proves the lemma. □

Remark. The converse of Lemma 4.4 is false. As Figure 6 illustrates, F_0 can be a face of F_1, F_1 a face of K, but F_0 cannot be a face of K. For a polytope, however, the converse of Lemma 4.4 is true (see Chapter II, Theorem 1.7).

Now, we will generalize Lemma 4.4.

4.5 Lemma. *If $F_1, \ldots, F_r$ are faces of a closed convex set K, then, $F := F_1 \cap \cdots \cap F_r$ is also a (possibly improper) face of K.*

PROOF. Since being a face is not affected by doing so, we may assume $0 \in F$ (unless $F = \emptyset$ in which case there is nothing to prove).

Let $H_i = \{x \mid \langle x, u_i \rangle = 0\}$ be a supporting hyperplane of K such that $F_i = K \cap H_i$, $i = 1, \ldots, r$. By possibly changing signs of some of the u_i, we can arrange

$$K \subset H_i^- = \{x \mid \langle x, u_i \rangle \leq 0\}, \quad i = 1, \ldots, r.$$

We set $u := u_1 + \cdots + u_r$. If necessary, we can replace u_1 by $2u_1$ so that $u \neq 0$ can always be assumed. We find

$$\langle x, u \rangle = \langle x, u_1 \rangle + \cdots + \langle x, u_r \rangle \leq 0 \qquad \text{for all } x \in K.$$

Therefore, $H := \{x \mid \langle x, u \rangle = 0\}$ is a supporting hyperplane of K. Moreover, $\langle x, u \rangle = 0$ is true if and only if $\langle x, u_1 \rangle = \cdots = \langle x, u_r \rangle = 0$. Hence,

$$x \in K \cap H \quad \text{if and only if} \quad x \in (K \cap H_1) \cap \cdots \cap (K \cap H_r) = F.$$

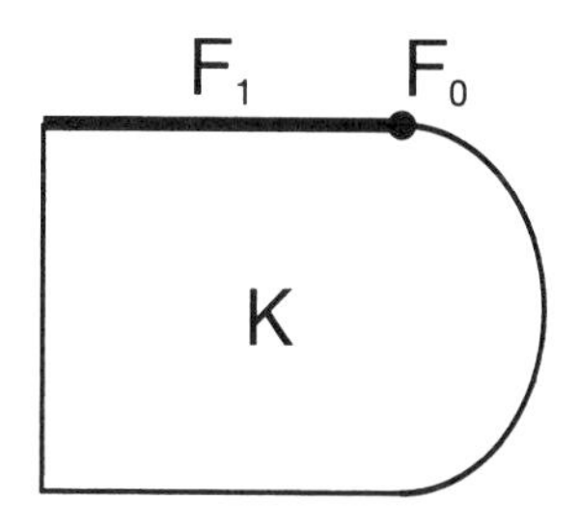

FIGURE 6.

□

4.6 Lemma.

(a) *Suppose F is a face of the closed convex set K and $x, \tilde{x} \in$ relint F. Then, any supporting hyperplane of K at x also contains $\tilde{x}$.*

(b) *If F, F' are faces of K and* (relint F) $\cap$ (relint F') $\neq \emptyset$, *then, $F = F'$.*

PROOF.

(a) The line segment $[x, \tilde{x}]$ is properly contained in a line segment $[y, \tilde{y}] \subset$ relint F. Should a supporting hyperplane at x not contain $\tilde{x}$ two of the points $\tilde{x}$ and y, $\tilde{y}$, would be separated, a contradiction.

(b) is a direct consequence of (a).

□

4.7 Definition. Let x be a point of the closed convex set K. We call

$$N(x) := -x + p_K^{-1}(x)$$

the *normal cone* of K at x.

4.8 Lemma. *$N(x)$ is a closed convex cone; it consists of* 0 *and all outer normals of K in x. If $x \in$* int *K, then, $N(x) = \{0\}$.*

PROOF. First, note that $N(x)$ is, indeed, a cone. From Lemmas 3.5 and 3.6, we deduce the second part of the lemma. $p_K^{-1}(x)$ and, hence, $-x + p_K^{-1}(x)$ is closed since p_K is continuous (Lemma 3.7). To show that $N(x)$ is convex, we arrange for $x = 0$ with a translation. Then, for $u, v \in N(0)$, we may assume $\langle K, u\rangle \leq 0$ and $\langle K, v\rangle \leq 0$, so that

$$\langle K, \lambda u + (1 - \lambda)v\rangle \leq 0 \qquad \text{for} \quad 0 \leq \lambda \leq 1;$$

hence, $\lambda u + (1 - \lambda)v \in N(0)$. □

4.9 Definition. Let σ be a cone. Then,

$$\check{\sigma} := \{y \mid \langle \sigma, y\rangle \geq 0\}$$

is called the *dual cone* of σ (Figure 7).

Lemma 4.8 implies Lemmas 4.10 and 4.11.

4.10 Lemma. *If σ is a cone with apex* 0, *then, $N(0) = -\check{\sigma}$ ($\check{\sigma}$ reflected in* 0*).*

4.11 Lemma. *Let F be a face of the closed convex set K. For $x, \tilde{x} \in$* relint *F $N(x) = N(\tilde{x})$.*

PROOF. This follows readily from Lemma 4.6. □

4.12 Definition. If F is face of a closed convex set K and $x \in$ relint F, then, $N(x)$ is denoted by $N(F)$ and is called the *cone of normals* of K in F.

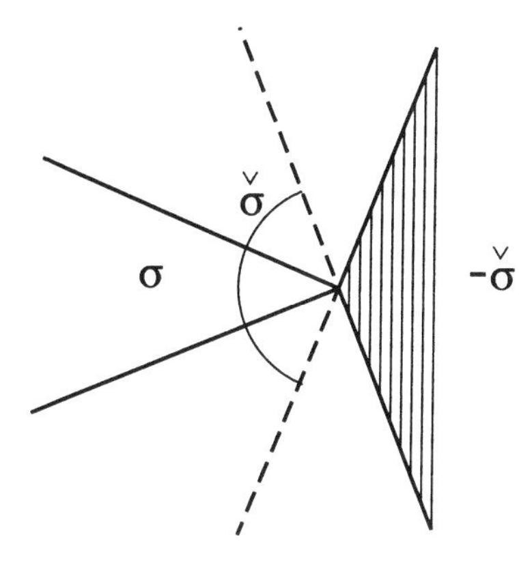

FIGURE 7.

4.13 Theorem. *Let K be a convex body in $\mathbb{R}^n$ and $x(F)$ one of the relative interior points of a face $F \neq \emptyset$ of K. Then, $\{\text{relint } N(x(F)) \mid F \text{ a face of } K\} = \{\text{relint } N(F) \mid F \text{ a face of } K\}$ is a partition (disjoint covering) of $\mathbb{R}^n$.*

PROOF. Let $0 \neq u \in \mathbb{R}^n$. Since K is bounded, there exists a hyperplane $H(\alpha, u) = \{z \mid \langle z, u\rangle = \alpha\}$ such that $K \subset H^-(\alpha, u)$. Put $H^- := \bigcap_\alpha H^-(\alpha, u)$, the intersection taken for all α, such that $K \subset H^-(\alpha, u)$. Clearly, H^- is again a closed half-space and $F := H \cap K \neq \emptyset$. For $x(F) \in \text{relint } F$, $u \in \text{relint } N(x(F))$; this is elementary in every plane passing through $x(F)$ and containing u; hence, it carries over the general situation. So, every $u \neq 0$ occurs in some cone relint $N(x(F))$. Also, the point 0 occurs in relint $N(x(K))$ since, for $x \in \text{relint } K$, the cone $N(x)$ is a linear space ($= \{0\}$ if $\dim K = n$).

Suppose $y \in \text{relint } N(x(F_1)) \cap \text{relint } N(x(F_2))$. Then, $p_K(y + x(F_1)) = x(F_1)$ and $p_K(y + x(F_2)) = x(F_2)$ so that, by Lemma 3.5, the supporting hyperplanes in $x(F_1)$ and $x(F_2)$ coincide. This implies $F_1 = F_2$. □

4.14 Definition. $\Sigma(K)$ denotes the set of all cones $N(F)$ and is called the *fan* of K (see Figure 8).

Exercises

1. Let K be convex and closed, int $K \neq \emptyset$, and let L be an affine subspace such that $L \cap \text{int } K = \emptyset$, $L \cap K \neq \emptyset$. Show that there exists a supporting hyperplane of K which contains L.
2. For a planar convex body K, write $F_i \longrightarrow F_j$ if F_i is a face of F_j, $\dim F_i = i$, $\dim F_j = j$, and construct valid diagrams like $F_0 \nearrow^{F_1} \searrow K$, $F_0 \longrightarrow K$. Call the diagram maximal if it can admit no further faces or arrows (in Figure 6, for example,

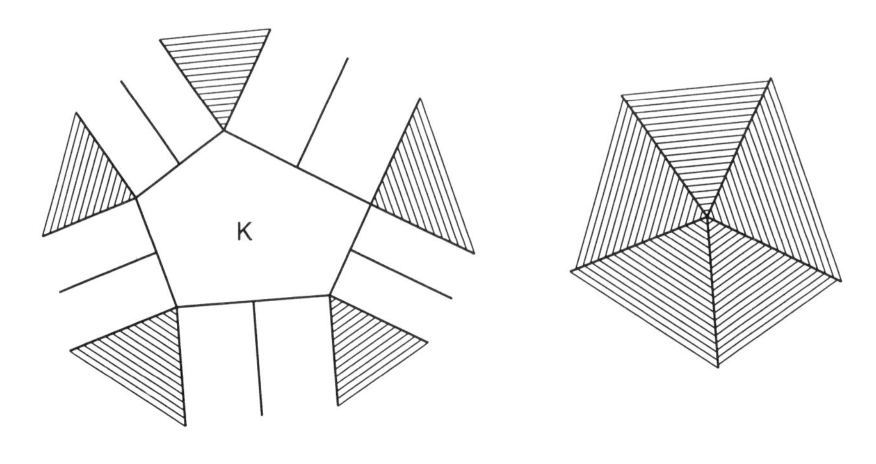

FIGURE 8.

$F_0 \longrightarrow F_1 \longrightarrow K$ is maximal). Show, by examples, that, for $\dim K = 3$, there exist maximal diagrams of the following types:

$$F_0 \longrightarrow K, \quad F_0 \longrightarrow F_1 \longrightarrow K, \quad \begin{array}{ccc} & F_1 & \\ \nearrow & & \searrow \\ F_0 & \longrightarrow & K \end{array}, \quad F_0 \longrightarrow F_2 \longrightarrow K,$$

$$\begin{array}{ccc} & F_2 & \\ \nearrow & & \searrow \\ F_0 & \longrightarrow & K \end{array}, \quad F_0 \longrightarrow F_1 \longrightarrow F_2 \longrightarrow K, \quad F_0 \longrightarrow \begin{array}{ccc} & F_2 & \\ \nearrow & & \searrow \\ F_1 & \longrightarrow & K \end{array},$$

$$\begin{array}{ccccc} & & F_2 & & \\ & \nearrow & \uparrow & \searrow & \\ F_0 & \longrightarrow & F_1 & \longrightarrow & K. \end{array}$$

3. Characterize convex polytopes which have the same fan.
4. For $0 \leq k < n$, call x a *k-boundary point* of the closed convex set K if $\dim N(x) = n - k$. Show (by using the nearest point map) that
   a. K possesses only countably many 0-boundary points, and
   b. the set of 1-boundary points can be covered by countably many rectifiable arcs (that is, images of line segments under Lipschitz maps).

5. Support function and distance function

Now we will generalize the linear function $h_{\{a\}} := \langle a, \cdot \rangle$ for arbitrary compact subsets K of $\mathbb{R}^n$:

5.1 Definition. Let $K \subset \mathbb{R}^n$ be a nonempty convex body. The map

$$h_K : \mathbb{R}^n \to \mathbb{R} \quad \text{defined by } u \mapsto \sup_{x \in K} \langle x, u \rangle$$

is called the *support function* of K (Figure 9a).

The next statement is an obvious consequence of the definition.

5.2 Lemma. *If $K + a$ is a translate of the convex body K, then,*

$$h_{K+a}(u) = h_K(u) + \langle a, u\rangle \qquad \textit{for all } u \in \mathbb{R}^n .$$

Example 1. For $n = 1$, set $K = [c, d]$. Then (compare Figure 9a for $c = -2$, $d = 1$),

$$h_{[c,d]}(u) = \begin{cases} \langle d, u\rangle & \text{for } u \geq 0 \\ \langle c, u\rangle & \text{for } u \leq 0. \end{cases}$$

5.3 Lemma.

(a) *For every fixed nonzero $u \in \mathbb{R}^n$, the hyperplane*

$$(*) \qquad H_K(u) := \{x \mid \langle x, u\rangle = h_K(u)\},$$

is a supporting hyperplane of K (Figure 9b).

(b) *Every supporting hyperplane of K has a representation of the form* $(*)$.

PROOF.

(a) Since K is compact and $\langle\cdot, u\rangle$ is continuous, for some $x_0 \in K$,

$$\langle x_0, u\rangle = h_K(u) = \sup_{x\in K}\langle x, u\rangle.$$

For an arbitrary $y \in K$, it follows that $\langle y, u\rangle \leq \langle x_0, u\rangle$; hence, $K \subset H_K^-(u)$. This proves (a).

(b) Let $H = \{x \mid \langle x, u\rangle = \langle x_0, u\rangle\}$ be a supporting hyperplane of K at x_0. We choose $u \neq 0$ such that $K \subset H^-$. Then, $\langle x_0, u\rangle = \sup_{x\in K}\langle x, u\rangle = h_K(u)$ which implies (b).

□

5.4 Definition. A function $f : \mathbb{R}^n \to \mathbb{R}$ is said to be *convex* if, for all $0 \leq \lambda \leq 1$ and $x, y \in \mathbb{R}^n$,

$$f(\lambda x + (1 - \lambda)y) \leq \lambda f(x) + (1 - \lambda)f(y).$$

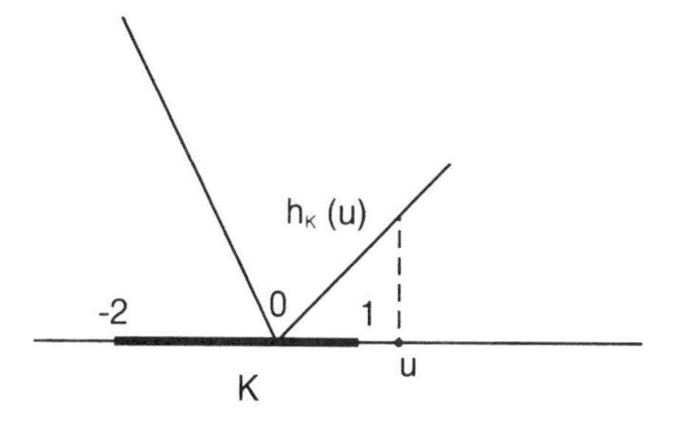

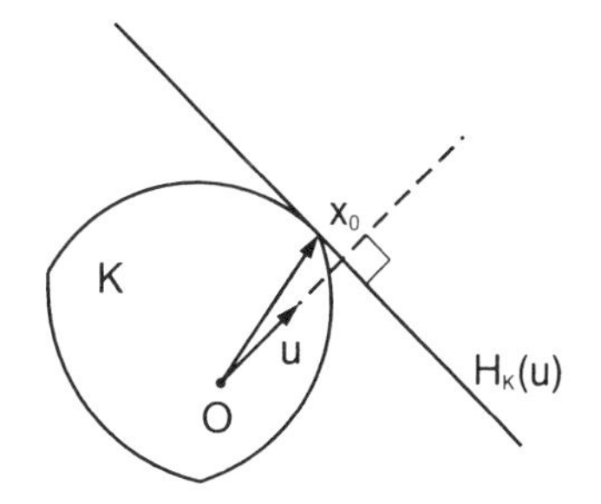

FIGURE 9a,b.

Note that if f is convex and L is an affine subspace of $\mathbb{R}^n$, then, $f|_L$ is also convex.

Example 2. For $n = 1$ and $x, y \in \mathbb{R}$, the graph $\Gamma(f)$ of a convex function f lies "below" the line-segment $[(x, f(x)), (y, f(y))]$ in $\mathbb{R}^2$. Hence for convex f, if $a \leq -1 < b < 0$, $f(b) = 1$, and $f(0) = 0$, then, $(a, f(a)$ and $(-b, -f(-b))$ are "above" the line through $(b, 1)$ and $(0, 0)$, so that $f(a) \geq -\frac{1}{b}$ and $f(b) \geq -f(-b)$.

5.5 Definition. A function $f : \mathbb{R}^n \to \mathbb{R}$ is called *positive homogeneous* if, for any $\lambda \geq 0$ and $x \in \mathbb{R}^n$,

$$f(\lambda x) = \lambda f(x).$$

5.6 Lemma. *A positive homogeneous function $f : \mathbb{R}^n \longrightarrow \mathbb{R}$ is convex if and only if*

(1) $$f(x + y) \leq f(x) + f(y) \quad \textit{for all } x, y \in \mathbb{R}^n .$$

PROOF. Let the positive homogeneous function f be convex. Then (1) follows from

$$\tfrac{1}{2} f(x + y) = f\left(\tfrac{1}{2}x + \tfrac{1}{2}y\right) \leq \tfrac{1}{2} f(x) + \tfrac{1}{2} f(y).$$

Conversely, if (1) holds for f, then, for $0 \leq \lambda \leq 1$,

$$f(\lambda x + (1 - \lambda)y) \leq f(\lambda x) + f((1 - \lambda)y) = \lambda f(x) + (1 - \lambda) f(y),$$

so f is convex. □

5.7 Lemma.

(a) *A function $f : \mathbb{R}^n \to \mathbb{R}$ is convex if and only if, for every convex combination $x = \lambda_0 x_0 + \cdots + \lambda_n x_n$, $\lambda_0 \geq 0, \ldots, \lambda_n \geq 0$, $\lambda_0 + \cdots + \lambda_n = 1$ of points $x_0, \ldots, x_n$*

(1) $$f(x) \leq \lambda_0 f(x_0) + \cdots + \lambda_n f(x_n).$$

(b) *Every convex function $f : \mathbb{R}^n \to \mathbb{R}$ is continuous.*

(c) *$f : \mathbb{R}^n \to \mathbb{R}$ is convex if and only if $\Gamma^+(f) := \{(x, \xi) \mid x \in \mathbb{R}^n, \xi \in \mathbb{R}, f(x) \leq \xi\}$ is a closed and convex subset of $\mathbb{R}^{n+1}$.*

(d) *A positive homogeneous function $f : \mathbb{R}^n \to \mathbb{R}$ is convex if and only if $\Gamma^+(f)$ is a closed convex cone.*

PROOF.

(a) If (1) is true we obtain, for $x_1 = \cdots = x_n$ (using $1 - \lambda_0 = \lambda_1 + \cdots + \lambda_n$),

$$f(\lambda_0 x_0 + (1 - \lambda_0)x_1) \leq \lambda_0 f(x_0) + (1 - \lambda_0) f(x_1),$$

so that f is convex.

If, conversely, f is convex, we proceed by induction and assume that f satisfies (1) (with n replaced by $n - 1$) on each $(n - 1)$-dimensional affine subspace of $\mathbb{R}^n$. Then, for $\lambda_0 < 1$ and $y := (1-\lambda_0)^{-1}(\lambda_1 x_1 + \cdots + \lambda_n x_n) = (\lambda_1 + \cdots + \lambda_n)^{-1}(\lambda_1 x_1 + \cdots + \lambda_n x_n)$, we find

$$\begin{aligned} f(\lambda_0 x_0 + \cdots + \lambda_n x_n) &= f(\lambda_0 x_0 + (1-\lambda_0)y) \\ &\le \lambda_0 f(x_0) + (1-\lambda_0) f(y) \\ &\le \lambda_0 f(x_0) + (\lambda_1 + \cdots + \lambda_n)\left(\sum_{i=1}^{n} (\lambda_1 + \cdots + \lambda_n)^{-1}\lambda_i f(x_i)\right) \\ &= \lambda_0 f(x_0) + \lambda_1 f(x_1) + \cdots + \lambda_n f(x_n), \end{aligned}$$

so that (1) follows.

(b) Given a point x_0 in $\mathbb{R}^n$, we consider a regular n-simplex $T := \operatorname{conv}\{x_1, \ldots, x_{n+1}\}$ which possesses x_0 as center of gravity and for which $\|x_1 - x_0\| = \cdots = \|x_{n+1} - x_0\| = 1$. We set $d := \max\{|f(x_1) - f(x_0)|, \ldots, |f(x_{n+1}) - f(x_0)|\}$. Let x lie in a δ_0-neighborhood $U_{\delta_0}(x_0)$ of x_0 such that $U_{\delta_0}(x_0) \subset T$. Since T is covered by the n-simplices $T_i := \operatorname{conv}\{x_0, x_1, \ldots, x_{i-1}, x_{i+1}, \ldots, x_{n+1}\}$, $i = 1, \ldots, n+1$, we may assume x to lie in one of the T_i, say in T_{n+1}, $x = \lambda_0 x_0 + \lambda_1 x_1 + \cdots + \lambda_n x_n$, $\lambda_0 \ge 0, \ldots, \lambda_n \ge 0$, $\lambda_0 + \cdots + \lambda_n = 1$. Clearly, $\lambda_i < \delta_0 \le 1$, $i = 1, \ldots, n$. We may assume $f(x) \ge 0$ in T (up to adding a constant). Given $\varepsilon > 0$, we choose $\delta := \frac{\varepsilon}{n(d+1)}$ and obtain (using (a) and assuming $\delta \le \delta_0$)

$$\begin{aligned} |f(x) - f(x_0)| &\le |\lambda_0 f(x_0) + \cdots + \lambda_n f(n) - f(x_0)| \\ &= |\lambda_1(f(x_1) - f(x_0)) + \cdots + \lambda_n(f(x_n) - f(x_0))| \\ &\le (\lambda_1 + \cdots + \lambda_n)d < n\delta(d+1) = \varepsilon. \end{aligned}$$

Therefore, f is continuous.

(c) Let f be convex. Given $(x, \xi), (y, \eta) \in \Gamma^+(f)$, for $0 \le \alpha \le 1$,

$$f(\alpha x + (1-\alpha)y) \le \alpha f(x) + (1-\alpha) f(y) \le \alpha\xi + (1-\alpha)\eta; \text{ hence,}$$
$$\alpha(x, \xi) + (1-\alpha)(y, \eta) = (\alpha x + (1-\alpha)y, \alpha\xi + (1-\alpha)\eta) \in \Gamma^+(f).$$

Therefore, $\Gamma^+(f)$ is convex. From (b), it readily follows that $\Gamma^+(f)$ is also closed. The arguments may be reversed.

(d) If f is positive homogeneous and convex, the closed set $\Gamma^+(f)$ is a cone. If, conversely, $\Gamma^+(f)$ is a closed and convex cone, f is homogeneous and, by (c), convex.

□

Remarks.

(1) By Carathéodory's theorem, in (a) we may choose x to be a convex combination of an arbitrary number of points.

(2) If, in the definition of a convex function, $\mathbb{R}^n$ is replaced by a closed convex subset of $\mathbb{R}^n$, (b) and (c) need no longer be true. Example: Let the subset be

the closed unit ball B of $\mathbb{R}^n$, and let $f(x) = 0$ for $x \in \operatorname{int} B$, $f(x) = 1$ for $x \in \partial B$.

5.8 Lemma. *The support function h_K of a convex body K is positive homogeneous and convex.*

PROOF. Let $\lambda \geq 0$. We find

$$h_K(\lambda u) = \sup_{x \in K} \langle x, \lambda u \rangle = \lambda \sup_{x \in K} \langle x, u \rangle = \lambda h_K(u).$$

Hence, h_K is positive homogeneous.

From $\langle x, u \rangle \leq h_K(u)$, $\langle x, v \rangle \leq h_K(v)$ for all $x \in K$, we obtain

$$\langle x, u + v \rangle \leq h_K(u) + h_K(v) \qquad \text{for all } x \in K.$$

Hence

$$h_K(u + v) = \sup_{x \in K} \langle x, u + v \rangle \leq h_K(u) + h_K(v).$$

Therefore, by Lemma 5.6, h_K is convex. □

5.9 Lemma. *h_K is linear on each cone of the fan $\Sigma(K)$ of K.*

PROOF. All points u in a fixed cone σ of $\Sigma(K)$ have the same nearest point $x_0 := p_K(u)$. As in the proof of Lemma 5.3 (b), we, thus, obtain

$$h_K|_\sigma = \langle x_0, \cdot \rangle|_\sigma.$$

□

5.10 Definition. Let K be an n-dimensional convex body in $\mathbb{R}^n$, and let $0 \in \operatorname{int} K$. The map

$$d_K : \mathbb{R}^n \to \mathbb{R}$$

defined by

$$d_K(\lambda \bar{x}) := \lambda, \qquad \text{for } \bar{x} \in \partial K \text{ and } \lambda \geq 0,$$

is called the *distance function* of K.

We show that d_K is well-defined (part (b) of the following lemma).

5.11 Lemma. *Let K be an n-dimensional convex body in $\mathbb{R}^n$.*

(a) *If a line g intersects ∂K in three different points, then, g is contained in a supporting hyperplane of K, so, in particular, $g \cap \operatorname{int} K = \emptyset$.*

(b) *Any ray emanating from a point in* int *K intersects ∂K in one and only one point.*

PROOF.

(a) Let $A, B, C \in g \cap \partial K$, and let B lie between A and C. We consider a supporting hyperplane $H = \{x \mid \langle x, u \rangle = c\}$ of K in B. If H did not contain both A and C, it would separate these points properly, which contradicts the definition of a supporting hyperplane.

(b) Let $y \in \text{int } K$, σ be a ray emanating from y, and h be the line that contains σ. The intersection $h \cap K$ is a convex body, hence, a line segment $[y_0, y_1]$. Either y_0 or y_1 equals $\sigma \cap \partial K$.

□

5.12 Lemma. *The distance function d_K is positive homogeneous and convex.*

PROOF. By definition, d_K is positive homogeneous.

To prove convexity, let $d_K(x) = \lambda, d_K(y) = \mu$. If $\lambda = 0$ or $\mu = 0$, then, $x = 0$ or $y = 0$, and there is nothing to prove. So let $\lambda \neq 0$, $\mu \neq 0$. For $\delta := \frac{\mu}{\lambda+\mu}$, we obtain $(1-\delta)\bar{x} + \delta\bar{y} \in K$, for $\lambda\bar{x} = x$, $\mu\bar{y} = y$, hence,

$$1 \geq d_K((1-\delta)\bar{x} + \delta\bar{y}) = d_K(\frac{\lambda}{\lambda+\mu}\bar{x} + \frac{\mu}{\lambda+\mu}\bar{y}) = d_K(\frac{1}{\lambda+\mu}(x+y))$$
$$= \frac{1}{\lambda+\mu} d_K(x+y),$$

hence, $d_K(x+y) \leq \lambda + \mu = d_K(x) + d_K(y)$. So d_K is convex by Lemma 5.6. □

5.13 Definition. A convex body K is called *centrally symmetric* if it is mapped onto itself by a reflection in a point c (which assigns to each $x = c + (x - c)$ the point $c - (x - c) = 2c - x$). We call c the *center* of K.

From the above lemmas, we derive Theorem 5.14.

5.14 Theorem. *Let K be a centrally symmetric convex body with $0 \in \text{int } K$ as its center. Then, d_K defines a norm on the vector space $\mathbb{R}^n$, that is, a map*

$$d_K = \|\cdot\| : \mathbb{R}^n \to \mathbb{R}$$

satisfying, for all $x, y \in \mathbb{R}^n$ and $\lambda \in \mathbb{R}$,

(a) $\|x\| = 0$ *if and only if* $x = 0$,

(b) $\|\lambda x\| = |\lambda| \cdot \|x\|$,

(c) $\|x + y\| \leq \|x\| + \|y\|$.

Example 3. The "maximum norm" in $\mathbb{R}^2$ is of the form

$$d_K(x) := \max\{|x_1|, |x_2|\}$$

where $x = (x_1, x_2)$ and K is the square with vertices $(1, 1)$, $(1, -1)$, $(-1, 1)$, $(-1, -1)$.

Example 4. The so-called *Manhattan norm* $d_{K'}(x) := |x_1| + |x_2|$ where K' is the square with vertices $(1, 0)$, $(0, 1)$, $(-1, 0)$, $(0, -1)$.

In the following section we shall see how the norms in Examples 3 and 4 are related to each other.

Exercises

1. Determine explicitly the support functions for the following convex bodies in $\mathbb{R}^2$.
   a. the unit disc,
   b. conv $\{(1, 0), (0, 1), (-1, 0), (0, -1)\}$, and
   c. the line segment $\{(-1, 0), (1, 0)\}$.
2. The support function h_K is linear if and only if K is a point.
3. Show explicitly that $d_K, d_{K'}$, in Examples 3 and 4, are norms.
4. Characterize those convex bodies K for which $d_K(x + y) = d_K(x) + d_K(y)$ implies that x and y are multiples.

6. Polar bodies

We will consider the polarity π in $\mathbb{R}^n$ with respect to the unit sphere $S := \{x \mid \langle x, x\rangle = 1\}$. It assigns to every affine subspace W of $\mathbb{R}^n$ with $0 \notin W$ a subspace $\pi(W)$ of $\mathbb{R}^n$ of dimension $n - 1 - \dim W$: If $0 \neq u$ is a point in $\mathbb{R}^n$, then,

$$\pi(u) = H_u := \{x \mid \langle x, u\rangle = 1\}.$$

If the affine subspaces U and V which generate W are not parallel and if W does not contain 0, then, $\pi(W) = \pi(U) \cap \pi(V)$. Note that $\pi \circ \pi$ is the identity.

The exceptional role of the point 0 can be avoided by going over to the projective extension of $\mathbb{R}^n$ by adding a "hyperplane at infinity", H_∞. Then, $\pi(0) = H_\infty$. That will be needed, for example, in Lemma 3.

6.1 Definition. Let $0 \in \operatorname{int} K$, where K is a convex body. Then, for $u \neq 0$, the half-spaces H_u^- which contain 0 and, for $H_0^- := \mathbb{R}^n$,

$$K^* := \bigcap_{u \in K} H_u^-$$

is called the *polar body* of K.

Clearly, $0 \in \operatorname{int} K^*$ and $K^* = \bigcap_{u \in \partial K} H_u^-$, since $0 \in \operatorname{int} K$.

Example 1. As three-dimensional examples, in Figure 10 we consider pairs of platonic solids with center at 0 and the sphere S inscribed in the outer body, hence, circumscribing the inner body (shaded).

6.2 Definition. We will represent the points of $\mathbb{R}^n \cup H_\infty$ by the one-dimensional subspaces of $\mathbb{R}^{n+1}$ such that the points of H_∞ are spanned by vectors $(0, \ldots, 0, \xi)$, $\xi \neq 0$. Then, a linear transformation of $\mathbb{R}^{n+1}$ up to multiplication by a nonzero factor is called a *projective transformation* of $\mathbb{R}^n \cup H_\infty$. It is called *permissible* with respect to the convex body $K \subset \mathbb{R}^n \cup H_\infty$, if H_∞ is mapped onto a hyperplane disjoint from K.

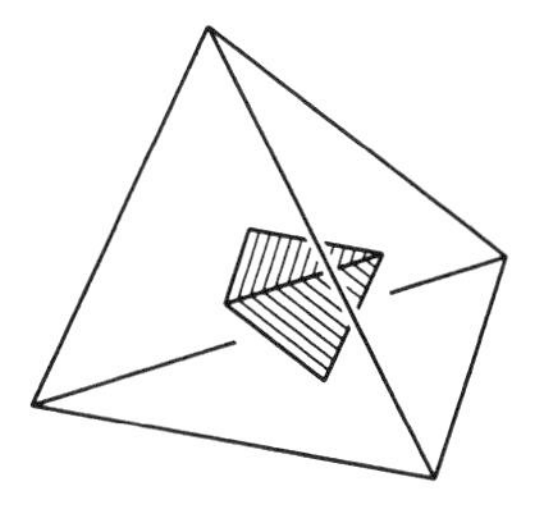
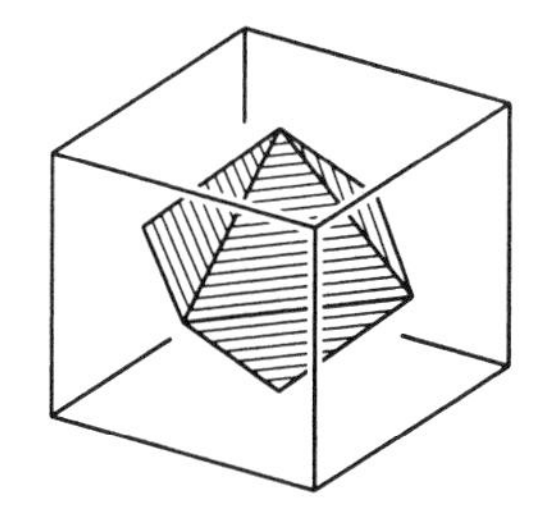
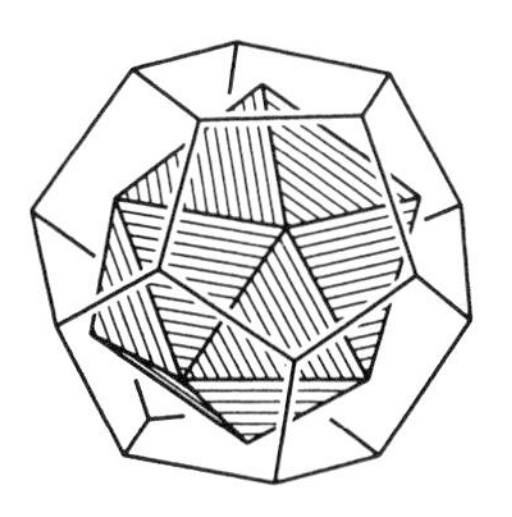

FIGURE 10. Left: Tetrahedron. Middle: Octahedron and cube. Right: Icosahedron and dodecahedron. (Peatonic solids).

6.3 Lemma. *If the convex body K is so translated to $\tau(K)$ that 0 remains in the interior, then, $(\tau(K))^*$ is obtained from K^* by a permissible projective transformation.*

PROOF. This follows from general facts on projective transformations. □

6.4 Theorem. *Let K be a convex body with $0 \in \operatorname{int} K$. Then,*

(*a*)
$$K^{**} = K;$$

(b) *The distance function of K equals the support function of K^*, and, conversely,*

$$d_K = h_{K^*}, \qquad d_{K^*} = h_K.$$

PROOF.

(a) By definition of H_u, for every $u \neq 0$ of K,

$$H_u^- = \{x \mid \langle u, x\rangle \leq 1\}$$

Therefore, (using the obvious notation $\langle K, x\rangle \leq 1$)

$$K^* = \{x \mid \langle K, x\rangle \leq 1\} \quad \text{and} \quad K^{**} = \{y \mid \langle K^*, y\rangle \leq 1\}.$$

If $y \in K$, then, the definition of K^* yields $\langle y, K^*\rangle \leq 1$ and, thus, $K \subset K^{**}$. Suppose $K \neq K^{**}$. Then, let $x \in K^{**} \setminus K$. For

$$x' := p_K(x) \quad \text{and} \quad u := \frac{x - x'}{\langle x', x - x'\rangle},$$

Lemma 3.5 yields

$$x \in H_u^+ \setminus H_u, \quad \text{but also } K \subset H_u^-,$$

whence $u \in K^*$. Since $x \in K^{**}$, it follows that $\langle u, x\rangle \leq 1$, i.e., $x \in H_u^-$, a contradiction.

Before showing part (b) we prove two lemmas.

6.5 Lemma. *Let K_1, K_2 be convex bodies such that $0 \in \operatorname{int} K_1$ and $K_1 \subset K_2$. Then, $K_2^* \subset K_1^*$.*

PROOF. If $y \in K_2^*$, then, $\langle K_2, y\rangle \leq 1$, hence, in particular, $\langle K_1, y\rangle \leq 1$. This implies $y \in K_1^*$. □

6.6 Lemma. *If $x \in \partial K$, $0 \in \operatorname{int} K$, then, H_x is a supporting hyperplane of K^*.*

PROOF. We know that $K^* = \bigcap_{x \in \partial K} H_x^-$. For every $x \in \partial K$, there exists a $\beta_x \in \mathbb{R}_{\geq 1}$ such that $H_{\beta_x x}$ is a supporting hyperplane of K^*. Thus, $\tilde{K} := \operatorname{conv}(\{\beta_x x \mid x \in \partial K\})$ includes K, and we obtain

$$\tilde{K}^* = \bigcap_{y \in \partial \tilde{K}} H_y^- = \bigcap_{x \in \partial K} H_{\beta_x x}^- \supset K^* = \bigcap_{x \in \partial K} H_x^-.$$

Since, obviously, $H_{\beta_x x}^- \subset H_x^-$, we find that $\beta_x = 1$ for every $x \in \partial K$. □

PROOF OF (B). in Theorem 6.4: Let $u \in \mathbb{R}^n \setminus \{0\}$. We may assume $u \in \partial K$, hence, $d_K(u) = 1$. By Lemma 6.6, H_u is a supporting hyperplane of K^*, and we obtain $h_{K^*}(u) = 1$ from Lemma 5.3. □

Example 2. In Figure 11 we illustrate the cones (in the notation of 5.7)

$$\Gamma^+(d_K) = \Gamma^+(h_{K'}) \text{ and } \Gamma^+(d_{K'}) = \Gamma^+(h_K)$$

for K and $K' = K^*$ of Examples 3 and 4 of section 5, where we obtain

$$\Gamma^+(d_K) = \operatorname{pos}(K + e), \quad \Gamma^+(d_{K'}) = \operatorname{pos}(K' + e)$$

for $e = (0, 0, 1) \in \mathbb{R}^3$.

Theorem 6.4 implies

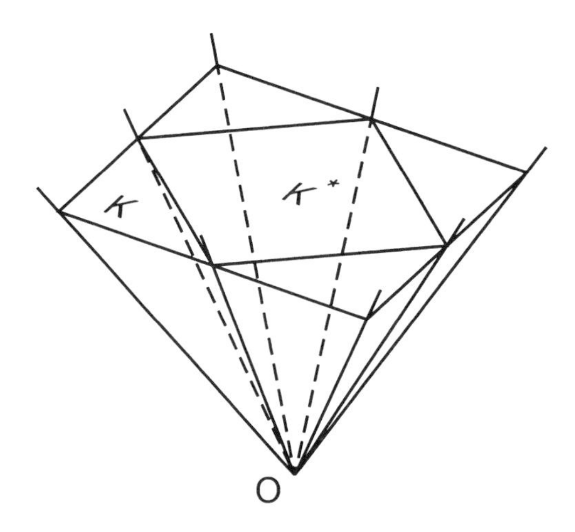

FIGURE 11.

6.7 Theorem. *Let K be a convex body in $\mathbb{R}^n$ with $0 \in \operatorname{int} K$. Set $K_+ := \Gamma^+(d_K) \subset \mathbb{R}^{n+1}$ (see Lemma 5.7) and $H := \{(x, 1) \mid x \in \mathbb{R}^n\}$. Then,*
(1) *∂K_+ is the graph of d_K in $\mathbb{R}^{n+1}$.*
(2) *$K_+ \cap H$ is a translate of K.*
(3) *$K_+^* \cap H$ is a translate of K^*.*
(4) *K_+, K_+^* are cones with apex 0 in $\mathbb{R}^{n+1}$.*

6.8 Theorem. *Every positive homogeneous and convex function $h : \mathbb{R}^n \to \mathbb{R}$ is the support function $h = h_K$ of a unique convex body K (whose dimension is possibly less than n).*

PROOF. Let us write $\mathbb{R}^n = U \oplus U^\perp$, where U is the maximal linear subspace of $\mathbb{R}^n$ on which h is linear. Then, there exists $a \in U$ such that, for $(x, x') \in U \oplus U^\perp$,

$$(*) \qquad h(x, x') = \langle x, a \rangle + h|_{U^\perp}(x').$$

Moreover, $\Gamma^+(h|_{U^\perp})$ is a cone with apex 0 in $U^\perp \oplus \mathbb{R}$ (see Lemma 5.7). Thus, there exists some $b \in U^\perp$ such that the hyperplane $H := \{(y, \langle y, b \rangle) | y \in U^\perp\}$ in $U^\perp \oplus \mathbb{R}$ intersects $\Gamma^+(h|_{U^\perp})$ only in the apex. Now the set

$$K_0 + (0, 1) := (U^\perp \times \{1\}) \cap \Gamma^+(h|_{U^\perp} - \langle \cdot, b \rangle)$$

is a convex body and, by Lemma 5.2, $h|_{U^\perp} - \langle \cdot, b \rangle$ the support function of $K_0 - b$. Finally, $(*)$ and Lemma 5.2 yield that h is the support function of $K := K_0 - b + a$. □

Exercises

1. Find explicitly the polar bodies of straight prisms and pyramids in $\mathbb{R}^3$ with regular polygons as bases.
2. Call an n-dimensional convex body K *strictly convex*, if ∂K does not contain a line segment, and *differentiable*, if there exists only one supporting hyperplane in each $x \in \partial K$. Show that K is strictly convex if and only if K^* is differentiable.
3. Let $\dim K < n$, and let $0 \in \operatorname{relint} K$. Use the definition for K^* as in the text.

   a. How is K^* obtained from the polar body of K relative to aff K?
   b. Is $K^{**} = K$?

4. Let K be an unbounded closed convex set, $\dim K = n$, and let $0 \in \operatorname{int} K$. We set $K^* := \bigcap_{u \in K} H_u^-$ where $H_0^- := \mathbb{R}^n$.

   a. Show that K^* is a convex body.
   b. Must $K^{**} = K$?

II

Combinatorial theory of polytopes and polyhedral sets

1. The boundary complex of a polyhedral set

We will turn now to the specific properties of convex polytopes or, briefly, polytopes. They have been introduced in I.1 as convex hulls of finite point sets in $\mathbb{R}^n$. Our first aim is to show that, equivalently, convex polytopes can be defined as bounded intersections of finitely many half-spaces. (This fact is of particular relevance in linear optimization).

1.1 Theorem. *Each polytope possesses only finitely many faces; they, too, are polytopes.*

PROOF. Let $P = \operatorname{conv}\{x_1, \ldots, x_r\}$, and let $F := P \cap H$ be a face where $H = \{x \mid \langle x, a\rangle = \alpha\}$ is a supporting hyperplane of P such that $P \subset H^-$. We may assume

$$x_1, \ldots, x_s \in H; \quad x_{s+1}, \ldots, x_r \in \operatorname{int} H^-$$

and find

$$\begin{aligned} \langle x_i, a\rangle &= \alpha && \text{for} \quad i = 1, \ldots, s, \\ \langle x_i, a\rangle &= \alpha - \beta_i, \quad \beta_i > 0 && \text{for} \quad i = s+1, \ldots, r. \end{aligned}$$

Then, for

$$x = \lambda_1 x_1 + \cdots + \lambda_r x_r, \qquad \lambda_1 + \cdots + \lambda_r = 1, \quad \lambda_j \geq 0, \quad j = 1, \ldots, r,$$

$$\langle x, a\rangle = \sum_{i=1}^{r} \lambda_i \langle x_i, a\rangle = \sum_{i=1}^{r} \lambda_i \alpha - \sum_{i=s+1}^{r} \lambda_i \beta_i = \alpha - \sum_{i=s+1}^{r} \lambda_i \beta_i.$$

Therefore, $x \in H$ if and only if $\sum_{i=s+1}^{r} \lambda_i \beta_i = 0$, which, in turn, is equivalent to $\lambda_{s+1} = \cdots = \lambda_r = 0$. So, x is a convex combination of $x_1, \ldots, x_s$. Hence $H \cap P = \operatorname{conv}\{x_1, \ldots, x_s\}$ is a polytope.

Since only finitely many convex hulls of elements of $\{x_1, \ldots, x_r\}$ exist, the theorem follows. □

1.2 Theorem (Krein-Milman theorem). *Each polytope P is the convex hull of its vertices, that is,*

$$P = \operatorname{conv}(\operatorname{vert} P).$$

PROOF. Trivially, $\operatorname{conv}(\operatorname{vert} P) \subset P$. For the opposite inclusion, we may assume that $P = \operatorname{conv}\{x_1, \dots, x_r\}$ and $x_i \notin \operatorname{conv}\{x_1, \dots, x_{i-1}, x_{i+1}, \dots, x_r\} =: P_i$ for $1 \leq i \leq r$. Let $q_i := p_{P_i}(x_i)$ be the image of x_i under the nearest point map p_{P_i} with respect to P_i. By I, Lemma 5.3, the hyperplane H_i through q_i with normal $x_i - q_i$ is a supporting hyperplane of P_i. We translate H_i by adding $x_i - q_i$ and so obtain a supporting hyperplane H_i' of P for which

$$\{x_i\} = H_i' \cap P$$

(otherwise, as is seen from the proof of Theorem 1.1, $F = H_i' \cap P$ would contain some $x_j \neq x_i$). Therefore x_i is a vertex of P. This implies $P \subset \operatorname{conv}(\operatorname{vert} P)$. Hence, the theorem follows. □

Convention:

If we write $P = \operatorname{conv}\{x_1, \dots, x_r\}$, we assume, if not otherwise stated, $x_1, \dots, x_r$ to be the vertices of P, $\operatorname{vert} P = \{x_1, \dots, x_r\}$.

1.3 Definition. The intersection of finitely many closed half-spaces in $\mathbb{R}^n$ is called a *polyhedral set.*

1.4 Theorem. *Every polytope P is a bounded polyhedral set.*

PROOF. We may assume $\operatorname{aff} P = \mathbb{R}^n$. Let $F_i := P \cap H_i$ be the facets of P ($(n-1)$-dimensional faces), and let $P \subset H_i^-$, $i = 1, \dots, s$.

Obviously, P is contained in

$$\bigcap_{i=1}^{s} H_i^- =: P'.$$

Suppose $x_0 \in P' \setminus P$. Consider the union $\mathcal{A}$ of all affine subspaces of $\mathbb{R}^n$ spanned by x_0 and at most $n-1$ vertices of P. Since $\mathcal{A}$ has no interior points, there exists

$$x \in (\operatorname{int} P) \setminus \mathcal{A}.$$

The line segment $[x, x_0]$ is not contained in $\mathcal{A}$ and intersects ∂P in a point y. Since ∂P is the union of all (proper) faces of P (I, Theorem 3.9), y is contained in a face F. From $\dim F < n-1$ would follow $x \in \mathcal{A}$, a contradiction. Therefore F is a facet, say F_0', and $y \in \operatorname{relint} F$. But, then, $\operatorname{aff} F_0$ would be one of the hyperplanes H_i, $i \in \{1, \dots, s\}$, and so $x_0 \notin P'$, a contradiction to the initial assumption. □

1.5 Theorem. *Every bounded polyhedral set is a polytope.*

PROOF. We will proceed by induction on $\dim P$, $P := H_1^- \cap \dots \cap H_s^-$. Let us assume that each of the (proper) faces $F_j := H_j \cap P$ is a polytope. Replacing $\mathbb{R}^n$

by aff P we may assume that P is of maximal dimension. Obviously,

$$\operatorname{conv}(\bigcup_{j=1}^{s} F_j) \subset P;$$

it suffices, thus, to show the opposite inclusion for int P. For $x \in \operatorname{int} P$, fix a ray σ emanating from x not parallel to any H_j for $j = 1, \ldots, s$. Then, by I, Lemma 5.11, $\sigma \cap \partial P$ consists of one point x_σ. Since $\partial P \subset \bigcup_{j=1}^{s} F_j$, the point x_σ is contained in a face, say F_{j_σ}. The analogous statement holds for the ray opposite to σ. Since $x \in [x_\sigma, x_\tau]$, we find $x \in \operatorname{conv}(F_{j_\sigma} \cup F_{j_\tau})$, and, then,

$$\text{(1)} \qquad \operatorname{int} P \subset \operatorname{conv}(\bigcup_{j=1}^{s} F_j).$$

□

We may summarize Theorems 1.4 and 1.5 as follows:

polytopes = bounded polyhedral sets

1.6 Corollary. *Any affine subspace L of $\mathbb{R}^n$ intersects a given polyhedral set (polytope) P in a polyhedral set (polytope).*

We are now ready to prove the converse of I, Lemma 4.4, in the case of polytopes.

1.7 Theorem. *Let P be a polyhedral set. If F_1 is a face of P and F_0 is a face of F_1, then, F_0 is a face of P.*

PROOF. First, let P be bounded, that is, a polytope P, and vert $P =: \{x_1, \ldots, x_m\}$. We may assume that $x_1 = 0 \in F_0 \neq F_1$. There are linearly independent u_0, u_1 such that, for $H_i := \{x \mid \langle x, u_i \rangle = 0\}$, $i = 0, 1$,

$$F_0 = H_0 \cap F_1, \quad F_1 \subset H_0^-,$$
$$F_1 = H_1 \cap P, \quad P \subset H_1^-.$$

We denote by $x_2, \ldots, x_s$ the vertices of $P \setminus F_1$, by $x_{s+1}, \ldots, x_t$ those of $F_1 \setminus F_0$. For $i = 2, \ldots, s$, there exist points u_i such that

$$H_i := \operatorname{lin}(\{x_i\} \cup (H_0 \cap H_1)) = \{x \mid \langle x, u_i \rangle = 0\}.$$

All u_i lie in the plane $(H_0 \cap H_1)^\perp$; hence, we may assume that $F_1 \subset \bigcap_{i=2}^{s} H_i^-$ and that all u_i, considered as points, lie on the line g through u_0 and u_1 (Figure 1):

$$u_i = u_0 + \alpha_i(u_1 - u_0), \qquad i = 2, \ldots, s.$$

The u_i's even lie on the ray of g emanating from u_1 and including u_0, since $\alpha_i \in \mathbb{R}_{<1}$. From $x_j \in H_i^-$, for $j \in \{s+1, \ldots, t\}$, we see that

$$0 > \langle x_j, u_i \rangle = (1 - \alpha_i)\langle x_j, u_0 \rangle.$$

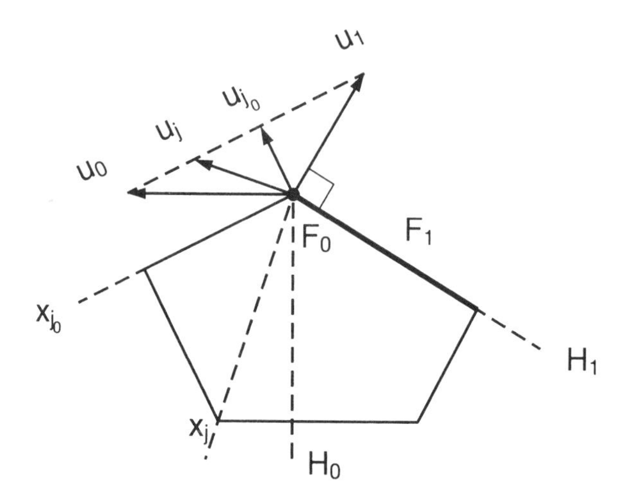

FIGURE 1.

Since $F_1 \subset H_0^-$ implies $\langle x_j, u_0 \rangle < 0$, $(1 - \alpha_i) > 0$. Hence, there exists a point $u \in g$ separating u_1 from $\{u_2, \ldots, u_s\}$ properly, that is,

$$u = \lambda_i u_1 + (1 - \lambda_i) u_i, \qquad \text{for some } 0 < \lambda_i < 1, i = 2, \ldots, s.$$

The hyperplane $H := \{x \mid \langle x, u \rangle = 0\}$ is a supporting hyperplane of P with $H \cap P = F_0$. For $x_j \in F_1$, we obtain

$$\langle x_j, u \rangle = \lambda_i \langle x_j, u_1 \rangle + (1 - \lambda_i) \langle x_j, u_i \rangle = (1 - \lambda_i) \langle x_j, u_i \rangle \leq 0,$$

since $F_1 \subset H_i^-$. Thus, $\langle x_j, u \rangle = 0$ if and only if $x_j \in F_0 \subset H_0 \cap H_1$. For $x_i \in \operatorname{vert} P \setminus F_1$, $\langle x_i, u_1 \rangle < 0$ and, thus,

$$\langle x_i, u \rangle = \lambda_i \langle x_i, u_1 \rangle + (1 - \lambda_i) \langle x_i, u_i \rangle = \lambda_i \langle x_i, u_1 \rangle < 0,$$

which implies $P \subset H^-$ and $P \cap H = F_0$.

If P is not a polytope, we choose a sufficiently large n-simplex S so that int S intersects each face of P (in particular, vert $P \subset$ int S). Then, all bounded faces of P are contained in int S. If F is an unbounded face of P, we find that $F = P \cap H$, H a supporting hyperplane of P, if and only if $F \cap S = P \cap S \cap H$. Each face F of P intersects $P \cap S$ in a face $F' := P \cap S \cap F$ of $P \cap S$ such that $\dim F = \dim F'$. So, the theorem readily follows from its validity for $P \cap S$. □

1.8 Theorem. *Any proper face of a polyhedral set P is a face of a facet of P.*

PROOF. Let $F = H \cap P$ be a face of P. We proceed in three steps:

(1) Every $x \in F$ lies in a facet of P: We may assume $\dim P = n$ and $\dim F \leq n - 2$. For some $y \in$ int P, we consider the union M of all affine subspaces spanned by y and an at most $(n-2)$-dimensional face of P. By Theorem 1.1, M is included in the union of finitely many hyperplanes. Therefore, in any

neighborhood of x, there exists a point $z \notin M$. The ray emanating from y and containing z intersects the boundary of P in a point $z_0 \notin M$, hence in a facet F_0 of P. Since z_0 can be arbitrarily close to x, we see that $x \in F_0$.

(2) The face F is included in a facet: Let $x \in \operatorname{relint} F$. By (1), $x \in F_0$ for some facet F_0. Since F_0 is a polytope, $x \in \operatorname{relint} F'$ for a face F' of F_0. By Theorem 1.7, F' is a face of P. Hence, by I, Lemma 4.6(b), $F = F'$, so that $F \subset F_0$.

(3) If F is included in a facet F_0, then, F is a face of F_0, by I, Lemma 4.4.

□

1.9 Theorem. *Let* $\dim F^j = j$, $j = 0, \ldots, n$, *and let* $F^i \subset F^k$ *for faces* F^i, F^k *of the polyhedral set* P, $i < k$. *Then, there exist faces* $F^{i+1}, \ldots, F^{k-1}$ *of* P *such that*

$$F^i \subset F^{i+1} \subset \cdots \subset F^{k-1} \subset F^k.$$

PROOF. We use induction on k. For $k = i + 1$, there is nothing to prove. So, let $k > i + 1$. Relative to aff F^k, the face F^i is contained in a facet F^{k-1} of F^k (Theorem 1.8). By applying the induction hypothesis, the theorem follows. □

1.10 Theorem. *Each* ($\dim P - 2$)*-face of a polyhedral set* P *is the intersection of precisely two facets of* P.

PROOF. We may assume that $\dim P = n$. By step (3) of the proof of Theorem 1.8, F is a face of each of finitely many facets $F_1, \ldots, F_s$ containing F. Essentially, we reduce the problem to the case $n = 2$.

In appropriate coordinates, we may decompose $\mathbb{R}^n$ as $\mathbb{R}^{n-2} \oplus E$ with $F \subset \mathbb{R}^{n-2}$ to be a neighborhood of 0 relative to $\mathbb{R}^{n-2}$. For every $x \in \operatorname{relint} F$ and every j, we obtain pairwise different lines l_{xj} in E, given by $\{x\} + E \cap F_j$. If $s = 1$, then $\dim[P \cap (\{x\} + E)] = 1$, which contradicts $\dim P = n$. If $s \geq 3$, by considering three lines, say l_{x1}, l_{x2}, l_{x3}, we may assume $l_{x3} \subset \operatorname{conv}(l_{x1}, l_{x2})$. Since that inclusion holds for all points in F near x, we find $F_3 \subset \operatorname{conv}(F_1 \cup F_2)$, and F_2 is not a facet of P. □

1.11 Theorem. *Let* P *be a polyhedral set, and fix* $j, k \in \mathbb{Z}_{>0}$ *with* $j \leq k < \dim P$. *Then, every* j*-face* F^j *is an intersection of* k*-faces of* P. *In particular,* F^j *is an intersection of facets of* P.

PROOF. By Theorem 1.9, there exists a chain

$$F^j \subset \cdots \subset F^{k-1} \subset F^k \subset F^{k+1}.$$

If $k = j + 1$, the theorem follows from Theorem 1.10. If $k > j + 1$, we apply Theorem 1.11 and see that $F^j = F_1^{j+1} \cap F_2^{j+1}$ is an intersection of facets relative to F^{j+2}. Again, if $k > j + 2$, we represent F_1^{j+1}, F_2^{j+1} as intersections of $(j+2)$-faces. Continuing in this way, the theorem follows. (As an illustration, see Figure 2.)

□

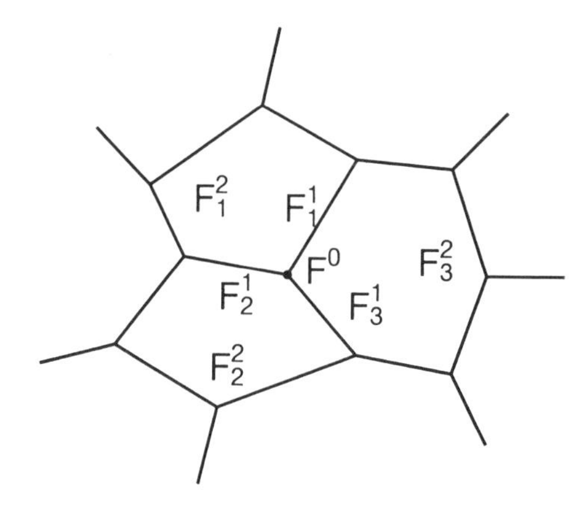

FIGURE 2. $F^0 = F_1^1 \cap F_2^1 = (F_1^2 \cap F_3^2) \cap (F_1^2 \cap F_2^2) = F_1^2 \cap F_2^2 \cap F_3^2$

Frequently, theorems on convex polytopes and, more generally, polyhedral sets refer only to the inclusion properties of their sets of faces. It is, therefore, useful to introduce the following notions.

1.12 Definition. The set of all proper faces of a polyhedral set P (without P and $\emptyset$) is denoted by $\mathcal{B}_0(P)$. $\mathcal{B}(P) := \mathcal{B}_0(P) \cup \{\emptyset\}$ is called *the boundary complex* $\mathcal{B}(P)$ of P, $\mathcal{B}_0(P)$ the *reduced boundary complex* of P. A bijective, inclusion-preserving map

$$\psi : \mathcal{B}(P) \quad \to \quad \mathcal{B}(P'),$$

where P, P' are polyhedral sets, is called a *combinatorial isomorphism* (or *equivalence*); when such a map exists, P and P' are called *combinatorially isomorphic*:

$$P \approx P'.$$

Equivalence classes under combinatorial isomorphisms are also said to be *types* of polyhedral sets (or polytopes).

It is readily seen that $\approx$ does satisfy the conditions of an equivalence relation (reflexivity, symmetry, transitivity).

Clearly, under an affine or permissible projective transformation, any polytope is mapped onto a combinatorially equivalent one. The converse is not true, as can be verified by pentagons in $\mathbb{R}^2$.

The exercises below, although expressed for individual polyhedral sets, are, in fact, concerned with types of polyhedral sets.

One of the basic invariants of a polytope under combinatorial isomorphisms is the vector defined as follows.

1.13 Definition. The number of k-faces of a polytope (or a polyhedral set) P is denoted by $f_k(P)$, $k = 0, \ldots, n-1$, and

$$f(P) := (f_0(P), \ldots, f_{n-1}(P))$$

is called the *f-vector* of P. For the improper faces $\emptyset$ and P, we set $f_{-1}(P) = 1$, $f_n(P) = 1$.

$\{f_0(P), \ldots, f_{n-1}(P)\}$ cannot be an arbitrary set of natural numbers. They must satisfy, for example, the "Euler relation" (to be discussed in III, 3).

Exercises

1. A polyhedral set $P \neq \emptyset$ in $\mathbb{R}^n$, $n \geq 1$, does not possess a vertex if and only if it contains a line.
2. Let $x \in \operatorname{relint} F$, F a face of the polyhedral set P in $\mathbb{R}^n$. Then, x is a k-boundary point of P (see I, 4, Exercise 4) if and only if $\dim F = k$.
3. If $V \subset \operatorname{vert} P$, where P is a polytope, then, $\operatorname{conv} V$ is a face of P if and only if $\operatorname{aff} V \cap \operatorname{conv}[(\operatorname{vert} P) \setminus V] = \emptyset$.
4. Let $1 \leq k \leq n$ and $x_1, \ldots, x_k \in \operatorname{vert} P$, P an n-dimensional polytope in $\mathbb{R}^n$. Then, there exists an $(n-k)$-face F of P such that $F \cap \{x_1, \ldots, x_k\} = \emptyset$.

2. Polar polytopes and quotient polytopes

In I, 6, we introduced polar bodies of full-dimensional convex bodies. Now we investigate special features of polar polytopes. In particular, we define polar faces. Subsequently, we introduce quotient polytopes and their relation to polar polytopes.

Example 1. Let p in Figure 10 of chapter I be a vertex of one of the "inner" polytopes P. Then, the polar plane $\pi(p)$ of p intersects P^* in the facet

$$F^* := P^* \cap \pi(p).$$

In the case of the octahedron, we obtain a square, in the case of the icosahedron, a pentagon. If q is an adjacent vertex of p, the polar face $[p, q]^*$ of $[p, q]$ is again a line segment

$$[p, q]^* = P^* \cap \pi(\operatorname{aff}[p, q]) = P^* \cap \pi(p) \cap \pi(q).$$

Example 2. Suppose $[p, q]$ is such an edge of a 4-dimensional polytope P that $[p, q]$ is the intersection of three facets of P which are combinatorially isomorphic to triangular prisms. Also suppose that p and q are contained each in only one more facet which is a simplex (Figure 3a). Then, P^* contains two facets p^*, q^*, and $[p, q]^* := p^* \cap q^* = P^* \cap \pi(\operatorname{aff}[p, q])$ is a triangle (Figure 3b).

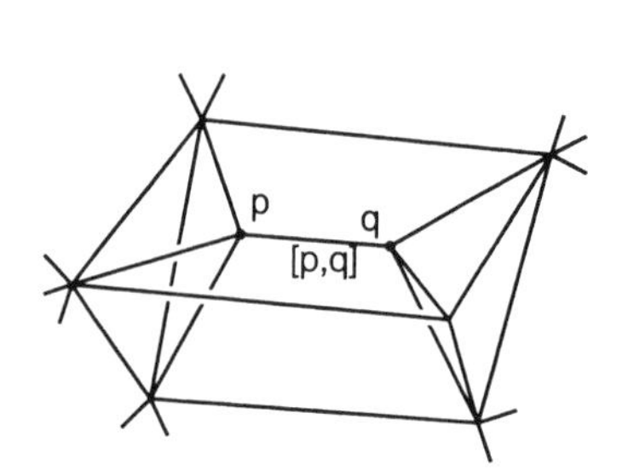

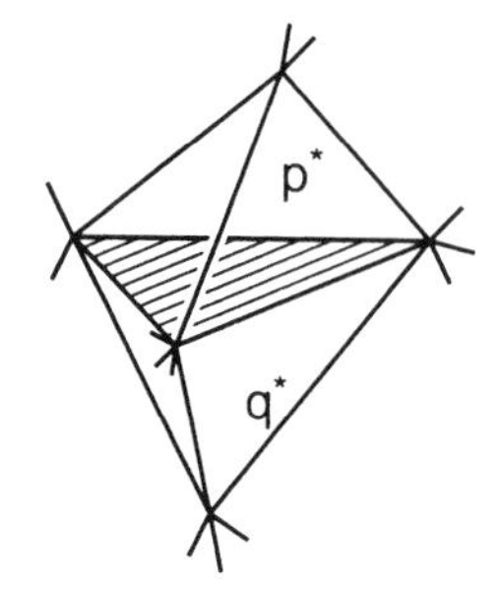

FIGURE 3a,b.

In general, we will prove Theorem 2.1.

2.1 Theorem. *Let P be an n-dimensional polytope in $\mathbb{R}^n$, $0 \in \operatorname{int} P$, and let P^* be the polar polytope under the polarity π.*

(a) *If F is a proper face of P, then,*

$$F^* := P^* \cap \pi(\operatorname{aff} F)$$

is a face of P^* and

$$\dim F^* = n - 1 - \dim F.$$

(b) *The assignment $F \mapsto F^*$, for all $F \in \mathcal{B}_0(P)$, induces a bijective and inclusion-reversing map*

$$\varphi : \quad \mathcal{B}_0(P) \quad \longrightarrow \quad \mathcal{B}_0(P^*)$$

between the reduced boundary complexes of P and P^*.

(c) *$F^{**} = F$.*

(d) *$P^* = \bigcap_{u \in \operatorname{vert} P} H_u^-$ (see I, 6.1.); in particular, P^* is an n-dimensional polytope with $0 \in \operatorname{int} P^*$.*

PROOF. We may assume that $F = \operatorname{conv}\{v_0, \ldots, v_r\}$ is a proper face of $P = \operatorname{conv}\{v_0, \ldots, v_s\}$.

(d) The inclusion $P^* = \bigcap_{u \in \partial P} H_u^- \subset \bigcap_{i=0}^{s} H_{v_i}^-$ is obvious. For the opposite inclusion, it suffices to verify that $x \in H_u^-$ for every $x \in \bigcap H_{v_i}^-$ and every $u \in \partial P$. By Theorem 1.2, there exists a representation

$$u = \lambda_0 v_0 + \cdots + \lambda_s v_s, \qquad \text{where } \lambda_0 + \cdots + \lambda_s = 1 \text{ and all } \lambda_i \geq 0.$$

Thus, $x \in H_u^-$ follows from

$$\langle x, u \rangle = \langle x, \lambda_0 v_0 + \cdots + \lambda_s v_s \rangle \leq (\lambda_0 + \cdots + \lambda_s) \cdot 1 = 1.$$

This implies, in particular, that P^* is a polyhedron. The assumption $0 \in \operatorname{int} P$ implies that there exists an n-simplex Δ with $0 \in \operatorname{int} \Delta \subset P$. As a subset of Δ^*, the polyhedron P^* is bounded and, thus, a polytope by Theorem 1.5.

(a) The subset F^* of P^* is a face. Since $\pi(\text{aff } F) = \bigcap_{i=0}^r H_{v_i}$, we obtain that

$$F^* = P^* \cap \pi(\text{aff } F) = P^* \cap \bigcap_{i=0}^{r} H_{v_i} = \bigcap_{i=0}^{r} (P^* \cap H_{v_i})$$

is a face by I, Lemma 4.5.

One inequality is obvious because of $F^* \subset \pi(\text{aff } F)$ and the dimension formula:

$$\dim F^* \leq \dim \pi(\text{aff } F) = n - 1 - \dim F.$$

There remains the inequality "$\geq$". By Theorem 1.11, there exist facets F_j such that $F = \bigcap_{j=0}^t F_j$. Since $0 \in \text{int } P$, we can find vectors $u_j \neq 0$ with $F_j = P \cap H_{u_j}$. We may assume that $t = \text{codim } F$; then, the u_j are affinely independent. If we can show that $u_j \in F_j^*$ ($\subset F^*$), for every j, then,

$$\dim F^* \geq t - 1 = n - 1 - \dim F$$

follows. Thus, let F itself be a facet with $F = P \cap H_u$ and $P \subset H_u^-$. Then,

$$F^* = P^* \cap \pi(\text{aff } F) = P^* \cap \pi(H_u) = \bigcap_{i=0}^{s} H_{v_i}^- \cap \bigcap_{x \in H_u} H_x$$

yields $u \in F^*$, since $u \in H_x$ holds for every $x \in H_u$ and since $v_i \in P \subset H_u^-$ implies $u \in H_{v_i}^-$.

(c) I, Theorem 6.4 implies

$$F = P \cap F = P \cap \pi(\pi(\text{aff } F)) = P^{**} \cap \pi(\text{aff } F^*) = F^{**}.$$

(b) Evidently, φ is inclusion-reversing, and, by (c), $\varphi^2 = \text{id}$, which implies the desired statement.

□

2.2 Definition. The face F^* of P^* in Theorem 2.1 is called the *polar face* of F.

We remark that many "linear" properties of P carry over to P^* since P^* is defined by the projective concept of polarity. In the example of a "bipyramid" P above, the polar body is a prism and, hence, has parallel edges joining "top" and "bottom". If we slightly disturb one of the vertices of P, this is no longer true, though the new polytope is combinatorially isomorphic to P. Even if P undergoes only a permissible projective transformation, P^* will not remain a prism; however, P^* will also undergo a permissible projective transformation so that its parallel edges are mapped onto edges whose affine hulls intersect in a point. In general (I, Lemma 6.3,

2.3 Lemma. *If P is mapped onto P' by a permissible projective transformation, then, P'^* is also the image of P^* under a suitable projective transformation.*

By the following theorem, polarity is given a combinatorial frame:

2.4 Theorem. *If P and P' are combinatorially equivalent, then, so are P^* and P'^*:*

$$P \approx P' \quad \textit{implies} \quad P^* \approx P'^*.$$

PROOF. This follows from Theorem 2.1(b) and the definition of combinatorial equivalence. □

So, we may consider P^* as either a polytope or a type of a polytope.

One of the most important combinatorial invariants is the f-vector of a polytope (see Definition 1.13). As a direct consequence of Theorem 2.1,

2.5 Theorem. *For each n-dimensional polytope P,*

$$f_k(P^*) = f_{n-k-1}(P), \qquad -1 \leq k \leq n.$$

In the example of platonic solids (see Figure 10 in I, 6), φ maps the 8 vertices, 12 edges, and 6 facets of a cube onto the 8 facets, 12 edges, and 6 vertices of an octahedron, respectively. In the case of a dodecahedron, the 20 vertices, 30 edges, and 12 facets are mapped onto the 20 facets, 30 edges, and 12 vertices of an icosahedron, respectively. If P is a simplex, P^* is also a simplex, and, hence, there are as many vertices as there are facets of a simplex.

As we have seen in Corollary 1.6, the intersection of an affine space U with a polytope P is again a polytope. We consider a special choice of U:

2.6 Lemma. *Let F be a proper face of the n-polytope $P \subset \mathbb{R}^n$. Then, an affine subspace U of $\mathbb{R}^n$ can be chosen so that the following conditions are satisfied.*

(a) $\operatorname{aff}(U \cup F) = \mathbb{R}^n$.

(b) *If F' is a face of P and $F \underset{\neq}{\subset} F'$, then $U \cap \operatorname{relint} F' \neq \emptyset$.*

(c) *If F' is a face of P and $F \not\subset F'$, then, $U \cap F' = \emptyset$.*

(d) $\dim F + \dim(U \cap P) = n - 1$.

PROOF. Let H be such a supporting hyperplane of P that $F = P \cap H$, and let $x \in \operatorname{relint} F$. We choose, in x an affine subspace, U_0 of H complementary to $\operatorname{aff} F$ (relative to H). Let $y \in \operatorname{int} P$. Then, for a sufficiently small $\delta > 0$,

$$U := U_0 + \delta(y - x)$$

has properties (a), (b), and (c) whereas (d) is a consequence of (a) and (b). □

2.7 Definition. Let P, F, U be given as in Lemma 2.6. Then, $P/F := P \cap U$ is called a *face figure* or a *quotient polytope* of P with respect to F. For $F = p \in \operatorname{vert} P$, we also say that P/p is a *vertex figure* (Figure 4).

Example 3. The vertex figures of

octahedra	are	quadrangles,
cubes	are	triangles,

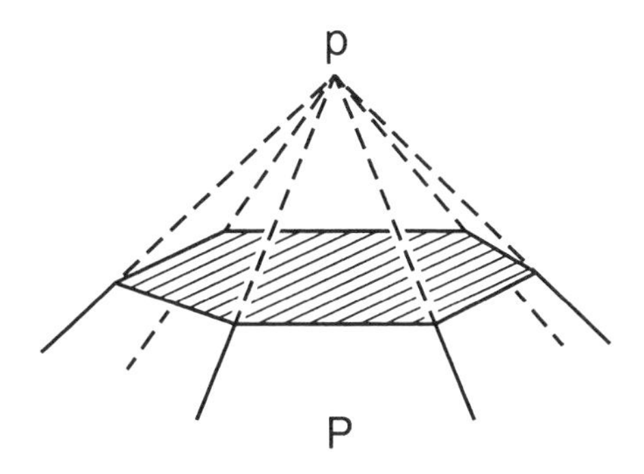

FIGURE 4.

icosahedra	are	pentagons, and
dodecahedra	are	triangles.

2.8 Theorem. *Let F be a proper face of the polytope P, and let P^* be the polar polytope of P with respect to a polarity π. For an affine subspace U of $\mathbb{R}^n$, let π_U denote the restriction of π to U, and set $\pi_U(P/F) := \pi_U(\text{aff }(P/F))$.*

(a) *For any face figure P/F of F,*

$$\pi_U(P/F) \approx F^*.$$

(b) *Any two face figures of P with respect to F are combinatorially isomorphic. Therefore, we can consider P/F to be the equivalence class of these face figures.*

PROOF. Let $k := \dim F$, so that $\dim U = n - k - 1$ for U as in 2.6. For any face G which contains F properly, we set

$$G \underset{\varphi}{\longmapsto} \pi_U(G \cap U).$$

If $g := \dim(G \cap U)$, so that $g = \dim G - k - 1$, then, by the construction in Lemma 2.6,

$$\dim \varphi(G) = (n - k - 1) - g - 1 = n - (g + 1 + k) - 1 = n - \dim G - 1.$$

Therefore $\dim \varphi(G) = \dim G^*$. The mapping φ is inclusion-reversing as is the polarity π. So, by

$$\varphi(G) \quad \longmapsto \quad G^* \in \mathcal{B}(F^*),$$

we obtain an isomorphism

$$\mathcal{B}_0(\pi_U(P/F)) \quad \longrightarrow \mathcal{B}_0(F^*),$$

whence (a) follows. Note that, by Lemma 2.6(c), only faces G including F properly contribute to $\mathcal{B}(\pi_U(P/F))$.

Part (b) of the theorem is a consequence of (a). □

Example 4. If F is a k-face of a simplex P, then, P/F is an $(n-k-1)$-simplex.

Example 5. If F is a facet of any n-polytope P, then, P/F is a point.

The mappings constructed in the proof of Theorem 2.8 easily yield the following result:

2.9 Theorem. *Let $[F, P]$ be the set of proper faces of P which properly contain the face F. Then, there are bijective mappings*

$$[F, P] \xrightarrow{\psi} \mathcal{B}_0(P/F) \xrightarrow{\varphi} \mathcal{B}_0(\pi_U(P/F)) \xrightarrow{\chi} \mathcal{B}_0(F^*),$$

where φ reverses and ψ, χ preserve inclusion.

Remark. The mapping ψ lowers dimension by $\dim F$, whereas φ reverses dimension according to Lemma 2.6(d) and χ is a combinatorial isomorphism.

Exercises

1. Find examples of 3-polytopes and 4-polytopes P other than simplices for which
$$f_k(P^*) = f_k(P),$$
$k = 0, 1, 2$ if $n = 3$; $k = 0, 1, 2, 3$ if $n = 4$.
2. Find examples of pairs of 3-polytopes which have the same sets of vertex figures but are not combinatorially equivalent.
3.
   a. Place on each facet F of a dodecahedron (I, Figure 10) a flat pentagonal pyramid conv $(F \cup \{p\})$, so that again a convex polytope is obtained and "old" edges are also edges of the new polytope P. Describe P^* and draw a diagram.
   b. How does P^* change for an n-dimensional P if a facet F of P is replaced by conv $(F \cup \{p\})$ so that, for the new polytope P', vert $P' = \{p\} \cup$ vert P?
4. Let P be an n-polytope, G a face of P, F a face of G. Suppose G/F to be defined relative to aff G. Show that the representations of G/F, P/F, and P/G can so be chosen that $(P/F)/(G/F)$ is defined (relative to aff(P/F)). Prove
$$(P/F)/(G/F) \approx P/G.$$

3. Special types of polytopes

There are two large classes of polytopes which are of great importance in the sequel:

3.1 Definition. If a polytope P has only simplices as proper faces, it is called *simplicial*.

For such P, we call P^* *simple*.

A more convenient characterization of simple polytopes is given by the following lemma:

3.2 Lemma. *An n-dimensional polytope Q is simple if and only if any vertex p of Q is contained in precisely n edges of Q.*

PROOF. Under the polarity π, a vertex p of Q corresponds to a facet F_p of Q^*, and the edges of Q containing p to the facets of F_p. Thus, the number of those edges is precisely the number of facets of F_p. Since Q^* is simplicial if and only if each of its facets F_p has precisely n facets (see the remark following I, 14), the lemma is evident. □

Now we will present some special types of polytopes.

3.3 Definition. If Q is an $(n-1)$-polytope in $\mathbb{R}^n$ and $p \notin \operatorname{aff} Q$, we call $P := \operatorname{conv}(\{p\} \cup Q)$ a *pyramid* with *basis* Q (or *over* Q) and with *apex* p. If Q, in turn, is a pyramid over an $(n-2)$-polytope, we say P is a *2-fold pyramid*. Inductively, P is called a *k-fold pyramid* over R if it is a pyramid whose basis is a $(k-1)$-fold pyramid over R, $1 \leq k \leq n$.

3.4 Theorem. *If P is a k-fold pyramid over R, then, P^* is a k-fold pyramid over the polar p^* of the apex p of P, $1 \leq k \leq n-1$.*

PROOF. Let Q be the basis of P (as a 1-fold pyramid). The polar face Q^* is a vertex of P^*, and p^* is a facet of P^*. If q is a vertex of P^* other than Q^*, $q = G^*$ for some facet G of P which is different from Q and, hence, contains p. Therefore, q is contained in p^*, and, hence, by I, Lemma 4.4, a vertex of p^*. This implies $P^* = \operatorname{conv}(Q^* \cup p^*)$ so that P^* is again a pyramid.

If $k > 1$, we apply the same arguments to Q instead of P. Continuing in this way, the theorem is proved. □

3.5 Definition. Let Q be an $(n-1)$-polytope, and let I be a line segment such that

$$(\operatorname{relint} Q) \cap (\operatorname{relint} I) = \{q\},$$

is a single point. Then, $P := \operatorname{conv}(Q \cup I)$ is called a *bipyramid* with *basis* Q (or *over* Q). P is said to be a *k-fold bipyramid* over R, $1 \leq k \leq n$, if, in the case $k > 1$, it is a bipyramid over a $(k-1)$-fold bipyramid over R.

3.6 Definition. Let Q be an $(n-1)$-polytope, and let Q' be a translate of Q not contained in aff Q. Then,

$$P := \operatorname{conv}(Q \cup Q')$$

is called a *prism*. P is said to be a *k-fold prism*over R, $1 \leq k \leq n$, if, in the case $k > 1$, it is a prism over a $(k-1)$-fold prism over R.

3.7 Theorem. *If P is a k-fold bipyramid, P^* is, up to a permissible projective transformation, a k-fold prism, and conversely* (Figure 5).

PROOF. First, let $k = 1$. Up to a translation, we can assume $q = 0$ (in the definition of a bipyramid). The translation induces a permissible projective transformation (see I, 6). By an appropriate linear transformation (shear transformation), we can arrange I perpendicular to aff Q. Let $I = [a, b]$. Now the polar faces of the $(n-2)$-faces of Q are line segments perpendicular to aff Q, with end points on the facets a^* and b^* of P^*, so that we obtain a prism. For $k > 1$, we again proceed inductively. □

3.8 Definition. An n-fold bipyramid is also called an *n-crosspolytope*, an n-fold prism is also called an *n-cube*.

2-crosspolytopes are convex quadrangles, 3-crosspolytopes are octahedra. n-cubes (also called *spars* or *parallelepipeds*) are affine images of ordinary n-cubes.

A particular case of Theorem 3.7 is

3.9 Corollary. *If P is an n-crosspolytope, then, up to a permissible projective transformation, P^* is an n-cube, and conversely.*

3.10 Definition. The curve $\{x(t) = (t, t^2, \ldots, t^n) \mid t \in \mathbb{R}\}$ is called a moment curve. . Let $x(t_1), \ldots, x(t_v)$ be different points on it, where $v > n$. Then,

$$C(v, n) := \operatorname{conv}\{x(t_1), \ldots, x(t_v)\}$$

is called a *cyclic polytope* (Figure 6).

If, in $\mathbb{R}^3$, every pair of two vertices of a polytope P are joined by an edge of P, it is readily seen that P is a simplex. The polytopes $C(v, n)$ for $n \geq 4$ show

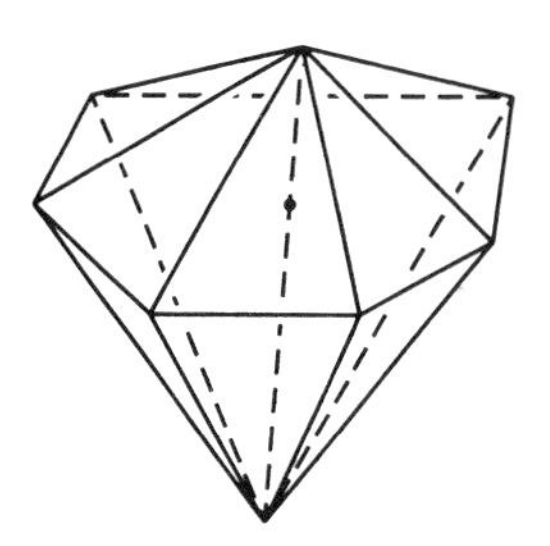

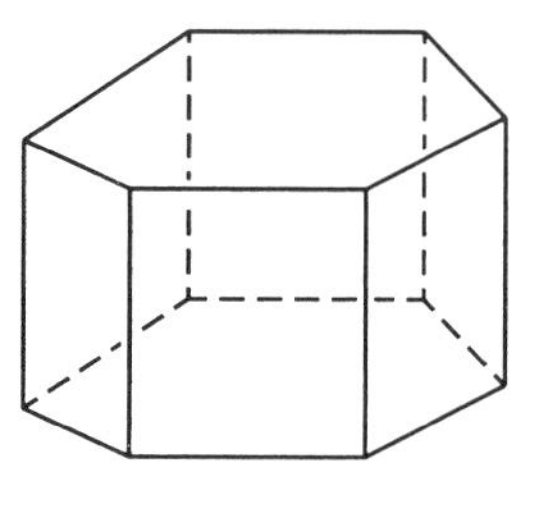

FIGURE 5.

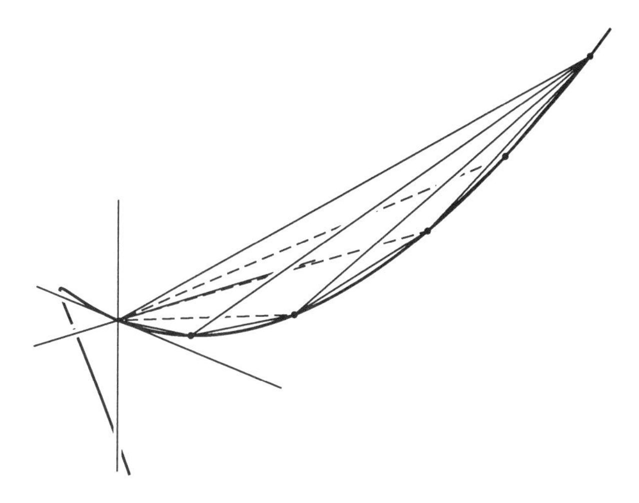

FIGURE 6.

that this is no longer true in $\mathbb{R}^n$. The importance of cyclic polytopes lies, more generally, in their richness of faces.

3.11 Theorem. *The cyclic polytopes are simplicial.*

PROOF. First, we show that every $n + 1$ of the vertices of $C(v, n)$ are affinely independent. For pairwise different points $t_0, \ldots, t_n$,

$$D := \begin{vmatrix} 1 & t_0 & t_0^2 & \ldots & t_0^n \\ 1 & t_1 & t_1^2 & \ldots & t_1^n \\ \vdots & & & & \vdots \\ 1 & t_n & t_n^2 & \ldots & t_n^n \end{vmatrix} = \prod_{0 \le i < j \le n} (t_j - t_i) \neq 0.$$

Hence $x(t_0), \ldots, x(t_n)$ are affinely independent.

Thus, each proper face F of $C(v, n)$ includes, at most, n vertices of $C(v, n)$, and those are affinely independent, i.e., F is a simplex. □

3.12 Theorem. *For $2 \le 2k \le n$, every subset of* vert $C(v, n)$ *containing k points is the vertex set of a $(k - 1)$-face of $C(v, n)$; hence,*

$$f_{k-1}(C(v, n)) = \binom{v}{k} \quad \textit{for} \quad 0 \le k \le \left[\frac{n}{2}\right].$$

PROOF. For $t_1 < \cdots < t_k$, write the real polynomial in t

$$p(t) := \prod_{i=1}^{k} (t - t_i)^2$$

as

$$p(t) = \beta_0 + \beta_1 t + \cdots + \beta_{2k} t^{2k}.$$

Consider the vector

$$b := (\beta_1, \beta_2, \dots, \beta_{2k}, 0, \dots, 0) \in \mathbb{R}^n$$

as normal vector of the hyperplane

$$H = \{x \in \mathbb{R}^n \mid \langle x, b\rangle = -\beta_0\}.$$

Then, for $1 \le i \le k$,

$$\langle x(t_i), b\rangle = \beta_1 t_i + \cdots + \beta_{2k} t_i^{2k} = p(t_i) - \beta_0 = -\beta_0,$$

hence, $x(t_i) \in H$ for $i = 1, \dots, k$.

For any further vertex $x(t_j)$,

$$\langle x(t_j), b\rangle = -\beta_0 + p(t_j) = -\beta_0 + \prod_{i=1}^{k}(t_j - t_i)^2 > -\beta_0,$$

so H is a supporting hyperplane of $C(v, n)$, and, by Theorem 3.11,

$$H \cap C(v, n) = \operatorname{conv}\{x(t_1), \dots, x(t_k)\}$$

is a $(k-1)$-face of $C(v, n)$. □

3.13 Theorem (Gale's evenness condition). *Let V_n be a set of n vertices of $C(v, n)$. Then, V_n is the vertex set of a facet of $C(v, n)$ if and only if all elements of* vert $C(v, n) \setminus V_n$ *are pairwise separated on the moment curve by an even number of elements of V_n.*

PROOF. Let $V_n = \{x(t_1), \dots, x(t_n)\}$ where

$$t_1 < \cdots < t_n,$$

and let $V :=$ vert $C(v, n)$. Furthermore, set

$$q(t) := \prod_{i=1}^{n}(t - t_i) = \sum_{j=0}^{n} \gamma_j t^j,$$

$c := (\gamma_1, \dots, \gamma_n)$, and

$$H := \{x \mid \langle x, c\rangle = -\gamma_0\} \quad \subset \mathbb{R}^n.$$

Then,

$$\langle x(t_i), c\rangle = \sum_{j=1}^{n} \gamma_j t_i^j = q(t_i) - \gamma_0 = -\gamma_0,$$

hence, $x(t_1), \dots, x(t_n)$ lie in H. For all other t

$$q(t) = \langle x(t), c\rangle + \gamma_0 \neq 0.$$

Therefore V_n is the vertex set of a face of $C(v, n)$ if and only if $V \setminus V_n$ lies on one side of H. This, in turn, is equivalent to $q(t)$ changing its sign an even number of times while, increasing t, we move from one vertex of $V \setminus V_n$ to another. This proves the theorem. □

Theorem 3.13 and Theorem 3.11 imply Theorem 3.14.

3.14 Theorem. *In $\mathbb{R}^n$, any two cyclic polytopes with v vertices are combinatorially equivalent.*

This justifies interpreting the notation $C(v, n)$ as the equivalence class of cyclic polytopes of dimension n with v vertices.

Remark. In III, 7., we shall see that in $\mathbb{R}^n$, for each k among all polytopes with $v > n$ vertices, the cyclic polytopes have a maximal number of k-faces.

Exercises

1. If P is a pyramid over Q,

$$f_j(P) = f_j(Q) + f_{j-1}(Q), \quad 0 \leq j \leq n-1.$$

For a k-fold pyramid over R,

$$f_j(P) = \sum_{i=0}^{\min\{k, j+1\}} \binom{k}{i} f_{j-i}(R), \quad 0 \leq j \leq n-1.$$

2. For an n-crosspolytope P and the n-cube P^*,

$$f_j(P) = 2^{j+1}\binom{n}{j+1}, \quad f_j(P^*) = 2^{n-j}\binom{n}{j}, \quad 0 \leq j \leq n-1.$$

3. Let R be an $(n-k)$-polytope in $\mathbb{R}^n$, $0 < k < n$, and let $0 \in \operatorname{relint} R$. Call R^* (I, 6, Exercise 3) an *infinite k-fold prism*, and show in which way it can be considered as the limit of (a) a k-fold prism, (b) a k-fold bipyramid.
4. Justify the face structure of $C(6, 3)$, as shown in Figure 6, by applying Gale's evenness condition.

4. Linear transforms and Gale transforms

The transformations, which we will introduce now, are useful tools for the investigation of polytopes (and polyhedral cones). (Their origin is from linear programming theory). In particular, they will be used for the classification of types of polytopes in section 6. Our point of view for linear algebraic aspects is close to that in P.R. Halmos' book "Finite Dimensional Vector Spaces".

First, we will present a coordinate-free definition. In the next section, the consideration of diagrams will be made more concrete by using matrices.

Let V be a v-dimensional real vector space and let V^* be the dual of V, that is, the linear space consisting of all linear functionals $x^* : V \longrightarrow \mathbb{R}$. For a real linear space U, let $L : V \to U$ be a linear map and L^* the dual map defined by

$L^*(u^*) := u^* \circ L$ for every $u^* \in U^*$. We write

$$V \xrightarrow{L} U, \qquad V^* \xleftarrow{L^*} U^*.$$

We fix a basis $b_1, \ldots, b_v$ of V and the dual basis $b_1^*, \ldots, b_v^*$ of V^*, so that

$$b_i^*(b_j) = \delta_{ij} := \begin{cases} 1 & \text{for } i = j, \\ 0 & \text{for } i \neq j. \end{cases}$$

For every subset M of V, there is a vector space

$$M^\perp := \{x^* \in V^* \mid x^*(M) = 0\}.$$

For $m \in V$, we write $m^\perp$ instead of $\{m\}^\perp$.

Now, we will introduce the notion of a short, exact sequence. Assume we are given a sequence

$$(1) \qquad 0 \longrightarrow W \underset{L_2}{\longrightarrow} V \underset{L_1}{\longrightarrow} U \longrightarrow 0$$

of linear maps and vector spaces. Then, we call (1) a *short, exact sequence* if

$$L_1 \text{ is surjective, } L_2 \text{ is injective, and } \ker L_1 = \operatorname{im} L_2.$$

Then, we may interpret W as a linear subspace of V and L_2 as the injection $W \hookrightarrow V$. Since we deal with vector spaces, by an easy exercise in linear algebra, the exactness of the sequence (1) is equivalent to the existence of a direct sum decomposition $V \cong W \oplus U$, or, more precisely, to the existence of homomorphisms

$$(1a) \qquad W \underset{M_2}{\overset{L_2}{\rightleftarrows}} V \underset{M_1}{\overset{L_1}{\rightleftarrows}} U,$$

such that (using the convention $(L \circ M)(\cdot) = L(M(\cdot))$ for the composition of maps)

$$L_1 \circ L_2 = 0,\ M_2 \circ M_1 = 0,\ L_1 \circ M_1 = \mathrm{id}_U,\ M_2 \circ L_2 = \mathrm{id}_W,\ M_1 \circ L_1 + L_2 \circ M_2 = \mathrm{id}_V.$$

4.1 Lemma. *The dual sequence*

$$(2) \qquad 0 \longleftarrow W^* \underset{L_2^*}{\longleftarrow} V^* \underset{L_1^*}{\longleftarrow} U^* \longleftarrow 0$$

of (1) *is also exact.*

PROOF. (1a) yields the existence of homomorphisms

$$(1b) \qquad W^* \underset{L_2^*}{\overset{M_2^*}{\rightleftarrows}} V^* \underset{L_1^*}{\overset{M_1^*}{\rightleftarrows}} U^*,$$

such that

$$L_2^* \circ L_1^* = 0,\ M_1^* \circ M_2^* = 0,\ M_1^* \circ L_1^* = \mathrm{id}_{U^*},\ L_2^* \circ M_2^* = \mathrm{id}_{W^*},\ L_1^* \circ M_1^* + M_2^* \circ L_2^* = \mathrm{id}_{V^*}.$$

□

We now concern ourselves with finite families $\{x_1, \ldots, x_v\}$ of (not necessarily different) elements whose linear hull is U. We fix an ordering and denote the associated finite sequence by

$$X := (x_1, \ldots, x_v).$$

If a basis of U is fixed, we write the coordinates of the x_j as column vectors and we also denote the resulting matrix $(x_1 \cdots x_v)$ by X. Further, we fix a v-dimensional vector space V with basis $b_1, \ldots, b_v$, that is, $V = \mathbb{R}\, b_1 \oplus \ldots \oplus \mathbb{R}\, b_v \cong \mathbb{R}^v$. The finite sequence X now defines a surjective linear map

$$L_1 : V \cong W \oplus U \quad \longrightarrow \quad U, \qquad \sum_{j=1}^{v} \alpha_j b_j \mapsto \sum_{j=1}^{v} \alpha_j x_j.$$

4.2 Definition. For the dual basis $b_1^*, \ldots, b_v^*$ of the basis $b_1, \ldots, b_v$ of V, we set

$$\bar{x}_i := L_2^*(b_i^*), \qquad i = 1, \ldots, v,$$

and call the finite sequence

$$\bar{X} := (\bar{x}_1, \ldots, \bar{x}_v)$$

a *linear transform* of the sequence X. x_i and $\bar{x}_i$ will be referred to as the *components* of X, $\bar{X}$.

Note that the components of $\bar{X}$ span W^*.

4.3 Lemma. *If $\bar{X}$ is a linear transform of X, then X is a linear transform of $\bar{X}$.*

PROOF. We will apply Lemma 4.1 twice, and use the elementary facts $U = U^{**}$, $L_1 = L_1^{**}$, and so on:

$$\begin{array}{ccccc} W & \xrightarrow[L_2]{} & V & \xrightarrow[L_1]{} & U \\ W^* & \xleftarrow[L_2^*]{} & V^* & \xleftarrow[L_1^*]{} & U^* \\ W = W^{**} & \xrightarrow[L_2 = L_2^{**}]{} & V = V^{**} & \xrightarrow[L_1 = L_1^{**}]{} & U = U^{**}. \end{array}$$

Now, we find, for the elements of the transform of $\bar{X}$,

$$L_1^{**}(b_i^{**}) = L_1(b_i) = x_i \quad \text{for} \quad i = 1, \ldots, v.$$

□

Example 1. Let U, W be subspaces of $V = \mathbb{R}^4 = U \oplus W$ where $U = \text{lin}\{e_1, e_2\}$, $W = \text{lin}\{e_3, e_4\}$, $e_1, \ldots, e_4$ the canonical basis of $\mathbb{R}^4$. We write the coordinates of the elements of U, V, W as column vectors and those of the elements of V^*, U^*, W^* as row vectors, respectively. Applying a dual vector of V^*, U^*, or W^* to a vector of V, U, W, respectively, can be carried out as (matrix) multiplication of a row vector and a column vector.

Let

$$X = (x_1 x_2 x_3 x_4) = \left(\begin{array}{cc|cc} 1 & 0 & 1 & 2 \\ 0 & 1 & 2 & 1 \end{array}\right)$$

$$(b_1 b_2 b_3 b_4) = \left(\begin{array}{cc|cc} 1 & 0 & 1 & 2 \\ 0 & 1 & 2 & 1 \\ \hline 0 & 0 & 1 & 0 \\ 0 & 0 & 0 & 1 \end{array}\right).$$

L_1, L_2 are given as matrices

$$L_1 = \left(\begin{array}{cc|cc} 1 & 0 & 0 & 0 \\ 0 & 1 & 0 & 0 \end{array}\right) = (\, E_2 \mid O \,)$$

$$L_2 = \left(\begin{array}{cc} 0 & 0 \\ 0 & 0 \\ \hline 0 & 0 \\ 0 & 1 \end{array}\right) = \left(\begin{array}{c} O \\ \hline E_2 \end{array}\right).$$

Then,

$$\begin{pmatrix} b_1^* \\ b_2^* \\ b_3^* \\ b_4^* \end{pmatrix} = \left(\begin{array}{cc|cc} 1 & 0 & -1 & -2 \\ 0 & 1 & -2 & -1 \\ \hline 0 & 0 & 1 & 0 \\ 0 & 0 & 0 & 1 \end{array}\right),$$

$L_1(b_i) = x_i$, $i = 1, \ldots, 4$ and $\operatorname{im} L_2 = \ker L_1$ are readily checked. We obtain (Figure 7)

$$\begin{pmatrix} \bar{x}_1 \\ \bar{x}_2 \\ \bar{x}_3 \\ \bar{x}_4 \end{pmatrix} = \left(\begin{array}{c} L_2^*(b_1^*) \\ L_2^*(b_2^*) \\ \hline L_2^*(b_3^*) \\ L_2^*(b_4^*) \end{array}\right) = \left(\begin{array}{c} b_1^* \\ b_2^* \\ \hline b_3^* \\ b_4^* \end{array}\right) L_2 = \left(\begin{array}{cc} -1 & -2 \\ -2 & -1 \\ \hline 1 & 0 \\ 0 & 1 \end{array}\right).$$

4.4 Definition. By the *rank* of a finite sequence X of vectors of U, we mean the dimension of the linear hull of the components of X,

$$\operatorname{rank} X := \dim \operatorname{lin} X.$$

4.5 Lemma. *If $X = (x_1, \ldots, x_v)$ is a sequence of points in U which generate U, then*

$$v = \operatorname{rank} X + \operatorname{rank} \bar{X}.$$

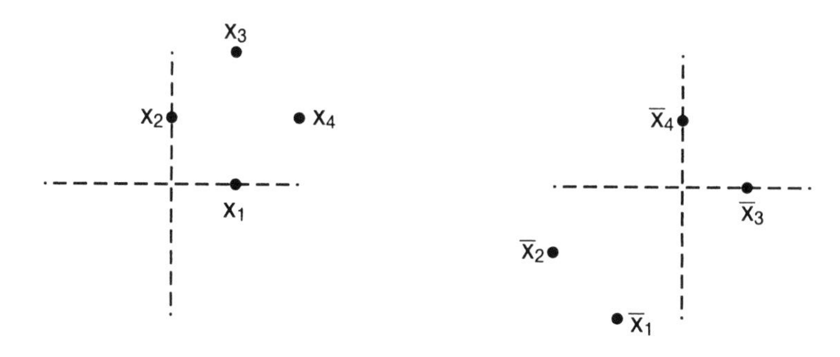

FIGURE 7.

PROOF. Since we assume X to span U, rank $X = \dim U$. In the same way we obtain rank $\bar{X} = \dim W^* = \dim W$. Thus,

$$v = \dim V = \dim W + \dim U = \text{rank } \bar{X} + \text{rank } X.$$

□

The following lemma is obvious:

4.6 Lemma.

(a) *If $L_U : U \longrightarrow U$ is a bijective linear map, $\bar{X}$ is a linear transform of $L_U(X)$ as well.*

(b) *If $L_{W^*} : W^* \longrightarrow W^*$ is a bijective linear map, $L_{W^*}(\bar{X})$ is a linear transform of X as well.*

In particular, for $\lambda \in \mathbb{R}_{\neq 0}$ and linear transform $\bar{X}$, the sequence $\lambda\bar{X}$ is a linear transform, too.

4.7 Definition. We call $\mathcal{L}(X) := \ker L_1 \subset V$ the *space of linear dependencies* (or *linear relations*) of X. It is convenient to write $\alpha \in \mathcal{L}(X) \subset V = \mathbb{R}\, b_1 \oplus \cdots \oplus \mathbb{R}\, b_v$ as column vector $\alpha = (\alpha_1, \ldots, \alpha_v)^t$ with respect to the basis $(b_1, \ldots, b_v)$. Such a linear dependency is called an *affine dependency* if

$$\alpha_1 + \cdots + \alpha_v = 0.$$

So, $\alpha = (\alpha_1, \ldots, \alpha_v)^t$ is a linear dependency for X if and only if $\Sigma\alpha_j x_j = 0$. Note that $\dim \mathcal{L}(X) = v - \dim U$. There is an interpretation of $\mathcal{L}(X)$ also via $\bar{X}$: Consider $\bar{X}$ as the linear mapping

$$\bar{X} : W \to V = \bigoplus_{i=1}^{v} \mathbb{R}\, b_i, \quad w \mapsto \sum_{i=1}^{v} \bar{x}_i(w) b_i;$$

then,

$$\mathcal{L}(X) = \text{im } \bar{X}.$$

4.8 Lemma. $(\alpha_1, \ldots, \alpha_v)^t \in \mathcal{L}(X)$ *if and only if there exists a vector* $a \in W$ *such that*

$$\alpha_i = \bar{x}_i(a) \qquad \textit{for } i = 1, \ldots, v.$$

PROOF. $(\alpha_1, \ldots, \alpha_v)^t \in \mathcal{L}(X) = \ker L_1 = \operatorname{im} L_2$ is equivalent to

$$\alpha_1 b_1 + \cdots + \alpha_v b_v = L_2(a) \qquad \text{for some } a \in W. \tag{3}$$

Applying b_i^* to both sides of (3), we obtain that, equivalently,

$$\alpha_i = b_i^*(L_2(a)) = [L_2^*(b_i^*)](a) = \bar{x}_i(a), \qquad i = 1, \ldots, v.$$

□

4.9 Lemma. *For* $i \in \{1, \ldots, v\}$, $x_i \notin \operatorname{lin}(X \setminus \{x_i\})$ *if and only if* $\bar{x}_i = 0$.

PROOF. $x_i \notin \operatorname{lin}(X \setminus \{x_i\})$ is equivalent to saying that, in all $(\alpha_1, \ldots, \alpha_v)^t \in \mathcal{L}(X), \alpha_i = 0$. This, in turn, is expressed (see Lemma 4.8) by the condition $\bar{x}_i(a) = 0$ for all $a \in W$, so that $\bar{x}_i$ is the zero map. □

For a subsequence Y of X, we set

$$\widetilde{Y} := (\bar{x}_i \mid x_i \notin Y).$$

4.10 Lemma. *The set of components of* Y *is linearly independent if and only if*

$$\operatorname{lin} \widetilde{Y} = \operatorname{lin} \bar{X}.$$

PROOF. Since linear independence does not depend on order, we may assume that $X = (Y, Z)$, so that

$$\bar{X} = (\tilde{Z}, \tilde{Y}) \qquad \text{and} \qquad \mathcal{L}(Y) \times \mathcal{L}(Z) \subset \mathcal{L}(X).$$

Since $\operatorname{lin} \tilde{Y} \subset \operatorname{lin} \bar{X} = W^*$, we may replace the condition $\operatorname{lin} \tilde{Y} = W^*$ by (4).

$$\begin{aligned} \text{For every } a \in W, \text{ the equation } \tilde{Y}(a) = 0 \text{ implies } 0 &= \bar{X}(a) \\ &= (\tilde{Z}(a), \tilde{Y}(a)). \end{aligned} \tag{4}$$

On the other hand, Y is linearly independent if and only if $\mathcal{L}(Y) = 0$, see 4.7.

If $\mathcal{L}(Y) = 0$, then, $\tilde{Z}(a) = 0$ for every $a \in W$ and (4) holds, see 4.8. If $\mathcal{L}(Y) \neq 0$, then, there exists some nonzero $(\alpha, 0) \in \mathcal{L}(Y) \times \mathcal{L}(Z) \subset \mathcal{L}(X)$, and thus, by 4.8, some $a \in W$ with $(\alpha, 0) = \bar{X}(a)$, so that (4) does not hold. □

Lemma 4.10 implies the following:

4.11 Lemma. Y *is a basis of* $\operatorname{lin} X$ *if and only if* $\widetilde{Y}$ *is a basis of* $\operatorname{lin} \bar{X}$.

Now, we will apply Lemma 4.10 to polyhedral cones introduced in I, 1. Let

$$\sigma := \operatorname{pos} X$$

be the cone determined by a finite sequence X in U. In order to study faces of the cone σ, we introduce a notion relating faces to their generating elements in X:

4.12 Definition. A (possibly empty) subsequence Y of X is called a face of X if there exists a hyperplane H in U such that

(a) $H \cap X = Y$, and

(b) H is a supporting hyperplane of $\sigma = \text{pos } X$.

The partially ordered set of faces of X (including $\emptyset$) is said to be the *face complex* $\mathcal{B}(X)$ of X.

In the circumstances above, $H \cap \sigma = \text{pos}(X \cap H) = \text{pos } Y$ is a face of σ; note that pos $\emptyset = \{0\}$. If $0 \in X$, then, 0 is contained in every face of X.

One of the main objectives with using linear transforms is the characterization of faces Y by properties of the corresponding subsequences $\tilde{Y}$ of $\bar{X}$.

4.13 Definition. If a subsequence Y of a finite sequence X in the vector space U is a face of X, then, $\widetilde{Y}$ is called a *coface* of X. The set of cofaces of X, partially ordered by inclusion, is called the *coface complex* of X.

4.14 Theorem. *A subsequence $\widetilde{Y}$ of $\bar{X}$ is a coface if and only if*

$$0 \in \text{relint pos } \widetilde{Y} \tag{5}$$

(or, equivalently, $0 \in \text{relint conv } \widetilde{Y}$*).*

In Example 1 above, $(\bar{x}_1, \bar{x}_3, \bar{x}_4)$, $(\bar{x}_2, \bar{x}_3, \bar{x}_4)$ and $(\bar{x}_1, \bar{x}_2, \bar{x}_3, \bar{x}_4)$ are exactly the cofaces, whereas (x_1), (x_2) and $\emptyset$ are faces; in the case of $\emptyset$, as a supporting hyperplane of σ, choose H, for which $H \cap \sigma = \{0\}$.

Proof of Theorem 4.14. For $\alpha \in \mathbb{R}^v$, we use the notation $\langle \alpha, \bar{X} \rangle := \Sigma_{j=1}^{v} \alpha_j \bar{x}_j$. Let Y be a subsequence of $X = (x_1, \dots, x_v)$, say $Y = (x_1, \dots, x_r)$. Then, $\tilde{Y}$ satisfies (5) if and only if there exists an $\alpha \in 0 \times \mathbb{R}^{v-r}_{>0}$ such that $\langle \alpha, \bar{X} \rangle = 0$. By Lemma 4.3 and Lemma 4.8, we may replace the condition "$\langle \alpha, \bar{X} \rangle = 0$" by requiring the existence of $a^* \in U^*$ such that $\alpha = X(a^*)$. Thus, we have the following equivalent statements:

(a) $\tilde{Y}$ satisfies (5).

(b) There exists $a^* \in U^*$ such that $X(a^*) \in 0 \times \mathbb{R}^{v-r}_{>0}$.

(c) There exists $a^* \in U^*$ such that $H := \{u \in U \mid a^*(u) = 0\}$ satisfies $x_1, \dots, x_r \in H$ and $x_{r+1}, \dots, x_v \in H^+ \setminus H$.

(d) $\widetilde{Y}$ is a coface of X.

□

Linear transforms of a particular kind will be used in 5.4 for the characterization of polytopes. First, we will make the following observation:

4.15 Lemma. *A linear transform $\bar{X}$ of X satisfies $\bar{x}_1 + \cdots + \bar{x}_v = 0$ if and only if the components of X lie in a hyperplane H of U for which $0 \notin H$.*

PROOF. Lemma 4.3 yields that X is a linear transform of $\bar{X}$. By Lemma 4.8, there exists $a^* \in U^*$ such that the points of X lie in $H := \{x \mid a^*(x) = 1\}$ if and only if $(1, \ldots, 1) \in \mathcal{L}(\bar{X})$, that is, if and only if $\bar{x}_1 + \cdots + \bar{x}_v = 0$. □

This may motivate the following construction: Let X be such that its components generate U. We identify U (as an affine space) with a hyperplane H in a linear space $\hat{U}$ where $0 \notin H$; hence, $\dim \hat{U} = \dim U + 1$. Then, X determines a finite sequence $X_{\hat{U}} = (\hat{x}_1, \ldots, \hat{x}_v)$ in $\hat{U}$ with components that generate $\hat{U}$.

4.16 Definition. A linear transform $\bar{X}_{\hat{U}} = (\bar{\hat{x}}_1, \ldots, \bar{\hat{x}}_v)$ of $X_{\hat{U}}$ is called a *Gale transform* (or sometimes an *affine transform*) of $X = (x_1, \ldots, x_v)$.

Remark. In appropriate coordinates of $\hat{U}$, we may always assume that $H = U \times \{1\}$.

Note one advantage of considering Gale transforms instead of linear transforms with respect to U: Whereas the components of $\bar{X}$ lie in W^*, those of $\bar{X}_{\hat{U}}$ lie in a lower-dimensional vector space $W^*_{\hat{U}}$; moreover, since $\bar{\hat{x}}_1 + \ldots + \bar{\hat{x}}_v = 0$, according to 4.15 $\bar{X}_{\hat{U}}$ is completely determined by only $v - 1$ of its elements.

Example 2. In Example 1, replace the elements x_3, x_4 by $\frac{1}{3}x_3, \frac{1}{3}x_4$, respectively. Then the linear transform obtained by the same calculation as in Example 1 is

$$\begin{pmatrix} \bar{x}_1 \\ \bar{x}_2 \\ \bar{x}_3 \\ \bar{x}_4 \end{pmatrix} = \begin{pmatrix} L_2^*(b_1^*) \\ L_2^*(b_2^*) \\ L_2^*(b_3^*) \\ L_4^*(b_4^*) \end{pmatrix} = \begin{pmatrix} -\frac{1}{3} & -\frac{2}{3} \\ -\frac{2}{3} & -\frac{1}{3} \\ 1 & 0 \\ 0 & 1 \end{pmatrix},$$

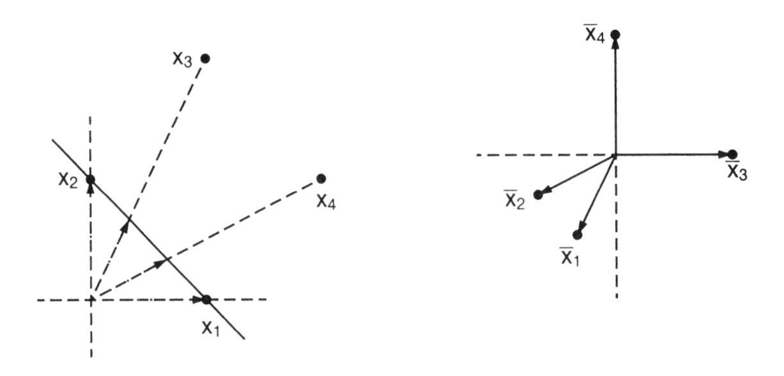

FIGURE 8.

and it is readily verified that $\bar{x}_1 + \bar{x}_2 + \bar{x}_3 + \bar{x}_4 = 0$ (see Figure 8).

Exercises

1.
   a. Let $V = U \oplus W$ where $U \cong \mathbb{R}^2$, $W \cong \mathbb{R}^3$. Proceed, as in Example 1, to find a linear transform of

$$(x_1 x_2 x_3 x_4 x_5) = \left(\begin{array}{cc|ccc} 1 & 0 & 2 & 3 & 1 \\ 0 & 1 & 3 & 3 & 1 \end{array}\right).$$

   b. Find the analog to Example 2.

2. Use Examples 1 and 2 to illustrate the proof of Lemma 4.7.
3. Consider a regular octahedron K that lies in the hyperplane $\{x = (\xi_1, \ldots, \xi_4) \mid \xi_4 = 1\}$ of $\mathbb{R}^4$.

   a. Find a linear transform for the vertices of the octahedron.
   b. Determine all faces of vert K.

4. Let $X = X_1 \cup X_2$ with $X_1 \cap X_2 = \emptyset$, and let $U = U_1 \oplus U_2$ where $U_1 := \operatorname{lin} X_1$, $U_2 := \operatorname{lin} X_2$. Then, $\bar{X} = \bar{X}_1 \cup \bar{X}_2$, and $\operatorname{lin} \bar{X}_1$, $\operatorname{lin} \bar{X}_2$ are complementary subspaces of $\operatorname{lin} \bar{X}$ (possibly $\operatorname{lin} \bar{X}_1 = \{0\}$ or $\operatorname{lin} \bar{X}_2 = \{0\}$).

5 Matrix representation of transforms

Examples 1 and 2 in the preceding section can be generalized so as to provide a method of calculating transforms by the use of matrices.

Let $U = \operatorname{lin} X = \mathbb{R}^n$ where $X = (x_1, \ldots, x_v)$, and let $e_1, \ldots, e_n$ be the canonical basis of $\mathbb{R}^n$. Up to renumbering, we may assume that $x_1 = e_1, \ldots, x_n = e_n$. Then, writing the elements of U as columns and denoting the transposed terms of a matrix $(\cdot)$ by $(\cdot)^t$,

$$\begin{aligned} x_{n+1} &=: (x_{n+1,1}, \ldots, x_{n+1,n})^t \\ &\vdots \\ x_v &=: (x_{v,1}, \ldots, x_{v,n})^t. \end{aligned}$$

In short, we can express $x_1, \ldots, x_v$ as coordinate columns

$$X = (x_1 \cdots x_v) = \left(\begin{array}{c|c} E_n & x_{n+1} \cdots x_v \end{array}\right).$$

We set $V := \mathbb{R}^n \oplus \mathbb{R}^{v-n}$ and define a basis (of column vectors) by

$$(b_1 \cdots b_v) := \left(\begin{array}{c|c} E_n & x_{n+1} \cdots x_v \\ \hline O & E_{v-n} \end{array}\right).$$

Now,

$$L_1 := \left(E_n \mid O \right)$$

satisfies $L_1(b_j) = x_j, j = 1, \ldots, v$. Furthermore, for $W = 0 \times \mathbb{R}^{v-n}$, we choose

$$L_2 := \left(\frac{O}{E_{v-n}} \right)$$

Then, multiplying from the left with L_1, L_2, we obtain an exact sequence

$$0 \longrightarrow W \underset{L_2}{\longrightarrow} V \underset{L_1}{\longrightarrow} U \longrightarrow 0.$$

5.1 Theorem. *Let $X = (x_1, \ldots, x_v)$ be a finite sequence in $\mathbb{R}^n$ such that $x_1, \ldots, x_n$ is the canonical basis. Then, a linear transform is obtained by the rows of the matrix*

$$\bar{X} = \begin{pmatrix} \bar{x}_1 \\ \vdots \\ \bar{x}_v \end{pmatrix} = \left(\frac{-x_{n+1} \cdots - x_v}{E_{v-n}} \right);$$

the columns of $\bar{X}$ form a basis of $\mathcal{L}(X)$.

PROOF. For the row vectors b_i^* defined by

$$\begin{pmatrix} b_1^* \\ \vdots \\ b_v^* \end{pmatrix} := \left(\begin{array}{c|c} E_n & -x_{n+1} \cdots - x_v \\ \hline O & E_{v-n} \end{array} \right),$$

we calculate $b_i^* \cdot b_j$. In fact, for $i = 1, \ldots, n$ and $j > n$,

$$b_i^* \cdot b_j = (0 \cdots 0 \underset{\underset{i}{\uparrow}}{1} 0 \cdots 0 - x_{n+1,i} \cdots - x_{vi}) \begin{pmatrix} x_{j1} \\ \vdots \\ x_{jn} \\ 0 \\ \vdots \\ 1 \\ \vdots \\ 0 \end{pmatrix} \begin{matrix} \\ \\ \\ \\ \\ \leftarrow j \\ \\ \\ \end{matrix} = x_{ji} - x_{ji} = 0.$$

Thus, the columns of $\bar{X}$ lie in $\mathcal{L}(X)$; they are obviously linearly independent and, thus, comprise a basis for $\mathcal{L}(X)$. For $i = n+1, \ldots, v$ or for $j \leq n, b_i^* \cdot b_j = \delta_{ij}$ is trivially true. Hence, $\{b_1^*, \ldots, b_v^*\}$ is the dual basis for $\{b_1, \ldots, b_v\}$, and

$$L_2^* \begin{pmatrix} b_1^* \\ \vdots \\ b_v^* \end{pmatrix} = \begin{pmatrix} b_1^* \\ \vdots \\ b_v^* \end{pmatrix} \cdot L_2 = \left(\frac{-x_{n+1} \cdots - x_v}{E_{v-n}} \right) = \begin{pmatrix} \bar{x}_1 \\ \vdots \\ \bar{x}_v \end{pmatrix}.$$

□

Occasionally (for example, in the proof of the next theorem), it is useful to have the canonical basis vectors $e_1, \ldots, e_n$ as part of the linear transform. This is achieved as follows:

5.2 Theorem. *Let $X = (x_1, \cdots, x_v)$ be given such that $x_1, \ldots, x_n$ are linearly independent in $\mathbb{R}^n$, and let $a_1, \ldots, a_{v-n}$ be any basis of $\mathcal{L}(X)$. Then, the rows $\bar{x}_i$ of*

$$(a_1 \cdots a_{v-n}) =: \begin{pmatrix} \bar{x}_1 \\ \vdots \\ \bar{x}_v \end{pmatrix}$$

form a linear transform of X.

PROOF. Since $\mathcal{L}(X)$ and $\bar{X}$ do not change when we apply an automorphism to $\mathbb{R}^n$, we may assume that $x_1, \ldots, x_n$ is the canonical basis of $\mathbb{R}^n$. By 5.1, we may assume that the columns of

$$\bar{X} = \left(\frac{-x_{n+1} \cdots - x_v}{E_{v-n}} \right)$$

form a basis of $\mathcal{L}(X)$. Now, we consider the matrix $(a_1 \cdots a_{v-n})$ whose columns are also a basis of $\mathcal{L}(X)$. We write

$$\left(\frac{D_0}{D} \right) := (a_1 \cdots a_{v-n}),$$

where D is a $(v-n) \times (v-n)$-matrix, and we assert that D has rank $v-n$. For otherwise, we could find a nonzero vector $a \in \mathcal{L}(X)$ whose last $v-n$ coordinates are zero Then a represents a nontrivial vanishing linear combination of $x_1, \ldots, x_n$, contrary to the assumption that $x_1, \ldots, x_n$ are linearly independent.

Therefore we can write, with $D_1 := D_0 D^{-1}$,

$$\left(\frac{D_0}{D} \right) = \left(\frac{D_1}{E_{v-n}} \right) D.$$

Since the columns of $\left(\frac{D_1}{E_{v-n}} \right)$ also represent linear dependencies and, thus, form a basis of $\mathcal{L}(X)$, D_1 equals the upper part of $\bar{X}$:

$$D_1 = (-x_{n+1} \cdots - x_v).$$

Therefore, by Lemma 4.6, the rows of $(a_1 \cdots a_{v-n})$ form a linear transform of X. □

We now come to the analogous properties for Gale transforms. Let $X = (x_1, \ldots, x_v)$ be a finite sequence generating U and choose H and $\hat{U}$ according to Definition 4.16. We, then, introduce coordinates in $\hat{U}$ so that $H = U \times \{1\}$. Thus, $X_{\hat{U}} = (\hat{x}_1, \ldots, \hat{x}_v)$, where $\hat{x}_j = \binom{x_j}{1}$. This way, every *affine* dependency $a \in \mathcal{L}_U(X)$ is a linear dependency in $\mathcal{L}_{\hat{U}}(X_{\hat{U}})$.

5.3 Theorem. *Let* $X = (x_1, \ldots, x_v)$ *be a finite sequence in* U, *where* $x_1, \ldots, x_{n+1}$ *are affinely independent; let* $a_1, \ldots, a_{v-n-1} \in \mathcal{L}_U(X)$ *be affine dependencies that generate* $\mathcal{L}_{\hat{U}}(X_{\hat{U}})$. *Then, the rows of* $(a_1 \cdots a_{v-n-1})$ *form a Gale transform* $\bar{X}_{\hat{U}}$ *of* X.

PROOF. Since $\hat{x}_1, \ldots, \hat{x}_{n+1}$ are linearly independent, Theorem 5.2 yields that $\bar{X}_{\hat{U}}$ is a linear transform of $X_{\hat{U}}$. □

Now we turn to the most important case, namely, that of a finite sequence given by the vertex set of a polytope P,

$$X = \text{vert } P. \tag{3}$$

We wish to characterize vert P by properties of Gale transforms of X. First, we consider two examples:

Example 1. We consider a triangular prism P in $\mathbb{R}^3$ and wish to find a Gale transform of $X :=$ vert P (see Figure 9). Since $v = 6$ and $n = 3$, the elements of a Gale transform lie in a two-dimensional space. By Theorem 5.3, it is sufficient to find two independent affine dependencies of $x_1, \ldots, x_6$. Since $x_4 - x_1, x_5 - x_2$, and $x_6 - x_3$ coincide,

$$(a_1 a_2)^t = \begin{pmatrix} -1 & 1 & 0 & 1 & -1 & 0 \\ -1 & 0 & 1 & 1 & 0 & -1 \end{pmatrix}.$$

By Theorem 5.3, the rows of $(a_1 a_2)$ provide a Gale transform.

Example 2. In the case of a pyramid in $\mathbb{R}^3$ with rectangular basis, we find, analogously to the first example, a single affine dependency

$$(1 \quad -1 \quad -1 \quad 1 \quad 0)^t .$$

The Gale transform, thus obtained, lies in $\mathbb{R}^1$. Two pairs of its points coincide.

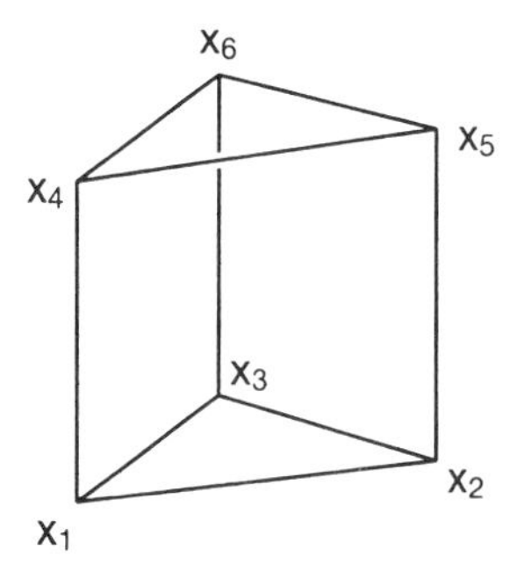

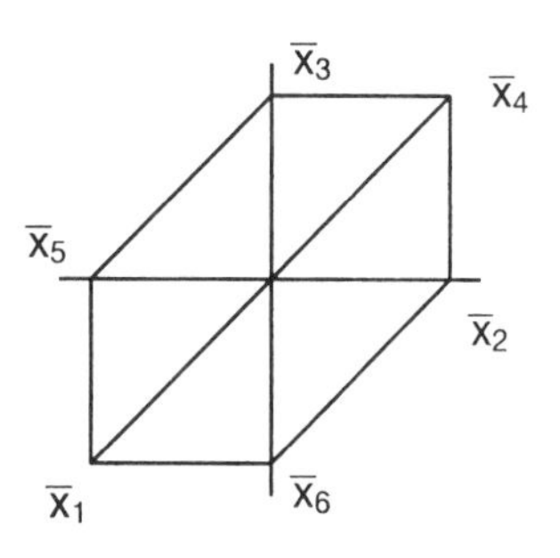

FIGURE 9.

5.4 Theorem. *A finite sequence X in $U = \text{aff}\, X$ consists of all points of the vertex set of a polytope P in U if and only if one (and, thus, every) Gale transform $\bar{X}_{\hat{U}}$ of X satisfies the following condition:*

(4) *For every hyperplane H in $\text{lin}\, \bar{X}_{\hat{U}}$ which contains 0, each of the associated half-spaces H^+ and H^- includes at least two points of $\bar{X}_{\hat{U}}$ in its interior.*

PROOF. In appropriate coordinates for $\hat{U}$, we may identify U with $U \times \{1\} \subset \hat{U}$. First, suppose that X comes from a polytope. Then, each element $\hat{x}_i$ of $X_{\hat{U}}$ is a face. By Theorem 4.14, $\bar{X}_{\hat{U}} \setminus \{\bar{\hat{x}}_i\}$ satisfies

$$0 \in \text{relint pos}(\bar{X}_{\hat{U}} \setminus \{\bar{\hat{x}}_i\}). \tag{5}$$

Let H be a hyperplane in $\text{lin}\, \bar{X}_{\hat{U}}$ with $0 \in H$. Let k^+, k^- be the number of elements $\bar{\hat{x}}_i$ in $\text{int}\, H^+$, $\text{int}\, H^-$, respectively. If $k^+ = k^- = 0$, $H = \text{lin}\, \bar{X}_{\hat{U}}$ so that H is not a hyperplane. $k^+ \leq 1$ and $k^- > 0$ or $k^- \leq 1$ and $k^+ > 0$ imply a contradiction to (5). So, (4) follows.

Conversely, if (4) holds, then, (5) readily follows (since, otherwise, 0 would lie on the boundary of $\sigma := \text{pos}(\bar{X}_{\hat{U}} \setminus \{\bar{\hat{x}}_i\})$, so that, for a supporting hyperplane H of σ with $0 \in H$ and $\sigma \subset H^+$, the half-space H^- would not contain two of the $\bar{\hat{x}}_i$'s in its interior). Therefore, by Theorem 4.14, $x_i \in \text{vert}\, P$ for $i = 1, \ldots, v$. □

5.5 Theorem. *A sequence $X = (x_1, \ldots, x_v)$ in $\mathbb{R}^n$ is the vertex set of a k-fold pyramid with apexes $x_1, \ldots, x_k$, if and only if, for a Gale transform $\bar{X}_{\hat{U}}$ of X,*

$$\bar{\hat{x}}_1 = \cdots = \bar{\hat{x}}_k = 0.$$

PROOF. A vertex x_i of a pyramid is an apex if and only if it is not affinely dependent on the other vertices. This is equivalent to the fact that, in every affine dependency of X, the ith coefficient is zero. Going over to $\mathbb{R}^n \times \{1\} \subset \mathbb{R}^n \times \mathbb{R} =: \hat{U}$, it just means the ith row in

$$\bar{X}_{\hat{U}} = (a_1 \cdots a_{v-n-1}),$$

is the zero vector. □

Example 3. The vertex set $X = (x_1, \ldots, x_{n+1})$ of an n-simplex in $\mathbb{R}^n$ is characterized by $\bar{\hat{x}}_1 = \cdots = \bar{\hat{x}}_n = 0$ for a Gale transform $\bar{X}_{\hat{U}}$ of X. This also follows directly from $\dim \text{lin}\, \bar{X}_{\hat{U}} = n + 1 - n - 1 = 0$ by Lemma 4.5.

Also, in the above Example 2 of a pyramid, it is seen that $\bar{\hat{x}}_5 = 0$.

The following lemma, which we shall need in V, Theorem 4.8 comments further on the relation between linear and Gale transforms, the latter being a linear transform with respect to a $(\dim U + 1)$-dimensional vector space $\hat{U}$.

5.6 Lemma. *If $\bar{X} = (\bar{x}_1, \ldots, \bar{x}_v)$ is a linear transform of the sequence $X = (x_1, \ldots, x_v)$ in U and if $b := \bar{x}_1 + \cdots + \bar{x}_v \neq 0$, then, we obtain a Gale transform of X by projecting $\bar{X}$ in direction b onto a one-codimensional linear subspace of* $\text{lin}\, \bar{X}$ *which does not contain b.*

PROOF. Up to a bijective linear transformation of $\mathcal{L}(X)$, we may assume that $(1, \dots, 1)^t$ is one of the basis vectors $a_1, \dots, a_{v-n}$ introduced in Theorem 5.2, say $a_{v-n} = (1, \dots, 1)^t$. So, we can write $\bar{x}_i = (x_i', 1)$, $i = 1, \dots, v$, and $b = (x_1' + \cdots + x_v', v)$. Furthermore, we may assume, up to a bijective linear transformation of $\operatorname{lin} \bar{X}$ which leaves $\operatorname{lin}\{(x_1', 1), \dots, (x_v', 1)\}$ invariant, that $b = (0, v)$.

Since $a_1, \dots, a_{v-n-1}$ are linearly independent they provide a basis of $\mathcal{L}(X)$ (relative to $\hat{U}$). Hence, $X' := (x_1', \dots, x_v')$ is a linear transform of X. Since $x_1' + \cdots + x_v' = 0$, X' is a Gale transform of X. □

Exercises

1. Find a Gale transform of a bipyramid over a regular pentagon.
2. Find a Gale transform of a four-dimensional crosspolytope.
3. Let X be an arbitrary finite set. Describe how a Gale transform $\bar{X}$ is obtained from a Gale transform of vert(conv X).
4. Let X consist of different points, and let X be centrally symmetric. Then, we cannot always find a Gale transform which is also centrally symmetric. (Consider, for example, a regular octahedron).

6 Classification of polytopes

Polytopes are called combinatorially isomorphic if their boundary complexes are isomorphic. There is no hope of classifying all polytopes under combinatorial isomorphism. However, partial classifications are possible. Gale transforms are especially helpful if the restriction is made by assuming "small" numbers of vertices.

6.1 Definition. Two sequences of points $\bar{X} = (\bar{x}_1, \dots, \bar{x}_v)$, $\bar{X}' = (\bar{x}_1', \dots, \bar{x}_v')$ are called *isomorphic* if they have isomorphic coface complexes (see Definitions 4.12 and 4.13). Every sequence isomorphic to a Gale transform $\bar{X}$ of X is called a *Gale diagram* of X.

By 4.14, $\bar{X}$ and $\bar{X}'$ are isomorphic if and only if there exists a bijection between their families of components such that, for each subsequence $\bar{Z}$ of $\bar{X}$ and the corresponding subsequence $\bar{Z}'$ of $\bar{X}'$,

$$0 \in \operatorname{relint} \operatorname{pos} \bar{Z} \qquad \text{if and only if} \qquad 0 \in \operatorname{relint} \operatorname{pos} \bar{Z}'.$$

Gale diagrams are more general than Gale transforms. Any Gale transform is a Gale diagram, but not conversely. For example, the condition of Lemma 4.15 will, in general, be violated if we proceed from a Gale transform to a Gale diagram (for nontrivial examples, see Lemma 6.3).

From the definitions and Theorem 4.14 we have the following two lemmas:

6.2 Lemma. *Two n-polytopes P, P' are combinatorially equivalent if and only if there exist isomorphic Gale diagrams of* $X :=$ vert P, $X' :=$ vert P'.

6.3 Lemma. *Let* $\bar{X} = (\bar{x}_1, \ldots, \bar{x}_v)$ *be a Gale transform, and let* $\beta_i > 0$, $i = 1, \ldots, v$. *Then,* $\bar{X}$ *and* $\bar{X}' := (\beta_1\bar{x}_1, \ldots, \beta_v\bar{x}_v)$ *are isomorphic Gale diagrams.*

Now, we prove a first classification theorem ($[u]$ denotes the integer part of u, that is, the largest integer $\leq u$).

6.4 Theorem. *There are precisely* $[\frac{1}{4}n^2]$ *combinatorial types of n-polytopes with* $n + 2$ *vertices.*

PROOF. Let P be an n-polytope, and let $X =$ vert P have $n + 2$ elements. We may assume that $P \subset \mathbb{R}^n \times \{1\} \subset U := \mathbb{R}^{n+1}$. Then, every Gale transform $\bar{X}$ lies in a one-dimensional space. By Lemma 6.3, we can replace $\bar{X}$ by a Gale diagram whose components are all either -1, 0 or 1. By condition (4) in Theorem 5.4, the classification problem reduces to finding all partitions of a set of $n + 2$ elements into three subsets M_{-1}, M_0, M_1 such that M_{-1} and M_1 each have at least two elements.

Let the numbers of elements of M_{-1}, M_0, M_1 be $1 + r \geq 2$, $t \geq 0$, $1 + s \geq 2$, respectively. For reasons of symmetry, we may assume

$$r \leq s.$$

Since $r + t + s = n$, there are the following possibilities:

r	t	s
1	0	$n-1$
2	0	$n-2$
$\vdots$	$\vdots$	$\vdots$
$[n/2]$	0	$n-[n/2]$
1	1	$n-2$
$\vdots$	$\vdots$	$\vdots$
$[(n-1)/2]$	1	$n-1-[(n-1)/2]$
$\vdots$	$\vdots$	$\vdots$
1	$n-2$	1

The total number of possibilities is

$$\left[\frac{n}{2}\right] + \left[\frac{n-1}{2}\right] + \cdots + \left[\frac{n-(n-3)}{2}\right] + \left[\frac{n-(n-2)}{2}\right].$$

By distinguishing the cases $n = 2k$ and $n = 2k + 1$, we find this sum to be $[\frac{1}{4}n^2]$. □

Thus, for an explicit classification of such polytopes, it suffices to provide $[\frac{1}{4}n^2]$ nonequivalent polytopes; that is easy for small n:

Example 1. $n = 2 : [\frac{1}{4}n^2] = 1$. Every convex quadrangle is a representative.

Example 2. $n = 3$: Then, $[\frac{1}{4}n^2] = 2$. Two nonisomorphic 3-polytopes with five vertices are represented below. Figure 10a shows a bipyramid over a triangle, Figure 10b a quadrangular pyramid. The two possible triples $(r+1, t, s+1)$ are $(2, 0, 3)$ and $(2, 1, 2)$; by Theorem 5.5, the second one belongs to Figure 10b.

Example 3. $n = 4 : [\frac{1}{4}n^2] = 4$. Figure 11 illustrates the Gale diagrams in the proof of Theorem 6.4 and central projections of the respective polytopes into a facet F. Although we cannot visualize the polytope itself, we can visualize its projection into F. We "look through a three-dimensional window into the four-dimensional polytope" (more about this in III, 4).

Type (a) represents a double simplex. The three-dimensional "window" is $F := \operatorname{conv}\{x_1, x_2, x_3, x_4\}$; the base of the bipyramid, projected into F, is $\operatorname{conv}\{x_2, x_3, x_4, x_5\}$.

Type (b) represents the type of the cyclic polytope $C(6,4)$. Of course, a polytope combinatorially equivalent to a cyclic polytope need not be cyclic.

Types (c) and (d) are illustrated by projecting the pyramid into the basis of the pyramid. The possible triples are indicated in Figure 11; the values of t again follow immediately from Theorem 5.5.

For $n = 3$, one of the types in Theorem 6.4 is simplicial; for $n = 4$, two of them are. What is the general law behind these examples? In order to find an answer, we first characterize simplicial polytopes by a property of the associated Gale transforms.

6.5 Theorem. *An n-dimensional polytope $P = \operatorname{conv}\{x_1, \ldots, x_v\}$ is simplicial if and only if, for a Gale diagram $\bar{X}$ of $X = $ (vert P), the following condition holds:*
(1) *For any hyperplane H in* $\operatorname{lin}\bar{X}$, *with* $0 \in H$, $0 \notin \operatorname{relint}\operatorname{conv}(\bar{X} \cap H)$.

PROOF. Suppose $0 \in \operatorname{relint}\operatorname{conv}(\bar{X} \cap H)$ for some hyperplane H. Since $\dim P = n$ and P lies in a hyperplane of U which does not contain 0, we have $\dim U = \operatorname{rank} X = n + 1$, so, $\operatorname{rank}\bar{X} = v - \operatorname{rank} X = v - n - 1$ (Lemma 4.5).

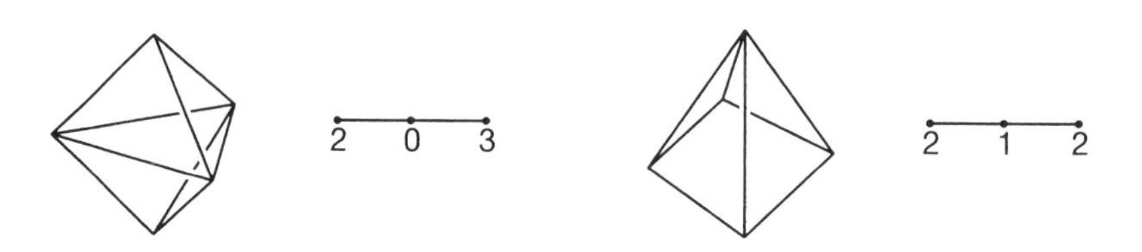

FIGURE 10a,b.

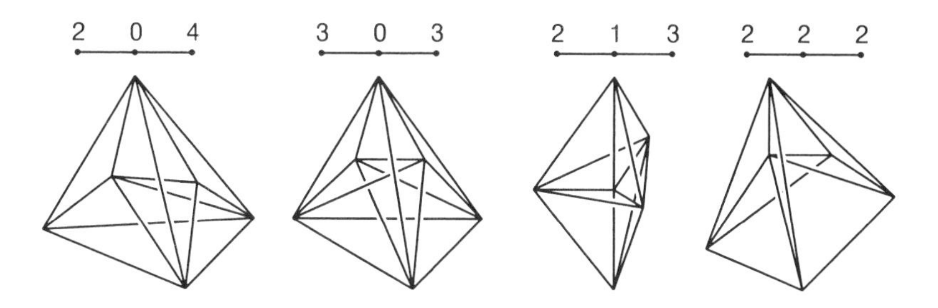

FIGURE 11a,b,c,d. (a) Double simplex. (b) Cyclic polytope $C(6, 4)$. (c) Pyramid over a double simplex. (d) 2-Fold pyramid.

Then, by Carathéodory's theorem (I, Theorem 2.3), for $k = \operatorname{rank}(H \cap \bar{X}) \leq v - n - 2$, there exist $k + 1 \leq v - n - 1$ affinely independent elements $\bar{x}_1, \ldots, \bar{x}_{k+1} \in \bar{X} \cap H$ such that $0 \in \operatorname{relint} \operatorname{conv} \{\bar{x}_1, \ldots, \bar{x}_{k+1}\}$. Therefore, $\{\bar{x}_1, \ldots, \bar{x}_{k+1}\}$ is a coface of X, and the corresponding (proper) face has at least $v - (v - n - 1) = n + 1$ vertices and cannot be a simplex.

The converse is also true. □

6.6 Theorem. *Of the combinatorial types of n-polytopes with $n + 2$ vertices, precisely $[\frac{n}{2}]$ are simplicial.*

PROOF. We may assume that $P \subset \mathbb{R}^n \times \{1\} \subset U = \mathbb{R}^{n+1}$; for $X = \operatorname{vert} P$, we obtain rank $\bar{X} = 1$. Thus, $H = \{0\}$ is the only possible hyperplane, as in Theorem 6.5. So P is simplicial if and only if in the table in the proof of 6.4, $t = 0$. □

It is also possible to calculate the exact number of types of n-polytopes with $n + 3$ vertices; the proof needs, however, a considerable amount of calculation (see the Appendix of this section). We restrict ourselves to 3-polytopes.

6.7 Theorem. *There exist precisely seven types of 3-polytopes with six vertices.*

PROOF. By Lemma 6.3 we can assume that a Gale diagram of $\bar{X}$ for $X = \operatorname{vert} P$ consists only of points on a circle with center 0 and possibly 0. From the definition of Gale diagrams and by Lemma 6.2, we can move any point of $\bar{X}$ on the circle without changing the type of P, as long as we observe the following rule:

(2) Never move into or beyond a point diametrically opposite to another point of $\bar{X}$.

So we can arrange all nonzero elements of $\bar{X}$ and their negatives to be vertices of a regular k-gon, $k \leq 12$.

We apply a further rule which does not change the type of P:

(3) If you can move a point of $\bar{X}$ on the circle into another one without violating (2), then, do so.

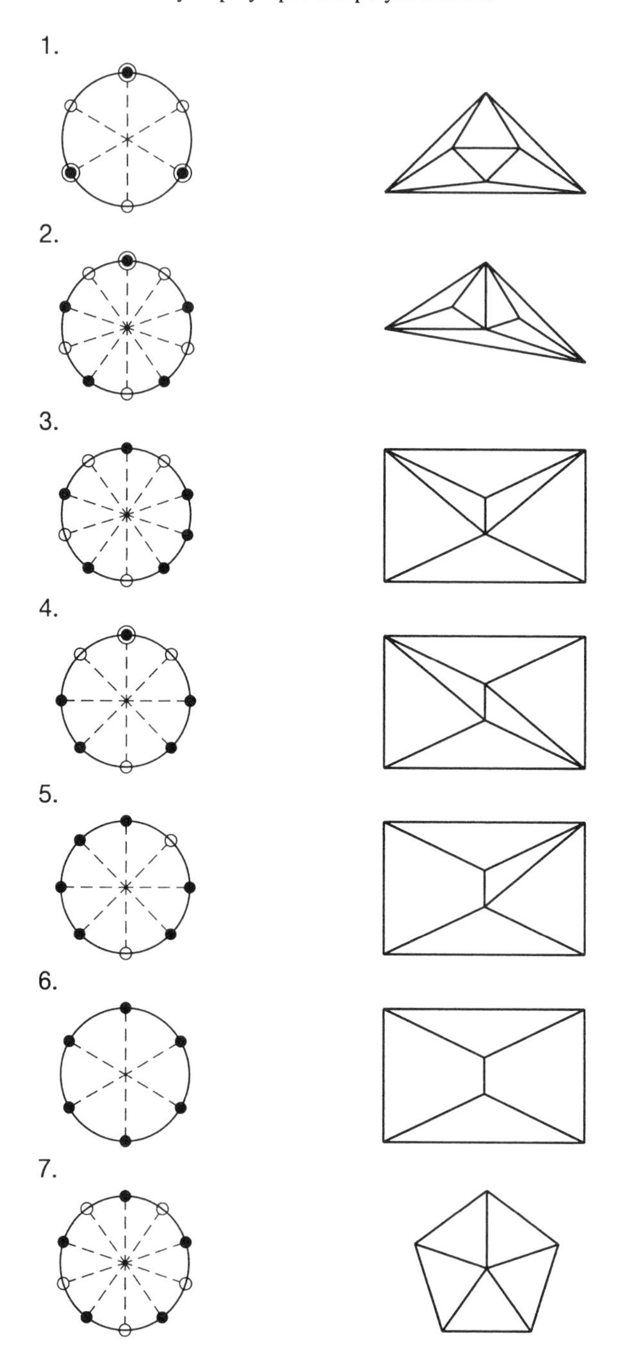

FIGURE 12. 1. Octahedron, 2. Simplex with two stacked simplices, 3. Pyramid with stacked simplex, 4. Prism with two faces bent, 5. Prism with one face bent, 6. Prism, 7. Pentagonal pyramid.

Rules (2) and (3) lead to standard diagrams which represent different types of polytopes.

Checking through all possibilities and applying Theorems 5.4 and 5.5, we obtain the list of different types of 3-polytopes with six vertices (in Figure 12). (By a little circle around a diagram point, we indicate two coinciding points of $\bar{X}$). □

Exercises

1. Determine all types of simplicial 4-polytopes with seven vertices.
2. Let $\bar{x}_1 = \bar{x}_2, \bar{x}_3 = \bar{x}_4, \bar{x}_5 = \bar{x}_6, \bar{x}_7 = \bar{x}_8$ be the vertices of a 3-simplex T, $0 \in \operatorname{int} T$. Describe a 3-polytope P with $X = \operatorname{vert} P = \{x_1, \ldots, x_8\}$ which has $\bar{X} = \{\bar{x}_1, \ldots, \bar{x}_8\}$ as a Gale transform.
3. If each facet of an n-polytope P has $v - 2$ vertices (v the number of vertices of P), then,

$$v \leq 2n.$$

4. Let P be an n-polytope, let $x \in P$, and let $\bar{X}$ be a Gale transform of $X := \{x\} \cup \operatorname{vert} P$. Let $H \ni 0$ be the hyperplane in $\operatorname{lin} \bar{X}$ with normal vector $\bar{x}$, and let π be the orthogonal projection onto H. Then, $\pi(\bar{X} \setminus \{\bar{x}\})$ is a Gale transform of $\operatorname{vert} P$.

III

Polyhedral spheres

1. Cell complexes

In the preceding chapter, we dealt with boundary complexes of convex polytopes. They consist of cell decompositions of topological spheres, the cells again being convex polytopes. If, however, any cell decomposition of a topological sphere is given, there need not exist a convex polytope with isomorphic (in the sense of inclusion of cells) boundary complex. We shall present counter-examples in section 4 below. In fact, one of the major unsolved problems in convex polytope theory is to find necessary and sufficient conditions for a cell-composed sphere to be isomorphic to the boundary complex of a polytope (*Steinitz problem*).

In this chapter, we will introduce the basic concepts for the topology-oriented part of polytope theory, including Euler and Dehn-Sommerville equations. We need "polyhedral" cells and cell complexes, rather than the general notion of cells and cell complexes. For that reason, we restrict our considerations to the explicit case of Definition 1.1, despite the slight inconvenience that, right from the beginning, we have to fix all (proper and improper) faces of cells. In cell complexes, $\emptyset$ is considered an (improper) face of a polytope, but not of a cone.

1.1 Definition. Let $\widetilde{F}$ be a k-dimensional polytope or a k-dimensional cone, $\{\widetilde{F}_i \mid i \in I\}$ the set of all its proper and improper faces, and $\varphi : \widetilde{F} \longrightarrow \varphi(\widetilde{F}) \subset \mathbb{R}^m$ a homeomorphism. Then,

$$F := \varphi(\widetilde{F}) \text{ is called a } k\text{-}cell \text{ in } \mathbb{R}^m \text{ with } faces\ \varphi(\widetilde{F}_i).$$

The cell F is called a

straight cell if $\varphi = \mathrm{id}_{\widetilde{F}}$,

simplex cell, if $\widetilde{F}$ is a simplex or a simplex cone,

spherical cell, if F (with its faces) is the intersection of the unit sphere and a cone.

Cones are always chosen to be straight.

Remark. A k-cell F comes from a polytope (as opposed to a cone) if and only if

$$\begin{cases} k \geq 1 & \text{and } F \text{ is compact,} \\ k = 0 & \text{and } \emptyset \text{ is a face of } F, \\ k = -1. \end{cases}$$

Remark. If $\{0\}$ is a cell, it must be said if it is on account of a cone or a polytope. In the former case, it has no proper faces. In the latter case, $\emptyset$ is a face of $\{0\}$.

1.2 Definition. A finite set $\mathcal{C}$ of cells in a Euclidean space is called a *cell complex* if the following two conditions are satisfied:
(a) If $F \in \mathcal{C}$ and F_0 is a face of F, then $F_0 \in \mathcal{C}$.
(b) If $F, F' \in \mathcal{C}$, then $F \cap F'$ is a common face of F and F'.

If all cells are simplex cells, we call $\mathcal{C}$ a *simplicial complex.* A cell complex consisting of spherical cells is said to be a *spherical complex.*

Remarks.
(1) Since $\emptyset$ is not a cone, it follows from (a) and (b) that either all or no cells of the cell complex come from cones.
(2) It should be noted that it is not sufficient to replace (b) in Definition 1.2 by "$F \cap F' \in \mathcal{C}$". If $\mathcal{C}$ consists of two triangles T, T' in $\mathbb{R}^2$ with their sides and if $T \cap T'$ is a 1-side of T properly contained in a 1-side of T', "$F \cap F' \in \mathcal{C}$" is always satisfied; $\mathcal{C}$ is, however, not a cell complex.

Clearly,

1.3 Lemma. *The boundary complex $\mathcal{B}(P)$ of a polytope P (including $\emptyset$ but excluding P) is a cell complex. It is a simplicial complex if and only if P is simplicial.*

1.4 Lemma. *The fan $\Sigma(P)$ of a polytope P (see* I, 4*) is a cell complex.*

As in the special case of boundary complexes $\mathcal{B}(P)$, we define the following:

1.5 Definition. Two cell complexes $\mathcal{C}, \mathcal{C}'$ (whose cells are not necessarily of the same type) are said to be *isomorphic*, $\mathcal{C} \approx \mathcal{C}'$, if there exists a bijective, inclusion-preserving map between them.

1.6 Lemma. *The fan $\Sigma(P)$ of an n-dimensional polytope P is isomorphic to the boundary complex $\mathcal{B}(P^*)$ of a polar polytope P^* of P (where $\{0\} \in \Sigma(P)$ corresponds to $\emptyset \in \mathcal{B}(P^*)$).*

PROOF. If v is a vertex of P, then, by I, Theorem 4.13, the normal cone $N(v)$ is an n-dimensional cone of $\Sigma(P)$, and $N(v) \cap H_v$ is the polar facet v^* of v (compare

II, Theorem 2.1(a)). We obtain an inclusion preserving bijection of vert P onto the set of facets of P^* which readily carries over to a bijection of $\Sigma(P)$ and $\mathcal{B}(P^*)$. □

Generalizing $\Sigma(P)$, we define the following:

1.7 Definition. A cell complex Σ consisting of cones in $\mathbb{R}^n$ with apex 0 is called a *fan*. The fan is called *polyhedral*, if all its cells are polyhedral, *simplicaial*, if each of its cones is a simplex cone, that is, the positive hull of linearly independent vectors. We say the fan is *complete* if its cones cover $\mathbb{R}^n$. We denote the set of all i-dimensional cones of Σ by $\Sigma^{(i)}$.

1.8 Lemma. *Let a cell complex $\mathcal{C}$ be given which is a fan, and let S be the unit sphere with center* 0. *Then,*

$$F \longmapsto F \cap S$$

for all cones $F \in \mathcal{C}$ induces an isomorphism of $\mathcal{C}$ onto a spherical cell complex.

1.9 Definition. If $\mathcal{C}$ is a cell complex, then, the point set $|\mathcal{C}| := \bigcup_{F \in \mathcal{C}} F$ is called the *support* of $\mathcal{C}$ or the *polyhedron* that underlies $\mathcal{C}$. If $|\mathcal{C}|$ is homeomorphic (that is, can be mapped bijectively, bicontinuously) to a k-sphere we call $\mathcal{C}$ a *polyhedral k-sphere* or shortly a *polyhedral sphere*.

Example 1. The boundary complex $\mathcal{B}(P)$ of an n-dimensional polytope is a polyhedral $(n-1)$-sphere.

Example 2. The intersection of a complete fan with the unit sphere is a polyhedral sphere.

1.10 Definition. A subset of a cell complex $\mathcal{C}$ is called a *subcomplex* of $\mathcal{C}$ if it is again a cell complex.

1.11 Definition. Let $\mathcal{C}$ be a cell complex, $F \in \mathcal{C}$.

$\operatorname{st}(F, \mathcal{C}) := \{F' \in \mathcal{C} \mid F \subset F'\}$ is called the *star* of F in $\mathcal{C}$.

$\overline{\operatorname{st}}(F, \mathcal{C}) := \{F'' \in \mathcal{C} \mid F'' \subset F' \in \operatorname{st}(F, \mathcal{C})\}$ is said to be the *closed star* of F in $\mathcal{C}$.

$$\operatorname{link}(F, \mathcal{C}) := \begin{cases} \{F' \in \overline{\operatorname{st}}(F, \mathcal{C}) \mid F' \cap F = \{0\}\} & \text{if } \mathcal{C} \text{ consists of cones,} \\ \{F' \in \overline{\operatorname{st}}(F, \mathcal{C}) \mid F' \cap F = \emptyset\} & \text{otherwise} \end{cases}$$

is called the *link* of F in $\mathcal{C}$.

Remarks.

(1) In general, $\operatorname{st}(F, \mathcal{C})$ is not a subcomplex of $\mathcal{C}$, $\overline{\operatorname{st}}(F, \mathcal{C})$ and $\operatorname{link}(F, \mathcal{C})$ are subcomplexes of $\mathcal{C}$.

(2) From the second (three-dimensional) example in Figure 1, it can be seen that, in general, $\operatorname{link}(F, \mathcal{C})$ is properly contained in $\overline{\operatorname{st}}(F, \mathcal{C}) \setminus \operatorname{st}(F, \mathcal{C})$.

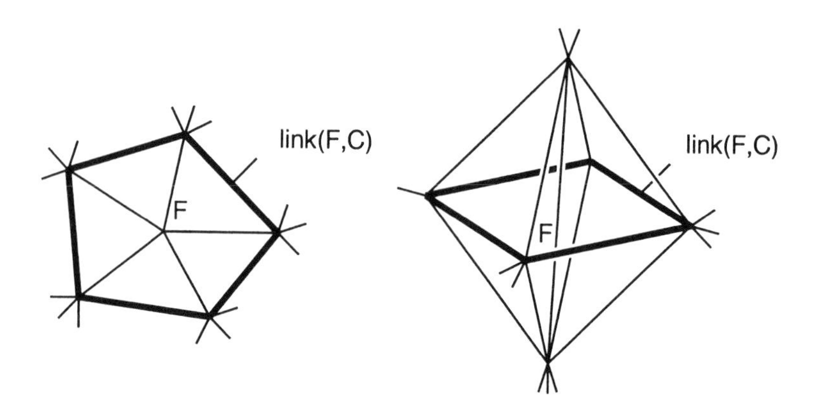

FIGURE 1.

Now, we will introduce a kind of multiplication for certain cell complexes $\mathcal{C}$ and $\mathcal{C}'$ in $\mathbb{R}^n$. To that end, let $F \in \mathcal{C}$ and $F' \in \mathcal{C}'$ be cells:

1.12 Definition. Assume that F, F' are cones with apex 0, so that $(\text{lin } F) \cap (\text{lin } F') = \{0\}$. Then, we call

$$F \cdot F' := \text{pos}(F \cup F')$$

the *join* of F and F'.

There is an analogous construction of $\mathcal{C}$ and $\mathcal{C}'$ for polytopes:

1.13 Definition. Let $\widetilde{F}$, $\widetilde{F}'$ be polytopes. If

$$\widetilde{F} \cdot \widetilde{F}' := \text{conv}(\widetilde{F} \cup \widetilde{F}')$$

is a $(\dim \widetilde{F} + \dim \widetilde{F}' + 1)$-polytope with $\widetilde{F}$ and $\widetilde{F}'$ as faces, then, we call $\widetilde{F} \cdot \widetilde{F}'$ the *join* of $\widetilde{F}$ and $\widetilde{F}'$. More generally, if there exists a homeomorphism

$$\varphi : \widetilde{F} \cdot \widetilde{F}' \longrightarrow \varphi(\widetilde{F} \cdot \widetilde{F}') \subset \mathbb{R}^n,$$

then, for $F := \varphi(\widetilde{F})$ and $F' := \varphi(\widetilde{F}')$, we call

$$F \cdot F' := \varphi(\widetilde{F} \cdot \widetilde{F}')$$

the *join* of F and F'. By convention, $\emptyset \cdot F' = F'$, $F \cdot \emptyset = F$, and $\emptyset \cdot \emptyset = \emptyset$.

Example 3. Each n-simplex T is the join $F \cdot F'$ of two of its faces F, F' such that $\text{vert } T = (\text{vert } F) \cup (\text{vert } F')$ and $(\text{vert } F) \cap (\text{vert } F') = \emptyset$.

For the construction of the join of the cell complexes $\mathcal{C}$ and $\mathcal{C}'$ in $\mathbb{R}^n$, we need an extra hypothesis (compare the remark after Lemma 1.15).

1.14 Definition. For $F \in \mathcal{C}$, assume that $F \cdot F'$ is defined for every $F' \in \mathcal{C}'$ and that, for every $F' \neq F'' \in \mathcal{C}'$, $\operatorname{relint}(F \cdot F') \cap \operatorname{relint}(F \cdot F'') = \emptyset$. Then, we call

$$F \cdot \mathcal{C}' := \{F \cdot F' \mid F' \in \mathcal{C}'\}$$

the *join* of F and $\mathcal{C}'$.

If $F \cdot \mathcal{C}'$ exists for each F of a cell complex $\mathcal{C}$ and if $\operatorname{relint}(F \cdot F') \cap \operatorname{relint}(G \cdot G') = \emptyset$ for any two different $F, G \in \mathcal{C}$ and any two different $F', G' \in \mathcal{C}'$, then, we call

$$\mathcal{C} \cdot \mathcal{C}' := \{F \cdot F' \mid F \in \mathcal{C}, F' \in \mathcal{C}'\}$$

the *join* of $\mathcal{C}$ and $\mathcal{C}'$.

1.15 Lemma. *The join $\mathcal{C} \cdot \mathcal{C}'$ of $\mathcal{C}$ and $\mathcal{C}'$ is a cell complex. It contains $\mathcal{C}$ and $\mathcal{C}'$.*

PROOF. For polytopes, $F \cdot \emptyset = \emptyset \cdot F = F$ and $\mathcal{C} \subset \mathcal{C} \cdot \mathcal{C}'$ as well as $\mathcal{C}' \subset \mathcal{C} \cdot \mathcal{C}'$. The defining properties 12 of a cell complex are readily verified. For cones, we proceed analogously. □

Remark. In the definition of $F \cdot \mathcal{C}'$ it is not sufficient to assume that $F \cdot F'$ is defined for all $F' \in \mathcal{C}'$. For example, let $\mathcal{C}' := \mathcal{B}(P)$ for a 2-polytope P in $\mathbb{R}^2$, and F a point in $\mathbb{R}^2 \setminus P$. Then, possibly, $F \cdot F'$ is defined for all $F' \in \mathcal{B}(P)$, but some of them overlap in (relative) interior points.

Example 4. Let P be a two-dimensional polytope in $\mathbb{R}^3$, and let $[p, p']$ be a line segment such that $\operatorname{relint} P \cap \operatorname{relint}[p, p']$ is a point. Then $\mathcal{B}(P) \cdot \{p, p', \emptyset\}$ is defined and is the boundary complex of a bipyramid.

Example 5. Let v be a vertex of a simplicial n-polytope P. Then,

$$\{v, \emptyset\} \cdot \operatorname{link}(v, \mathcal{B}(P)) = \overline{\operatorname{st}}(v, \mathcal{B}(P)).$$

Example 6. Let $\mathcal{C}, \mathcal{C}'$ be a regular r-gon and a regular s-gon, $r, s \geq 3$ in the ξ_1, ξ_2-plane and the ξ_3, ξ_4-plane of $\mathbb{R}^4$, respectively, both with centroid (=barycenter) 0. Then, $\mathcal{C} \cdot \mathcal{C}'$ is defined and is a polyhedral 3-sphere with $r \cdot s$ 3-cells.

1.16 Lemma. *Let F, F' be cones in $\mathbb{R}^n$ such that $F \cdot F'$ is defined.*

(a) $F \cdot F' = F + F'$ *(vector sum).*

(b) $F \cdot F'$ *is a cone with apex* 0.

(c) *If S is the unit sphere of $\mathbb{R}^n$, then,*

$$(F \cap S) \cdot (F' \cap S) = (F \cdot F') \cap S$$

for the spherical cells $F \cap S$, $F' \cap S$.

PROOF. (a) For $x \in F \cdot F'$, we may set

$$x = \lambda_1 y_1 + \cdots + \lambda_k y_k + \lambda_{k+1} y_{k+1} + \cdots + \lambda_m y_m$$

where $y_1, \ldots, y_k \in F$, $y_{k+1}, \ldots, y_m \in F'$, and $\lambda_1, \ldots, \lambda_m \geq 0$. Then, for $y := \lambda_1 y_1 + \cdots + \lambda_k y_k$, $y' := \lambda_{k+1} y_{k+1} + \cdots + \lambda_m y_m$, we have $x = y + y'$.

Hence, $F \cdot F' \subset F + F'$. The inclusion $F + F' \subset F \cdot F'$ is trivially true, so (a) follows.

(b) By definition, $F \cdot F'$ is a cone. Suppose $\{0\}$ is not an apex of $F \cdot F'$. Then, there exists $x \neq 0$ such that $x \in F \cdot F'$ and $-x \in F \cdot F'$. By (a), we have $x = y + y'$ and $-x = z + z'$ where $y, z \in F$ and $y', z' \in F'$. Now $0 = x - x = y + z + y' + z'$, so that $y + z = -(y' + z')$. By the assumption $(\text{lin } F) \cap (\text{lin } F') = \{0\}$, we obtain $y + z = 0 = y' + z'$, and, hence, $z = -y, z' = -y'$. Since 0 is an apex of F and of F', we find $z = y = z' = y' = 0$ and, therefore, $x = 0$, a contradiction.

(c) Since, by (a) and (b), $F \cdot F'$ is a cone with apex 0, we can find a hyperplane H which does not pass through 0 and intersects $F \cdot F'$ in a polytope $(F \cdot F') \cap H \neq \emptyset$. By central projection from 0, we see that $(F \cdot F') \cap H$ and $(F \cdot F') \cap S$ are homeomorphic. It is readily seen that $(F \cdot F') \cap H = (F \cap H) \cdot (F' \cap H)$, so that (c) follows. □

Exercises

1. If P is a k-fold bipyramid over R (see II, 3), there exists a k-cube Q such that $\mathcal{B}(P) = \mathcal{B}(R) \cdot \mathcal{B}(Q)$.
2. Let v be a vertex of a simplicial polytope P. Then $\text{link}(v, \mathcal{B}(P))$ is a polyhedral sphere isomorphic to $\mathcal{B}_0(Q) = \mathcal{B}(Q) \setminus \emptyset$ for some $(n-1)$-dimensional polytope Q.
3. Let C be a 4-cube. Then $\mathcal{B}(C)$ contains a subcomplex $\mathcal{C}_0$ such that $|\mathcal{C}_0|$ is homeomorphic to an ordinary torus ("tube").
4. Let v be a vertex of a polytope P. Define $\Sigma := \{\text{pos}(F - v) \mid F \in \mathcal{B}(P) \setminus \text{st}(v, \mathcal{B}(P))$ a proper face of $P\}$. Then, Σ is a fan, and $\Sigma \setminus \{0\} \approx [\mathcal{B}(P) \setminus \text{st}(v, \mathcal{B}(P))]$.

2. Stellar operations

The operations introduced now are of fundamental importance in the so-called piecewise linear (p.l.) topology. Although the present section can be considered part of that theory, our aim is an application of a special type of these operations in algebraic geometry ("blowing up" and "blowing down", Chapter VI).

2.1 Definition. Let $\mathcal{C}$ be a cell complex, $F \in \mathcal{C}$, $F \neq \emptyset$. For a point $p \in \text{relint } F$, we call the transition

$$\mathcal{C} \longrightarrow (\mathcal{C} \setminus \text{st}(F, \mathcal{C})) \cup p \cdot (\overline{\text{st}}(F, \mathcal{C}) \setminus \text{st}(F, \mathcal{C})) =: s(p; F)\mathcal{C}$$

a *stellar subdivision* (or *elementary subdivision*) of $\mathcal{C}$ *in direction* p. The inverse operation $s^{-1}(p; F)$ defined by $s^{-1}(p; F)(s(p; F)\mathcal{C}) = \mathcal{C}$ is said to be an *inverse stellar* (or *elementary*) *subdivision*.

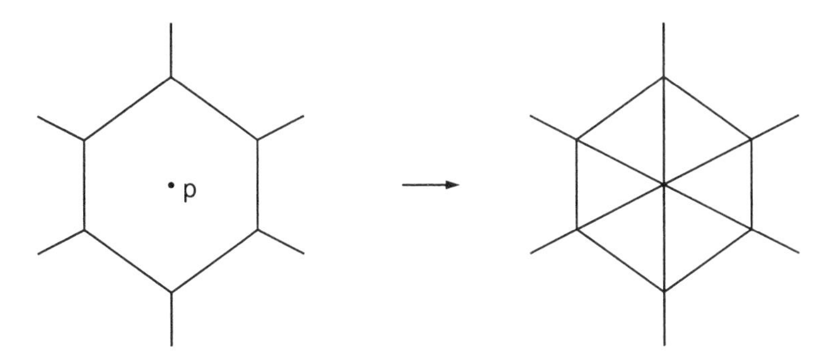

FIGURE 2.

Put more simply, we throw out the star of a face F and join a relative interior point of F to the boundary of the star. The result is called a stellar subdivision.

Example 1. If F is a point p, then, $s(p; \{p\})\mathcal{C}$ and $\mathcal{C}$ have the same 0-cells, though, in general, $s(p; \{p\})\mathcal{C} \neq \mathcal{C}$.

Example 2. If F is a cell of maximal dimension in $\mathcal{C}$, then, $\operatorname{st}(F, \mathcal{C}) = \{F\}$ and $s(p; F)\mathcal{C}$ splits only F and leaves all other cells of $\mathcal{C}$ unchanged (Figure 2).

Example 3. Let $\dim F = 1$ in a cell complex $\mathcal{C}$ whose cells have maximal dimension 2. Figures 3a, b, c illustrate the cases in which the number of 2-cells that contain the 1-cell F is 1, 2, or 3, respectively. In polyhedral 2-spheres, of course, only case (b) occurs.

A point $p \in |\mathcal{C}|$ determines $F \in \mathcal{C}$ by the condition "$p \in \operatorname{relint} F$". In the following example, p is chosen so that it determines different faces in different cell decompositions of a set.

Example 4. In Figure 4, the cell complex in the middle (six 3-simplices plus faces) is obtained by stellar subdivision of the right one (three 3-simplices plus faces) as well as of the left one (two 3-simplices plus faces).

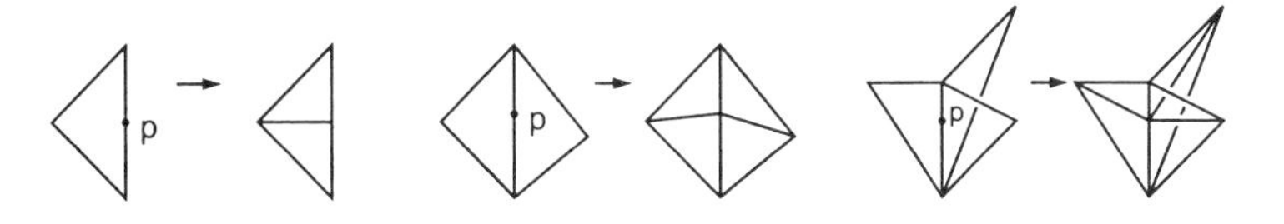

FIGURE 3a,b,c.

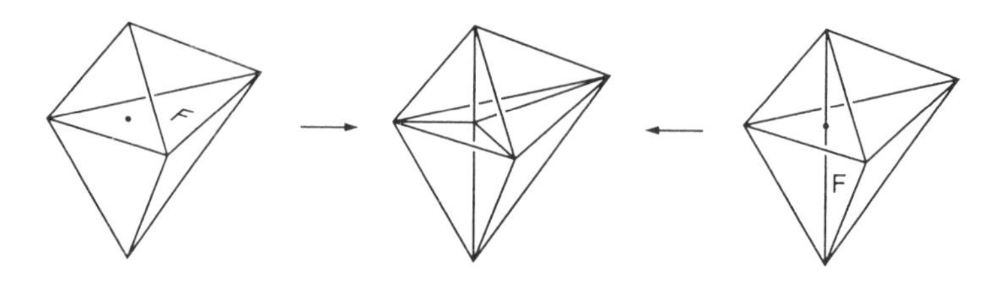

FIGURE 4.

The boundary complex of a polytope stellar subdivision can be described as follows:

2.2 Theorem. *Let P be a polytope and $s(p;\, F)\mathcal{B}(P)$ a stellar subdivision of its boundary complex. Then, there exists a polytope P' such that*

$$s(p;\, F)\mathcal{B}(P) \approx \mathcal{B}(P').$$

PROOF. We may assume $\dim P = n$. For $\dim F = n - 1$, we obtain a P' by placing a sufficiently "flat" pyramid onto F. Equivalently, we choose some point $q \in (\bigcap \operatorname{int} H_j^-) \setminus P$ where $H_1, \ldots, H_r$ are the affine hulls of those facets of P not equal to F and $P \subset H_j^-$, $j = 1, \ldots, r$, and, then, set $P' := \operatorname{conv}(\{q\} \cup P)$.

Let $0 \leq \dim F < n - 1$ and $p \in \operatorname{relint} F$. Furthermore, let $F = P \cap H_P(u)$, $H_P(u)$ a supporting hyperplane of P with outer normal u. We consider an $(n-1)$-face $\bar{F} \in \operatorname{st}(F, \mathcal{B}(P))$, $\bar{F} = P \cap H_P(v)$, and a face F_0 of $\bar{F}$ not in $\operatorname{st}(F, \mathcal{B}(P))$, $F_0 = P \cap H_P(w)$ (Figure 5). We choose some $p \in \operatorname{relint} F$. For $0 < \alpha < 1$, $F_0 = P \cap H_P(\alpha v + (1-\alpha)w)$ (we may assume $\langle u, v\rangle > 0$). If $q \in p + \mathbb{R}_{\geq 0}\, u$ is sufficiently close to p, we find some $\alpha \in (0, 1)$, such that $q \in H_P(\alpha v + (1 - \alpha)w) =: H(q, F_0)$.

We can assume $P \subset H_P^-(u) \cap H_P^-(v) \cap H_P^-(w)$. Then, also, $P \subset H^-(q, F_0)$. Now we consider the set of all $(n-2)$-faces of P. Let $F_0 = F_0^{(1)}, \ldots, F_0^{(s)}$ be all $(n-2)$-faces in $\overline{\operatorname{st}}(F, \mathcal{B}(P)) \setminus \operatorname{st}(F, \mathcal{B}(P))$. We can choose the same q for all these faces, so that $P \subset H^-(q, F_0^{(j)})$, $j = 1, \ldots, s$. Let $H_1, \ldots, H_r$ be the

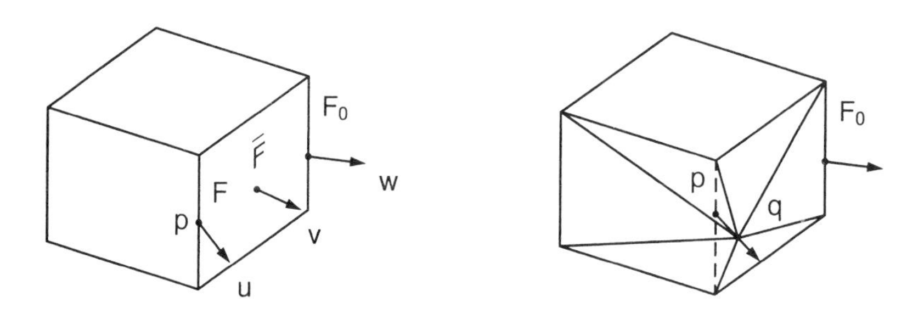

FIGURE 5.

affine hulls of all facets of P that do not lie in $\mathrm{st}(F, \mathcal{B}(P))$, where $P \subset H_j^-$, $j = 1, \ldots, r$. We set

$$P' := (\bigcap_{i=1}^{r} H_i^-) \cap (\bigcap_{j=1}^{s} H^-(q, F_0^{(j)})).$$

Clearly, $P' = \mathrm{conv}(\{q\} \cup P)$. Furthermore, $q \cdot F_0^{(j)}$ are facets of P', $j = 1, \ldots, s$.
The assignment

$$p \cdot F_0^{(j)} \longmapsto q \cdot F_0^{(j)} \qquad j = 1, \ldots, s$$

can readily be extended to an isomorphism

$$s(p; F)\mathcal{B}(P) \longrightarrow \mathcal{B}(P').$$

□

2.3 Definition. We say P' is obtained from P by *pulling up* a point p of the boundary of P.

A stellar subdivision of $\mathcal{B}(P)$ can also be characterized by a "cutting off" operation dual to "pulling up".

2.4 Theorem. *Let P^* be the polar polytope of P, and let F^* be the polar face of a proper face F of P. If the hyperplane H strictly separates* vert F^* *from* (vert P^*) $\setminus$ (vert F^*) *and if $F^* \subset H^+$, then, for an arbitrary $p \in$* relint F*,*

$$\mathcal{B}(P^* \cap H^-) \approx \mathcal{B}(P'^*),$$

where P' is defined according to Theorem 2.2.

Proof. We may assume $n \geq 2$. If $p_1, \ldots, p_m$ are the vertices of F, then, $p_1^*, \ldots, p_m^*$ are the facets of P^* that intersect in $F^* \in \mathcal{B}(P^*)$. The duals $F_0^{(i)*}$ of the $(n-2)$-faces $F_0^{(i)}$, defined as in the proof of Theorem 2.2, are line segments emanating from vertices of F^*. If, also, q is introduced, as in the proof of Theorem 2.2, the polar hyperplane H of q intersects each $F_0^{(i)*}$ in a point $q_i := (q \cdot F_0^{(i)})^*$, $i = 1, \ldots, s$. So, $\mathrm{conv}\{q_1, \ldots, q_s\} = q^*$ in $\mathcal{B}(P'^*)$ is a facet whose boundary complex consists of the polars G^* of faces $G \in \mathcal{B}(P')$ that contain q. Since the other faces of $P^* \cap H^-$ and P'^* are in a natural bijective and inclusion-preserving correspondence, the theorem follows. □

The following combination of stellar subdivisions can, on the one hand, be used to turn any cell complex into a simplicial one, and, on the other, be used to decompose a cell complex into cells of arbitrarily small diameters (as is needed in p.l. topology).

2.5 Definition. Let $\mathcal{C}$ be a cell complex whose cells of dimension i are denoted $F_1^i, \ldots, F_{f_i}^i, i = 0, \ldots, k$ (k the maximal dimension of a cell). For each F_j^i, we

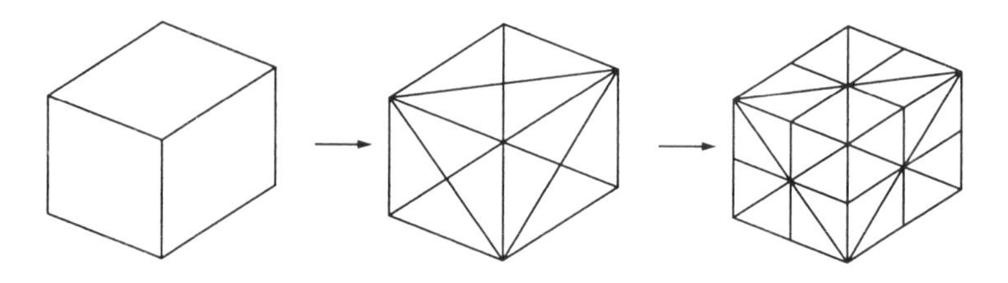

FIGURE 6.

choose a point $p^i_j \in \text{relint } F^i_j$, for example, the barycenter. We set $s^i_j := s(p^i_j;\, F^i_j)$. Then,

$$\beta(\mathcal{C}) := s^1_{f_1} \circ \cdots \circ s^1_1 \circ \cdots \circ s^k_{f_k} \circ \cdots \circ s^k_1(\mathcal{C})$$

is called a *barycentric subdivision* of $\mathcal{C}$.

It should be noted that s^i_j does not affect any cell $F \neq F^i_j$ of dimension $\leq i$, so that barycentric subdivisions are well-defined.

Example 5. Figure 6 illustrates $\beta(\mathcal{B}(C))$, C a cube, for the "visible" part of $\mathcal{B}(C)$.

Example 6. Figure 7 shows $\beta(\mathcal{C})$ for a cell complex consisting of three triangles, six edges, and four vertices.

Finally, we will mention the combinatorial aspect of a problem which will occupy us in Chapter V, 6, for special stellar subdivisions:

Problem. Given two polyhedral n-spheres $\mathcal{C}$, $\mathcal{C}'$ of the same dimension n, do there exist stellar subdivision $s_1, \ldots, s_p, s'_1, \ldots, s'_q$ such that from $\mathcal{C}, \mathcal{C}'$ the same combinatorial sphere $\mathcal{C}''$ is obtained as indicated in the following diagram?

$$\mathcal{C} \underset{s_1}{\longrightarrow} \cdots \underset{s_p}{\longrightarrow} \mathcal{C}'' \underset{s'_q}{\longleftarrow} \cdots \underset{s'_1}{\longleftarrow} \mathcal{C}'$$

The answer is yes for $n = 2$, and it seems also to be yes for $n > 2$ (c.f. Appendix III, 2.).

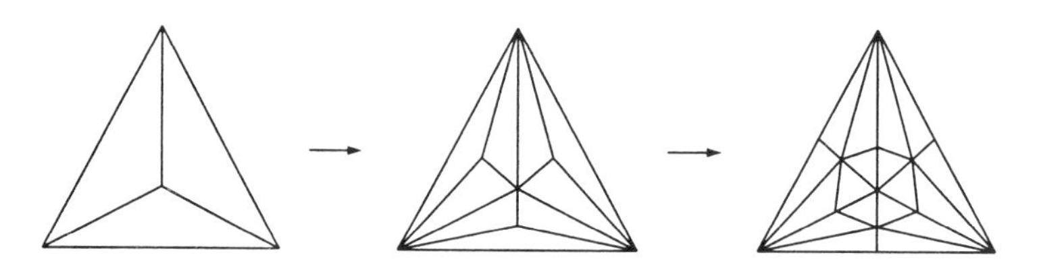

FIGURE 7.

Besides the barycentric subdivision, there is a decomposition of the boundary complex $\mathcal{B}(P)$ of a polytope P into a simplicial complex in which no new vertices occur:

2.6 Theorem. *Let $\mathcal{C}$ be a cell complex and let $Z_1, \ldots, Z_r$ be the 1-cells (in case $\mathcal{C}$ is a complex of cones) or 0-cells otherwise (in some order). For $v_i \in$ relint Z_i ($\{v_i\} = Z_i$ in case of 0-cells), $i = 1, \ldots, r$, we set $\mathcal{C}_0 := \mathcal{C}$ and $\mathcal{C}_i = s(v_i; Z_i)\mathcal{C}_{i-1}$, $i = 1, \ldots, r$. Then, $\mathcal{C}_r$ is a simplicial complex which has the same 1-cells (in case of a complex of cones) or 0-cells (in case of a compact complex) as $\mathcal{C}$ has. Furthermore, $|\mathcal{C}_r| = |\mathcal{C}|$.*

PROOF. By definition, $s(v_i; Z_i)$ does not add a 1-cell (in case of a cone complex) or a 0-cell (otherwise) to $\mathcal{C}_{i-1}$, $i = 1, \ldots, r$. Suppose $F \in \mathcal{C}_r$ is not a simplex cone or a simplex. Then, there exist $Z_j \subset F$, $Z_k \subset F$ such that $\operatorname{conv}(Z_j \cup Z_k)$ is not a cell of $\mathcal{C}_r$. But $s(v_j; Z_j)\mathcal{C}_{j-1}$ contains $\operatorname{conv}(Z_j \cup Z_k)$ (by definition), a contradiction. □

Now, we will turn to the problem of completing a noncomplete fan Σ (corresponding to problems of compactification in algebraic geometry). First, we introduce the following notion.

2.7 Definition. Let $\mathcal{C}_1, \mathcal{C}_2$ be cell complexes of polytopes or cones in $\mathbb{R}^n$ (possibly $\mathcal{C}_1$ consists of polytopes and $\mathcal{C}_2$ of cones). Then, we call $\mathcal{C}_1 \sqcap \mathcal{C}_2 := \{\sigma_1 \cap \sigma_2 \mid \sigma_1 \in \mathcal{C}_1, \sigma_2 \in \mathcal{C}_2\}$ the *intersection complex* of $\mathcal{C}_1$ and $\mathcal{C}_2$.

Clearly, $\mathcal{C}_1 \sqcap \mathcal{C}_2$ is a cell complex.

If $\mathcal{C}_1, \mathcal{C}_1$ are complete fans, $\mathcal{C}_1 \sqcap \mathcal{C}_2$ is again a complete fan.

However, if $\mathcal{C}_1$ consists of polytopes, so does $\mathcal{C}_1 \sqcap \mathcal{C}_2$, no matter if $\mathcal{C}_2$ consists of polytopes or cones.

2.8 Theorem. *Every fan Σ can be extended to a complete fan $\Sigma' \supset \Sigma$.*

PROOF. Let S be the unit sphere. We shall prove a somewhat stronger version of the theorem:

Given a noncomplete fan Σ in $\mathbb{R}^n$ and an $\varepsilon > 0$, we can find a fan Σ'' and a complete fan Σ' such that the following five conditions are satisfied.

(i) $\Sigma \subset \Sigma'' \subset \Sigma'$.

(ii) $\Sigma' \setminus \Sigma''$ and $\Sigma' \cap \Sigma''$ consist of simplex cones.

(iii) $\Sigma'' \setminus \Sigma$ consists of k-fold joins (Definition 1.12) $\tau \cdot \varrho^{(1)} \cdots \varrho^{(k)}$, $\tau \in \Sigma$, $\varrho^{(i)}$ a 1-cone, $i = 1, \ldots, k$, $1 \le k \le n-1$ and their faces containing at least one cone $\varrho^{(i)}$.

(iv) $|\Sigma| \setminus \{0\} \subset \operatorname{int} |\Sigma''|$.

(v) If $n \ge 2$ and $\sigma'' \in \Sigma'' \setminus \Sigma$, then, each point of $\sigma'' \cap S$ has distance less than ε from a cell of Σ.

We proceed by induction on n. For $n = 1$, Σ is either $\{\{0\}\}$ or $\{\{0\}, \mathbb{R}_{\geq 0}\}$ or $\{\{0\}, \mathbb{R}_{\leq 0}\}$ and $\Sigma' = \Sigma'' = \{\{0\}, \mathbb{R}_{\geq 0}, \mathbb{R}_{\leq 0}\}$.

For $n = 2$, the set $S \setminus |\Sigma|$ consists of finitely many arcs (without end points). If $\overline{pq}$ is such an arc, we choose two points p', q' on it with distances less than $\varepsilon/2$ from p, q, respectively. Then $\operatorname{pos}\{p, p'\}$ and $\operatorname{pos}\{q, q'\}$ are considered as elements of Σ'', and $\operatorname{pos}\{p', q'\}$ as an element of Σ'. We do the same for all arcs and split them if they are greater than or equal to π. We consider Σ as a subset of Σ'', also, Σ'' as a subset of Σ' and obtain fans satisfying (i) to (v).

Let $n > 2$, and let $\varrho := \mathbb{R}_{\geq 0}\, a, a = \varrho \cap S$, be a 1-cone of Σ for which $\operatorname{st}(\varrho, \Sigma)$ is not complete. (If $\Sigma = \{\{0\}\}$ we first extend Σ by a 1-cone). We consider the tangent hyperplane H of S at a and the fan $\Sigma_a := \{\operatorname{pos}(\sigma \cap H - a) \mid \sigma \in \operatorname{st}(\varrho, \Sigma)\}$ with origin a. By the induction hypothesis, there exist fans Σ''_a, Σ'_a such that conditions (i) to (v) hold. We construct from them an extension of Σ as follows.

If $\mathbb{R}_{\geq 0}(b - a)$ is a 1-cone of $\Sigma'_a \setminus \Sigma_a$, we assume $\|b - a\| < \varepsilon_a$ for some bound $\varepsilon_a \leq \varepsilon$ which depends only on a. Let $\varrho_a^{(i)} := \mathbb{R}_{\geq 0}(b^{(i)} - a) \in \Sigma''_a$, $\|b^{(i)} - a\| < \varepsilon_a, i = 1, \ldots, k$, and $\tau_a \in \Sigma_a$ be given, $\dim \tau_a = n - k - 1$, such that $\tau_a \cdot \varrho_a^{(1)} \cdots \varrho_a^{(k)} \in \Sigma''_a \setminus \Sigma_a$ is an $(n-1)$-cone. To τ_a there corresponds a unique $\tau \in \operatorname{st}(\varrho, \Sigma)$ such that $\tau_a = \operatorname{pos}(\tau \cap H - a)$. We set $\varrho^{(i)} := \mathbb{R}_{\geq 0}\, b^{(i)}$, $i = 1, \ldots, k$. Then, $\bar{\tau} := \tau \cdot \varrho^{(1)} \cdots \varrho^{(k)}$ is well-defined. If $b^{(i)}$ is not a rational vector, we replace it by a rational vector close to $b^{(i)}$ and denote the new vector again by $b^{(i)}$. From (ii) and (iii) applied to $\Sigma_a, \Sigma''_a, \Sigma'_a$, we see that changing the $b^{(i)}$ slightly, provides an isomorphic change of Σ''_a and Σ'_a whereas Σ_a remains unchanged. If ε_a is chosen small enough, the cone $\bar{\tau}$ intersects $|\Sigma|$ only in τ. Therefore, Σ together with $\bar{\tau}$ and the faces of $\bar{\tau}$ is again a fan $\Sigma(a, \tau)$.

Let $\widetilde{\tau}_a \cdot \widetilde{\varrho}_a^{(1)} \cdots \widetilde{\varrho}_a^{(l)} \in \Sigma''_a \setminus \Sigma_a$ be another such $(n-1)$-cone. We associate with it an n-cone $\bar{\widetilde{\tau}}$ analogously, and continue in this way until $\Sigma''_a \setminus \Sigma_a$ is exhausted. We obtain a fan $\Sigma(a)$.

Now let a simplex $(n-1)$-cone $\hat{\varrho}_a^{(1)} \cdots \hat{\varrho}_a^{(n-1)} \in \Sigma'_a \setminus \Sigma''_a$ be given, $\hat{\varrho}_a^{(i)} := \mathbb{R}_{\geq 0}(\hat{b}^{(i)} - a)$, $\|\hat{b}^{(i)} - a\| < \varepsilon_a, i = 1, \ldots, n-1$. As in the case of $b^{(i)}$ we may assume $\hat{b}^{(i)}$ to be rational. We consider the simplex n-cone $\hat{\tau} := \operatorname{pos}\{a, \hat{b}^{(1)}, \ldots, \hat{b}^{(n-1)}\}$ which, for sufficiently small ε_a, intersects Σ only in ϱ. Adding all such cones to $\Sigma(a)$ provides a cone $\Sigma''(a)$.

If a 1-cone of Σ different from ϱ exists on the boundary of $|\Sigma|$, we proceed in the same way as before with Σ replaced by $\Sigma''(a)$. Continuing in this way, we end up with a fan Σ''_0 which satisfies $|\Sigma| \setminus \{0\} \subset \operatorname{int} |\Sigma''_0|$.

The set $S \setminus |\Sigma''_0|$ is open relative to S. Its (topological) closure does not intersect $|\Sigma|$. We intend to cover $S \setminus |\Sigma''_0|$ by "small" spherical simplices. For this purpose, we introduce the following subdivision of S into spherical simplices.

Consider a cube C circumscribed to S, and apply barycentric subdivisons to $\mathcal{B}(C)$ successively until all simplices have diameter less than ε_0 for a given $\varepsilon_0 \leq \varepsilon$. The positive hulls of these simplices intersect S in spherical simplices with diameter less than ε_0. If ε_0 is chosen small enough, we can cover $S \setminus |\Sigma''_0|$ by spherical simplices which do not intersect Σ. Together with their faces, they provide a simplicial spherical complex $\mathcal{C}$. From $\mathcal{C}$ we obtain a simplical fan $\Sigma_{\mathcal{C}} := \{\text{pos}\, T \mid T \in \mathcal{C}\}$. It satisfies $|\Sigma_{\mathcal{C}}| \cap |\Sigma| = \{0\}$ and $|\Sigma_{\mathcal{C}}| \cup |\Sigma_0''| = \mathbb{R}^n$.

The last problem to be solved is turning the overlapping part of $\Sigma_{\mathcal{C}}$ and Σ_0'' into a cell complex. Let $\mathcal{B}_0$ consist of all cones of Σ_0'' which do not intersect int $|\Sigma_0''|$ (the "boundary complex" of Σ_0''), and let $\mathcal{B}_0' := \mathcal{B}_0 \sqcap \Sigma_{\mathcal{C}}$. We decompose the cones σ of $\Sigma_0'' \setminus \Sigma$ as follows. If $\sigma \in \mathcal{B}_0$, replace $\mathcal{B}(\sigma)$ (the boundary complex of σ) by $\mathcal{B}(\sigma) \sqcap \Sigma_{\mathcal{C}} \subset \mathcal{B}_0'$. If $\sigma \notin \mathcal{B}_0$, it intersects both Σ and $\mathcal{B}_0$. If $\sigma_0 := \sigma \cap |\mathcal{B}_0|$, we replace $\mathcal{B}(\sigma_0)$ by $\mathcal{B}(\sigma_0) \sqcap \Sigma_{\mathcal{C}}$. Having achieved this change for all cones $\sigma \in \Sigma_0'' \setminus \Sigma$, we obtain a new cell complex Σ_1'' from Σ_0''. Its cells, however, are not all polyhedral cones but polyhedral cones with subdivided faces. In order to obtain only polyhedral cones, we choose a point $p_\sigma \in \text{relint}\, \sigma$ for each $\sigma \in \Sigma_0'' \setminus (\Sigma \cup \mathcal{B}_0)$, put $\varrho_\sigma := \mathbb{R}_{\geq 0}\, p_\sigma$ and apply a stellar subdivision in the sense that $\text{st}(\sigma, \Sigma_1'')$ is replaced by the join of ϱ_σ and the (subdivided) boundary complex of $\text{st}(\sigma, \Sigma_1'')$. Hereby, we start with those σ which have an $(n-1)$-face σ_0 in $\mathcal{B}_0$ and continue with $\dim \sigma = n - 2$, and so on. We end up with a fan which we denote by Σ_2''.

Next we wish to decompose the remainder $\sigma' \setminus |\Sigma_2''|$ of each cone $\sigma' \in \Sigma_{\mathcal{C}} \setminus \Sigma_0''$ into polyhedral cones. If the elements of $\mathcal{C}$ have been chosen with sufficiently small diameters, the following condition holds:

(*) Each n-cone of Σ_0'', which intersects a given $\sigma' \in \Sigma_{\mathcal{C}}$, belongs to $\text{st}(\varrho, \Sigma_0'')$ for one and the same $\varrho \in \mathcal{B}_0$.

(Note that, although Σ_0'' has been decomposed into Σ_2'' above, we continue using it temporarily).

As in the earlier part of the proof, we associate with $\text{st}(\varrho, \Sigma_0'')$ an $(n-1)$-dimensional fan Σ_p in the tangent hyperplane H_0 of S at $p = \varrho \cap S$. By the induction hypothesis, we can extend Σ_p to a complete fan Σ_p^c. Because of (*), Σ_p^c induces a cell decomposition of $H_0 \cap (\sigma' \setminus \text{int}\, |\Sigma_0''|)$ from which we obtain (by taking positive hulls) a cell decomposition of $\sigma' \setminus \text{int}\, |\Sigma_0''| = \sigma' \setminus \text{int}\, |\Sigma_2''|$ into cells $\sigma_1', \ldots, \sigma_k'$ for some k. This cell decomposition preserves the cells of $\mathcal{B}_0'$ contained in σ' and is hence compatible with Σ_2''.

However, if we have accomplished all these decompositions, they may not induce the same decompositions of a common face τ of two of the cells. Let $\Sigma(\tau, 1), \ldots, \Sigma(\tau, r)$ be different decompositions of τ into cell complexes. Then, we consider $\Sigma(\tau, 1) \sqcap \cdots \sqcap \Sigma(\tau, r)$ and decompose the n-cones, which contain τ by stellar subdivisions, into polyhedral cones (as we have done when we decomposed Σ_1'' into Σ_2''). Proceeding in this way for all cones $\sigma' \in \Sigma_{\mathcal{C}}$ which intersect $\mathcal{B}_0'$, we obtain a decomposition of $(|\Sigma_{\mathcal{C}}| \setminus |\Sigma_2''|) \cup |\mathcal{B}_0'|$ into a cell complex of polyhedral cones Σ_0' which contains $\mathcal{B}_0'$ as a subcomplex of polyhedral cones.

Finally, we apply stellar subdivisions as in $\mathcal{C}$ to all cells of $\Sigma_0' \cup \Sigma_2''$ which are not simplex cones and which do not intersect $|\Sigma| \setminus \{0\}$. These stellar subdivisions do not affect Σ. Thus, we obtain from $\Sigma_0' \cup \Sigma_2''$ a cell complex of cones Σ' and denote by Σ'' the subcomplex of all cones of Σ' which have at least a 1-cone in Σ together with their faces. Now Σ' and Σ'' are readily seen to satisfy (i) to (v). □

Remark. Although, from Theorem 2.8, an algorithm can be deduced for completing a fan it will, in general, not be the most efficient one. For $n = 3$ and $\operatorname{pos} |\Sigma| = \mathbb{R}^3$, completion without additional 1-cones can be found, which is not true for higher dimensions (c.f. the Appendix to this section).

Exercises

1. Show that $\beta(\mathcal{C})$ is always simplicial.
2. Illustrate a solution of the above problem for Σ the boundary complex of a 3-simplex and Σ' the boundary complex of an octahedron.
3. Given a 3-simplex, do there exist three stellar subdivisions such that, in the resulting complex, each pair of vertices is joined by a 1-cell?
4. Dualize $\beta(\mathcal{C})$ according to Theorem 2.4, and apply to Examples 5 and 6.

3. The Euler and the Dehn-Sommerville equations

Generalizing the f-vector $f(P)$ of a polytope P (II, Definition 1.13), we call, for an arbitrary cell complex $\mathcal{C}$,

$$f(\mathcal{C}) := (f_0(\mathcal{C}), \dots, f_{n-1}(\mathcal{C}))$$

the *f-vector* of $\mathcal{C}$, where $f_j(\mathcal{C})$ denotes the number of j-cells of $\mathcal{C}$, $j = 0, \dots, n-1$.

If $f_j = f_j(P)$ are the numbers of j-faces of an n-polytope P, there is a basic relationship between them, known, for $n = 3$, as Euler's formula

$$f_0 - f_1 + f_2 = 2. \tag{1}$$

This equation is, in fact, true for any polyhedral 2-sphere; its proof is, however, quite difficult. We shall prove a generalization of (1) to n-dimensional spherical complexes obtained as intersections of the unit sphere with complete fans, which (up to isomorphisms) includes the case of boundary complexes of polytopes.

3.1 Theorem (Euler-Poincaré's theorem). *Let the spherical complex $\mathcal{C}$ be obtained by intersecting the cones of a complete fan in $\mathbb{R}^n$ with an $(n-1)$-sphere S^{n-1} about 0. Then,*

$$\sum_{j=0}^{n-1} (-1)^j f_j(\mathcal{C}) = 1 + (-1)^{n-1}. \tag{2}$$

PROOF. We use induction on n. For $n = 1$, $\mathcal{C}$ is a pair of points and $(-1)^0 \cdot 2 = 1 + 1$ so that (2) holds. Let (2) be true for S^{n-2} instead of S^{n-1}.

It is readily seen that there exists an $(n-2)$-dimensional linear subspace U of $\mathbb{R}^n$ which intersects S^{n-1} only in the relative interiors of several at least two-dimensional cells of $\mathcal{C}$, called cells of the *second kind*. All other cells are not intersected by U; we call them cells of the *first kind*.

To the set of vertices (0-cells) of a spherical cell F of $\mathcal{C}$, we assign its convex hull $\hat{F}$ in $\mathbb{R}^n$ (of possibly higher dimension than F if F is not a spherical simplex). Then, pos F = pos $\hat{F}$. If a is a point in a sufficiently small neighborhood of 0, the assignment

$$F = S^{n-1} \cap \operatorname{pos} F \mapsto S^{n-1} \cap \operatorname{pos}(\hat{F} - a)$$

is an isomorphism of $\mathcal{C}$ and a spherical complex $\mathcal{C}'$. We can choose $\mathcal{C}'$ such that for any vertex $x \in \mathcal{C}'$ we have $-x \notin \mathcal{C}'$. Keeping this in mind, we can choose U such that the following conditions are satisfied for any hyperplane $H \supset U$ and any $F \in \mathcal{C}$,

(i) $H \cap F$ is a spherical cell.
(ii) Either $H \cap F$ is empty or a vertex or we have $\dim(H \cap F) = -1 + \dim F$.
(iii) H contains at most one vertex of $\mathcal{C}$.

Let $H_p := \operatorname{lin}(\{p\} \cup U)$ be the hyperplane passing through the vertex p of $\mathcal{C}$. On a semi-circle κ consisting of points represented by unit normals of the hyperplanes $H \supset U$ we obtain a linear ordering of the H_p, say, $H_1, H_3, \ldots, H_{2f_0-1}$. We choose hyperplanes $H_2, H_4, \ldots, H_{2f_0}$ so as to make the unit normal of H_{2i} on κ lie properly between that of H_{2i-1} and that of H_{2i+1}, $i = 1, \ldots, f_0$; take $H_{2f_0+1} = H_1$. (The left side of Figure 8 illustrates the H_i for a spherical cell of the first kind, where $\mathcal{C}$ has five vertices; the right side refers to a cell of the second kind, $\mathcal{C}$ having seven vertices).

Let F^j be a j-face of $\mathcal{C}$, $j = 1, \ldots, n-1$, and let $\mathcal{P}_i := \{H_i \cap F \mid F \in \mathcal{C}\}$ be a complex of spherical cells on an $(n-2)$-sphere. We set

$$\Phi(F^j, \mathcal{P}_i) = \begin{cases} 0 & \text{if } |\mathcal{P}_i| \cap \operatorname{relint} F^j = \emptyset, \\ 1 & \text{otherwise.} \end{cases}$$

We claim that

$$\sum_{i=1}^{2f_0} (-1)^i \Phi(F^j, \mathcal{P}_i) = \begin{cases} 1 & \text{if } F^j \text{ is of the first kind,} \\ 0 & \text{if } F^j \text{ is of the second kind} \end{cases} \tag{3}$$

(see Figure 8).

Proof: If H_i, with i odd, intersects a face F^j of the first kind, the hyperplane H_{i+1} intersects relint F^j unless $H_i \cap F^j$ is the "last" vertex, that is, a vertex of F^j with maximal i. H_i itself also intersects relint F^j, except for the "first" and the "last" vertex of F^j. Therefore, in the alternating sum on the left side of (3), the number of even i, for which we have $\Phi(F^j, \mathcal{P}_i) = 1$, exceeds the number of odd i with $\Phi(F^j, \mathcal{P}_i) = 1$ by one. This proves the first part of (3).

If F^j is of the second kind, there is no "first" or "last" vertex of F^j, and each H_i intersects, by definition, relint F^j. Therefore, the second part of (3) follows.

Let g_j denote the number of all j-faces of the second kind.

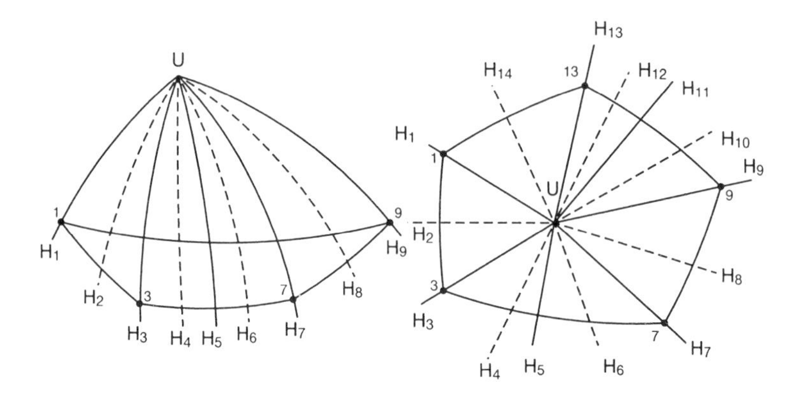

FIGURE 8.

We sum up (3) for all j-faces with a fixed j, and obtain

$$\sum_{j\text{-faces}} \sum_{i=1}^{2f_0} (-1)^i \Phi(F^j, \mathcal{P}_i) = f_j(\mathcal{C}) - g_j. \tag{4}$$

Now, summing over j, we find

$$\sum_{j=1}^{n-1} (-1)^j \sum_{j\text{-faces}} \sum_{i=1}^{2f_0} (-1)^i \Phi(F^j, \mathcal{P}_i) = \sum_{j=1}^{n-1} (-1)^j f_j(\mathcal{C}) - \sum_{j=1}^{n-1} (-1)^j g_j. \tag{5}$$

We set $\mathcal{C}_0 := \{F \cap U \mid F \in \mathcal{C}\}$ and apply the theorem inductively to $\mathcal{C}_0$. So, we obtain

$$\sum_{j=1}^{n-1} (-1)^j g_j = \sum_{j=1}^{n-1} (-1)^j f_{j-2}(\mathcal{C}_0) = 1 + (-1)^{n-3} = 1 + (-1)^{n-1}. \tag{6}$$

Looking further at the left side of (5), the main step is to change the order of summation. We claim that

$$\sum_{j\text{-faces}} \Phi(F^j, \mathcal{P}_i) = \begin{cases} f_0(\mathcal{P}_i) - 1 & \text{if} \quad j = 1 \text{ and } i \text{ odd}, \\ f_{j-1}(\mathcal{P}_i) & \text{otherwise.} \end{cases} \tag{7}$$

Proof: If i is even, each $(j-1)$-face of $\mathcal{P}_i$ is, by definition of H_i, the intersection of a j-cell F^j with H_i. The same is true if i is odd and $j \geq 1$. For odd i, let p_i be the vertex for which $H_i = \operatorname{lin}(\{p_i\} \cup U)$. It is not the intersection of H_i and relint F^1 of a 1-cell. All other vertices of $\mathcal{P}_i$ are such intersections. Therefore, (7) is true.

From (5), (6), (7) by applying the induction hypothesis to $\mathcal{P}_i$, we obtain

$$\sum_{j=1}^{n-1} (-1)^j \sum_{j\text{-faces}} \Phi(F^j, \mathcal{P}_i) \tag{8}$$

$$= \begin{cases} \sum_{j=1}^{n-1}(-1)^j f_{j-1}(\mathcal{P}_i) + 1 & \text{if } i \text{ is odd} \\ \sum_{j=1}^{n-1}(-1)^j f_{j-1}(\mathcal{P}_i) & \text{if } i \text{ is even} \end{cases}$$

$$= \begin{cases} -1 - (-1)^{n-2} + 1 = (-1)^{n-1} & \text{if } i \text{ is odd,} \\ -1 - (-1)^{n-2} = -1 + (-1)^{n-1} & \text{if } i \text{ is even.} \end{cases}$$

Now,

$$(9) \qquad \sum_{i=1}^{2f_0}(-1)^i \sum_{j=1}^{n-1}(-1)^j \sum_{j-\text{faces}} \Phi(F^j, \mathcal{P}_i)$$

$$= \sum_{i \text{ even}}(-1)^i((-1)^{n-1} - 1) + \sum_{i \text{ odd}}(-1)^i(-1)^{n-1} = -f_0(\mathcal{C}).$$

From (5), (6), and (9) we conclude that

$$\sum_{j=1}^{n-1}(-1)^j f_j(\mathcal{C}) - 1 - (-1)^{n-1} = -f_0(\mathcal{C}),$$

and, hence, we obtain (2). □

In terms of the f-vector Euler's theorem implies Theorem 3.2.

3.2 Theorem. *The f-vectors of polyhedral $(n-1)$-spheres, defined by complete fans in $\mathbb{R}^n$, lie in a hyperplane of $\mathbb{R}^n$, called Euler's hyperplane.*

If the spherical complex is simplicial, there are more relationships for the f-vectors. Before proving this, we introduce an analog to the quotient polytope P/F (F a face of the polytope P; see II, 2) for fans and spherical complexes.

3.3 Definition. Let τ be a cone of a fan Σ, and let U be the orthocomplement of the linear hull $\operatorname{lin} \tau$ of τ in $\mathbb{R}^n$. The fan, obtained from $\operatorname{st}(\tau, \Sigma)$ by perpendicular projection onto U, is called the *quotient fan* Σ/τ of Σ to τ. If $\mathcal{C} := \{\sigma \cap S^{n-1} \mid \sigma \in \Sigma\}$ and $F := \tau \cap S^{n-1}$ is substituted for Σ, τ, respectively, we obtain a spherical complex from Σ/τ called the *quotient complex* $\mathcal{C}/F$ (on the sphere $S^{n-2-\dim F}$).

We note that, for $\dim F = n-1$ or $\operatorname{st}(F, \mathcal{C}) = \{F\}$, the complex $\mathcal{C}/F$ is empty. It is readily checked that Σ/τ is well-defined. If Σ is complete and we set $r := \dim \tau$, then, Σ/τ is $(n-r)$-dimensional. Furthermore, we have

3.4 Lemma. *Let $\mathcal{C}$ be the $(n-1)$-dimensional spherical complex obtained from a complete fan in $\mathbb{R}^n$ according to Definition 3.3. For any k-dimensional cell F of $\mathcal{C}$, $k \leq n-2$, the quotient $\mathcal{C}/F$ is again complete, and, for the number $\tilde{f}_i(F)$ of i-dimensional cells of $\mathcal{C}$ that contain F,*

$$\tilde{f}_i(F) = f_{i-k-1}(\mathcal{C}/F), \qquad i = k, \ldots, n-1.$$

3.5 Theorem (Dehn-Sommerville equations). *Let $\mathcal{C}$ be the spherical complex obtained from a complete simplicial fan in $\mathbb{R}^n$ by intersecting its cones with S^{n-1}. Then,*

$$(E_k^n) \qquad \sum_{i=k}^{n-1}(-1)^i\binom{i+1}{k+1}f_i(\mathcal{C}) = (-1)^{n-1}f_k(\mathcal{C}), \qquad k=-1,\ldots,n-1.$$

Remark. For $k=-1$, we obtain Euler's theorem as a special case of the Dehn-Sommerville equations. For $k=n-1$, the result is trivial.

PROOF OF THEOREM 3.5. Let us fix a k-face F. By Lemma 3.4,

$$\tilde{f}_i(F) = f_{i-k-1}(\mathcal{C}/F).$$

Therefore,

$$\begin{aligned}\sum_{i=k}^{n-1}(-1)^i\tilde{f}_i(F) &= \sum_{i=k}^{n-1}(-1)^i f_{i-k-1}(\mathcal{C}/F) \underset{j:=i-k-1}{=} \sum_{j=-1}^{n-k-2}(-1)^{j+k+1}f_j(\mathcal{C}/F)\\ &= (-1)^{k+1}\sum_{j=-1}^{n-k-2}(-1)^j f_j(\mathcal{C}/F) = (-1)^{k+1}(-1)^{n-k-2}\\ &= (-1)^{n-1}\end{aligned}$$

by Euler's equation. Summation over all k-faces of $\mathcal{C}$ yields

$$(10) \qquad \sum_{k\text{-faces } F}\sum_{i=k}^{n-1}(-1)^i\tilde{f}_i(F) = \sum_{k\text{-faces } F}(-1)^{n-1} = f_k(\mathcal{C})\cdot(-1)^{n-1}.$$

Changing the order of summation on the left side of (10), we obtain

$$(11) \qquad \sum_{i=k}^{n-1}(-1)^i\sum_{k\text{-faces } F}\tilde{f}_i(F) = (-1)^{n-1}f_k(\mathcal{C}).$$

Since $\mathcal{C}$ is simplicial, any i-face of $\mathcal{C}$ contains $\binom{i+1}{k+1}$ k-faces. Therefore,

$$\sum_{k\text{-faces } F}\tilde{f}_i(F) = \binom{i+1}{k+1}f_i(\mathcal{C}).$$

So, we obtain the theorem from (11). □

Example. For $n=3$ and $k\le 1$, the formulas (E_k^n) read explicitly ($f_i := f_i(\mathcal{C})$):

$$(E_{-1}^3) \qquad -f_{-1}+f_0-f_1+f_2 = f_{-1},$$

$$(E_0^3) \qquad f_0-2f_1+3f_2 = f_0,$$

$$(E_1^3) \qquad -f_1+3f_2 = f_1.$$

As we remarked before, (E_{-1}^3) is Euler's equation (1). (E_0^3) and (E_1^3) provide the same equation

$$3f_2 = 2f_1,$$

which does not follow from Euler's equation.

3.6 Definition. For $f_i := f_i(\mathcal{C}), i = -1, \ldots, n-1$, we set

$$h_k = h_k(\mathcal{C}) := \sum_{i=0}^{k} (-1)^{k-i} \binom{n-i}{n-k} f_{i-1}, \qquad 0 \leq k \leq n$$

and call $h = (h_0, \ldots, h_n)$ the *h-vector* of $\mathcal{C}$.

3.7 Theorem. *The Dehn-Sommerville equations are equivalent to the equations*

$$h_k = h_{n-k}, \qquad 0 \leq k \leq n.$$

PROOF. We define two functions of the real variable t:

$$F(t) := \sum_{i=0}^{n} f_{i-1} t^i; \qquad H(t) := (1-t)^n F(\frac{t}{1-t}).$$

Then, $H(t) = \sum_{i=0}^{n} f_{i-1} t^i (1-t)^{n-i}$ is a polynomial in t and can be written as

$$H(t) = \sum_{k=0}^{n} h'_k t^k$$

with coefficients

$$h'_k = \sum_{i=0}^{k} (-1)^{k-i} \binom{n-i}{n-k} f_{i-1}.$$

Hence, $h'_k = h_k, k = 0, \ldots, n$.

If we calculate $F(t-1)$ and $F(-t)$, we see that the Dehn-Sommerville equations are equivalent to

$$F(t-1) = (-1)^n F(-t), \qquad \text{or}$$

$$H(t) = t^n H(t^{-1}). \tag{10}$$

Comparing coefficients in (10) shows the equivalence of (E^n_{k-1}) and $h_k = h_{n-k}$, $k = 0, \ldots, n$. □

In section 6 the h-vector will be given a geometric meaning in case $\mathcal{C}$ is the boundary complex of a simple n-polytope.

Exercises

1. Let $\mathcal{C}_1, \mathcal{C}_2$ be simplicial 2-spheres isomorphic to boundary complexes $\mathcal{B}(P_1)$, $\mathcal{B}(P_2)$ of simplicial polytopes P_1, P_2, respectively.

   Suppose $\Delta_1, \Delta'_1 \in \mathcal{C}_1$ and $\Delta_1 \cap \Delta'_1 = \emptyset$, also $\Delta_2, \Delta'_2 \in \mathcal{C}_2$ and $\Delta_2 \cap \Delta'_2 = \emptyset$ for 2-cells $\Delta_1, \Delta'_1, \Delta_2, \Delta'_2$. By identifying ("gluing together") of Δ_1 and Δ_2, also Δ_1' and Δ_2', and taking away the interior of $\Delta_1 = \Delta_2$, $\Delta_1' = \Delta_2'$ we obtain a simplicial complex $\mathcal{C}$ whose set is a torus-like surface.

a. Prove that

$$f_0(\mathcal{C}) - f_1(\mathcal{C}) + f_2(\mathcal{C}) = 0.$$

b. If we repeat the gluing together of disjoint triangles of $\mathcal{C}_1, \mathcal{C}_2$ so as to obtain closed surfaces with g "holes", what is the analogous formula of (a)?

2. Find the corresponding equation to (a) in Exercise 1, for a cell complex obtained from two simplicial $(n-1)$-spheres by gluing together two pairs of disjoint $(n-1)$-simplices.

3.
   a. Show directly that, for any simplicial spherical complex $\mathcal{C}$, the left side of (2) does not change if a stellar subdivision of $\mathcal{C}$ is applied.
   b. By stellar operations, prove the Euler- and the Dehn-Sommerville equations for all simplicial spheres obtained from the boundary complex of an n-simplex.

4. The hyperplanes in $\mathbb{R}^n$ with equations (E_k^n) intersect in an affine space of dimension $[\frac{1}{2}n]$.

4. Schlegel diagrams, n-diagrams, and polytopality of spheres

If a 3-polytope P is made of glass and, if we stand close enough to one of its facets F, we "see" the boundary complex $\mathcal{B}(P)$ as centrally projected into F. Figure 9 illustrates the case of platonic solids (see I, 6).

Because of the symmetries which platonic solids possess, all such projections for one P look alike. If we consider, however, a triangular prism, there are two different types of "windows", triangles and rectangles, and we obtain two types of projections (Figure 10).

The higher-dimensional analog is of some use in the investigation of polyhedral spheres. In particular, we can "visualize" a 4-polytope by "looking through" one of its three-dimensional faces.

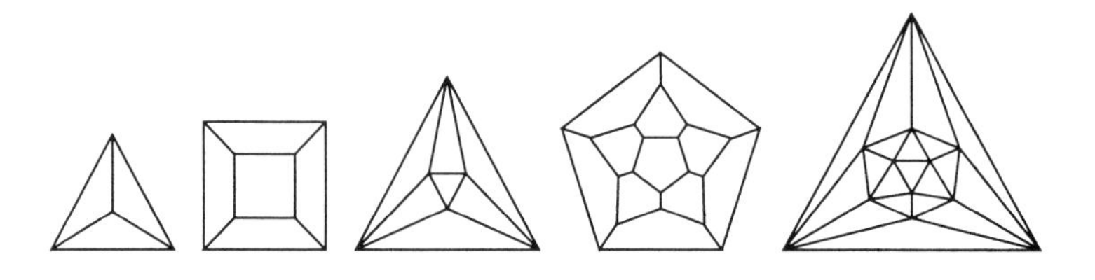

FIGURE 9. Tetrahedron, Cube, Octahedron, Dodecahedron, and Icosahedron.

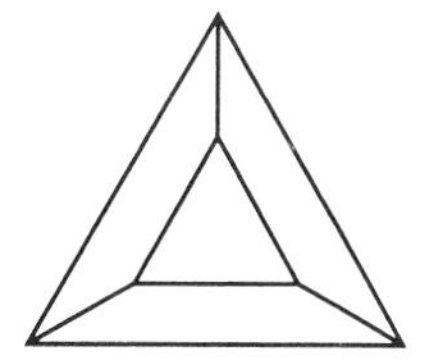
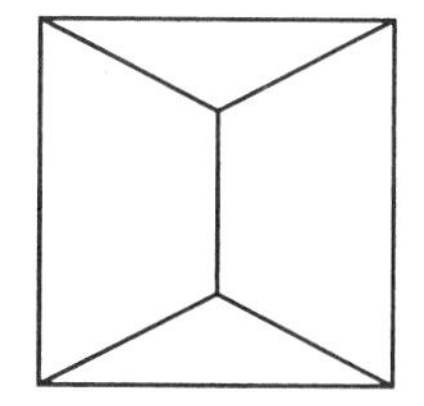

FIGURE 10.

4.1 Definition. Let $P := H_1^- \cap \cdots \cap H_r^-$ be an n-polytope with facets $F_1 := P \cap H_1, \ldots, F_r := P \cap H_r$, and let $p \in \text{int}\, H_1^+ \cap \text{int}\, H_2^- \cap \cdots \cap \text{int}\, H_r^-$. We consider the cell complex $\mathcal{C}$ obtained from $\mathcal{B}(P) \setminus \{F_1\}$ by central projection into F_1 with projection center p and call $\mathcal{C} \cup \{F_1\}$ a *Schlegel diagram* of P.

Figure 11 shows Schlegel diagrams of a 4-simplex and a 4-cube.

A natural question arises. Given any cell decomposition $\mathcal{C}$ of an $(n-1)$-polytope F, is $\mathcal{C} \cup \{F\}$ always a Schlegel diagram or at least isomorphic to a Schlegel diagram? To be more precise, we define the following:

4.2 Definition. Let $\mathcal{C}$ be a cell complex in $\mathbb{R}^{n-1}$ such that $|\mathcal{C}| = F$ is an $(n-1)$-polytope, $F_0 \in \mathcal{C}$ for each face F_0 of F and $F' \cap \partial F$ is a face of F for any $F' \in \mathcal{C}$. Then, $\mathcal{C} \cup \{F\}$ is said to be an *$(n-1)$-diagram. F* is called the *base* of the $(n-1)$-diagram.

So our question can be restated: Is any $(n-1)$-diagram equal or isomorphic to a Schlegel diagram and, hence, isomorphic to the boundary complex of an n-polytope?

As can be seen from V, 4, Example 1, an $(n-1)$-diagram need not be a Schlegel diagram though it is isomorphic to a Schlegel diagram. In fact, one of the most celebrated results of classical polytope theory is a theorem of E. Steinitz (1922)

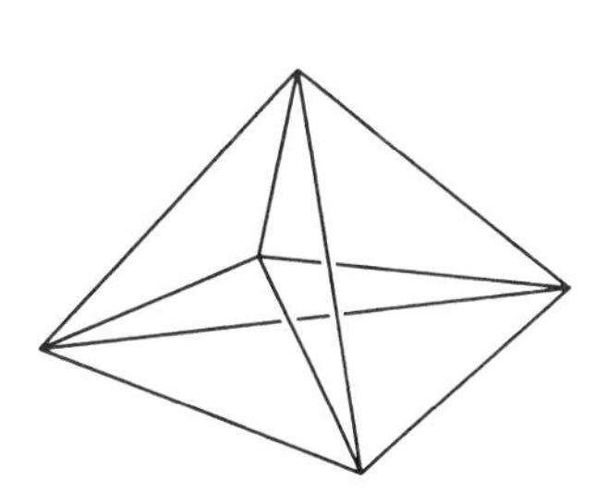
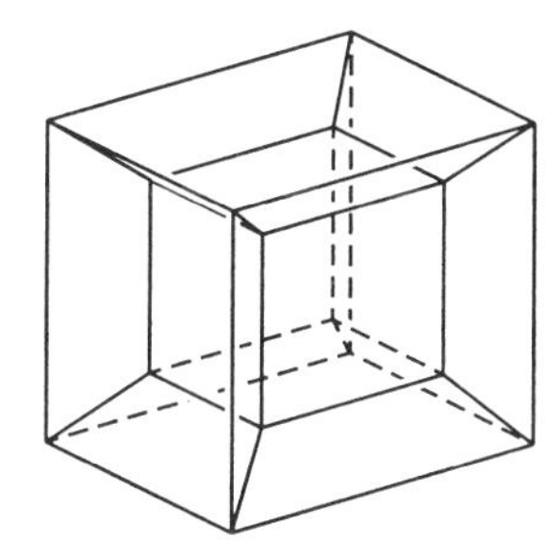

FIGURE 11.

saying that any cell decomposition of a 2-sphere, in particular, any 2-diagram, is isomorphic to a Schlegel diagram.

When Brückner (1893) used 3-diagrams in the classification of 4-polytopes with few vertices, he tacitly assumed that all 3-diagrams are isomorphic to Schlegel diagrams. It was not until 1965 that B. Grünbaum proved this to be false. Eventually, it turned out that one of the 3-diagrams found by Brückner (the "Brückner sphere", see below) was not isomorphic to a Schlegel diagram.

4.3 Definition. A polyhedral $(n-1)$-sphere is said to be *polytopal* if it is isomorphic to the boundary complex of an n-polytope.

So, a general problem can be formulated which is far from being solved:

Steinitz problem. Find necessary and sufficient conditions for a polyhedral $(n-1)$-sphere to be polytopal.

We present an example of a 3-diagram which is not isomorphic to a Schlegel diagram. It is simpler than the original one by Grünbaum.

4.4 Definition. Let $F := [1, 3, 5, 7]$ be a 3-simplex with vertices 1, 3, 5, 7, and let points $2, 4, 6 \in \operatorname{int} F$ be chosen such that the simplices $T_1 := [1, 2, 3, 4]$, $T_2 := [3, 4, 5, 6]$, $T_3 := [1, 2, 5, 6]$ satisfy

$$T_1 \cap T_2 = [3, 4] \qquad T_2 \cap T_3 = [5, 6] \qquad T_1 \cap T_3 = [1, 2]$$

(see Figure 12), and such that there exists a point $8 \in \operatorname{int}(\operatorname{conv}\{1, 2, 3, 4, 5, 6\} \setminus (T_1 \cup T_2 \cup T_3))$ from which the triangles [1, 2, 3], [2, 3, 4], [3, 4, 5], [4, 5, 6], [1, 2, 6], [1, 5, 6] can be "seen".

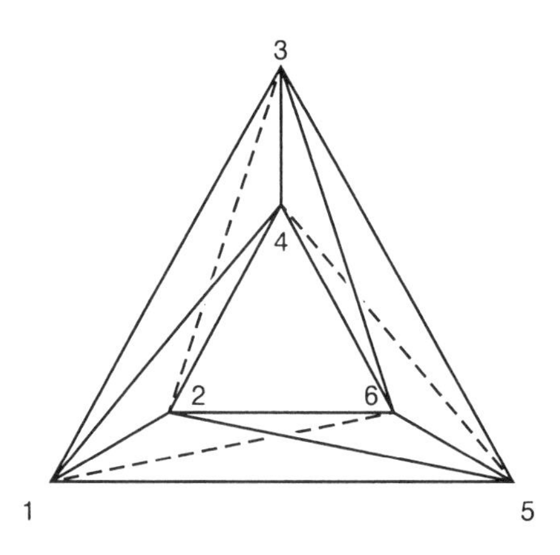

FIGURE 12.

The 3-diagram, having the following 3-simplices as 3-cells, is (up to an isomorphism) called the *Barnette sphere*:

$$T_1, T_2, T_3, [1, 2, 4, 7], [1, 3, 4, 7], [3, 4, 6, 7], [3, 5, 6, 7], [1, 2, 5, 7],$$
$$[2, 5, 6, 7], [2, 4, 6, 7], [1, 2, 3, 8], [2, 3, 4, 8], [3, 4, 5, 8], [4, 5, 6, 8],$$
$$[1, 2, 6, 8], [1, 5, 6, 8], [1, 3, 5, 8], [2, 4, 6, 8], \text{ and } F \text{ as the base.}$$

The defining properties of a polyhedral 3-sphere (Definition 1.9) are readily checked. In order to get a 3-diagram one may choose, for example, 2, 4, 6 as vertices of a triangle $\Delta \subset \text{int } F$ with sides parallel to $[1, 3], [3, 5], [1, 5]$, respectively, and then rotate Δ a small amount about the line that joins the barycenters of Δ and $[1, 3, 5]$.

4.5 Definition. In the Barnette sphere, let $p := [7, 8] \cap [2, 4, 6] \in \text{relint}[2, 4, 6]$. We apply successively $s(p; [2, 4, 6])$, $s^{-1}(p; [7, 8])$, and call the resulting 3-diagram a *Bruckner sphere*.

4.6 Theorem. *There does not exist a 3-diagram, with base T_i, $i = 1, 2$ or 3, that is isomorphic to the Barnette sphere. Hence the Barnette sphere is not polytopal.*

PROOF. The assignment $1 \mapsto 3 \mapsto 5 \mapsto 1$, $2 \mapsto 4 \mapsto 6 \mapsto 2$, $7 \mapsto 7$, $8 \mapsto 8$ induces an automorphism of the Barnette sphere, as does the assignment $1 \mapsto 5 \mapsto 3 \mapsto 1$, $2 \mapsto 6 \mapsto 4 \mapsto 2$, $7 \mapsto 7$, $8 \mapsto 8$. Therefore, it is sufficient to prove the theorem for $T_2 = [3, 4, 5, 6]$.

Suppose T_2 were the base of a 3-diagram of the Barnette sphere. Then, $7, 8 \in \text{int } T_2$ are such that each two of the simplices $[3, 4, 6, 7]$, $[3, 5, 6, 7]$, $[3, 4, 5, 8]$, $[4, 5, 6, 8]$ do not have an interior point in common (Figure 13). The set $T := T_2 \setminus ([3, 4, 6, 7] \cup [3, 5, 6, 7] \cup [3, 4, 5, 8] \cup [4, 5, 6, 8])$ is to be filled with the remaining cells of the Barnette sphere.

In any 3-diagram $\mathcal{C}$ we have, the following directly from the definitions:

(1) For any edge $[a, b]$, the polygonal path $\text{link}([a, b], \mathcal{C})$ can be projected onto a plane perpendicular to the line $\text{aff}[a, b]$ in which the projected path has no self-intersection.

In the Barnette sphere $\mathcal{S}$, since $\text{link}([1, 2], \mathcal{S}) = [3, 4] \cup [4, 7] \cup [7, 5] \cup [5, 6] \cup [6, 8] \cup [8, 3] =: \pi$, we must choose $1, 2 \in T$ such that (1) is satisfied.

However, the triangular paths $[3, 8] \cup [8, 6] \cup [6, 3]$ and $[4, 7] \cup [7, 5] \cup [5, 4]$ are linked as links of a chain. It is readily seen from Figure 13 that there does not exist a projection of π without self-intersection onto any plane in $\mathbb{R}^3$.

Therefore, T_2 is not the base of a 3-diagram of the Barnette sphere. Since each 3-face of a polytopal 3-sphere must carry a 3-diagram, the theorem follows. □

Remark. The proof of Theorem 4.6 is based on the fact that any polytopal $(n-1)$-sphere can be realized in each of its $(n-1)$-faces as an $(n-1)$-diagram. This condition, however, is not sufficient for polytopality, as Barnette and Schulz have shown by other examples (c.f. the Appendix to this section).

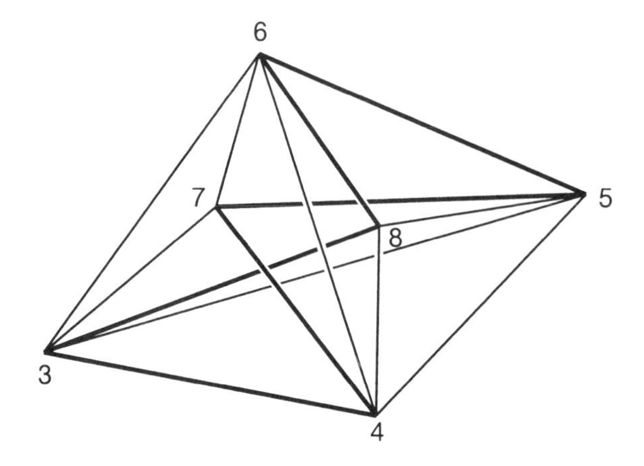

FIGURE 13.

As a direct consequence of Theorem 4.6, we have (in the terminology of section 1) the following theorem:

4.7 Theorem. *Not every (simplicial) complete fan is isomorphic to a fan spanned by the faces of a polytope P.*

Exercises

1. The Brückner sphere is not polytopal.
2. Stellar subdivisions preserve polytopality, inverse stellar subdivisions, in general, do not.
3. The Brückner sphere and the Barnette sphere can be turned into polytopal spheres by appropriate pairs of a stellar subdivision and an inverse stellar subdivision.
4. By using Gale transforms, show that any $(n-1)$-sphere with $n+2$ vertices is polytopal.

5. Embedding problems

Theorem 4.7 suggests the question: Which polyhedral spheres are isomorphic to spherical complexes spanning a complete fan in the sense of Lemma 1.8? For simplicial spheres, this is evidently equivalent to asking: When is a simplicial sphere realizable or "embeddable" as the boundary complex of a "starshaped" body?

In this section we clarify some basic questions on embeddability of polyhedral spheres. In the simplicial case, we have the following theorem:

5.1 Theorem. *Any simplicial complex $\mathcal{C}$ with v vertices is isomorphic to a subcomplex of a $(v-1)$-simplex T^{v-1}.*

PROOF. Any subset of the set of vertices of T^{v-1} is the set of vertices of a face of T^{v-1}. Hence, an arbitrary bijection of the set of vertices of $\mathcal{C}$ onto vert T^{v-1} can be extended to an isomorphism of $\mathcal{C}$ onto a subcomplex of $\mathcal{B}(T^{v-1})$. □

5.2 Definition. Given a cell complex $\mathcal{C}$, we call an injective map

$$\varphi : |\mathcal{C}| \longrightarrow \mathbb{R}^k$$

a *polyhedral embedding*, if it is continuous, has a continuous inverse on $\varphi(|\mathcal{C}|)$, and satisfies the following condition:

(a) For any cell $F \in \mathcal{C}$, $\varphi(F)$ is a polytope in $\mathbb{R}^k$.

If such a φ exists, we call $\mathcal{C}$ *polyhedrally embeddable*, or, briefly, *embeddable* into $\mathbb{R}^k$.

So, as a consequence of Theorem 5.1, we know that any simplicial complex is embeddable into a sufficiently high-dimensional $\mathbb{R}^k$. An analogous theorem for general cell complexes fails to be true even for polyhedral spheres:

5.3 Theorem. *There exists a polyhedral 3-sphere with eight vertices which is not embeddable into any $\mathbb{R}^k$.*

PROOF. We construct a 3-diagram which has the property of the theorem. Let $[1, 2, 3, 4]$ be a 3-simplex, and let 5, 6, 7 be chosen in $\operatorname{int}[1, 2, 3, 4]$ such that $[1, 2]$ and $[5, 6]$ are parallel, $[1, 3]$ and $[5, 7]$ are parallel, but $[2, 3]$ and $[6, 7]$ are not parallel (Figure 14).

We may assume $[2,7]$ to be an edge of the 3-polytope $P_0 := \operatorname{conv}\{1, 2, 3, 5, 6, 7\}$. We choose $8 \in \operatorname{int} P_0 \setminus [2, 3, 6, 7]$ such that $[2, 3, 6, 7, 8]$ and $[4, 5, 6, 7, 8]$ are double-tetrahedra. Let $\mathcal{S} := \mathcal{C} \cup \{[1, 2, 3, 4]\}$ be the 3-diagram with the base $[1, 2, 3, 4]$, and let the 3-cells of $\mathcal{C}$ be given as follows: $[1, 2, 3, 8]$, $[1, 2, 4, 5, 6]$, $[1, 3, 4, 5, 7]$, $[2, 3, 4, 7]$, $[2, 4, 6, 7]$, $[1, 2, 5, 6, 8]$, $[1, 3, 5, 7, 8]$, $[2, 3, 6, 7, 8]$, $[4, 5, 6, 7, 8]$.

Suppose $\mathcal{S}$ were embedded in $\mathbb{R}^k$, $k \geq 4$. We leave the notation unchanged. Then, the planes $\operatorname{aff}[1, 2, 5, 6]$ and $\operatorname{aff}[1, 3, 5, 7]$ lie in a 3-space U and intersect in the line $\operatorname{aff}[1, 5]$; therefore, 1, 2, 3, 5, 6, 7 lie in U. Since 2, 3, 6, 7 are affinely independent, the double-tetrahedron $[2, 3, 6, 7, 8]$ is also contained in U, so that $8 \in U$. Similarly, since 5, 6, 7, 8 are affinely independent, $[4, 5, 6, 7, 8] \subset U$, and, hence, $4 \in U$. Then, all cells of $\mathcal{S}$ would lie in $U \subset \mathbb{R}^k$.

For any vertex i, consider the (five or six) 3-polytopes of $\operatorname{st}(i; \mathcal{S})$. Each 2-face which contains i lies on precisely two of these polytopes, hence, not on the boundary of $\operatorname{st}(i; \mathcal{S})$. This implies i to be an interior point of $|\operatorname{st}(i; \mathcal{S})|$, $i = 1, \ldots, 8$.

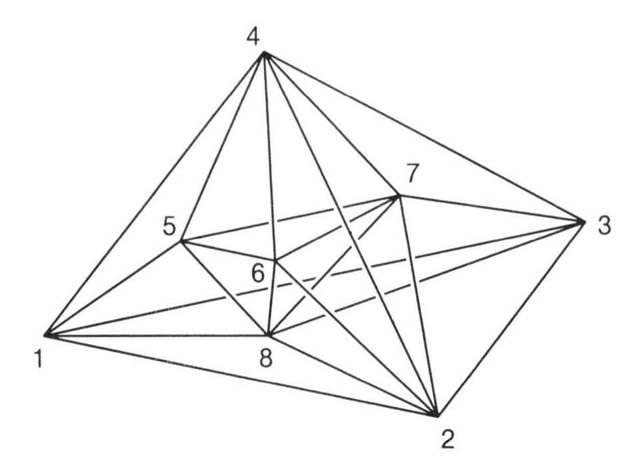

FIGURE 14.

If we consider, however, $P := \operatorname{conv} \mathcal{S}$, we see that $P = \operatorname{conv}\{1, \ldots, 8\}$, so that some i is not an interior point of $|\operatorname{st}(i; \mathcal{S})|$, a contradiction. □

5.4 Definition. We call a polyhedral $(n-1)$-sphere $\mathcal{S}$ *star-shaped* if it can be polyhedrally embedded into $\mathbb{R}^n$ so that there exists a point p with the property that each ray emanating from p meets $|\mathcal{S}|$ in one and only one point. p is said to be a *kernel point*. The embedding is, then, also called *star-shaped*.

Clearly, each polytopal sphere is also star-shaped. The converse, however, is not true:

Example. The Barnette sphere $\mathcal{S}$ is star-shaped. In its definition we have a realization of $\mathcal{S} \setminus \operatorname{st}(8, \mathcal{S})$ in $\mathbb{R}^3$. If 8 is placed in $\mathbb{R}^4 \setminus \mathbb{R}^3$ we join 8 to the faces of $\operatorname{link}(8, \mathcal{S})$ and obtain a pyramid. Its boundary complex is a star-shaped embedding of $\mathcal{S}$.

Theorem 5.3 shows that not every polyhedral sphere has a star-shaped embedding. This still remains true if we restrict ourselves to simplicial spheres:

5.5 Theorem. *There exists a simplicial* 3*-sphere with* 12 *vertices which has no star-shaped embedding.*

PROOF. We consider two copies $\mathcal{S}$, $\mathcal{S}'$ of a Barnette sphere with 3-faces $[3, 4, 5, 6]$, $[3', 4', 5', 6']$, respectively, that are not bases of 3-diagrams (Theorem 4.6). We glue $\mathcal{S}, \mathcal{S}'$ together by identifying i and i', $i = 3, 4, 5, 6$ and leaving away $\operatorname{relint}[3, 4, 5, 6] = \operatorname{relint}[3', 4', 5', 6']$. Let $\bar{\mathcal{S}}$ be the sphere thus obtained.

Suppose $\bar{\mathcal{S}}$ is realized in $\mathbb{R}^4$ as a star-shaped sphere. Then, the hyperplane aff[3, 4, 5, 6] splits $\bar{\mathcal{S}}$ into two parts $\mathcal{S}_1$, $\mathcal{S}_2$. If we alter the position of 0 by a small amount, it remains a kernel point. So, we can choose $0 \notin \text{aff}[3, 4, 5, 6]$ as a kernel point. The boundary of the cone pos[3, 4, 5, 6] separates $\mathbb{R}^4$ into two sets. One of them contains $\mathcal{S}_1$, the other $\mathcal{S}_2$. By central projection of either $\mathcal{S}_1$ or $\mathcal{S}_2$ onto [3, 4, 5, 6], we obtain a 3-diagram of the Barnette sphere, which contradicts Theorem 4.6. □

Star-shaped spheres provide examples for the following:

5.6 Definition. A polyhedral sphere $\mathcal{S}$ is called *fan-like* if there exists a complete fan Σ and an isomorphism

$$\Phi : \mathcal{S} \longrightarrow \Sigma \tag{1}$$

which associates, with each face $F \in \mathcal{S}$, the cone pos F (in particular, pos $\emptyset = \{0\}$).

If $\mathcal{S}$ is simplicial, the existence of Φ is equivalent to $\mathcal{S}$ being star-shaped. This is not true for non-simplicial $\mathcal{S}$.

5.7 Theorem.
(a) *The 3-sphere of Theorem 5.3 is fan-like but not embeddable and, hence, not star-shaped.*
(b) *The 3-sphere of Theorem 5.5 is embeddable but not fan-like.*

PROOF. (a) Let U be defined as in the proof of Theorem 5.3, and let $0 \in \text{relint}[1, 2, 3, 8]$. We set $U \subset \mathbb{R}^4$ and choose a vector $a \in \mathbb{R}^4 \setminus U$. Then, the assignment

$$i \mapsto \begin{cases} i - a & \text{for } i = 1, 2, 3, 4 \\ i + a & \text{for } i = 5, 6, 7, 8 \end{cases}$$

readily induces an isomorphism Φ of the given 3-sphere onto a fan-like 3-sphere in $\mathbb{R}^4$: its 3-faces, which are not simplices, are "bent" but have convex cones as positive hulls.

(b) By Theorem 5.5, the given 3-sphere is not fan-like. By Theorem 5.1 it is embeddable into $\mathbb{R}^{11}$. (An embedding into $\mathbb{R}^4$ can also readily be found). □

As a summary, we list the hierarchy of $(n-1)$-spheres we have obtained.

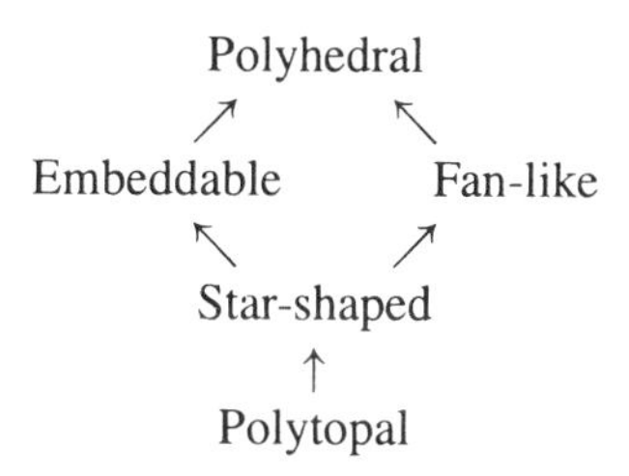

In chapter V, we shall further refine this list.

Exercises

1. The Brückner sphere is star-shaped.
2. Modify the 3-sphere of Theorem 5.3 by choosing [6, 7] parallel to [2, 3], so that the base [1, 2, 3, 4] is decomposed into a simplex, a double simplex ([4, 5, 6, 7, 8]), and four pyramids. The resulting 3-sphere is polytopal.
3. Generalize Theorem 5.3 to k-spheres for arbitrary $k \geq 3$.
4. We call a polyhedral 2-sphere simple if each of its vertices lies on precisely three 2-faces. If a simple 2-sphere is embedded in $\mathbb{R}^3$, it is automatically the boundary complex of a polytope.

6. Shellings

Now, we will introduce a further tool for the investigation of polyhedral $(n-1)$-spheres, in particular, of the boundary complexes of n-polytopes.

6.1 Definition. Let $F_1, \ldots, F_r$ be the $(n-1)$-cells of a polyhedral $(n-1)$-sphere $\mathcal{S}$. If the numbering can be chosen so that

$$F_i \cap (F_{i+1} \cup \cdots \cup F_r) \tag{1}$$

is a union of $(n-2)$-cells of $\mathcal{S}$, for any $i = 1, \ldots, r-1$, we call the numbering $F_1, \ldots, F_r$ a *shelling* of $\mathcal{S}$. If a shelling exists, we say $\mathcal{S}$ is *shellable*.

Remarks.
(a) It is known that nonshellable polyhedral spheres exist.
(b) There exist several definitions of "shelling" in the literature, but not all are equivalent. One of them is (2) below.

6.2 Lemma. *Let $F_1, \ldots, F_r$ be a numbering of the facets ($(n-1)$-cells) of a polyhedral $(n-1)$-sphere $\mathcal{S}$ such that the following is true:*

$$\begin{aligned} &F_1 \cup \cdots \cup F_i \text{ and} \\ &F_{i+1} \cup \cdots \cup F_r \text{ are topological } (n-1)\text{-balls}, \ i = 1, \ldots, r-1. \end{aligned} \tag{2}$$

Then, $\mathcal{S}$ is shellable.

PROOF. (2) implies that $(F_1 \cup \cdots \cup F_i) \cap (F_{i+1} \cup \cdots \cup F_r) = \partial(F_1 \cup \cdots \cup F_i) = \partial(F_{i+1} \cup \cdots \cup F_r)$ is a polyhedral $(n-2)$-sphere, hence composed of $(n-2)$-cells. Since $F_i \cap (F_{i+1} \cup \cdots \cup F_r)$ is the (topological) closure of $\partial(F_{i+1} \cup \cdots \cup F_r) \setminus \partial(F_1 \cup \cdots \cup F_{i-1})$, it is also composed of $(n-2)$-cells. □

Remark. The converse is not true. Consider a triangular prism in $\mathbb{R}^3$, choose F_3, F_4, F_5 as the rectangular sides and F_1, F_2 as the triangular faces.

There is a surprisingly simple proof for the shellability of polytopal spheres:

6.3 Theorem (Bruggesser-Mani shelling). *If P is a polytope, $\mathcal{B}(P)$ is shellable.*

PROOF. Let $\mathbb{R}^n = \operatorname{aff} P$. We choose $p \in \mathbb{R}^n \setminus P$ so close to a facet F of P that the central projection from p provides a Schlegel diagram of $\mathcal{B}(P)$ with base F. Let H be a hyperplane that strictly separates p and P. By a *permissible projective transformation* φ (see I, 6), we map H onto the hyperplane at infinity. We set

$$p' := \varphi(p), \quad F' := \varphi(F), \quad \text{and } P' := \varphi(P).$$

Clearly, shellability is an invariant under permissible projective transformations (as under any isomorphism of a given polyhedral sphere).

Let $T := \operatorname{conv}(\{p'\} \cup F')$. Then, $P' \subset T$. We choose a point $q' \in \operatorname{relint} F'$ such that $x_1' := [p', q'] \cap |\mathcal{B}(P') \setminus \{F'\}| \in \operatorname{relint} F_1'$ for some facet F_1' of P' (Figure 15).

If F_i is any facet of P not equal to F, then, aff F_i does not intersect the line segment $[p, q]$ for $q = \varphi^{-1}(q')$ (by the choice of p). Since relint$[p', q']$ is the image under φ of the complement of $[p, q]$ on the line pq, we find $[p', q'] \cap \operatorname{aff} F_i' \neq \emptyset$. The point of intersection lies in $[p', x_1']$. We set

$$x_i' := [p', x_1'] \cap \operatorname{aff} F_i', \qquad i = 2, \ldots, r-1,$$

where r is the number of facets of P'. By an appropriate choice of q', all points x_i' are different, $i = 1, \ldots, r-1$. Up to renumbering, we can assume x_j' to lie between x_{j-1}' and x_{j+1}', $j = 2, \ldots, r-2$. Then, we claim that

$$F_1', \ldots, F_{r-1}', F_r' := F' \quad \text{is a shelling of } \mathcal{B}(P').$$

The idea of the proof is as follows. We start a "space flight" from x_1' in direction p'. At any stage of the flight, we "see" an $(n-1)$-polytope obtained from P' by

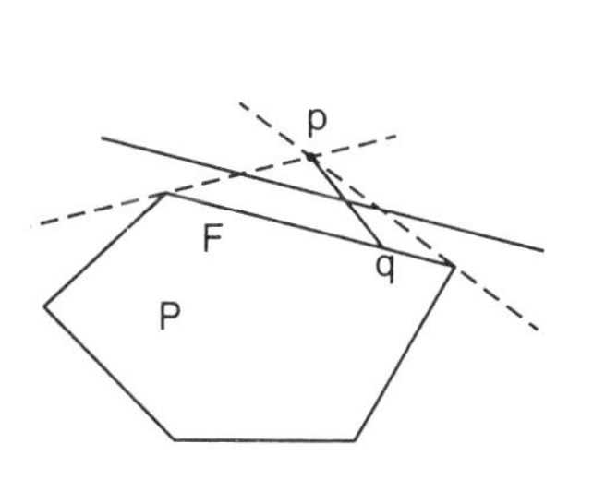

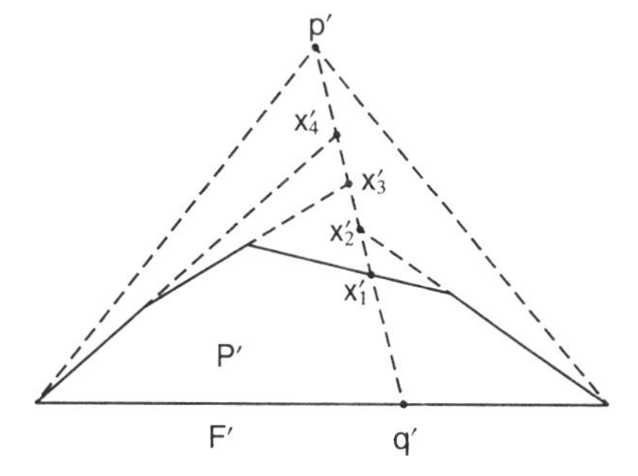

FIGURE 15.

central projection. Whenever we pass an x'_i, we "see" one new facet. $F'_1 \cup \cdots \cup F'_i$, then, is the "visible" part of $\partial P'$, and $F'_{i+1} \cup \cdots \cup F'_r$ is the "invisible" region of $\partial P'$. Both sets are topological $(n-1)$-balls, and we can apply Lemma 6.2.

The idea is made precise by the natural order of the x'_i on $[p', q']$, and

$$F_1 := \varphi^{-1}(F'_1), \ldots, F_r := \varphi^{-1}(F'_r)$$

provides the shelling we look for. □

The "space flight" in the proof of Theorem 6.3 can also be carried out without applying, first, the permissible projective transformation φ. The "space ship" must, then, pass, in the projective extension of $\mathbb{R}^n$, through "infinity" and return to the "planet polytope" from the opposite side.

We dualize a "space flight" in the following way. Let $x_i := \varphi^{-1}(x'_i)$ be the inverse images of $x'_1, \ldots, x'_{r-1}$ in the proof of Theorem 6.3. They lie on a line g which cuts the interior of P. Let ϱ_1, ϱ_2 be the two rays of which $g \setminus P$ is composed, and suppose $x_1, \ldots, x_s \in \varrho_1, x_{s+1}, \ldots, x_{r-1} \in \varrho_2$. Suppose $0 \in g \cap \operatorname{int} P$. We consider the polytopes

$$P_i := \operatorname{conv}(\{x_i\} \cup P), \qquad i = 1, \ldots, r-1.$$

If H_i is the polar hyperplane of x_i, then, the polar polytope P^* is split by H_i into two polytopes, one of which is P_i^*. In an appropriate orientation, we obtain

$$P_i^* = \begin{cases} P^* \cap H_i^- & \text{for } i = 1, \ldots, s, \\ P^* \cap H_i^+ & \text{for } i = s+1, \ldots, r-1, \end{cases}$$

The sequence

$$P_1^*, P_2^*, \ldots, P_s^*, P^* \cap H_{s+1}^-, \ldots, P^* \cap H_{r-1}^-$$

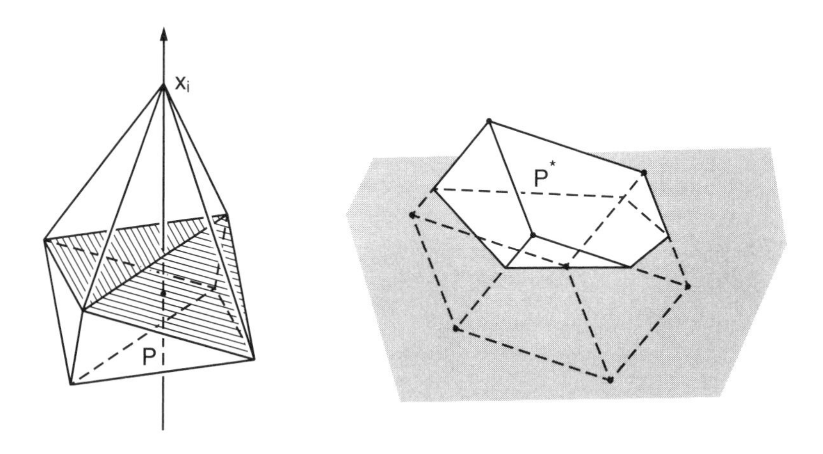

FIGURE 16.

can now be looked at as the part of P^* still under water if the "rock" P^* is successively "lifted" out of a lake with "surface" parallel to all the H_i (and perpendicular to g since $0 \in g \cap \operatorname{int} P$). In Figure 16, left side, three facets of P (hatched) can be "seen" from x_i. Their polar faces, which are vertices of P^*, are "above the water surface" (right side of Figure 16).

It is somewhat involved, however, to give a definition of a dual shelling purely in terms of cell complexes. This is why we have defined "shellable" more generally than it can be done by using (2). We can characterize shellability in a dualizable form:

6.4 Lemma. *Let $F_1, \ldots, F_r$ be the $(n-1)$-cells of a polyhedral $(n-1)$-sphere $\mathcal{S}$. Then, $\mathcal{S}$ is shellable if and only if the numbering of $F_1, \ldots, F_r$ can be chosen so that the following two conditions are satisfied for each F_i, $i = 1, \ldots, r-1$:*
(a′) $F_i \cap F_j$ is an $(n-2)$-cell for at least one $j > i$.
(b′) If F_i, F_j contain a cell $F' \in \mathcal{S}$ and $j > i$, then there is some $j' > i$ such that $F_i \cap F_{j'}$ is an $(n-2)$-cell of $\mathcal{S}$ which contains F'.

PROOF. This is readily verified from Definition 6.1. □

An abstract dualization (which formally interchanges 0-cells and $(n-1)$-cells and reverses inclusion) leads to the following notion.

6.5 Definition. Let $q_1, \ldots, q_s$ be the vertices of a polyhedral $(n-1)$-sphere $\mathcal{S}$. If the numbering can be chosen so that for each $i \in \{1, \ldots, s-1\}$,
(a) $[q_i, q_j]$ is a 1-cell of $\mathcal{S}$ for at least one $j > i$, and
(b) if $q_i, q_j \in F$ for a cell $F \in \mathcal{S}$ and $j > i$, then there exists $j' > i$ such that $[q_i, q_{j'}]$ is a face of F,
then, the numbering $q_1, \ldots, q_s$ is called a *dual shelling* of $\mathcal{S}$.

6.6 Lemma. *Let P be an n-polytope in $\mathbb{R}^n$ and $F_1, \ldots, F_r$ the shelling of $\mathcal{B}(P)$, constructed in the proof of Theorem* 6.3. *Then,*

$$F_1^*, \ldots, F_r^*$$

is a dual shelling of $\mathcal{B}(P^*)$. Here, the line $g = \mathbb{R} \cdot m$ is so chosen that no two vertices of P^* have the same foot on g.

6.7 Definition. For an n-polytope P in $\mathbb{R}^n$, we fix $m \in \mathbb{R}^n$, according to Lemma 6.6. For a face F of P^*, we say that F *culminates* in its vertex v if $\langle m, v\rangle = \max_{x \in F}\langle m, x\rangle$. We call v a *k-vertex* of P^* if the face of maximal dimension which culminates in v has dimension k. We denote the number of k-vertices of P^* by $\tilde{h}_k(P^*)$.

6.8 Theorem. *Let P be a simplicial polytope, P^* a polar polytope (which is simple). Then, $(\tilde{h}_0(P^*), \ldots, \tilde{h}_n(P^*))$ is the h-vector of P, and*
(3)

$$\tilde{h}_k(P^*) = h_k(P) = \sum_{i=0}^{k} (-1)^{k-i} \binom{n-i}{n-k} f_{i-1}(P) = \sum_{i=k}^{n} (-1)^{i-k} \binom{i}{k} f_i(P^*)$$

PROOF. If F culminates in a k-vertex v of P^* and has dimension k, the number of i-faces culminating in v is $\binom{k}{i}$ (since P^* is simple). Hence,

$$f_i(P^*) = \sum_{k=i}^{n} \binom{k}{i} \tilde{h}_k(P^*), \qquad i = 0, \ldots, n. \tag{4}$$

Resolving (4) for the $\tilde{h}_k(P^*)$ and applying II, Theorem 2.5, and III, Theorem 3.5 yields (3). □

Exercises

1. Find a shelling of the 4-cube (using a Schlegel diagram).
2. How many shellings of the boundary complex of a regular octahedron P exist up to a symmetry of P?
3. Find a shelling a) of the Barnette sphere , b) of the Brückner sphere.
4. Given the shelling of a polyhedral 2-sphere $\mathcal{S}$, find a shelling of $s(p;\, F)\mathcal{S}$ for any stellar subdivision $s(p;\, F)$ of $\mathcal{S}$.

7. Upper bound theorem

A fundamental question of polytope theory is that of an upper bound for the number of i-faces $f_i(P)$, $i = 1, \ldots, n-1$, provided the number $f_0(P)$ of vertices is prescribed. In particular, it is of importance in linear optimization. A conjecture by T. S. Motzkin [1957] was first proved by P. McMullen [1970]. There is a precise answer to the question, that is, the upper bounds are attained for cyclic polytopes. We remark that the result has been extended by R. Stanley [1975] to arbitrary simplicial spheres. Stanley uses tools of commutative algebra. The proof for polytopes we present here is due to N. Alon and G. Kalai [1985].

First we prove a combinatorial lemma. card A denotes the number of elements of a finite set A.

7.1 Lemma. *For a set of natural numbers $Q := \{1, \ldots, q\}$, let there exist h pairs of subsets A_i, B_i of Q and $m, s \in \mathbb{Z}_{\geq 0}$ such that*

$$\text{card } A_i \leq s \leq m \leq \text{card } B_i, \qquad i = 1, \ldots, h,$$

and

$$A_i \subset B_i \quad \textit{for } i = 1, \ldots, h, \qquad A_i \not\subset B_j \quad \textit{for } 1 \le i < j \le h. \tag{1}$$

Then

$$h \le \binom{q - m + s}{s}. \tag{2}$$

PROOF. We may assume card $A_i = s \ge 1, i = 1, \ldots, h$, since adding elements to A_i does not increase h. The proof uses a trick. It assigns a vector of an exterior product space to each set A_i and to each set $B_i \setminus A_i$, so that (2) follows from knowing the dimension of that space.

If $V = \mathbb{R}^{q-m+s}$, let $\bigwedge^s V$ be the vector space generated by all exterior products $x_1 \wedge \cdots \wedge x_s$ of vectors of V. As is known from linear algebra,

$$\dim(\textstyle\bigwedge^s V) = \binom{q - m + s}{s}. \tag{3}$$

We choose vectors $a_1, \ldots, a_q \in \mathbb{R}^{q-m+s}$ in general position, that is, any set of at most $q - m + s$ of them is linearly independent. For $i \in \{1, \ldots, h\}$, we set

$$y_i := \bigwedge_{j \in A_i} a_j \in \bigwedge^s V \qquad \bar{y}_i := \bigwedge_{k \in Q \setminus B_i} a_k \in \bigwedge^{q - \text{card } B_i}.$$

From (1) and properties of the exterior product, we find

$$y_i \wedge \bar{y}_i \neq 0 \qquad \text{for } i \in \{1, \ldots, h\},$$
$$y_i \wedge \bar{y}_j = 0 \qquad \text{for } 1 \le i < j \le h.$$

We claim that $y_1, \ldots, y_h$ are linearly independent in $\bigwedge^s V$. Suppose $\alpha_1 y_1 + \cdots + \alpha_h y_h = 0$, and, up to reordering, $\alpha_j \neq 0$ for $j = 1, \ldots, k, \alpha_{k+1} = \cdots = \alpha_h = 0$. Then $0 = 0 \wedge \bar{y}_k = (\alpha_1 y_1 + \cdots + \alpha_k y_k) \wedge \bar{y}_k = \alpha_k y_k \wedge \bar{y}_k$, which implies $\alpha_k = 0$, a contradiction.

So, we have found h linearly independent vectors in $\bigwedge^s V$, and, hence, by (3),

$$h \le \dim(\textstyle\bigwedge^s V) = \binom{q - m + s}{s}.$$

□

7.2 Definition. Let $\mathcal{C}$ be an $(m-1)$-dimensional simplicial complex. An $(s-1)$-face $F \in \mathcal{C}, s \le m$, is called *free* if it is contained in a unique face of maximal dimension $m - 1$. The transition

$$\mathcal{C} \longrightarrow \begin{cases} \mathcal{C} \setminus \operatorname{st}(F, \mathcal{C}) & \text{for } s > 0 \\ \mathcal{C} \setminus \{v\}, \text{v a 0-cell}, & \text{for } s = 0, m = 1 \end{cases}$$

for a free face F is called an *elementary* (s, m)*-collapse* . A *collapse process* on $\mathcal{C}$ is a sequence

$$\mathcal{C} = \mathcal{C}_0 \supset \mathcal{C}_1 \supset \cdots \supset \mathcal{C}_t \tag{4}$$

of simplicial complexes such that C_i is obtained from C_{i-1} by an elementary collapse, $i = 1, \ldots, t$.

Remark. The case $s = 0$, $m > 1$ does not occur for the following reason: The only case in which $\emptyset$ is contained in a unique face of maximal dimension is that in which C has only one vertex.

Example. If C consists of a 3-simplex and its faces, it can be turned into $\emptyset$ by one $(2, 4)$-collapse, two $(2, 3)$-collapses, three $(1, 2)$-collapses, and one $(0, 1)$-collapse. One may also start with a $(3, 4)$-collapse or the $(4, 4)$-collapse.

7.3 Lemma. *Let $s, m \in \mathbb{Z}_{\geq 0}$, $s \leq m$, and let the simplicial complex C have q vertices. The number h of all elementary (s', m')-collapses with $s' \leq s$, $m' \geq m$ in any collapse process on C, is at most $\binom{q-m+s}{s}$.*

PROOF. We consider a collapse process (4). When proceeding from C_{i-1} to C_i, let F_i be a free face and $M_i \supset F_i$ a face of maximal dimension $m - 1$, so that all faces which contain F_i are eliminated, $i = 1, \ldots, t$.

Among the pairs (M_i, F_i), we select those pairs $M_i' = \operatorname{conv}\{a_{i_1}, \ldots, a_{i_{m'}}\}$, $F_i' = \operatorname{conv}\{a_{i_1}, \ldots, a_{i_{s'}}\}$ for which $B_i := \{i_1, \ldots, i_{m'}\}$, $A_i := \{i_1, \ldots, i_{s'}\}$ satisfy

$$s' \leq s \quad \text{and} \quad m' \geq m.$$

Let h be the number of such pairs. By definition of an elementary collapse, the assumptions of Lemma 7.1 hold, so that the lemma follows. □

7.4 Theorem (Upper bound theorem).

Let C be a shellable simplicial $(n - 1)$-sphere or a polytopal sphere with v vertices. Then,

$$f_i(C) \leq f_i(C(v, n)), \qquad i = 1, \ldots, n - 1,$$

where $C(v, n)$ is a cyclic n-polytope with v vertices.

PROOF. In the case of a polytopal sphere, we may assume C to be simplicial, since splitting a face without introducing new vertices increases the numbers $f_i(C)$, for $i \geq 1$ (Theorem 2.6), and does not disturb C to be polytopal and, hence, shellable (Theorem 6.3). As in Definition 3.6, we introduce the h-vector $(h_0, h_1, \ldots, h_n)$ defined by

$$h_i := \sum_{j=-1}^{i-1} (-1)^{i-j-1} \binom{n - j - 1}{n - i} f_j(C), \qquad i = 0, \ldots, n.$$

We set

$$\tilde{h}_i := h_0 + \cdots + h_i \quad \text{for } i = 0, \ldots, n, \quad \tilde{h}_{-1} := 0.$$

Then,

$$h_i = \tilde{h}_i - \tilde{h}_{i-1}, \quad i = 0, \ldots, n.$$

As we have seen in II, Theorem 3.12, $f_i(C(v,n)) = \binom{v}{i+1}$ for $i = 1, \ldots, [\frac{n}{2}] - 1$. As a consequence, we claim that

$$f_j(\mathcal{C}) = \sum_{i=0}^{[\frac{n}{2}]-1} \left[\binom{n-i-1}{n-j-2} - \binom{i}{n-j-2}\right]\tilde{h}_i$$

$$+ \left[\binom{[\frac{n}{2}]}{n-j-1} + \binom{[\frac{n}{2}]+1}{n-j-1}\right]\tilde{h}_{[\frac{n}{2}]}, \text{ if } n \text{ is odd, and} \tag{5}$$

$$f_j(\mathcal{C}) = \sum_{i=0}^{[\frac{n}{2}]-1} \left[\binom{n-i-1}{n-j-2} - \binom{i}{n-j-2}\right]\tilde{h}_i$$

$$+ \binom{\frac{n}{2}}{n-j-1}\left(\tilde{h}_{\frac{n}{2}-1} + \tilde{h}_{\frac{n}{2}}\right), \text{ if } n \text{ is even.} \tag{6}$$

Proof of (5). We set $f_j := f_j(\mathcal{C})$.

$$f_j = \sum_{i=0}^{j+1} \binom{n-i}{n-j-1} h_i = \sum_{i=0}^{\frac{n-1}{2}} \binom{n-i}{n-j-1} h_i$$

$$+ \sum_{i=\frac{n+1}{2}}^{j+1} \binom{n-i}{n-j-1} h_{n-i} \text{ by Theorem 3.7}$$

$$= \sum_{i=0}^{\frac{n-1}{2}} \binom{n-i}{n-j-1} h_i + \sum_{i=n-j-1}^{\frac{n-1}{2}} \binom{i}{n-j-1} h_i$$

by change of index ($i \leftrightarrow n-i$)

$$= \sum_{i=0}^{\frac{n-1}{2}} \left[\binom{n-i}{n-j-1} + \binom{i}{n-j-1}\right]\tilde{h}_i$$

$$- \sum_{i=1}^{\frac{n-1}{2}} \left[\binom{n-i}{n-j-1} + \binom{i}{n-j-1}\right]\tilde{h}_{i-1}$$

$$= \sum_{i=0}^{\frac{n-1}{2}} \left[\binom{n-i}{n-j-1} + \binom{i}{n-j-1}\right]\tilde{h}_i$$

$$- \sum_{i=0}^{\frac{n-3}{2}} \left[\binom{n-i-1}{n-j-1} + \binom{i+1}{n-j-1}\right]\tilde{h}_i$$

$$= \sum_{i=0}^{\frac{n-3}{2}} \left\{\left[\binom{n-i}{n-j-1} - \binom{n-i-1}{n-j-1}\right]\right.$$

$$\left. - \left[\binom{i+1}{n-j-1} - \binom{i}{n-j-1}\right]\right\}\tilde{h}_i$$

$$+\left[\binom{\frac{n+1}{2}}{n-j-1}+\binom{\frac{n-1}{2}}{n-j-1}\right]\tilde{h}_{\frac{n-1}{2}},$$

which equals the right side of (5).

PROOF OF (6).

$$f_j=\sum_{i=0}^{j+1}\binom{n-i}{n-j-1}h_i=\sum_{i=0}^{\frac{n}{2}}\binom{n-i}{n-j-1}h_i$$

$$+\sum_{i=\frac{n}{2}+1}^{j+1}\binom{n-i}{n-j-1}h_{n-i} \text{ by Theorem 3.7}$$

$$=\sum_{i=0}^{\frac{n}{2}}\binom{n-i}{n-j-1}h_i+\sum_{i=n-j-1}^{\frac{n}{2}-1}\binom{i}{n-j-1}h_i$$

by change of index ($i \leftrightarrow n-i$)

$$=\sum_{i=0}^{\frac{n}{2}}\binom{n-i}{n-j-1}\tilde{h}_i-\sum_{i=1}^{\frac{n}{2}}\binom{n-i}{n-j-1}\tilde{h}_{i-1}$$

$$+\sum_{i=0}^{\frac{n}{2}-1}\binom{i}{n-j-1}\tilde{h}_i-\sum_{i=1}^{\frac{n}{2}-1}\binom{i}{n-j-1}\tilde{h}_{i-1}$$

$$=\sum_{i=0}^{\frac{n}{2}}\binom{n-i}{n-j-1}\tilde{h}_i-\sum_{i=0}^{\frac{n}{2}-1}\binom{n-i-1}{n-j-1}\tilde{h}_i$$

$$+\sum_{i=0}^{\frac{n}{2}-1}\binom{i}{n-j-1}\tilde{h}_i-\sum_{i=0}^{\frac{n}{2}-2}\binom{i+1}{n-j-1}\tilde{h}_i,$$

which equals the right side of (6).

From $0 \le i \le \frac{n}{2}-1$ we deduce $n-i-1 \ge \frac{n}{2} > i$, and, hence,

$$\binom{n-i-1}{n-j-2}-\binom{i}{n-j-2}\ge 0.$$

Therefore, all coefficients in (5) and (6) are nonnegative. Since

$$f_i(C(v,n))=\binom{v}{i+1}, \quad \text{for } i=1,\dots,\left[\frac{n}{2}\right]-1,$$

an easy calculation gives

$$h_i(C(v,i))=\binom{v-n+i-1}{i} \quad \text{and} \quad \tilde{h}_i(C(v,i))=\binom{v-n+i}{i}$$

for $C(v,i)$ $i=1,\dots,[\frac{n}{2}]$.

So, because of (5) and (6), it suffices to show that

$$\tilde{h}_i \leq \binom{v-n+i}{i}, \qquad \text{for } i = 1, \ldots, \left[\frac{n}{2}\right]. \tag{7}$$

For the proof of (7), we use the shellability of the simplicial sphere $\mathcal{C}$. If F is a face of $\mathcal{C}$, we set $\bar{F} := \mathcal{B}(F) \cup \{F\}$ (set of all faces of F including F). Let $F_1, \ldots, F_r$ be the *shelling*. We obtain

$$\bar{F}_k \cap \left(\bigcup_{i=k+1}^{r} \bar{F}_i \right) = \bigcup_{j=1}^{s_k} G_j^k, \qquad k = 1, \ldots, r-1,$$

where G_j^k, $j = 1, \ldots, s_k$, are distinct $(n-2)$-faces of $\mathcal{C}$. We set $\mathcal{C}_i := \bigcup_{k=i+1}^{r} \bar{F}_k$, $i = 0, \ldots, r$. In particular $\mathcal{C}_r = \emptyset$. For $i = 1, \ldots, r-1$, let

$$S_i := \bar{F}_i \setminus \left(\bigcap_{j=1}^{s_i} G_j^i \right), \qquad S_r := \emptyset.$$

S_i is readily seen to be a free face of $\mathcal{C}_{i-1}$. We obtain $\mathcal{C}_i$ from $\mathcal{C}_{i-1}$ by deleting S_i and all faces which contain S_i, that is, by an elementary $(\operatorname{card} S_i, n)$-collapse.

We denote by g_i the number of elementary (i, n)-collapses in the shelling $F_1, \ldots, F_r$, that is, the number of elements of $\{k \mid s_k = 1\}$. Since the number of j-faces deleted in an elementary (i, n)-collapse is $\binom{n-i}{j+1-i}$ and since $\mathcal{C}_r = \emptyset$,

$$f_j = \sum_{i=0}^{j+1} \binom{n-i}{j+1-i} g_i, \qquad \text{for } j = -1, \ldots, n-1.$$

Therefore, by definition of h_i and Theorem 6.8, $g_i = h_i$ for $i = 0, \ldots, n$. The number of all elementary (i', n)-collapses with $i' \leq i$ in the collapse process

$$\mathcal{C} = \mathcal{C}_0 \supset \mathcal{C}_1 \supset \cdots \supset \mathcal{C}_r$$

is $\tilde{h}_i = \sum_{j=0}^{i} g_j$. Hence, by Lemma 7.3, we obtain (7). □

Exercises

1. Describe explicitly the collapse process of $\mathcal{C} = \mathcal{B}(P)$ where P is the Barnette sphere with a shelling as found in 6, Exercise 2.
2. Let $f_i(\mathcal{B}(P)) = c_i(v, n)$, $i = 1, \ldots, n-1$, and let P be an n-polytope. For which v is P automatically a cyclic polytope?
3. Let P be a 3-polytope with v vertices. If P is decomposed into simplices whose vertices lie in vert P, we obtain a three-dimensional simplicial complex $\mathcal{C}$. Find upper bounds for $f_1(\mathcal{C})$, $f_2(\mathcal{C})$, $f_3(\mathcal{C})$ (as functions of v) which are sharp for $v \leq 6$.
4. Given a simplicial n-polytope P with $v \leq n+3$ vertices, find sharp lower bounds for $f_i(P)$, $i = 1, \ldots, n-1$.

IV

Minkowski sum and mixed volume

1. Minkowski sum

A fundamental operation for convex sets is the following (which can be defined for arbitrary sets in $\mathbb{R}^n$).

1.1 Definition. For any two sets K, L in $\mathbb{R}^n$, we call

$$K + L := \{x + y \mid x \in K, y \in L\}$$

the *Minkowski sum* or, briefly, the *sum* of K and L.

The sum of two triangles K, L in a plane is either a triangle, a quadrangle, a pentagon, or a hexagon (Figure 1).

Minkowski addition may increase the dimension, as is seen from the examples in Figure 2.

1.2 Lemma.

(a) *If τ denotes a translation, then, for any sets K, L in $\mathbb{R}^n$,*

$$\tau(K) + L = \tau(K + L) = K + \tau(L).$$

(b) *If K, L are both convex, closed convex or convex bodies, then, $K + L$ is convex, closed convex, or a convex body, respectively.*

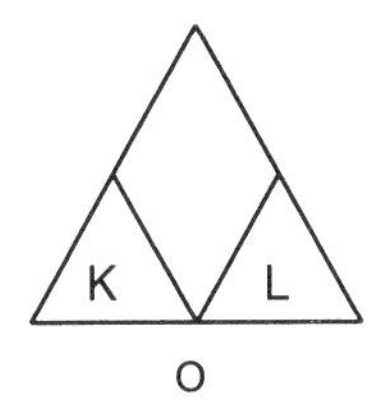

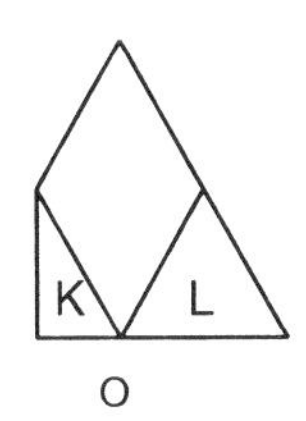

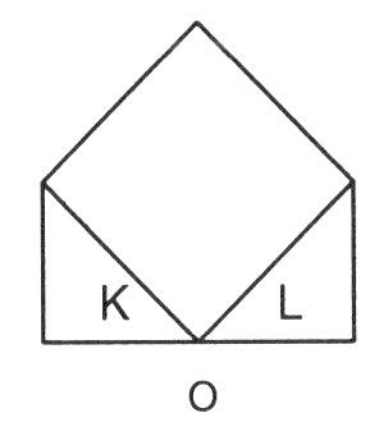

FIGURE 1a,b,c,d.

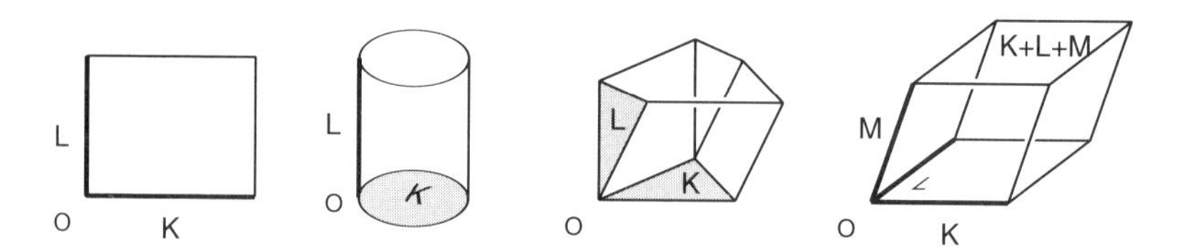

FIGURE 2a,b,c,d.

PROOF.
(a) τ is given by a translation vector t. So, the assertion follows from $(t + K) + L = t + (K + L) = K + (t + L)$.
(b) Let $x, x' \in K$, and $y, y' \in L$. Then, for $0 \leq \lambda \leq 1$,

$$\lambda(x + y) + (1 - \lambda)(x' + y') = \lambda x + (1 - \lambda)x' + \lambda y + (1 - \lambda)y' \in K + L,$$

if K and L are convex. The properties "closed" and "bounded" carry over from K and L to $K + L$ since addition is a continuous operation and maps pairs of bounded sets onto a bounded set.

□

Remark. Most considerations about Minkowski sums are invariant under translations, which is so because of Lemma 1.2(a). If we draw figures, we shall, in general, not mark the origin. It is often helpful to visualize Minkowski addition for convex sets as follows: Hold L in one of its points, p say, and move L around by translations such that p attains all points of K. Then, the translates of L cover $K + L$, that is, $K + L = \bigcup_{p \in K}(p + L)$. If $p \in L$, it can easily be shown that $K + L = K \cup \bigcup_{p \in \partial K}(p + L)$, provided $K \cap L \neq \emptyset$.

1.3 Definition. If λ is a real number and $K \subset \mathbb{R}^n$ is a set, then, we call $\lambda K := \{\lambda x \mid x \in K\}$ a *multiple* of K. If $\lambda_1, \ldots, \lambda_r \in \mathbb{R}$ and $K_1, \ldots, K_r$ are sets in $\mathbb{R}^n$, we call $\lambda_1 K_1 + \cdots + \lambda_r K_r$ a *linear combination* of $K_1, \ldots, K_r$.

Remark. λ may be negative. However, $(-1)K =: -K$ is not the negative of K with respect to Minkowski addition: In Figure 1d, $L = -K$, but $K + L = K + (-K)$ is a hexagon. For $\lambda \in \mathbb{Z}_{>0}$ the notion λK has two meanings, but, fortunately, for convex K both have the same value: $\underbrace{K + \cdots + K}_{\lambda \text{ times}} = \lambda K$.

From the definition of linear combinations and Lemma 1.2(b), we have the following lemma:

1.4 Lemma. *If $K_1, \ldots, K_r$ are convex and $\lambda_1, \ldots, \lambda_r$ are any real numbers, then, $\lambda_1 K_1 + \cdots + \lambda_r K_r$ is convex.*

A very useful property of the Minkowski sum is that it respects the additivity of the support functions:

1.5 Theorem.

(a) *If h_K, h_L are the support functions of the convex bodies K, L, respectively, then, $h_K + h_L$ is the support function of $K + L$,*

$$h_{K+L} = h_K + h_L.$$

(b) *If F is a face of $K + L$, then, there exist faces F_K, F_L of K, L, respectively, such that*

$$F = F_K + F_L.$$

In particular, each vertex of $K + L$ is the sum of vertices of K, L, respectively.

(c) *If K, L are polytopes, so is $K + L$.*

(d) *If K, L are lattice polytopes, so is $K + L$.*

PROOF.

(a) $h_{K+L}(u) = \sup_{x\in K, y\in L}\langle x + y, u\rangle = \sup_{x\in K}\langle x, u\rangle + \sup_{y\in L}\langle y, u\rangle = h_K(u) + h_L(u)$ for any $u \in \mathbb{R}^n \setminus \{0\}$.

(b) For any convex body C and a vector u pointing away from C, according to I, 5.3, let $H_C(u)$ be a supporting hyperplane of C. Then, $F = (K + L) \cap H_{K+L}(u)$. We set $F_K := K \cap H_K(u)$, $F_L := L \cap H_L(u)$. Then, $H_K(u)$, $H_L(u)$ and $H_{K+L}(u)$ are parallel hyperplanes, $H_{K+L}(u) = H_K(u) + H_L(u)$. Up to a translation of L, we may assume $H_K(u) = H_L(u)$. Then, we obtain

$$\begin{aligned} F &= (K + L) \cap H_{K+L}(u) = (K + L) \cap (H_K(u) + H_L(u)) \\ &= (K \cap H_K(u)) + (L \cap H_L(u)) = F_K + F_L. \end{aligned}$$

(c) is a consequence of (b), since the sum of two vertices is a vertex and each vertex of $K + L$ is obtained in this way.

(d) is also a consequence of (b).

□

1.6 Definition. Let K, L be convex bodies in $\mathbb{R}^n$ such that $(\operatorname{aff} K) \cap (\operatorname{aff} L)$ is a point. Then, we call $K + L$ the *direct sum* $K \oplus L$ of K and L.

Figures 2a, b, and d illustrate direct sums; the examples in Figures 1 and 2c are not direct sums. The polar body $(K + L)^*$ of a sum of convex bodies has, in general, no plausible interpretation in terms of K^*, L^*. Only in the case of direct sums do we present such an interpretation.

1.7 Definition. Let K, L be convex bodies such that $K \cap L = \{0\} \subset \operatorname{relint} K \cap \operatorname{relint} L$. Then, we set $K \circ L := \operatorname{conv}(K \cup L)$ and we say $K \circ L$ *splits* into K and L.

Examples are any convex quadrangle in the plane, which splits into two line segments, and any bipyramid in 3-space, in particular, an octahedron which splits into a square and a line segment.

1.8 Theorem. *Let K, L be polytopes in orthogonal subspaces of $\mathbb{R}^n$ such that* $0 \in \operatorname{relint} K \cap \operatorname{relint} L$, $\dim K + \dim L = n$, *and denote by $K^{(*)}$, $L^{(*)}$ the polar polytopes of K, L relative to* $\operatorname{lin} K$, $\operatorname{lin} L$, *respectively. Then,*

$$(K \oplus L)^* = K^{(*)} \circ L^{(*)}. \tag{1}$$

PROOF. Since $0 \in \operatorname{relint} K^{(*)} \cap \operatorname{relint} L^{(*)}$, the right side of (1) is defined. We shall prove $K \oplus L = (K^{(*)} \circ L^{(*)})^*$, which is equivalent to (1).

Let v be a vertex of $Q := K^{(*)} \circ L^{(*)}$. We denote the unit sphere with center 0 by S. Let H_v be the polar hyperplane of v with respect to S, and let $0 \in H_v^-$. Then, by II, Theorem 2,1(d), $Q^* := \bigcap_{v \in \operatorname{vert} Q} H_v^-$. Since each vertex of Q is either a vertex of $K^{(*)}$ or a vertex of $L^{(*)}$, we see that H_v is perpendicular either to $\operatorname{lin} K$ or to $\operatorname{lin} L$.

Let F be a facet of Q, $F = \operatorname{conv}\{v_1, \dots, v_r, v_{r+1}, \dots, v_s\}$ where $v_1, \dots, v_r \in \operatorname{vert} K^{(*)}$ and $v_{r+1}, \dots, v_s \in \operatorname{vert} L^{(*)}$ ($s \geq n$ since $\dim K^{(*)} = \dim K$ and $\dim L^{(*)} = \dim L$). Then, $\operatorname{conv}\{v_1, \dots, v_r\}$ is a facet of $K^{(*)}$ (relative to $\operatorname{lin} K^{(*)}$), so that $a_K := H_{v_1} \cap \cdots \cap H_{v_r} \cap \operatorname{lin} K$ is a point, and, hence, $H_{v_1} \cap \cdots \cap H_{v_r}$ is a translate g of $\operatorname{lin} L$. Similarly, $a_L := H_{v_{r+1}} \cap \cdots \cap H_{v_s} \cap \operatorname{lin} L$ is a point and $H_{v_{r+1}} \cap \cdots \cap H_{v_s}$ a translate h of $\operatorname{lin} K$. The intersection $g \cap h$ is the polar face $a := F^*$ of F (the pole of the hyperplane aff F). Hence, $a = a_K + a_L$ is true for any vertex a of Q^* and vertices $a_K \in \operatorname{vert} K^{(*)(*)} = \operatorname{vert} K$, $a_L \in \operatorname{vert} L^{(*)(*)} = \operatorname{vert} L$.

Since $K \oplus L$ is determined by the sums of vert K and vert L, $K \oplus L = Q^* = (K^{(*)} \circ L^{(*)})^*$ follows. □

Remark. Theorem 1.8 provides a new proof of II, Theorem 3.7. If P is a k-fold bipyramid over R, then P^* is obtained from $R^{(*)}$ by adding k line segments

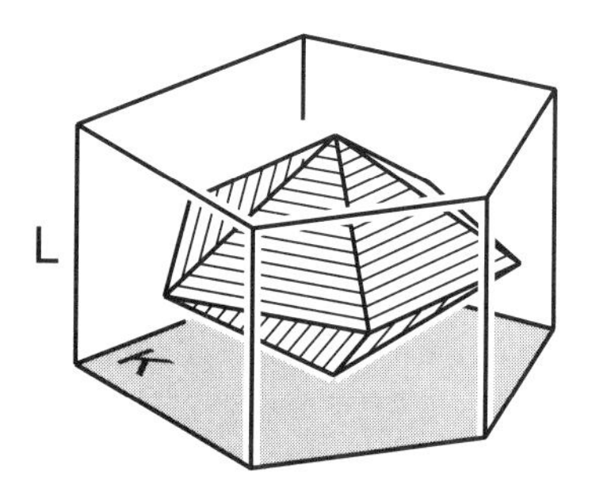

FIGURE 3.

$I_1, \ldots, I_k$, thus

$$P^* = R^{(*)} \oplus I_1 \oplus \cdots \oplus I_k.$$

Example. In $\mathbb{R}^3$ consider a bipyramid (II, Definition 3.5 with $K = Q$, $L = I$). Figure 3 illustrates K, L in translated positions.

Exercises

1.
   a. If K or L is not convex, $K + L$ may or may not be convex.
   b. If K is not convex and not closed, $K + L$ may be open, closed, or none of both.
   c. If K and $K + L$ are convex bodies and L is convex, then L is a convex body.
2. If K, L are convex bodies and if K has differentiable boundary, that is, K has a unique supporting hyperplane in each point of ∂K, then, $\partial(K + L)$ is differentiable.
3.
   a. Find two 3-simplices in $\mathbb{R}^3$ whose Minkowski sum has 16 vertices.
   b. Prove $f_0(P + Q) \leq f_0(P) \cdot f_0(Q)$ for any pair of polytopes in $\mathbb{R}^n$ ($f_0(\cdot) =$ number of vertices).
4. In $\mathbb{R}^4$ let Q be a convex r-gon and R a convex s-gon such that $Q \circ R$ exists. Determine the f-vector of $Q \circ R$.

2. Hausdorff metric

Now, we will introduce a distance function which turns the set $\mathcal{K}$ of all convex bodies in $\mathbb{R}^n$ into a metric space. Here, we use the Minkowski addition of balls.

2.1 Definition. Let B be the unit ball with center 0, and K a convex body in $\mathbb{R}^n$. Then, we call, for $\lambda \geq 0$,

$$K + \lambda B$$

the λ-*parallel body* of K.

2.2 Definition. Given two convex bodies K, L in $\mathbb{R}^n$, let $d(K, L)$ denote the smallest λ for which L lies in the λ-parallel body of K and K in the λ-parallel body of L, so that

$$d(K, L) := \inf\{\lambda \mid K + \lambda B \supset L \text{ and } L + \lambda B \supset K\}.$$

We call $d(K, L)$ the *Hausdorff distance* of K and L.

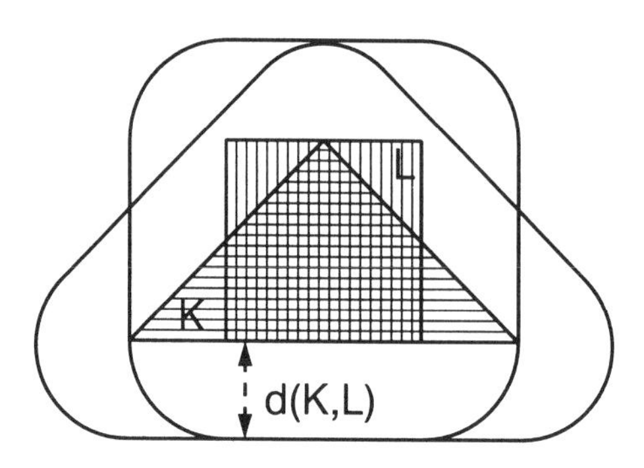

FIGURE 4.

2.3 Theorem. *Let $\mathcal{K}$ denote the set of all convex bodies in $\mathbb{R}^n$. The Hausdorff distance is a metric on $\mathcal{K}$, that is, for all $K, L, M \in \mathcal{K}$,*
(1) $d(K, L) \geq 0$.
(2) $d(K, L) = 0$ *if and only if* $K = L$.
(3) $d(K, L) = d(L, K)$.
(4) $d(K, L) \leq d(K, M) + d(M, L)$ *(triangle inequality).*

Example 1. In $\mathbb{R}^2$, let $K := \text{conv}\{2e_1, -2e_1, 2e_2\}$, $L := \text{conv}\{e_1, -e_1, e_1 + 2e_2, -e_1 + 2e_2\}$. As is seen from Figure 4, $d(K, L) = 1$.

PROOF OF THEOREM 2.3.
(1) and (3) are true by definition of d.
(2) If $d(K, L) = 0$, $K + 0 \cdot B \supset L$, and $L + 0 \cdot B \supset K$; hence, $K = L$. Clearly, $d(K, K) = 0$.
(4) We set $r := d(K, M)$, $s := d(M, L)$, and $t := d(K, L)$. Then, $K + rB \supset M$, $M + rB \supset K$ and $M + sB \supset L$, $L + sB \supset M$ imply

$$\begin{array}{rcccl} K + (r+s)B & \supset & M + sB & \supset & L, \\ L + (r+s)B & \supset & M + rB & \supset & K, \end{array}$$

hence, $r + s \geq t$.

□

Remark. If we define $d_0(K, L) := \inf_{x \in K, y \in L} \|x - y\|$, then, d_0 is not a metric since (2) is, in general, violated.
Also $d_1(K, L) := \sup_{x \in K, y \in L} \|x - y\|$ does not define a metric as is seen, for example, by choosing $K = L$ as a line segment.

2.4 Lemma. *For any two convex bodies K, L and a ball B,*

$$d(K, L) = d(K + B, L + B).$$

PROOF. We will proceed in several steps.

(a) Let x be a point such that $x + B \subset K + B$. We assert $x \in K$. Suppose, otherwise, $x \notin K$. Let $x' \in K$ be the point nearest to x ($x' = p_K(x)$ according to I, Definition 3.2). Let

$$H = \{y \mid \langle y, u \rangle = \langle x', u \rangle\}$$

be the supporting hyperplane through x' with $K \subset H^-$ (according to I, 35). Then, the point in which $\langle \cdot, u \rangle$ takes the maximal value on $x + B$ does not belong to $x + B$, a contradiction to $x + B$ being a closed set.

(b) Let $K + B \subset L + B$. We show $K \subset L$. In fact, $p + B \subset L + B$ for every $p \in K$ is, by (a), equivalent to $p \in L$ for every $p \in K$.

(c) Proof of the lemma. For every $\lambda > 0$, we have the following equivalent statements:

(c.1) $d(K, L) \leq \lambda$.

(c.2) $K + \lambda B \supset L$ and $L + \lambda B \supset K$.

(c.3) $K + \lambda B + B \supset L + B$ and $L + \lambda B + B \supset K + B$.

(c.4) $d(K + B, L + B) \leq \lambda$.

□

2.5 Theorem. *Let K be any convex body. There exists a sequence of polytopes $P_j \subset K$ which converges to K with respect to the Hausdorff metric, thus,*

$$P_1, P_2, \ldots \quad \longrightarrow \quad K.$$

PROOF. Given any $\varepsilon > 0$, assign to each $x \in \partial K$ the open ball B_x with center x and radius ε. Then, $\{B_x\}$ is an open covering of ∂K. Since ∂K is compact, there exists a finite subcovering $\{B_{x_1}, \ldots, B_{x_q}\}$ of ∂K. Let $P := \operatorname{conv}\{x_1, \ldots, x_q\}$. Then, $P + \varepsilon B \supset B_{x_1} \cup \cdots \cup B_{x_q} \supset \partial K$; hence, $P + \varepsilon B$ being convex,

$$P + \varepsilon B \supset K.$$

Trivially, $P \subset K \subset K + \varepsilon B$, hence, $d(P, K) \leq \varepsilon$. Therefore, a sequence

$$P_1, P_2, \ldots \quad \longrightarrow \quad K$$

can be found. □

2.6 Theorem. *Let $K, K_1, K_2, \ldots$ be convex bodies, and let $0 \in \operatorname{int} K$. If*

$$K_1, K_2, \ldots \quad \longrightarrow \quad K, \qquad \textit{then,} \qquad K \cap K_1, K \cap K_2, \ldots \quad \longrightarrow \quad K.$$

PROOF. Let B_0 be a ball with center 0 and radius $r_0 > 0$, such that $2B_0 \subset K$. Then, $d(B_0, K) \geq r_0$. Since, for $K_1, K_2, \ldots \longrightarrow K$, $K_i + \varepsilon_i B \supset K \supset 2B_0$ for appropriate $\varepsilon_1, \varepsilon_2, \ldots \longrightarrow 0$, there exists an i_0 such that

$$K_i \supset B_0 \qquad \text{for } i \geq i_0.$$

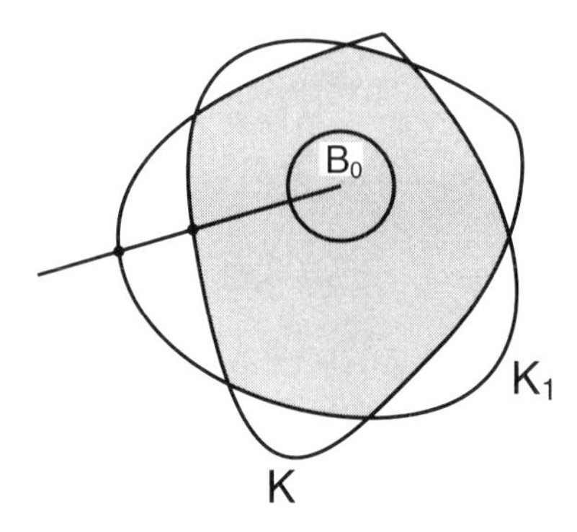

FIGURE 5.

The same is true for the $K_i \cap K$.

Let ϱ be any ray emanating from 0, and denote the points in which ϱ intersects $\partial K, \partial K_i$ by p^ϱ, p_i^ϱ, respectively, (Figure 5). Then,

$$K = \bigcup_{\text{all}\varrho} [0, p^\varrho] \qquad K_i = \bigcup_{\text{all}\varrho} [0, p_i^\varrho],$$

and

$$K \cap K_i = \bigcup_{\text{all}\varrho} [0, p^\varrho] \cap [0, p_i^\varrho].$$

But $[0, p^\varrho] \cap [0, p_i^\varrho]$ equals $[0, p^\varrho]$ or $[0, p_i^\varrho]$. We set

$$\delta_i := \sup_\varrho \| p_i^\varrho - p^\varrho \|.$$

Then,

$$(K_i \cap K) + \delta_i B \supset K, \quad \text{and} \quad K + \delta_i B \supset K_i \cap K.$$

At the same time,

$$K_i + \delta_i B \supset K, \quad \text{and} \quad K + \delta_i B \supset K_i.$$

Since $K_1, K_2, \ldots \longrightarrow K$, we obtain $\delta_1, \delta_2, \ldots \longrightarrow 0$, and, hence, $K \cap K_1, K \cap K_2, \ldots \longrightarrow K$. □

Remark. If K has no interior points, the theorem is false: Take, for example a sequence of disjoint line segments $[x_i, y_i]$ in $\mathbb{R}^1$ whose endpoints x_i, y_i converge to 0. Then, $[x_1, y_1], [x_2, y_2], \ldots \longrightarrow \{0\} =: K$, but $K \cap [x_i, y_i] = \emptyset$, $i = 1, 2, \ldots$

2.7 Theorem. *Let K be a convex body, and $0 \in \operatorname{int} K$. Then,*

$$K_1, K_2, \ldots \longrightarrow K \qquad \textit{if and only if}$$
$$K_1^*, K_2^*, \ldots \longrightarrow K^* \qquad \textit{for the polar bodies with respect to the unit sphere.}$$

PROOF. Let $K_1, K_2, \ldots \longrightarrow K$. By Theorem 2.6, $K \cap K_1, K \cap K_2, \ldots \longrightarrow K$, so we simply let $K_i \subset K$, $i = 1, 2, \ldots$

As in the proof of Theorem 2.6, we introduce ϱ, p^ϱ, p_i^ϱ and find

$$\sup_{\text{all } \varrho} \|p_1^\varrho - p^\varrho\|, \sup_{\text{all } \varrho} \|p_2^\varrho - p^\varrho\|, \ldots \quad \longrightarrow \quad 0. \tag{1}$$

Let H^ϱ, H_i^ϱ be the polar hyperplanes of p^ϱ, p_i^ϱ, respectively, $i = 1, 2, \ldots$, and let $0 \in H^{\varrho-}$, $0 \in H_i^{\varrho-}$. Then, by definition of K^*, K_i^*,

$$K^* = \bigcap_{\text{all } \varrho} H^{\varrho-} \qquad K_i^* = \bigcap_{\text{all } \varrho} H_i^{\varrho-}.$$

Suppose $K_1^*, K_2^*, \ldots \not\rightarrow K^*$. Then, considering $K^* \subset K_i^*, i = 1, 2, \ldots$, we can choose a sequence of points $q_i \in (\partial K_i^*) \setminus K^*$ such that $\|q_1 - p_{K^*}(q_1)\|, \|q_2 - p_{K^*}(q_2)\|, \ldots \not\rightarrow 0$.

Let $H^{(i)}$ be the supporting hyperplane of K^* in $p_{K^*}(q_i)$, and let H_i be the supporting hyperplane of K_i^* which has the same outer normal as $H^{(i)}$ has (Figure 6). Then, for the distance α_i between $H^{(i)}$ and H_i, $\alpha_i \geq \|q_i - p_{K^*}(q_i)\|$, and, hence,

$$\alpha_1, \alpha_2, \ldots \quad \not\rightarrow \quad 0. \tag{2}$$

We consider the poles p^{ϱ_i}, $p_i^{\varrho_i}$ of $H^{(i)}$, H_i, respectively; they are the intersections of the same ray ϱ_i with ∂K, ∂K_i, respectively. Since $\alpha_i \leq d(K_i, K)$, it now follows that $K_1, K_2, \ldots \not\rightarrow K$, a contradiction.

Therefore $K_1, K_2, \ldots \longrightarrow K$ implies $K_1^*, K_2^*, \ldots \longrightarrow K^*$. The converse is also true since $K_i^{**} = K_i$, and $K^{**} = K$. □

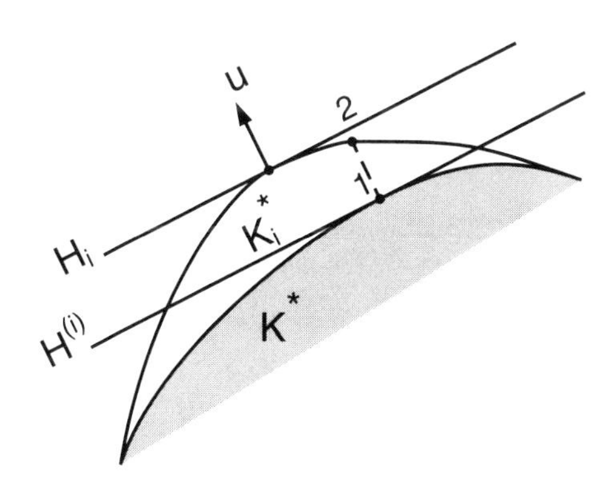

FIGURE 6. $2 = q_i$, $1 = p_{K^*}(q_i)$

2.8 Theorem. *Given a convex body, for any one of the following conditions, there exists a convergent sequence $P_1, P_2, \ldots \longrightarrow K$ of polytopes satisfying it:*
(a) *All P_i are simplicial.* ,
(b) *All P_i are simple.* ,
(c) *All P_i are inscribed into K.*
(d) *All P_i are circumscribed to K.*

PROOF.
(a) Given any sequence $P_1, P_2, \ldots \longrightarrow K$ (see Theorem 2.5), we need only slightly disturb the vertices of each P_i in a sufficiently small neighborhood to obtain simplicial polytopes.
(b) follows from (a) if Theorem 2.7 is applied to the polar of K relative to the affine hull of K.
(c) is implied by the proof of Theorem 2.5.
(d) is obtained from (c) by applying Theorem 2.7 to a polar body of K relative to aff K.

□

2.9 Theorem. *Let $K, K_1, K_2, \ldots$ be convex bodies, and let $B_0 = r_0 B$ be a ball with radius $r_0 > 0$. If*

$$B_0 + K_1, B_0 + K_2, \ldots \quad \longrightarrow \quad K,$$

then, there exists a convex body K_0 such that $K = B_0 + K_0$.

PROOF. This is a consequence of Lemma 2.4. □

2.10 Theorem (Blaschke's selection theorem). *Let $\{K_\alpha\}_{\alpha \in I}$ be a set of infinitely many convex bodies, all contained in the unit ball B. Then, there exists an infinite sequence of different elements of this set, say $K_1, K_2, \ldots$, which converges (in the Hausdorff metric) to a convex body K also contained in B; thus,*

$$K_1, K_2, \ldots \quad \longrightarrow \quad K.$$

PROOF. If we add to each K_α the ball $B_0 := 3B$, we obtain

$$0 \in B \subset \operatorname{int}(K_\alpha + B_0) \subset 4B.$$

Therefore, by Theorem 2.9, it suffices to find a convergent sequence among the $K_\alpha + B_0$. Again, we write K_α instead of $K_\alpha + B_0$ and assume $B \subset \operatorname{int} K_\alpha \subset 4B$ for all $\alpha \in I$. So, in particular, all K_α and the approximating polytopes we shall use are full-dimensional.

We consider rays $\varrho^{(1)}, \ldots, \varrho^{(r)}$ emanating from 0, and the points $p_\alpha^{(j)} := \varrho^{(j)} \cap \partial K_\alpha$, $\alpha \in I$, $j = 1, \ldots, r$ (Figure 7). If the rays are chosen appropriately, we obtain

$$B \subset P_\alpha := \operatorname{conv}\{p_\alpha^{(1)}, \ldots, p_\alpha^{(r)}\}, \quad \alpha \in I.$$

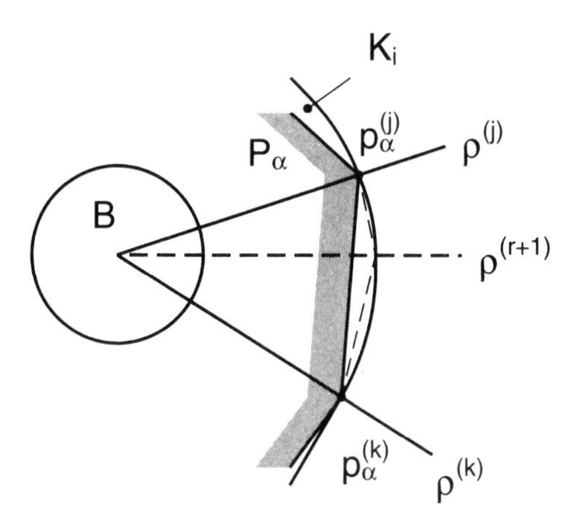

FIGURE 7.

Each set $\{p_\alpha^{(j)}\}$, $\alpha \in I$, j fixed, is contained in a closed line segment $S_i \subset \varrho^{(i)}$. Therefore, the points

$$(p_\alpha^{(1)}, \dots, p_\alpha^{(r)}) \in S_1 \times \cdots \times S_r$$

(topological product space, or a parallelotope in $\mathbb{R}^r$) lie in a compact set, and, hence, we can pick from them a convergent sequence, say

$$(p_1^{(1)}, \dots, p_1^{(r)}), (p_2^{(1)}, \dots, p_2^{(r)}), \dots \quad \longrightarrow \quad (p^{(1)}, \dots, p^{(r)})$$

such that $p_r^{(1)} \neq p_s^{(1)}$ for $r \neq s$. Then, obviously, for $P := \operatorname{conv}\{p^{(1)}, \dots, p^{(r)}\}$, we obtain a Hausdorff convergence

$$P_1, P_2, \dots \quad \longrightarrow \quad P$$

where $P_r \neq P_s$ for $r \neq s$. Now we consider a further ray $\varrho^{(r+1)}$ emanating from 0, and set $p_i^{(r+1)} := \partial K_i \cap \varrho^{(r+1)}$; $i = 1, 2, \dots$ Again, there is a convergent subsequence of $p_1^{(r+1)}, p_2^{(r+1)}, \dots$ which we denote by $p_{11}^{(r+1)}, p_{12}^{(r+1)}, \dots$,

$$p_{11}^{(r+1)}, p_{12}^{(r+1)}, \dots \quad \longrightarrow \quad p^{(r+1)}. \tag{3}$$

We set $P_{1i} := \operatorname{conv}\{p_{1i}^{(1)}, \dots, p_{1i}^{(r)}, p_{1i}^{(r+1)}\}$ (where $p_{11}^{(j)}, p_{12}^{(j)}, \dots$ is, for $j = 1, \dots, r$, the subsequence of $p_1^{(j)}, p_2^{(j)}, \dots$ with the indices according to (3)), and

$$P_{(1)} := \operatorname{conv}(P \cup \{p^{(r+1)}\}).$$

Then,

$$\begin{array}{c} P_{11}, P_{12}, \dots \quad \longrightarrow \quad P_{(1)}, \qquad \text{with vert } P_{1j} \subset \partial K_{1j} \\ \text{for a subsequence } K_{11}, K_{12}, \dots \text{ of } K_1, K_2, \dots \end{array} \tag{4}$$

We continue choosing rays $\varrho^{(r+k)}$ emanating from 0 such that, for a subsequence $P_{k1}, P_{k2}, \dots$ of (4), using an obvious notation,

$$P_{k1}, P_{k2}, \dots \longrightarrow P_{(k)} := \operatorname{conv}(P_{(k-1)} \cup \{p^{(r+k)}\}),$$

$k = 2, 3, \ldots$, vert $P_{kj} \subset \partial K_{kj}$ ($K_{k1}, K_{k2}, \ldots$ a subsequence of $K_{k-1,1}, K_{k-1,2}, \ldots$).

Let ε'_k be the infimum of all $\varepsilon > 0$ which permit an open covering of $S := \partial(4B)$ by $r + k$ open ε-balls centered at $q_j = S \cap \varrho^{(j)}$, and set $\varepsilon_k := 2\varepsilon'_k$. We can choose the rays $\varrho^{(r+k)}$ such that $\varepsilon_1, \varepsilon_2, \ldots, \varepsilon_k, \ldots \longrightarrow 0$.

Since $P_{(1)} \subset P_{(2)} \subset \cdots$ the set $P_0 := P_{(1)} \cup P_{(2)} \cup \cdots$ is convex. Furthermore, P_0 is bounded, so $K := \bar{P}_0$ (closed hull) is a convex body. We see now that

$$P_{ii} + \varepsilon_i B \supset K, \qquad \text{and (trivially)} \quad K + \varepsilon_i B \supset P_{ii}.$$

Also,

$$P_{ii} + \varepsilon_i B \supset K_{ii}, \qquad \text{and (trivially)} \quad K_{ii} + \varepsilon_i B \supset P_{ii}$$

where $P_{rr} \neq P_{ss}$ for $r \neq s$. Hence the diagonal sequence consists of different elements and converges:

$$K_{11}, K_{22}, \ldots \quad \longrightarrow \quad K$$

□

The following equivalence relationship for polytopes defines classes of polytopes which will be needed in the following two sections and in V, section 5.

2.11 Definition. Two n-polytopes P and Q are called *strictly combinatorially isomorphic* if there exists a bijective, inclusion-preserving map $\varphi : \mathcal{B}(P) \longrightarrow \mathcal{B}(Q)$ such that the following condition is satisfied

(5) For any face $F \in \mathcal{B}(P)$, the affine hulls aff F and aff $\varphi(F)$ are translates of each other.

About general combinatorial isomorphism compare II, Definition 1.12.

2.12 Lemma. *Two n-polytopes P, Q are strictly combinatorially isomorphic if and only if $\Sigma(P) = \Sigma(Q)$ (compare* I, *Definition 4.14).*

PROOF. $\Sigma(P)$ consists of the normal cones $N(x)$ of P, as introduced in I, Definition 4.7. Hereby, $N(x) = N(x')$ for $x, x' \in$ relint F for any face $F \in \mathcal{B}(P)$ (I, Lemma 4.11). Since aff F is a translate of $(N(x))^{\perp}$, the lemma readily follows. □

2.13 Lemma. *If P and Q are strictly combinatorially isomorphic, then, P and $P + Q$ are also strictly combinatorially isomorphic.*

PROOF. This is a direct consequence of Lemma 2.12. □

2.14 Theorem. *Let $K_1, \ldots, K_r$ be convex bodies. Then, there exist convergent sequences of polytopes*

$$P_{i1}, P_{i2}, \ldots, \quad \longrightarrow \quad K_i; \qquad i = 1, \ldots, r \tag{6}$$

such that $P_{1j}, \ldots, P_{rj}$, are, for any fixed j, pairwise strictly combinatorially isomorphic.

PROOF. The existence of sequences (6) is guaranteed by Theorem 2.5. If $k := \dim K_i < n$ for some $i \in \{1, \ldots, r\}$, we consider, first, a sequence (6) relative to aff K_i, and, then, choose an $(n-k)$-fold pyramid over P_{ij} contained in $P_{ij} + \varepsilon_j B$ for sufficiently small $\varepsilon_j > 0$. Suppose $\varepsilon_{ij} := d(P_{ij}, K_i)$ if $d(P_{ij}, K_i) > 0$, $i = 1, \ldots, r$; $j = 1, 2, \ldots$, otherwise $\varepsilon_{ij} := \frac{1}{j}$. Then, we set $\tilde{P}_{ij} := P_{ij} + \varepsilon_{1j} P_{1j} + \cdots + \varepsilon_{rj} P_{rj}$ and obtain pairwise strictly combinatorially isomorphic polytopes which satisfy $\tilde{P}_{i1}, \tilde{P}_{i2}, \ldots \longrightarrow K_i, i = 1, \ldots, r$. □

Exercises

1. Among all squares inscribed in a unit circle, find two which have largest possible Hausdorff distance. What is the value of this distance?
2. We can extend the definition of Hausdorff distance to closed convex sets, allowing the distance to become infinite. Find necessary and sufficient conditions for two unbounded polyhedral sets to have finite Hausdorff distance.
3. Let K be a convex body and $L = \{p\}$ be a single point. Then, $d(K, L) = \sup_{x \in K} \|x - p\|$.
4. Given a two-dimensional convex body K, find a sequence of convex bodies $K_1, K_2, \ldots \longrightarrow K$ such that each K_i has a twice differentiable support function. (Hint: Approximate K by polytopes P_i, and "round off" the vertices of each P_i by arcs cut out of curves of the type $x_2 = c x_1^r, r \in \mathbb{Z}_{\geq 3}$).

3. Volume and mixed volume

We assume that it is known from calculus that any n-dimensional convex body K possesses an n-dimensional volume $V(K)$. A map Φ of a set $M \subset \mathbb{R}^n$ onto a set $M' \subset \mathbb{R}^n$ is called a *Lipschitz map* if there exists a constant $c \geq 0$ such that

$$\|\Phi(x) - \Phi(y)\| \leq c \cdot \|x - y\|$$

for any pair $x, y \in M$. Clearly such a Φ is uniformly continuous on M.

If a set M can be dissected into finitely many $(n-1)$-polytopes, its $(n-1)$-volume is the sum of the $(n-1)$-volumes of the polytopes. We further assume that it is known from calculus that, if M has finite $(n-1)$-volume and is mapped under a Lipschitz map onto a set M', then M' has finite $(n-1)$-volume.

3.1 Lemma. *If K is an n-dimensional convex body, then, ∂K has (finite) $(n-1)$-volume.*

PROOF. Since K is bounded, we find a simplex Δ which contains K. The nearest point map p_K being Lipschitz ($c = 1$; see I, Lemma 3.7), we obtain the lemma from

$$p_K(\partial\Delta) = \partial K.$$

□

Volume and Minkowski addition are linked as follows.

Theorem (H. Minkowski).

Let $K_1, \ldots, K_r$ be convex bodies in $\mathbb{R}^n$, and $\lambda_i \geq 0$, $i = 1, \ldots, r$.

Then, $V(\lambda_1 K_1 + \cdots + \lambda_r K_r)$ is a homogeneous polynomial of degree n in $\lambda_1, \ldots, \lambda_r$,

$$(1) \qquad V(\lambda_1 K_1 + \cdots + \lambda_r K_r) = \sum_{\varrho_1, \ldots, \varrho_n = 1}^{r} V(K_{\varrho_1}, \ldots, K_{\varrho_n}) \lambda_{\varrho_1} \cdots \lambda_{\varrho_n},$$

the summation being carried out independently over the ϱ_i, $i = 1, \ldots, n$.

3.3 Definition. Arranging the coefficients on the right side of (1) such that $V(K_{\pi(\varrho_1)}, \ldots, K_{\pi(\varrho_n)}) = V(K_{\varrho_1}, \ldots, K_{\varrho_n})$ for any permutation π of $\varrho_1, \ldots, \varrho_n$, we call $V(K_{\varrho_1}, \ldots, K_{\varrho_n})$ the (n-dimensional) *mixed volume* of $K_{\varrho_1}, \ldots, K_{\varrho_n}$.

PROOF OF THEOREM 3.2. We shall first carry out the proof for polytopes and deduce Lemma 3.4, Lemma 3.5, Lemma 3.6, and Theorem 3.7 for polytopes. Then, Lemma 3.8 extends all statements to arbitrary convex bodies.

We apply induction with respect to the dimension n. For $n = 1$, the K_i are intervals $[p_i, q_i]$ or points $p_i = [p_i, p_i]$, $p_i \leq q_i$. We find

$$\begin{aligned}
\lambda_1 K_1 + \cdots + \lambda_r K_r &= [\lambda_1 p_1 + \cdots + \lambda_r p_r, \lambda_1 q_1 + \cdots + \lambda_r q_r] \\
V(\lambda_1 K_1 + \cdots + \lambda_r K_r) &= \lambda_1 q_1 + \cdots + \lambda_r q_r - (\lambda_1 p_1 + \cdots + \lambda_r p_r) \\
&= \lambda_1 (q_1 - p_1) + \cdots + \lambda_r (q_r - p_r) \\
&= \lambda_1 V(K_1) + \cdots + \lambda_r V(K_r)
\end{aligned}$$

which is a polynomial of degree 1 in $\lambda_1, \ldots, \lambda_r$.

Suppose the theorem is true for $n-1$ instead of n. First, we assume $K_1, \ldots, K_r$ to be polytopes.

Let $u \neq 0$ be a fixed normal direction. We consider the supporting hyperplanes $H_{K_j}(u)$ and the faces

$$F_j := F_j(u) := H_{K_j}(u) \cap K_j, \quad j = 1, \ldots, r.$$

We set

$$\begin{aligned}
K_\Lambda &:= \lambda_1 K_1 + \cdots + \lambda_r K_r \qquad (\Lambda := (\lambda_1, \ldots, \lambda_r)), \\
F_\Lambda(u) &:= H_{K_\Lambda}(u) \cap K_\Lambda \qquad \text{(Figure 8)}.
\end{aligned}$$

Then

$$F_\Lambda(u) = \lambda_1 F_1(u) + \cdots + \lambda_r F_r(u).$$

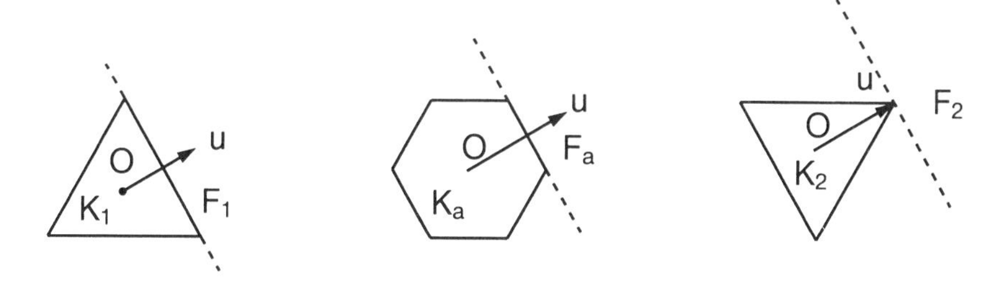

FIGURE 8.

Volumes do not change under translations, so for every given u we can assume that all F_i lie in the hyperplane $H_{K_\Lambda}(u)$ and apply the induction hypothesis to $v(F_\Lambda(u))$. Furthermore, $0 \in K_\Lambda$ can be arranged. If $\dim K_\Lambda < n$, the assertion of the theorem follows by inductive assumption. So, let $\dim K_\Lambda = n$, and let $u_1, \ldots, u_m$ be all outer facet normals of K_Λ, $\|u_i\| = 1$, $i = 1, \ldots, m$. We decompose K_Λ into pyramids with apex 0 over the facets and obtain ($v(\cdot)$ denoting $(n-1)$-dimensional volume, h_{K_Λ} the support function of K_Λ)

$$\begin{aligned} V(K_\Lambda) &= \frac{1}{n} \sum_{j=1}^{m} h_{K_\Lambda}(u_j) v(F_\Lambda(u_j)) \\ &= \frac{1}{n} \sum_{j=1}^{m} \Big(\lambda_1 h_{K_1}(u_j) + \cdots + \lambda_r h_{K_r}(u_j)\Big) v(F_\Lambda(u_j)). \end{aligned}$$

Since, by inductive assumption, $v(F_\Lambda(u_j))$ is a homogeneous polynomial of degree $n-1$ in $\lambda_1, \ldots, \lambda_r$, $V(K_\Lambda)$ is a homogeneous polynomial of degree n in the same variables. This proves the theorem for polytopes.

Before we finish the proof, we collect some basic properties of mixed volumes for polytopes where we write $K_1, \ldots, K_n$ instead of $K_{i_1}, \ldots, K_{i_n}$, $1 \le i_1 \le \cdots \le i_n \le r$.

3.4 Lemma. $V(K_1, \ldots, K_n) = V(K_{\pi(1)}, \ldots, K_{\pi(n)})$ *for any permutation* π *of* $(1, \ldots, n)$.

PROOF. This follows from the definition of mixed volume. □

3.5 Lemma. $V(K, \ldots, K) = V(K)$ *is the volume of* K.

PROOF. The proof is immediate with the equation $\lambda^n V(K) = V(\lambda K) = \lambda^n V(K, K, \ldots, K)$ where the first equality is true since V is an n-dimensional volume, and the second a consequence of Theorem 3.2. □

Example 1. For planar convex bodies K_1, K_2 we have $V(K_1 + K_2) = V(K_1) + V(K_2) + 2V(K_1, K_2)$. In Figure 9a, b and c the volume of the hatched areas is $2V(K_1, K_2)$.

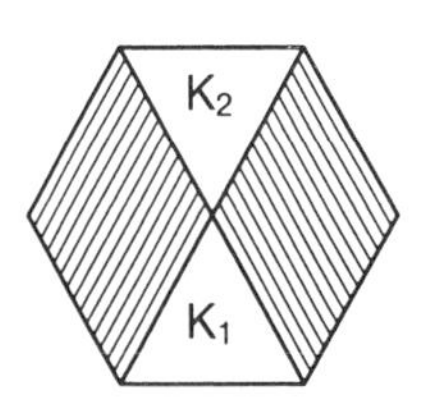

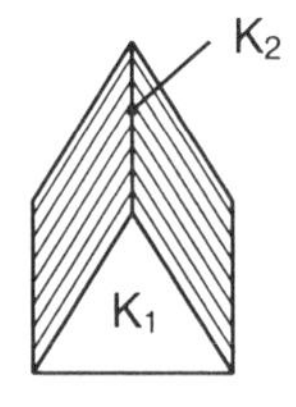

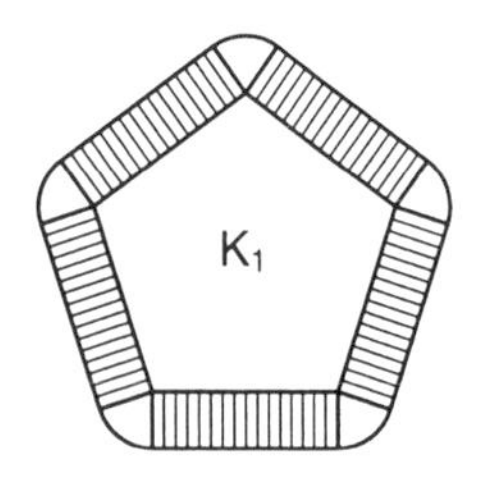

FIGURE 9a,b,c.

3.6 Lemma. *$V(\lambda_1 K_1 + \lambda_1' K_1', K_2, \dots, K_n) = \lambda_1 V(K_1, K_2, \dots, K_n) + \lambda_1' V(K_1', K_2, \dots, K_n)$ for any $\lambda_1 \geq 0$, $\lambda_1' \geq 0$.*

PROOF. by Theorem 3.2. □

3.7 Theorem. *$V(K_1, \dots, K_n)$ can be expressed as a linear combination of volumes of sums of the K_i, $i \in \{1, \dots, n\}$, that is,*

$$\begin{aligned} n!V(K_1, \dots, K_n) =& V(K_1 + \cdots + K_n) \\ & - \sum_{i=1}^{n} V(K_1 + \cdots + K_{i-1} + K_{i+1} + \cdots + K_n) \\ & + \sum_{i<j} V(K_1 + \cdots + K_{i-1} + K_{i+1} + \cdots + K_{j-1} \\ & + K_{j+1} + \cdots + K_n) - + \cdots \\ & + (-1)^{n-2} \sum_{i<j} V(K_i + K_j) + (-1)^{n-1} \sum_{i=1}^{n} V(K_i). \end{aligned}$$

In particular, for $n = 3$,

$$\begin{aligned} 6V(K_1, K_2, K_3) =& V(K_1 + K_2 + K_3) - V(K_2 + K_3) - V(K_1 + K_3) \\ & - V(K_1 + K_2) + V(K_1) + V(K_2) + V(K_3). \end{aligned}$$

PROOF. Because of Lemma 3.4 and Lemma 3.6, we can treat the equation of the theorem in the same way as the equation

$$\begin{aligned} n!a_1 \cdots a_n =& (a_1 + \cdots + a_n)^n \\ & - \sum_{i=1}^{n} (a_1 + \cdots + a_{i-1} + a_{i+1} + \cdots a_n)^n \\ & + - \cdots + (-1)^{n-2} \sum_{i<j} (a_i + a_j)^n + (-1)^{n-1} \sum_{i=1}^{n} a_i^n \end{aligned} \tag{2}$$

in elementary algebra.

To verify (2) we proceed as follows. First, we look at all terms in which a_1^n occurs. In the first term, we find it once, in the second $n-1$ times, in the third $\binom{n-1}{2}$ times, and so on. Altogether, it occurs

$$1-(n-1)+\binom{n-1}{2}-+\cdots+(-1)^{n-3}\binom{n-1}{n-3} \\ +(-1)^{n-2}(n-1)+(-1)^{n-1}=(1-1)^{n-1}=0$$

times. In the same way, we proceed for $a_1^{n-1}a_2$, and so on. The only term which occurs only once on the right side of (2) is $n!a_1\cdots a_n$ (in the first term). All others cancel out. □

3.8 Lemma. *The mixed volume $V(K_1,\ldots,K_n)$ depends continuously on $K_1,\ldots,K_n$ (in the Hausdorff metric), hence, Lemmas* 3.4 *to* 3.6 *and Theorem* 3.7 *extend to arbitrary convex bodies.*

PROOF. This follows from Theorem 3.7 since the ordinary volume and Minkowski sum depend continuously on the polytopes. □

The Proof of Theorem 3.2 is now readily completed by applying Lemma 3.8.

Example 2. Let $S := \operatorname{conv}\{0, e_1, e_2, e_3\}$ be a simplex in $\mathbb{R}^3$, and $I_1 := [0, e_1]$, $I_2 := [0, e_2]$ line segments. We wish to calculate $V(S, I_1, I_2)$. By Theorem 3.7,

$$6V(S,I_1,I_2)=V(S+I_1+I_2)-V(S+I_1)-V(S+I_2) \\ -V(I_1+I_2)+V(S)+V(I_1)+V(I_2).$$

Clearly $V(I_1)=V(I_2)=V(I_1+I_2)=0$ since $V(\cdot)$ denotes three-dimensional volume. For reasons of symmetry, $V(S+I_1)=V(S+I_2)$. We know $V(S)=\frac{1}{6}$, so we must only calculate $V(S+I_1)$ and $V(S+I_1+I_2)$. We

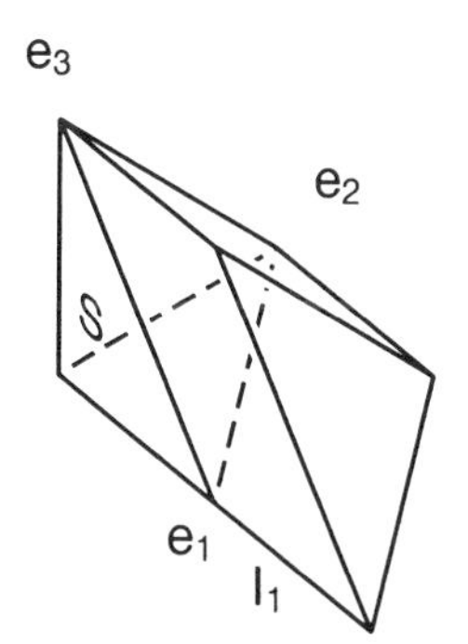

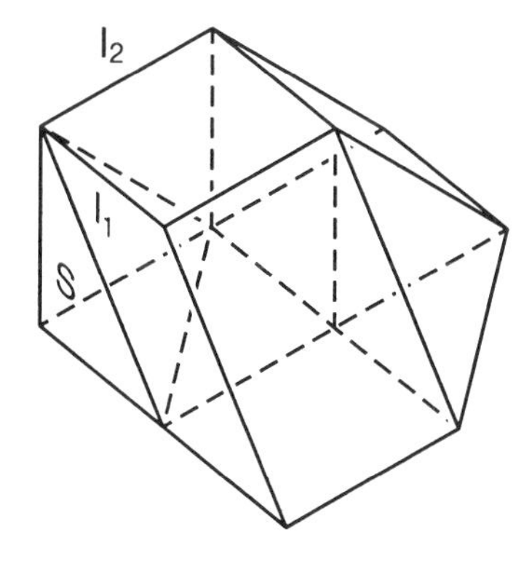

FIGURE 10.

obtain $S + I_1$ from S by joining a triangular prism to it (see Figure 10) whose volume is evidently $\frac{1}{2}$. Therefore, $V(S + I_1) = V(S + I_2) = \frac{2}{3}$.

From Figure 10, it is readily seen that $V(S + I_1 + I_2) = 2\frac{1}{6}$. Therefore $6V(S, I_1, I_2) = 2\frac{1}{6} - \frac{4}{3} + \frac{1}{6} = 1$, hence, $V(S, I_1, I_2) = \frac{1}{6}$.

3.9 Theorem. *If $K_1, \dots, K_n$ are lattice polytopes in $\mathbb{R}^n$, then, $n!V(K_1, \dots, K_n)$ is an integer.*

PROOF. By Theorem 1.5(d), sums of the K_i are lattice polytopes. But lattice polytopes can be dissected into lattice simplices. Each lattice simplex can, by a volume-preserving affine transformation, be mapped onto a simplex with integral coordinates on the coordinate axes of $\mathbb{R}^n$; therefore, each has volume $\frac{k}{n!}$, k an integer. Hence, the theorem follows from Theorem 3.7. □

Exercises

1. The circumference of a planar convex polytope P has length $2V(P, B)$, where B is the unit disc.
2. Let S be the simplex as in Example 2, and let I_1 again be $[0, e_1]$. Find $V(S, -S, I_1)$.
3. Let K be any convex body, I a line segment. Then, $V(I, K, \dots, K)$ can be interpreted as volume of a cylinder. (Hint: $V(I, \dots, I, K, \dots, K) = 0$ if I occurs at least twice).
4. A sum $I_1 + \dots + I_r$ of finitely many line segments is called a zonotope. If $I_j = [a_j, b_j]$, we call $b_j - a_j$ a spanning vector of I_j, $j = 1, \dots, r$. Express $V(I_1 + \dots + I_r)$ by determinants of spanning vectors.

4. Further properties of mixed volumes

Though Minkowski sums have a plausible geometric meaning, this is not so for mixed volumes. We shall characterize mixed volumes in a way which, at least in a number of cases, displays them in a geometrical light. Our main aim will be presented in the next section.

First, we let P be a polytope and B the unit ball. We introduce angular measures in the following way.

4.1 Definition. Let F be a proper face of P. For $x \in \operatorname{relint} F$, we set

$$\Theta_F := p_P^{-1}(x) \cap (P + B) - x$$

and call Θ_F the *outer angle* of P in F (or in x). If $\dim \Theta_F (= \dim(\operatorname{aff} \Theta_F)) = k$, we say $v_k(\Theta_F)$ is the *angular measure* of P in F (or in x), where $v_k(\cdot)$ denotes k-dimensional volume.

4.2 Lemma.

$$\binom{n}{k} V(\underbrace{B, \ldots, B}_{k}, \underbrace{P, \ldots, P}_{n-k}) = \sum_{(n-k)-\text{faces } F} v_k(\Theta_F) v_{n-k}(F).$$

PROOF. For $k = 0$, we have $F = P$ as an improper face. Then, we have $\Theta_F = \{0\}$, and we set $v_0(\{0\}) = 1$. Then, $V(P, \ldots, P) = V(P) = v_n(P)$ is the volume of P (see Lemma 3.5).

We consider $P + \lambda B, \lambda \geq 0$. By Theorem 3.2,

$$\begin{aligned}(1) \quad V(P + \lambda B) =& V(P) + \lambda\binom{n}{1} V(B, P, \ldots, P) \\ &+ \lambda^2 \binom{n}{2} V(B, B, P, \ldots, P) \\ &+ \cdots + \lambda^{n-1}\binom{n}{n-1} V(B, \ldots, B, P) + \lambda^n V(B).\end{aligned}$$

Conversely, we decompose $P + \lambda B$ as follows.

$k = 0$: $F = P$ is the "inner" part of $P + \lambda B$, and has volume $V(P)$.

$k = 1$: Let F be a facet, and consider "above" F the prism with basis F and height λ. The volume of the union of all such prisms is $\lambda \sum v_{n-1}(F)$ summed over all facets F.

$k = 2$: Consider an $(n - 2)$-face F of P. Above F, we have a "log" (compare Figure 11)

$$\bigcup_{p \in \text{relint } F} [p_P^{-1}(p) \cap (P + \lambda B)].$$

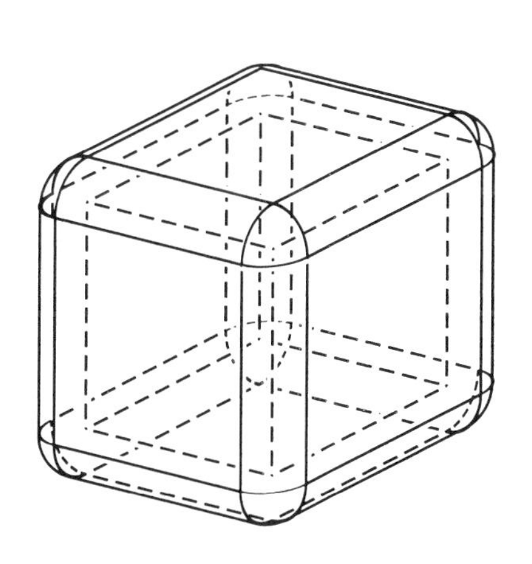

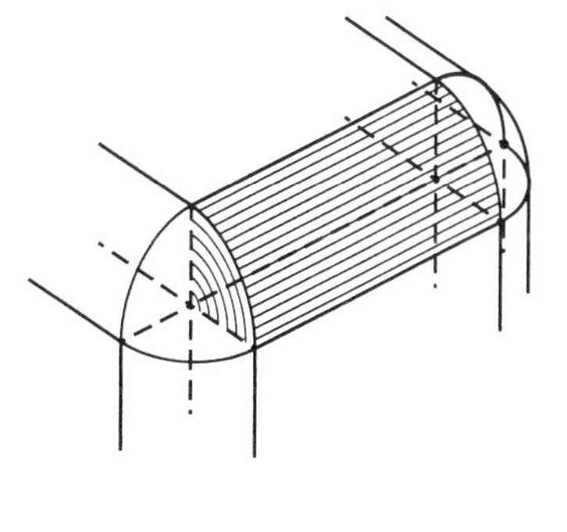

FIGURE 11.

The total volume of all such pieces of $P + \lambda B$ is

$$\lambda^2 \sum_{F \text{ an } (n-2)-\text{face}} v_2(\Theta_F) v_{n-2}(F).$$

We continue in this way and obtain, finally,

$k = n - 1$: For any edge F of P, we find, "above," F a piece of $P + \lambda B$ whose volume is $\lambda^{n-1} v_{n-1}(\Theta_F) v_1(F)$, so, altogether

$$\lambda^{n-1} \sum_{F \text{ an edge}} v_{n-1}(\Theta_F) v_1(F).$$

Now,

$$(2) \qquad V(P + \lambda B) = V(P) + \lambda \sum_{(n-1)-\text{faces } F} v_1(\Theta_F) v_{n-1}(F) + \cdots + \lambda^{n-1} \sum_{1-\text{faces } F} v_{n-1}(\Theta_F) v_1(F) + \lambda^n V(B).$$

By comparing coefficients in (1) and (2), we obtain the lemma. □

Example 1. Let P be a triangular prism with vertices $0, e_1, e_2, e_3, e_1 + e_3, e_2 + e_3$. It is readily seen that $3V(B, P, P) = 3 + \sqrt{2}$ and $3V(B, B, P) = (2 + \frac{1}{2}\sqrt{2})\pi$.

Lemma 4.2 can be generalized such that B is replaced by any convex body Q. We introduce outer angles with respect to Q which depend on the shape of Q and also on the choice of $0 \in Q$. Before we introduce generalized outer angles, we define the following notions.

4.3 Definition. Let Q be a convex body in $\mathbb{R}^n$, $0 \in \operatorname{int} Q$, and let M be a set of vectors in $\mathbb{R}^n$. Then, we call the point set

$$\Gamma_M^Q := \bigcup_{u \in M \setminus \{0\}} (Q \cap H_Q(u))$$

the *M-shadow boundary* of Q. We say Γ_M^Q is *sharp* if $(x + M^\perp) \cap Q$ is a single point for each $x \in \Gamma_M^Q$. If M is a subspace of $\mathbb{R}^n$, we call Γ_M^Q the *shadow boundary of Q in direction $M^\perp$*.

The intuitive meaning of "shadow boundary" is best illustrated for M being an $(n-1)$-dimensional subspace in $\mathbb{R}^n$. Under "illumination" of Q by rays of light in the direction of a spanning vector of the line $M^\perp$, the shadow boundary Γ_M^Q consists of those points on ∂Q in which the rays of light "touch" Q.

Example 2. Let Q be a cube in $\mathbb{R}^3$, M a two-dimensional subspace of $\mathbb{R}^3$. In general, Γ_M^Q is a hexagon on ∂Q (1234561 in Figure 12). If we choose P to be the octahedron Q^* and $F = [a, b]$ is an edge of P (compare Figure 12), then, $\Gamma_{N(F)}^Q = F^*$ (polar face of F on Q).

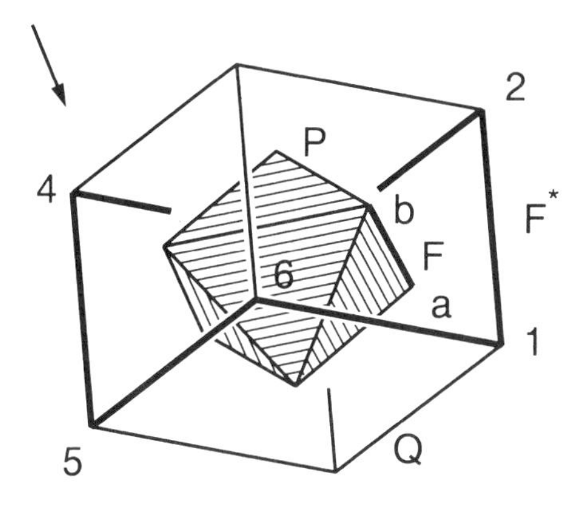

FIGURE 12.

Example 3. If $Q = B$ is the unit ball and U is an r-dimensional subspace of $\mathbb{R}^n$, then, Γ_U^Q is an r-dimensional great sphere on ∂B. For a k-dimensional face F of any polytope P, the set $\Gamma_{N(F)}^Q$ is a spherical cell on ∂B of dimension $n - k - 1$, and

$$\operatorname{conv}(\Gamma_{N(F)}^Q \cup \{0\}) = \Theta_F.$$

Example 4. Let $Q = P$ be an n-polytope, and let M be a hyperplane. Then, Γ_M^Q is sharp if M is not perpendicular to the affine hull of an at least one-dimensional face of Q.

4.4 Definition. Let U_F denote the linear subspace of $\mathbb{R}^n$ which is a translate of aff F (F some polytope). We say two polytopes P, Q are in *skew position* if, for any pair $F := P \cap H_P(u)$, $G := Q \cap H_Q(u)$ of faces defined by the same outer normal u, we have

$$U_F \cap U_G = \{0\}.$$

Clearly,

4.5 Lemma. *If P and Q are in skew position, then, for any proper face F of P the $N(F)$-shadow boundary $\Gamma_{N(F)}^Q$ of Q is sharp ($N(F)$ the cone of normals; see I, Definition 4.12).*

Now we show the following lemma:

4.6 Lemma. *Let P, Q be n-polytopes in $\mathbb{R}^n$. Then, for any $\varepsilon > 0$, there exists a polytope Q' such that*
(1) *$d(Q, Q') < \varepsilon$, and*

(2) *P and Q' are in skew position.*

PROOF. Let $G_1, \dots, G_r$ be those faces of Q for which faces $F_1, \dots, F_r$ of P exist, respectively, such that $U_{F_i} \cap U_{G_i} \neq \{0\}, i = 1, \dots, r$. Up to renumbering, let $\dim G_1 = \cdots = \dim G_{r_1} =: k_1, \dots, \dim G_{r_s} = \cdots = \dim G_r =: k_s$ and $k_1 > \cdots > k_s \geq 1$. From each G_i, we choose a point

$$q_i \in \operatorname{relint} G_i, \quad i = 1, \dots, r.$$

Successively, we pull up q_i to a point q_i' in the sense of III, Definition 2.3, where $\|q_i' - q_i\| < \varepsilon$ can be assumed, $i = 1, \dots, r$.

First, we pull up the q_i with $\dim G_i = k_1$, then those q_j with $\dim G_j = k_2$, and so on. Since $k_1 > \cdots > k_s$, in the ith pulling up the faces $G_{i+1}, \dots, G_r$ are not affected. So, as a result of the r pulling ups, we obtain a polytope Q' for which $d(Q, Q') < \varepsilon$.

If G_j is contained in a nonsharp $N(F)$-shadow boundary $\Gamma^Q_{N(F)}$ ($F = F_j$ or some other face F of P) such that a translate of G_j is contained in U_F, then, $\Gamma^{Q'}_{N(F)}$ no longer contains a point of G_j, and, for $q_j \in \operatorname{relint} G_j$, the vertex q_j' of $\hat{\Theta}^{Q'}_F$ satisfies $(U_F + q_j') \cap Q' = \{q_j'\}$. This readily implies (2). □

4.7 Lemma. *If we set, for the points q_i' in the proof of Lemma 4.6,*

$$q_i(t) := q_i + t(q_i' - q_i), \qquad 0 \leq t \leq 1, \quad i = 1, \dots, r,$$

then, for $Q(t) := \operatorname{conv}(Q \cup \{q_1(t), \dots, q_r(t)\})$, $\mathcal{P} := \{q_1, \dots, q_r\}$, the $N(F)$-shadow boundary $\Gamma^{Q(t)}_{N(F)}$ is sharp, provided $0 < t \leq 1$, and the set

$$\hat{\Theta}^{Q,\mathcal{P}}_F := \lim_{t \to 0} \Gamma^{Q(t)}_{N(F)} \subset \Gamma^Q_{N(F)}$$

is a topological ball of dimension, at most, $n - \dim F - 1$.

PROOF. Each $t\varepsilon$ with $0 < t \leq 1$ can be chosen instead of ε so as to provide a sharp shadow boundary of the "pulled up" polytope $Q(t)$. Each two such shadow boundaries represent subcomplexes of $\mathcal{B}(Q(t))$ which are isomorphic. Since the vertices of these subcomplexes converge to distinct points as $t \to 0$, the lemma follows. □

4.8 Definition. We call a set $\mathcal{P} := \{q_1, \dots, q_r\}$ as chosen in Lemma 4.7 a *pulling set* and say $\hat{\Theta}^{Q,\mathcal{P}}_F$ is a *local $\mathcal{P}$-shadow boundary* or *local shadow boundary* $\hat{\Theta}^Q_F$ if it does not depend on $\mathcal{P}$. We set

$$\bar{\Theta}^{Q,\mathcal{P}}_F := \{sy \mid 0 \leq s \leq 1, \ y \in \hat{\Theta}^{Q,\mathcal{P}}_F\}.$$

Furthermore, let E_F be the linear space which is an orthogonal complement of aff F, and let π_F be the orthogonal projection onto E_F. Then,

$$\Theta^{Q,\mathcal{P}}_F := \pi_F(\bar{\Theta}^{Q,\mathcal{P}}_F)$$

is called the *outer $(Q, \mathcal{P})$-angle* of P in F (or in a point of relint F). We set $\Theta^Q_F := \Theta^{Q,\mathcal{P}}_F$ if it does not depend on $\mathcal{P}$ and say Θ^Q_F is the *outer Q-angle* of P in F.

Figure 13 provides an illustration for a set $\bar{\Theta}_F^{Q,\mathcal{P}}$ and its projection to $\Theta_F^{Q,\mathcal{P}}$. E_F is shown in a translated position (so it can be better visualized). Like a "Chinese fan", $\bar{\Theta}_F^{Q,\mathcal{P}}$ is composed of triangles with one side on $\hat{\Theta}_F^{Q,\mathcal{P}}$ which is a polygonal arc from 1 to 2. If P and Q are in a skew position, $\hat{\Theta}_F^{Q,\mathcal{P}} = \hat{\Theta}_F^{Q}$ and $\Theta_F^{Q,\mathcal{P}} = \Theta_F^{Q}$ do not depend on the choice of $\mathcal{P}$.

Clearly, $\Theta_F^{Q,\mathcal{P}}$ has finite k-dimensional volume $v_k(\Theta_F^{Q,\mathcal{P}})$ for $k = n - \dim F$.

4.9 Lemma. *Let P, Q be n-polytopes. Then,*

$$(3)\qquad \binom{n}{k} V(\underbrace{Q,\ldots,Q}_{k},\underbrace{P,\ldots,P}_{n-k}) = \sum_{(n-k)-\text{faces } F \text{ of } P} v_k(\Theta_F^{Q,\mathcal{P}})v_{n-k}(F).$$

PROOF. As in the case $Q = B$ (Lemma 4.2), we consider a split of $P + \lambda Q$ into pieces "above" the faces of the "inner" part P of $P + \lambda Q, 0 < \lambda \leq 1$. However, in this case, we cannot use the nearest point map for describing these pieces. Instead, we observe that each of them is bounded by two translates of $\Theta_F^{\lambda Q,\mathcal{P}}$ and translates of F (compare Figure 13). For the volume of the piece, we obtain

$$v_k(\Theta_F^{\lambda Q,\mathcal{P}})v_{n-k}(F) = \lambda^k v_k(\Theta_F^{Q,\mathcal{P}})v_{n-k}(F).$$

Now we apply the same arguments as in the proof of Lemma 4.2 and obtain Lemma 4.9. □

Example 5. Let P be a unit cube and εP^* an octahedron (P^* polar to P, compare Example 2), $\varepsilon > 0$. As an illustration, replace each circular arc in Figure 11 by a line segment with the same endpoints. Now, we readily obtain

$$\binom{3}{1} V(\varepsilon P^*, \varepsilon P^*, P) = 12 \cdot \frac{1}{2}\varepsilon^2 = 6\varepsilon^2,$$

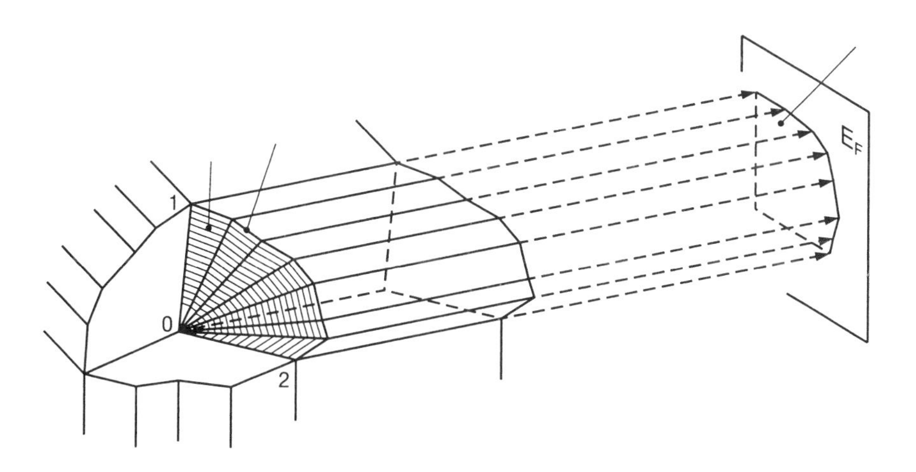

FIGURE 13.

hence, $V(P^*, P^*, P) = 2$.

Lemma 4.9 is the starting point for proving the following three theorems.

4.10 Theorem. *Let K be any convex body, and let $P_1, \ldots, P_{n-1}$ be polytopes. Then, ($h_K(\cdot)$ support function, compare* I, *5.1)*

$$nV(K, P_1, \ldots, P_{n-1}) = \sum_{\|u\|=1} h_K(u) v(F_1, \ldots, F_{n-1}) \tag{4}$$

where $F_i := P_i \cap H_{P_i}(u)$, u an outer facet normal of $P = P_1 + \cdots + P_{n-1}$, $F = F_1 + \cdots + F_{n-1}$ a facet of P, and $v(\cdot, \ldots, \cdot)$ denotes $(n-1)$-dimensional mixed volume.

PROOF. We set $P := \lambda_1 P_1 + \cdots + \lambda_{n-1} P_{n-1}$, $\lambda_i \geq 0$, $i = 1, \ldots, n-1$. For a given $\varepsilon > 0$, we choose a polytope Q such that $d(K, Q) < \varepsilon$ and P, Q are in skew position (Lemma 4.5). For any facet F of P, the outer Q-angle Θ_F^Q is a line segment perpendicular to aff F and has length $h_Q(u)$ if u is the outer unit normal of F. If we apply Theorem 3.2 to both sides of (3) in Lemma 4.9 and compare coefficients, we obtain (4) in the case $K = Q$. Since mixed volumes and support functions depend continuously on the respective convex bodies, the theorem follows. □

4.11 Theorem. *Let $Q, P_1, \ldots, P_{n-k}$ be polytopes, and let $\mathcal{P}$ be a pulling set for $P_1 + \cdots + P_{n-k}$. Then,*

$$\binom{n}{k} V(\underbrace{Q, \ldots, Q}_{k}, P_1, \ldots, P_{n-k}) = \sum_F v_k(\Theta_F^{Q,\mathcal{P}}) v_{n-k}(F_1, \ldots, F_{n-k}) \tag{5}$$

where $F = F_1 + \cdots + F_{n-k}$ is an arbitrary $(n-k)$-face of $P = P_1 + \cdots + P_{n-k}$, F_i a face of P_i, $i = 1, \ldots, n-k$, and $v_{n-k}(\cdot, \ldots, \cdot)$ the $(n-k)$-dimensional mixed volume.

PROOF. We compare coefficients after replacing P in (3) by $\lambda_1 P_1 + \cdots + \lambda_{n-k} P_{n-k}$, and apply continuity arguments. □

Remark. Theorem 4.11 may be extended to convex bodies K instead of polytopes Q after Θ_F^K has been defined by a limit. We do not carry this out.

4.12 Theorem. *Let $K, K', K_1, \ldots, K_{n-1}$ be convex bodies, and let $K \subset K'$. Then,*

$$V(K, K_1, \ldots, K_{n-1}) \leq V(K', K_1, \ldots, K_{n-1}). \tag{6}$$

PROOF. Given any $\varepsilon > 0$, we choose polytopes $P_1, \ldots, P_{n-1}$ such that $d(K_i, P_i) < \varepsilon, i = 1, \ldots, n-1$.

Let $0 \in K \subset K'$. Then, $h_K(u) \leq h_{K'}(u)$ for all $u \in \mathbb{R}^n$. Hence, (6) is implied by (4) in the case of polytopes $P_1, \ldots, P_{n-1}$. By continuity arguments, Theorem 4.12 follows. □

4.13 Theorem. *Let $K_1, \dots, K_n$ be convex bodies. Then,*

(a) $V(K_1, \dots, K_n) \geq 0$,

(b) $V(K_1, \dots, K_n) > 0$ *if and only if each K_i contains a line segment $I_i = [a_i, b_i]$ such that $b_1 - a_1, \dots, b_n - a_n$ are linearly independent.*

PROOF.

(a) If we replace each K_i by one of its points, p_i, $i = 1, \dots, n$, and apply Theorem 4.12, we obtain (a) from $V(p_1, \dots, p_n) = 0$ (which follows, for example, by Theorem 3.7).

(b) From Theorem 3.7, we obtain

$$n!V(I_1, \dots, I_n) = V(I_1 + \cdots + I_n) = |\det(b_1 - a_1, \dots, b_n - a_n)|,$$

and by (6), we know $V(K_1, \dots, K_n) \geq V(I_1, \dots, I_n)$. So, $|\det(b_1 - a_1, \dots, b_n - a_n)| > 0$ implies $V(K_1, \dots, K_n) > 0$.

Conversely, let $V(K_1, \dots, K_n) > 0$, and suppose $|\det(b_1 - a_1, \dots, b_n - a_n)| = 0$ for arbitrary $a_i, b_i \in K_i$, $i = 1, \dots, n$. Then, the linear hull of $(K_1 - K_1) \cup \cdots \cup (K_n - K_n)$ has dimension, at most, $n - 1$, and, up to translations, $K_i \subset H$, $i = 1, \dots, n$. Then, $K_{i_1} + \cdots + K_{i_r} \subset H$ and $V(K_{i_1} + \cdots + K_{i_r}) = 0$ for arbitrary index sets $\{i_1, \dots, i_r\} \subset \{1, \dots, n\}$. By Theorem 3.7, this implies $V(K_1, \dots, K_n) = 0$, a contradiction.

□

Finally, we prove a property of two-dimensional mixed volume needed in the following section.

4.14 Theorem. *Let A, B be planar convex bodies, and suppose there exist parallel lines $g \neq h$ which both support A and B. Then,*

$$2V(A, B) \geq V(A) + V(B). \tag{7}$$

PROOF. We can assume A, B to be strictly combinatorially isomorphic 2-polytopes, since a brief calculation shows that the inequality (7) is true if and only if it is true for $A_\lambda := \lambda A + (1 - \lambda)B$, $B_\mu := \mu A + (1 - \mu)B$, $0 < \lambda < \mu < 1$. ($A = [1, 2, 3, 4, 5, 6, a]$, $B = [1', 2', 3', 4', 5', 6', b]$ in Figure 14.)

Consider g, h to be "vertical" lines (Figure 14), and let κ_A, κ_B be the "upper" polygonal arcs on $\partial A, \partial B$, respectively, having endpoints on g and h. By a translation with traces g, h we move B so that κ_B is "below" κ_A. Replacing A, B by A_λ, B_μ, respectively, we can arrange, for sufficiently small $\mu - \lambda > 0$, that the following is true:

(c) Let a, b be vertices of κ_A, κ_B, respectively, $a, b \notin g \cup h$, which correspond to each other under strict combinatorial isomorphism of A_λ, B_μ. Also let g', h' be the lines through a, b, respectively, parallel to g, h. Then, the quadrangle bounded by $g', h', \kappa_{A_\lambda}$ and κ_{B_μ} is contained in A_λ (compare Figure 14).

Again we write A, B instead of A_λ, B_μ, respectively. From

$$V(A + B) = 2V(A, B) + V(A) + V(B),$$

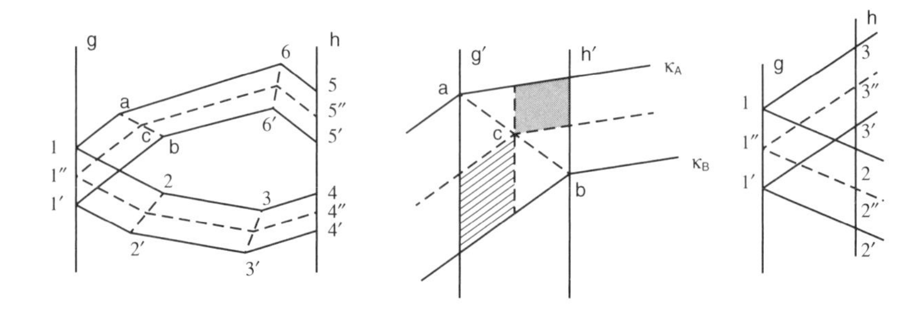

FIGURE 14a,b,c.

we find that

$$2V(A, B) - V(A) - V(B) = V(A + B) - 2V(A) - 2V(B).$$

It will be more convenient to consider

$$V(A, B) - \tfrac{1}{2}V(A) - \tfrac{1}{2}V(B) = 2V(\tfrac{1}{2}(A + B)) - V(A) - V(B) \tag{8}$$

since $\kappa_{\frac{1}{2}(A+B)}$ has vertices $i'' = \frac{1}{2}(i + i')$, $i = 1, \dots, 6$ (see Figure 14a,b).

If we consider the triangles (or trapezoids) of A, B, $\frac{1}{2}(A + B)$ as illustrated in Figure 14c, we see that

$$2V(\Delta(\tfrac{1}{2}(A + B))) - V(\Delta(A)) - V(\Delta(B)) = 0.$$

In Figure 14b, an elementary argument shows that the volume of the hatched parallelogram is larger than that of the dotted parallelogram (or equal if they are degenerate). By a reflection in a line perpendicular to g, h and interchanging A and B, we obtain an analogous statement for a, b on the "lower" parts of ∂A, ∂B, respectively. If we sum the differences of such volumes for all vertices of A, B, we find that the right hand side of (8) is nonnegative. This proves the theorem. □

Exercises

1. Let $p_1, p_2, \dots$ be a sequence of points in $\mathbb{R}^n$ converging to a point p, such that all points of the sequence are vertices of $P := \operatorname{conv}\{p_1, p_2, \dots\} \cup \{p\}$. Then, we call P a *pseudopolytope*. Find a pseudopolytope P in $\mathbb{R}^3$ which has a sharp shadow boundary of infinite length.
2. Let $S := \operatorname{conv}\{0, e_1, e_2, e_3\}$; e_1, e_2, e_3 be the canonical basis of $\mathbb{R}^3$. Calculate $V(S, S, -S)$ with the aid of Lemma 4.11.
3. If K, L are convex bodies and $I = [a, b]$ a line segment in R^3, then, for the orthogonal projection π onto $(\operatorname{aff} I)^{\perp}$,

$$V(K, L, I) = v_2(\pi(K), \pi(L)) \cdot \|a - b\|.$$

4. In Theorem 4.14, let A and B be polytopes. Then, equality in (7) holds if and only if A, B are obtained, one from the other, by "telescopic extension in direction g", that is, by adding a line segment parallel to g (or a point).

5 Alexandrov–Fenchel's inequality

We now state one of the fundamental theorems in the classical theory of convex bodies. For a proof, we refer to Schneider [1993].

5.1 Theorem (Alexandrov–Fenchel's inequality). *Let $K, L, K_1, \ldots, K_{n-2}$ be convex bodies in $\mathbb{R}^n$. Then,*

$$\text{(AF)} \qquad \begin{aligned} V(K, L, K_1, \ldots, K_{n-2})^2 \geq & V(K, K, K_1, \ldots, K_{n-2}) \\ & \times V(L, L, K_1, \ldots, K_{n-2}). \end{aligned}$$

Let $\mathcal{C} := \{K_1, \ldots, K_{n-2}\}$, so that $V(K, L, \mathcal{C}) := V(K, L, K_1, \ldots, K_{n-2})$. We discuss some useful variations of Theorem 5.1.

5.2 Theorem. *For $K_\lambda := \lambda K + (1-\lambda)L$, $L_\mu := \mu K + (1-\mu)L$, $0 \leq \lambda \leq 1$, $0 \leq \mu \leq 1$,*

$$\text{(1)} \quad \begin{aligned} & V(K_\lambda, L_\mu, \mathcal{C})^2 - V(K_\lambda, K_\lambda, \mathcal{C})V(L_\mu, L_\mu, \mathcal{C}) \\ & \qquad = (\lambda - \mu)^2[V(K, L, \mathcal{C})^2 - V(K, K, \mathcal{C})V(L, L, \mathcal{C})], \end{aligned}$$

so that (AF) *is true for K, L if and only if it is true for K_λ, L_μ with $\lambda \neq \mu$.*

PROOF. We apply the linearity rules, Lemma 3.6 and Lemma 3.4, and verify (1) by elementary calculation. □

We notice that (AF) is trivially true if $V(K, K, \mathcal{C}) = 0$ or $V(L, L, \mathcal{C}) = 0$. So we may assume $V(K, K, \mathcal{C}) > 0$ and $V(L, L, \mathcal{C}) > 0$. Up to a homothety of L, we can arrange the side condition

$$\text{(2)} \qquad V(K, K, \mathcal{C}) = V(L, L, \mathcal{C})$$

so that (AF) becomes equivalent to

$$\text{(3)} \qquad V(K, L, \mathcal{C}) \geq V(K, K, \mathcal{C}).$$

5.3 Theorem. *Let $K, L, K_1, \ldots, K_{n-2}$ be polytopes, $\mathcal{C} = \{K_1, \ldots, K_{n-2}\}$, and let $V(K, K, \mathcal{C}) > 0$, $V(L, L, \mathcal{C}) > 0$. We choose pulling sets $\mathcal{P}_K, \mathcal{P}_L, \mathcal{P}_{K+L}$ for $K, L, K+L$, respectively, such that the outer angles $\Theta_F^{K,\mathcal{P}_K}$, $\Theta_F^{L,\mathcal{P}_L}$, $\Theta_F^{K+L,\mathcal{P}_{K+L}}$ are defined for each $(n-2)$-face F of $P := K_1 + \cdots + K_{n-2}$ (Definition 4.8). Then,* (2) *is equivalent to*

$$\text{(4)} \qquad \sum \left[v_2(\Theta_F^{K,\mathcal{P}_K}) - v_2(\Theta_F^{L,\mathcal{P}_L}) \right] v_{n-2}(F_1, \ldots, F_{n-2}) = 0$$

and (3) *is true if and only if*

$$\sum \left[v_2(\Theta_F^{K+L,\mathcal{P}_{K+L}}) - 4v_2(\Theta_F^{K,\mathcal{P}_K}) \right] v_{n-2}(F_1, \ldots, F_{n-2}) \geq 0 \tag{5}$$

is satisfied.

PROOF. By Theorem 4.3, we may replace (2) by

$$\sum \left[v_2(\Theta_F^{K,\mathcal{P}_K}) - v_2(\Theta_F^{L,\mathcal{P}_L}) \right] v_{n-2}(F_1, \ldots, F_{n-2}) = 0. \tag{2'}$$

Furthermore, by using (2), we obtain

$$\binom{n}{2} V(K+L, K+L, \mathcal{C}) = \sum v_2(\Theta_F^{K+L,\mathcal{P}_{K+L}})\, v_{n-2}(F_1, \ldots, F_{n-2})$$
$$= 2\binom{n}{2}[V(K, L, \mathcal{C}) + V(K, K, \mathcal{C})],$$

so that, if (2), or equivalently (2′), is valid, (3) becomes equivalent to

$$\begin{aligned} V(K,L,\mathcal{C}) - V(K,K,\mathcal{C}) &= \frac{1}{2}\binom{n}{2}^{-1} \sum \left[v_2(\Theta_F^{K+L,\mathcal{P}_{K+L}}) - 4v_2(\Theta_F^{K,\mathcal{P}_K}) \right] \\ &\quad \times v_{n-2}(F_1, \ldots, F_{n-2}) \geq 0. \end{aligned} \tag{3'}$$

Each of the sums is carried out for all $(n-2)$-faces F of $P = K_1 + \cdots + K_{n-2}$. □

5.4 Corollary. *If K, L in Theorem 5.3 are replaced, according to Theorem 5.2, by appropriate strictly combinatorially isomorphic polytopes (denoted again K, L) and if $P = K_1 + \cdots + K_{n-2}$ is also strictly combinatorially isomorphic to K, L, then, we may choose $\Theta_F^{K,\mathcal{P}_K}$, $\Theta_F^{L,\mathcal{P}_L}$, $\Theta_F^{K+L,\mathcal{P}_{K+L}}$ such that*

$$\Theta_F^{K+L,\mathcal{P}_{K+L}} = \Theta_F^{K,\mathcal{P}_K} + \Theta_F^{L,\mathcal{P}_L} \tag{6}$$

and (5) *in Theorem* 5.3 *becomes equivalent to*

$$\sum \left[v_2(\Theta_F^{K,\mathcal{P}_K}, \Theta_F^{L,\mathcal{P}_L}) - v_2(\Theta_F^{K,\mathcal{P}_K}) \right] v_{n-2}(F_1, \ldots, F_{n-2}) \geq 0 \tag{5'}$$

where the sum is taken for all $(n-2)$-faces of P.

PROOF. In Theorem 5.2, if the term $|\lambda - \mu|$ is chosen small enough and $0 \in (\operatorname{int} K) \cap (\operatorname{int} L)$, then, given any facet G of L, we may find a ray ϱ_G emanating from 0 which intersects relint G, relint G', and relint G'' where G', G'' are the facets of K, $K+L$, respectively, which correspond to G under strict combinatorial isomorphisms. Let $\mathcal{P}_L$ consist of all $\varrho_G \cap G$ and an arbitrarily chosen point on each $(n-2)$-face of K, also $\mathcal{P}_K$ consist of all $\varrho_G \cap G'$ and points on $(n-2)$-faces. Then, $\mathcal{P}_{K+L}$ may be chosen as $\{a + b \mid a \in \mathcal{P}_K, b \in \mathcal{P}_L\}$, and $\Theta_F^{K,\mathcal{P}_K}$, $\Theta_F^{L,\mathcal{P}_L}$, $\Theta_F^{K+L,\mathcal{P}_{K+L}}$ are quadrangles such that the sides containing 0 as an endpoint have an intersection different from $\{0\}$. Then, (6) is satisfied, and (5′) follows from Theorem 3.2. □

Remark. For $n = 2$, $V(K, K, \mathcal{C}) = V(K) = V(L) = V(L, L, \mathcal{C})$ in Corollary 0, and (5′) becomes equivalent to Theorem 4.14.

We now turn to the question: When does equality hold in (AF)?

Clearly a sufficient condition is K and L to be homothetic: If $K = a + tL$, $t > 0$, then, $V(K, L, \mathcal{C}) = V(a + tL, L, \mathcal{C}) = tV(L, L, \mathcal{C})$, so that equality in (AF) follows. However, homothety of K and L is not necessary.

Example 1. In $\mathbb{R}^3$, let $I_j := [a_j, b_j]$, $j = 1, 2, 3$, be line segments such that $b_1 - a_1, b_2 - a_2, b_3 - a_3$ are linearly independent. Then, for $K := I_1 + I_3$, $L := I_2 + I_3$, $K_1 := I_1 + I_2$, an easy calculation shows that equality holds in (AF).

Example 2. Let K_1 be a 3-simplex in $\mathbb{R}^3$, and let K, L be obtained from K_1 by chopping off the vertices but leaving at least one point of each edge remaining (Figure 15). Then, (2′) and equality in (3′) are trivially true, so that equality in (AF) follows.

Remark. The problem to find necessary and sufficient conditions for $K, L, \mathcal{C}$ under which equality in (AF) holds is currently unsolved (see the Appendix to this section).

We present a special result:

5.5 Theorem. *Let $K_1, \ldots, K_{n-2}$ be strictly combinatorially isomorphic n-polytopes, and let K, L be n-polytopes such that $K + L$ is in skew position to a K_{j_0} (and, hence, to each K_j, $j = 1, \ldots, n-2$). Then, equality in* (AF) *holds if and only if, up to a homothety of L, we have*

(S) *(Schneider's condition) If $H_{K_{j_0}}(u)$ is a supporting hyperplane of K_{j_0} such that $F = K_{j_0} \cap H_{K_{j_0}}(u)$ is an $(n-2)$-face of K_{j_0}, then,*

$$H_K(u) = H_L(u).$$

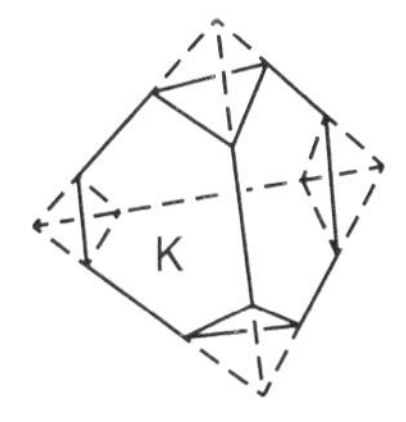

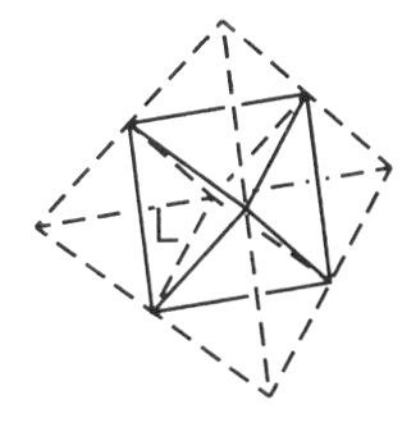

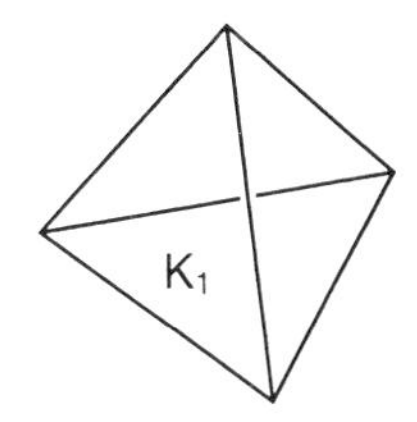

FIGURE 15.

PROOF. We can express (S) in terms of Θ_F^K, Θ_F^L. Consider the set $\hat{\Theta}_F^K$ (see Definition 4.8). By the assumptions of the theorem and by using (1) in Lemma 5.4, we may suppose K, L to be strictly combinatorially isomorphic and both in skew position to P. Then $\hat{\Theta}_F^K$ is a polygonal arc. Both end points of $\hat{\Theta}_F^K$ are also end points of further local shadow boundaries. In fact, there is a one-to-one–correspondence between the set of local shadow boundaries and the set of edges of the dual polytope P^* of P. It provides a homeomorphism between the union of all local shadow boundaries of K and the union of all (closed) edges of P^* (which carries the edge graph of P^*). Since the edge graph of P^* is connected, so is the union of all local shadow boundaries. Hence, from the definition of Θ_F^K, (and, analoguously, Θ_F^L), (S) is equivalent to the following condition:

(S′) For each $(n-2)$-face F of $P = K_1 + \cdots + K_{n-2}$,

$$\Theta_F^K = \Theta_F^L.$$

It even suffices to show the following:

(S″) Let F be an $(n-2)$-face of P, and let a_K, a_L be parallel edges of Θ_F^K, Θ_F^L, respectively, which do not contain 0. Then, the lengths α_K, α_L of a_K, a_L, respectively, are equal.

To deduce (S″) from equality in (AF), we introduce the following operations. Let $G = K \cap H_K(u)$ be a face of K. Also, let H be a hyperplane parallel to $H_K(u)$ which separates vert G from (vert K) \ (vert G), such that vert $G \subset \operatorname{int} H^+$, (vert K) \ (vert G) $\subset H^-$. Let ε be the distance of H and $H_K(u)$. Then, we set

$$c(u, \varepsilon)K := K \cap H^-$$

and call $c(u, \varepsilon)$ a *cut operation* (Figure 16).

Suppose, in (S″),

$$\alpha_K < \alpha_L. \tag{7}$$

Let G_K, G_L be the inverse images of a_K, a_L on the local shadow boundaries $\hat{\Theta}_F^K$, $\hat{\Theta}_F^L$, respectively. Then G_K, G_L are also parallel line segments. Let $G_K = K \cap H_K(u)$, $G_L = L \cap H_L(u)$, $F = P \cap H_P(u)$.

We apply cut operations $c(u, \varepsilon), c(u, \varepsilon')$ to K, L, respectively. As a consequence, trapezoids are cut off from Θ_F^K, Θ_F^L. Also other outer K-angles and L-angles are,

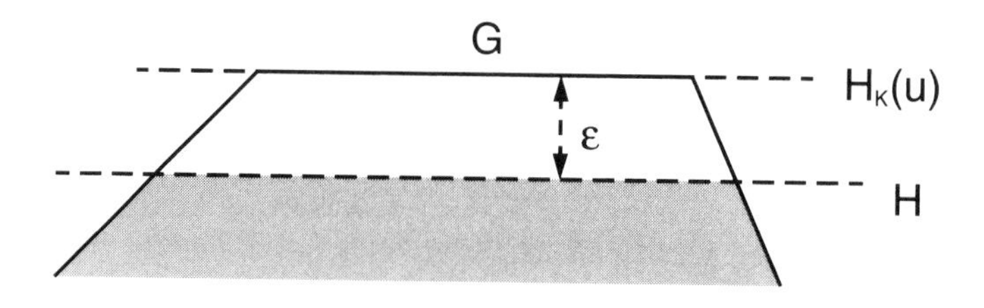

FIGURE 16.

in general, affected. We denote them by $\Theta_F^K =: \Theta_{F(1)}^K, \dots, \Theta_{F(r)}^K$ and $\Theta_F^L =: \Theta_{F(1)}^L, \dots, \Theta_{F(r)}^L$.

From each $\Theta_{F(i)}^K$, an area is cut off which can be split according to Figure 17.

One part is a trapezoid T_K^i with upper base length $c_i \alpha_K$ and altitude $d_i \varepsilon$ for constants c_i, d_i, $0 < c_i \leq 1$, $0 < d_i \leq 1$, $c_1 = d_1 = 1$. The remaining part (shaded in Figure 17) can be split into triangles $t_K^{i,1}, \dots, t_K^{i,s_i}$ (as indicated in Figure 17) with areas $e_{ij} \cdot \varepsilon^2$, e_{ij} constants, $i = 1, \dots, r$; $j = 1, \dots, s_i$.

We proceed analogously for $\Theta_{F(i)}^L$. From the strict combinatorial isomorphism of K, L, we see that the trapezoids T_L^i have upper base lengths $c_i \alpha_L$ and altitudes $d_i \varepsilon'$ for the same constants c_i, d_i as above. Also, the corresponding triangles $t_L^{i,j}$ have areas $e_{ij} \varepsilon'^2$, $i = 1, \dots, r$, $j = 1, \dots, s_i$.

Hence, the change in (2′), as a consequence of $c(u, \varepsilon)$ applied to K and $c(u, \varepsilon')$ applied to L, amounts to

$$(\alpha_K \varepsilon - \alpha_L \varepsilon') \sum_{i=1}^{r} c_i d_i + (\varepsilon^2 - \varepsilon'^2) \sum_{i=1}^{r} \sum_{j=1}^{s_i} e_{ij}. \tag{8}$$

Since $\alpha_K < \alpha_L$, we find $\varepsilon > \varepsilon'$ (both sufficiently small) such that (8) vanishes. Then, (2′) is established for $c(u, \varepsilon)K$, $c(u, \varepsilon')L$ instead of K, L, respectively.

Now we investigate the change in (3′). Since T_K^i and T_L^i are strictly combinatorially isomorphic,

$$T_{K+L}^i = T_K^i + T_L^i.$$

Let the lower base lines of T_K^i, T_L^i have lengths $c_i \alpha_K + \mathring{c}_i \varepsilon$, $c_i \alpha_L + \mathring{c}_i \varepsilon'$, respectively, for appropriate constants $\mathring{c}_i$, $i = 1, \dots, r$. Then, by a brief calculation (Figure 18),

$$\begin{aligned} &v_2(T_{K+L}^i) - 2v_2(T_K^i) - 2v_2(T_L^i) \\ &\quad = 4[v_2(\tfrac{1}{2}(T_K^i + T_L^i)) - \tfrac{1}{2} v_2(T_K^i) - \tfrac{1}{2} v_2(T_L^i)] \\ &\quad = 4c_i d_i (\varepsilon - \varepsilon')(\alpha_L - \alpha_K) - 8\, \mathring{c}_i c_i (\varepsilon - \varepsilon')^2. \end{aligned}$$

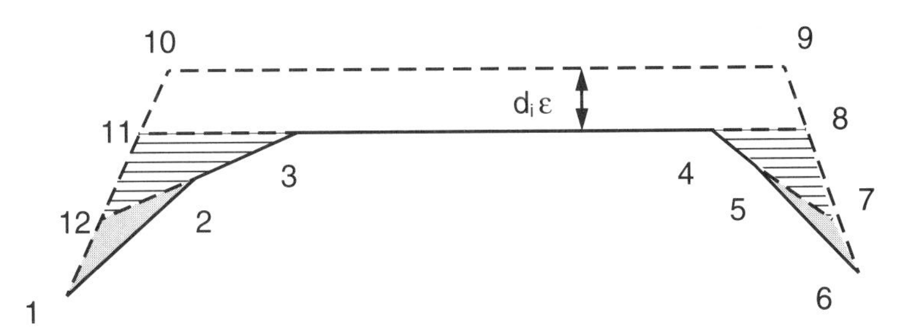

FIGURE 17. $T_K^i = [8, 9, 10, 11]$ $t_K^{i,2} = [1, 2, 12]$ $t_K^{i,4} = [5, 6, 7]$
$t_K^{i,1} = [3, 11, 12]$ $t_K^{i,3} = [4, 7, 8]$

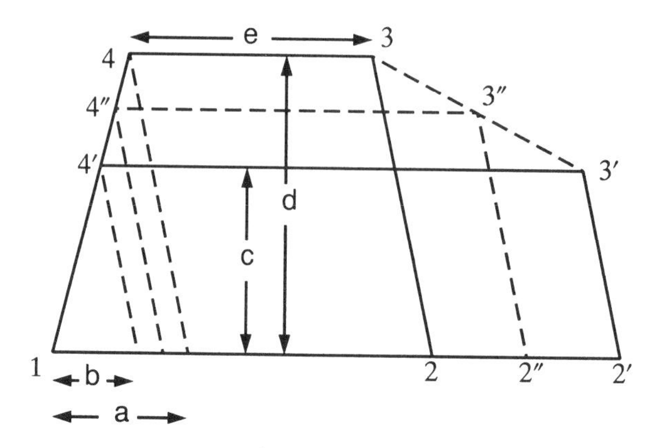

FIGURE 18. $T^i_K = [1, 2, 3, 4]$ $T^i_{K+L} = [1, 2'', 3'', 4'']$ $\overset{\circ}{c}_i \cdot \varepsilon' = b$ $c_i \cdot \varepsilon = d$
$T^i_L = [1, 2', 3', 4']$ $\overset{\circ}{c}_i \cdot \varepsilon = a$ $c_i \cdot \varepsilon' = c$ $d_i \cdot \alpha_k = e$

In the equation above, the quantity $8\, \overset{\circ}{c}_i c_i (\varepsilon - \varepsilon')^2$ stems from the homothetic triangles with base lengths $\overset{\circ}{c}_i \varepsilon$, $\overset{\circ}{c}_i \varepsilon'$, $\frac{1}{2}\, \overset{\circ}{c}_i (\varepsilon + \varepsilon')$ which are parts of T^i_K, T^i_L, T^i_{K+L}, respectively. In the same way, for the triangles $t^{i,j}_K$, $t^{i,j}_L$, $t^{i,j}_{K+L}$, we find that

$$v_2(t^{i,j}_{K+L}) - 2v_2(t^{i,j}_K) - 2v_2(t^{i,j}_L) = 8e_{ij}(\varepsilon - \varepsilon')^2.$$

Now the total change in (3′) as implied by $c(u, \varepsilon)$, $c(u, \varepsilon')$ is

$$(9) \qquad 4(\varepsilon - \varepsilon')(\alpha_L - \alpha_K) \sum_{i=1}^{r} c_i d_i - 8(\varepsilon - \varepsilon')^2 \sum_{i=1}^{r} \Big(\overset{\circ}{c}_i c_i + \sum_{j=1}^{s_i} e_{ij}\Big).$$

For sufficiently small $\varepsilon, \varepsilon'$ ($\varepsilon > \varepsilon'$), the term (9) becomes positive. Since it must be subtracted from (3′), we obtain a contradiction to (AF).

Therefore, $\alpha_K = \alpha_L$, and we obtain (S″) and, hence, (S′), as a consequence of equality in (AF).

If, conversely, (S′) is true, the unions of local shadow boundaries of K and L (see beginning of the proof) are translates of each other, so that, by (2′), (3′) equality in (AF) holds.

This completes the proof of Theorem 5.5. □

Exercises

1. Deduce from (AF) Minkowski's "Inequality of the first kind":

$$V(K, \ldots, K, L) \geq V(K)^{(n-1)/n} V(L)^{1/n}.$$

2. Let I_1, I_2, I_3, I_4 be line segments in $\mathbb{R}^4$ such that the difference vectors of their end points are linearly independent. Show that equality in (AF) holds for

$$K := 2I_1 + 2I_2 + I_3 + I_4, \quad L := I_1 + I_2 + 2I_3 + 2I_4,$$
$$K_1 := I_1 + I_3, \quad K_2 := I_2 + I_4,$$

but (S) is not true.
3. In $\mathbb{R}^3$, let $K := \operatorname{conv}\{e_1, e_2, e_3\}$, $L := \operatorname{conv}\{e_1, e_2, 2e_3\}$, $K_1 := -K$. Show that

$$V(K, L, K_1)^2 > V(K, K, K_1)V(L, L, K_1).$$

4. In $\mathbb{R}^3$, let K_1 be a line segment. Characterize equality in (AF).

6 Ehrhart's Theorem

Let P be a lattice polytope, that is, a polytope in $\mathbb{R}^n$ whose vertices all lie in the (canonical) lattice $\mathbb{Z}^n$. We denote by $G(P)$ the number of lattice points contained in P. A similar theory can be developed for $G(P)$ as for volumes and mixed volumes. Here we restrict ourselves to a special question: How does $G(kP)$ depend on k for natural numbers k? The answer is given by Ehrhart's theorem. We shall need Ehrhart's theorem in the last section of chapter VIII. It turns out to be equivalent to the so-called Riemann–Roch–Hirzebruch theorem for torus-invariant invertible sheaves on smooth, compact, and projective toric varieties.

First we consider a special case:

6.1 Lemma. *Let $S := \operatorname{conv}\{0, a_1, \ldots, a_n\}$ be an n-simplex, where $a_1, \ldots, a_n \in \mathbb{Z}^n$. There exist constants $\beta_1, \ldots, \beta_n \in \mathbb{Z}_{\geq 0}$ such that, for any natural number k, the number of lattice points in kS is*

$$G(kS) = \binom{n+k}{n} + \beta_1\binom{n+k-1}{n} + \cdots + \beta_n\binom{k}{n}.$$

PROOF. Any lattice point $y \in kS$ has a unique representation

$$(1) \qquad y = x + \alpha_1 a_1 + \cdots + \alpha_n a_n, \qquad \alpha_i \in \mathbb{Z}_{\geq 0}, \quad i = 1, \ldots, n,$$

where x lies in the half-open interval (parallelepiped)

$$A := \{t_1 a_1 + \cdots + t_n a_n \mid 0 \leq t_i < 1, i = 1, \ldots, n\}.$$

Let H_i be the hyperplane which contains $ia_1, \ldots, ia_n$, for any fixed $i \in \{0, \ldots, k\}$. We determine the number of lattice points in $H_j \cap (kS)$, $j = 1, \ldots, k$, and those in the "layers" bounded by H_{i-1} and H_i for $i = 1, \ldots, n$. Then, the lemma is established by using (1).

First, let $x = 0$ in (1). Then the combinations (1) which lie in H_j satisfy

$$(2) \qquad \alpha_1 + \cdots + \alpha_n = j, \qquad 0 \leq \alpha_i \leq j, \quad i = 1, \ldots, n.$$

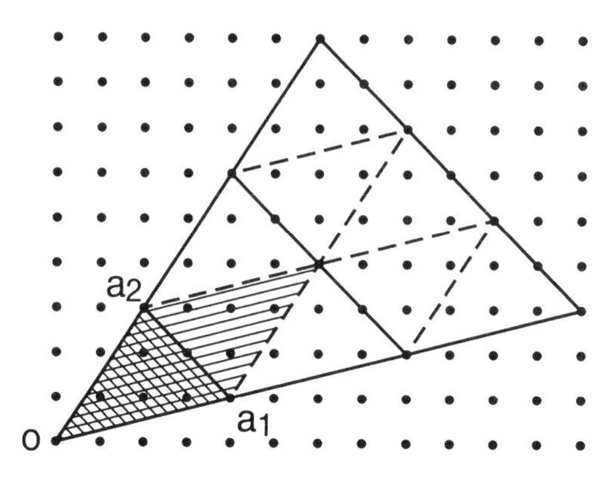

FIGURE 19. $S = [0, a_1, a_2]$, $A = [0, a_1, a_2, a_1 + a_2] \setminus ([a_1, a_1 + a_2] \cup [a_2, a_1 + a_2])$

We claim that there are precisely $\binom{n+j-1}{n-1}$ partitions of the type included in (2). In fact, for $n = 1$, there is only $1 = \binom{j}{0}$ possibility. Suppose that the claim is true for $n - 1$ instead of n. Then, there are $\binom{n+j-\alpha_n-2}{n-2}$ partitions

$$\alpha_1 + \cdots + \alpha_{n-1} = j - \alpha_n$$

of $j - \alpha_n$. By summing, we find that the number of solutions of (2) is

(3)
$$\sum_{\alpha_n=0}^{j} \binom{n+j-\alpha_n-2}{n-2} = \binom{n-2}{n-2} + \cdots + \binom{n+j-3}{n-2} + \binom{n+j-2}{n-2}.$$

By another induction (assuming the claim to hold for $j - 1$ instead of j) and by using the evident formula

$$\binom{p}{q} + \binom{p}{q+1} = \binom{p+1}{q+1}, \qquad (p \leq q),$$

we see that the right hand side of (3) amounts to

$$\binom{n+j-2}{n-1} + \binom{n+j-2}{n-2} = \binom{n+j-1}{n-1},$$

as claimed.

Now, we sum over the j, and use an analogous calculation by which we find the total number of solutions of (1) for $x = 0$,

$$\sum_{j=0}^{k} \binom{n+j-1}{n-1} = \binom{n+k}{n}.$$

If x lies properly between H_0 and H_1 or on H_1, the number of solutions of (1) reduces to $\binom{n+k-1}{n}$. Similarly, we find $\binom{n+k-i}{n}$ possibilities for each x properly between H_{i-1} and H_i or on H_i, $i \in \{1, \ldots, n\}$.

Let β_i denote the number of lattice points in A properly between H_{i-1} and H_i or on H_i, $i \in \{1, \dots, n\}$. In the case $k < n$, $\beta_j = 0$ for $j \in \{k+1, \dots, n\}$. Then, we obtain the formula of the lemma. □

Example 1.. For $a_1 = (4, 1)$, $a_2 = (2, 3)$ we see, with the aid of Figure 19, that $\beta_1 = 5$, $\beta_2 = 4$, and, hence, $G(kS) = \binom{2+k}{2} + 5\binom{1+k}{2} + 4\binom{k}{2} = 5k^2 + 2k + 1$. In particular, setting $k = 3$, gives $G(3S) = 52$.

We call the set $S_0 := S \setminus (\operatorname{conv}\{0, a_1, \dots, a_{n-1}\} \cup \dots \cup \operatorname{conv}\{0, a_2, \dots, a_n\})$ the *pseudo-simplex* associated with S. By $G(M)$, we denote generally the number of lattice points contained in the set M.

6.2 Lemma. *$G(kS_0)$ is a polynomial in $k \in \mathbb{Z}_{\geq 0}$.*

PROOF. We will use induction on the dimension of S_0. For $\dim S_0 = 0$, there is nothing to prove. So let the lemma be true for the pseudosimplices of dimension less than n.

We consider the proper faces $F^{(1)}, \dots, F^{(s)}$ of S which contain 0 and satisfy $0 < \dim F^{(i)} < n, i = 1, \dots, s$. Then, $S \setminus S_0 = \{0\} \cup F_0^{(1)} \cup \dots \cup F_0^{(s)}$ is a disjoint union. By inductive assumption, $G(k(S \setminus S_0)) = 1 + G(kF_0^{(1)}) + \dots + G(kF_0^{(s)})$ is a polynomial in k. Hence, by Lemma 6.1, $G(kS_0) = G(kS) - 1 - G(kF_0^{(1)}) - \dots - G(kF_0^{(s)})$ is also a polynomial in k. □

Theorem (Ehrhart's theorem). *Let P be a lattice polytope in $\mathbb{R}^n$. Then, $G(kP)$ is a polynomial in $k \in \mathbb{Z}_{\geq 0}$.*

PROOF. We can assume 0 to be a vertex of P. First, we decompose all faces of P which do not contain 0 into simplicial complexes whose 0-cells are the given vertices (see III, Theorem 2.6). Then, we join each simplex, thus obtained, to 0 resulting in a decomposition of P into a simplicial complex $\mathcal{C}$ whose 0-cells are the vertices of P. The cells $S^{(1)}, \dots, S^{(r)}$ of $\mathcal{C}$, which contain 0 and are $\neq \{0\}$, have the following property that

$$P = \{0\} \cup S_0^{(1)} \cup \dots \cup S_0^{(t)}$$

is a disjoint union. Now the theorem follows from Lemma 6.2. □

Remark. If $G(kP) = \sum_{i=0}^{n} \gamma_i k^i, \gamma_i \in \mathbb{Z}_{\geq 0}$, is the Ehrhart polynomial according to Theorem 6.3 and if $\overset{\circ}{P} := \operatorname{relint} P$, then, by a proof similar to that of Ehrhart's theorem, one can show *Ehrhart's reciprocity law*:

$$G(k\overset{\circ}{P}) = (-1)^{\dim P} \sum_{i=0}^{n} \gamma_i (-k)^i .$$

Exercises

1. In $\mathbb{R}^3$, let $a_1 := (4, 1, 1)$, $a_2 = (2, 3, 2)$, $a_3 = (3, 2, 1)$. Find $G(kP)$ for $P := \operatorname{conv}\{0, a_1, a_2, a_3\}$. Calculate explicitly $G(2P)$, $G(3P)$, $G(4P)$.
2. Let C be the unit cube (with 2^n lattice points). Verify $G(kC) = (k+1)^n$ by Lemma 6.1.
3. Let $\overset{\circ}{G}(P)$ denote the number of lattice points on the boundary of a lattice polytope P. Find a formula for $\overset{\circ}{G}(kS)$ analogous to that in Lemma 6.1.
4. 4. For the simplex S, as in Lemma 6.1, prove ($V(\cdot)$ denoting volume)

$$n!V(S) = \sum_{j=0}^{n} \binom{n}{j} (-1)^j G((n-j)S).$$

7 Zonotopes and arrangements of hyperplanes

7.1 Definition. A Minkowski sum

$$Z := S_1 + \cdots + S_r$$

of line segments $S_1, \ldots, S_r$ in $\mathbb{R}^n$ is called a *zonotope*. If each n of the line segments point in linearly independent directions we say the zonotope is *independent*. .

The name "zonotope" is motivated by the "zones" obtained in the case $n \geq 3$ and independent Z as the union of all line segments in the boundary of Z which are parallel to one S_i, $i \in \{1, \ldots, r\}$ (Figure 20a). If Z is not independent, one can choose "zones" as indicated in Figure 20b.

There are many interesting features of zonotopes and applications to other parts of geometry. We select only a few which we shall need in Chapter VII.

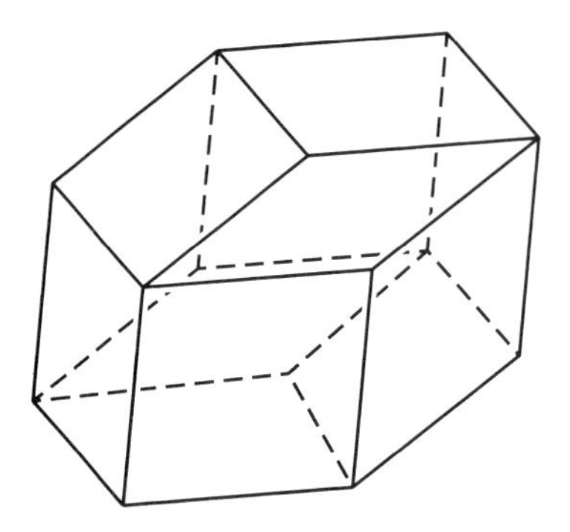

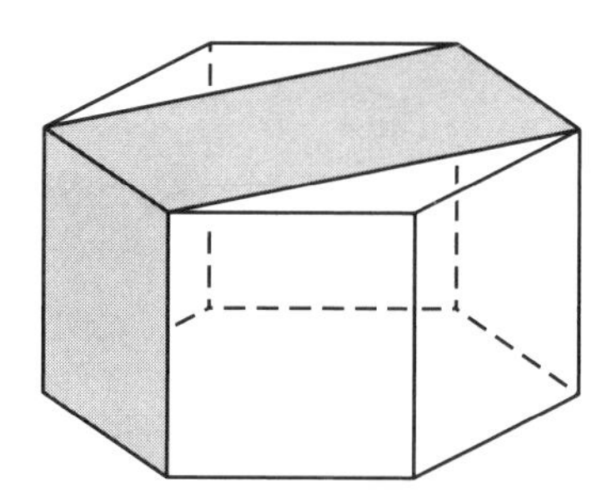

FIGURE 20a,b.

7.2 Theorem. *For $n = 2$, a polytope Z is a zonotope if and only if it is centrally symmetric.*

PROOF. Let Z be a zonotope. Up to translations, we can assume that all S_i have midpoints in 0, $i = 1, \ldots, r$. So, Z is readily seen to be centrally symmetric.

Conversely, if Z is a planar and centrally symmetric polytope, say, a $2k$-gon $Z = [a_1, \ldots, a_{2k}]$, consider the parallelogram $[a_1, a_2, a_{k+1}, a_{k+2}]$ and the polytopes $Z_1 := [a_1, a_{k+2}, \ldots, a_{2k}, a_1]$, $Z_2 := [a_2, a_3, \ldots, a_{k+1}, a_2]$ (Figure 21).

Then the $(2k-2)$-gon $Z_1 \cup (Z_2 + a_1 - a_2) =: Z_0$ is again a centrally symmetric polytope and

$$Z = Z_0 + [a_1, a_2].$$

By induction, we can assume Z_0 to be a zonotope. Therefore Z is also a zonotope. □

Remark. Centrally symmetric polytopes in dimensions higher than two need not be zonotopes as is seen from a regular octahedron in $\mathbb{R}^3$ which cannot be a zonotope because of the following theorem:

7.3 Theorem. *Each face of a zonotope Z is again a zonotope.*

PROOF. Let F be a facet of Z and H be a supporting hyperplane of Z for which $F = Z \cap H$ is true. We may assume $0 \in \operatorname{vert} F$ and $S_i = [0, b_i]$, $i = 1, \ldots, r$. $[0, b_i]$ is either contained in H or satisfies $[0, b_i] \cap H = \{0\}$, say, $[0, b_i] \subset H$ for $i = 1, \ldots, s$. Then $F = S_1 + \cdots + S_s$ readily follows. □

7.4 Definition. A finite set of $(n-1)$-dimensional subspaces of $\mathbb{R}^n$ is called a *hyperplane arrangement* of $\mathbb{R}^n$. The cones into which $\mathbb{R}^n$ is split by the hyperplanes are said to be the *cells* of the hyperplane arrangement.

Remark. It is often useful to consider the projective ($(n-2)$-dimensional) hyperplanes defined by the hyperplanes of the hyperplane arrangement. For example,

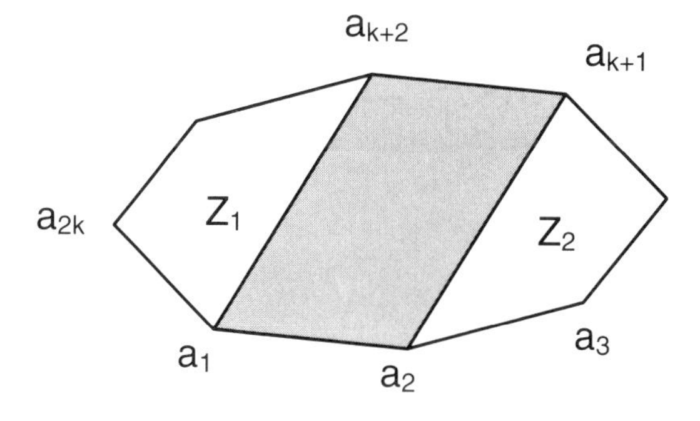

FIGURE 21.

for $n = 3$, plane arrangement can well be visualized by line arrangements in the real projective plane.

The following theorem refers to complete fans as defined in III, 1.

7.5 Theorem. *Let $Z = S_1 + \cdots + S_r$ be an n-dimensional zonotope in $\mathbb{R}^n$, and let 0 be the center of symmetry of Z. Then, the complete fan Σ defined by projecting the faces of Z^* consists of the cells of a hyperplane arrangement.*

Any hyperplane arrangement containing at least n linearly independent hyperplanes is obtained in this way.

PROOF. Let H_i be the $(n-1)$-subspace of $\mathbb{R}^n$ orthogonal to aff $S_i, i = 1, \ldots, r$. Each face F of Z which contains a translate of S_i—we call it an S_i-face—has an affine hull orthogonal to H_i; hence, F^* is contained in H_i. If Z is projected orthogonally onto H_i, we obtain a polytope Z_1 whose faces are projections of S_i-faces. The polar faces of Z_1 relative to H_i are the F^*, F an S_i-face. Therefore, H_i carries a complete fan contained in Σ.

The one-dimensional space g spanned by a vertex v of Z^* is the intersection of all hyperplanes H_j which are spanned by $(n-2)$-faces of Z^* containing v. g splits into two rays which are the one-dimensional cones, and pos $F^* = \text{pos}\{v_1, \ldots, v_p\}$ if $\{v_1, \ldots, v_p\} = \text{vert } F^*$.

Conversely, if a hyperplane arrangement $\{H_1, \ldots, H_r\}$ is given such that not all H_i are linearly dependent, $i = 1, \ldots, r$, then, $\mathbb{R}^n \setminus (H_1 \cup \cdots \cup H_r)$ consists of open cones whose closures σ are cones with apex 0. If a_i is a normal vector of H_i, $i = 1, \ldots, r$, then, for $S_i := [-a_i, a_i]$ we obtain a zonotope $Z = S_1 + \cdots + S_r$, and Z^* spans a complete fan which is readily seen to have the above cones σ as n-cones. □

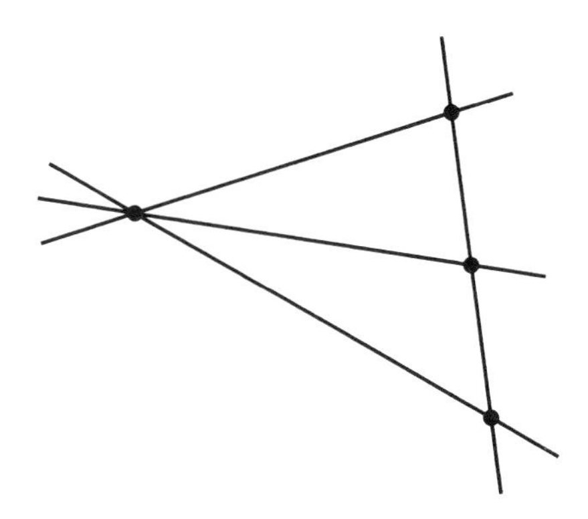

FIGURE 22.

Example 1. If Z is a hexagonal prism with centrally symmetric basis, hence, Z^* a hexagonal bipyramid, the projective line arrangement corresponding to it is of the type illustrated in Figure 22.

Exercises

1. A zonotope Z is independent if and only if aff S_i and aff S_j are not parallel for $i \neq j$, and each proper face of Z is a spar (affine image of a cube).
2. Let P be a 3-polytope whose facets are centrally symmetric. Then, P is a zonotope. (Hint: Show that any edge of P determines a "zone"; apply induction on r).
3. Find simplicial arrangements of 6,7,8 or 9 lines in the real projective plane which do not correspond to bipyramids.
4. Determine the f-vector of an arbitrary three-dimensional, independent zonotope.

V

Lattice polytopes and fans

1. Lattice cones

In I, 1, we introduced the (polyhedral) cone σ as the positive hull of finitely many vectors $x_1, \ldots, x_k$.

$$\sigma = \operatorname{pos}\{x_1, \ldots, x_k\} = \mathbb{R}_{\geq 0}\, x_1 + \cdots + \mathbb{R}_{\geq 0}\, x_k \tag{1}$$

It has an apex if there exists a supporting hyperplane H such that $H \cap \sigma = \{0\}$ (compare I, Definition 1.9). That property has an elementary geometric interpretation:

Remark. Let H and H' be parallel hyperplanes in $\mathbb{R}^n$ with $0 \in H$. Let σ be a polyhedral cone with $\sigma \cap H' \neq \emptyset$. Then, $H \cap \sigma = \{0\}$ if and only if $H' \cap \sigma$ is bounded.

PROOF. Assume there is a $0 \neq x \in H \cap \sigma$. Then, $\operatorname{pos} x \subset H \cap \sigma$ is unbounded. If $H \neq H'$, then, choose an $a \in H' \cap \sigma$; then $a + \operatorname{pos} x \subset H' \cap \sigma$ is unbounded, a contradiction.

Now assume that there is an unbounded sequence $a_1, a_2, \ldots$ in $H' \cap \sigma$. The sequence $\|a_1\|^{-1}a_1, \|a_2\|^{-1}a_2, \ldots$ of unit vectors has a cluster point $a \in \sigma$. Since H and H' are parallel, it is easy to see that $a \in H$, contrary to $H \cap \sigma = \{0\}$.

The following theorem enables us to apply results on polytopes to cones:

1.1 Theorem. *The following properties of an, at least, two-dimensional polyhedral cone σ are equivalent.*

(a) *σ has apex* 0.

(b) *There exists a hyperplane $H' \not\ni 0$ such that $H' \cap \sigma$ is a polytope.*

(c) *There exists a polytope P of dimension* $\dim \sigma - 1$ *such that*

$$\sigma = \operatorname{pos} P.$$

PROOF. Suppose (a) to be true: Let H be a hyperplane with $H \cap \sigma = \{0\}$. We choose a hyperplane H' parallel to H such that $H' \cap \sigma \neq \emptyset$. Then, $H' \cap \sigma$ is a polyhedral set, which is bounded according to the above Remark. This proves (b).

Now, we deduce (c) from (b). Set $P := H' \cap \sigma$, and suppose there is a vector $x \in \sigma \setminus \operatorname{pos} P$. Since $\dim \sigma > 1$, there exists a vector $a \in H' \cap \sigma$ such that x, a are linearly independent. Going over to the plane generated by 0, x, and a, we may assume that $n = 2$. Then, $H' \cap \sigma$ bounded implies $H \cap \sigma = \{0\}$ for H as in the above remark; in particular, $x \notin H$. But, then, $\operatorname{pos} x \cap H'$ is not empty, though $\operatorname{pos} x \cap \operatorname{pos} P = \{0\}$.

Let (c) be true. Since $\dim \operatorname{aff} P < \dim \sigma$, we obtain $0 \notin \operatorname{aff} P$. Hence there exists a hyperplane H' including $\operatorname{aff} P$ but not 0. Let H be the hyperplane parallel to H' through 0. Then $H \cap \sigma = \{0\}$ by the above remark. □

Remark. If $\sigma = \mathbb{R}$, then, (b) holds, but neither (a) nor (c). So the condition $\dim \sigma \geq 2$ in Theorem 1.1 is essential.

As a consequence of Theorem 1.1, we obtain the following theorem from Theorems 1.4 and 1.5 in II applied to $(n-1)$-polytopes:

1.2 Theorem. *Every polyhedral cone with apex* 0 *is a polyhedral set*

$$H_1^+ \cap \cdots \cap H_r^+, \tag{2}$$

where

$$H_1 \cap \cdots \cap H_r = \{0\}. \tag{3}$$

Conversely, any polyhedral set (2) *satisfying* (3) *is a polyhedral cone with apex* 0.

In this chapter, we restrict our attention mainly to the case where $x_1, \ldots, x_k$ are lattice points:

1.3 Definition. The points $x \in \mathbb{Z}^n \subset \mathbb{R}^n$ are called *lattice vectors*. In (1), if the vectors $x_1, \ldots, x_k$ are lattice vectors, then, we call σ a *lattice cone* (or *rational polyhedral cone*). Similarly, we say that $\operatorname{conv}\{x_1, \ldots, x_r\}$ is a *lattice polytope* if $x_1, \ldots, x_r$ are lattice points.

Remark. The condition "lattice vector" in the definition of a lattice cone may be replaced by "vector with rational coordinates", since multiplying all coordinates with their common denominator yields a lattice vector. To a lattice cone $\sigma \subset \mathbb{R}^n$ corresponds a "rational" cone $\sigma_{\mathbb{Q}} := \sigma \cap \mathbb{Q}^n$ such that $\sigma = \bar{\sigma}_{\mathbb{Q}}$. Thus, for lattice cones, we may go over to $\mathbb{Q}^n$ instead of $\mathbb{Z}^n$ and profit from the theory of $\mathbb{Q}$-vector spaces, which is easier than that of $\mathbb{Z}$-modules.

Most properties of lattice cones we consider are not invariant under combinatorial isomorphisms (see definition in II, 1), not even under general linear transformations. Nevertheless, combinatorial facts are often a useful starting point.

For example, if a classification is to be achieved, we first look at the "coarse" combinatorial classification, then specify further. There is a special group of linear transformations which leaves invariant all properties of cones to be discussed.

1.4 Definition. A linear transformation $L : \mathbb{R}^n \longrightarrow \mathbb{R}^n$ and the matrix representing it, with respect to the canonical basis, are called *unimodular* if $L(\mathbb{Z}^n) = \mathbb{Z}^n$.

Unimodular transformations can be characterized as follows.

1.5 Lemma. *A linear transformation $L : \mathbb{R}^n \longrightarrow \mathbb{R}^n$ is unimodular if and only if it is represented by a matrix A with integral entries and* $\det A = \pm 1$.

PROOF. Let A be unimodular. Evidently, A is invertible as a real matrix. From $A(\mathbb{Z}^n) \subset \mathbb{Z}^n$, we infer that A has integral entries: the columns of A are the image of the canonical basis vectors e_j. In the same way, A^{-1} is integral. Thus, A is invertible over $\mathbb{Z}$; hence, $\det A$ is a unit in $\mathbb{Z}$. The converse is obvious. □

Example. The linear transformation of $\mathbb{R}^2$ represented by

$$A = \begin{pmatrix} 1 & 1 \\ 0 & 1 \end{pmatrix},$$

a "shear-transformation" with $\{te_1 \mid t \in \mathbb{R}\}$ as axis, is unimodular.

A cone need not have an apex. However, we can represent every cone in the following way.

1.6 Lemma.

(a) *Any n-dimensional cone σ is the vector sum of a cone σ_0 with apex* 0 *and a linear space U; thus,*

$$\sigma = \sigma_0 + U,$$

where

$$\dim \sigma_0 + \dim U = n.$$

(b) *If σ is a lattice cone, σ_0 and U can also be chosen to be lattice cones.*

PROOF.

(a) The set $U := \sigma \cap (-\sigma)$ is the maximal linear subspace of $\mathbb{R}^n$ included in σ. We can choose the generators $x_1, \dots, x_k$ of σ so that $x_1, \dots, x_r \in U$, $x_j \notin U$ for $j = r+1, \dots, k$, and no x_j, $j = r+1, \dots, k$, is a positive linear combination of other elements of $\{x_1, \dots, x_k\}$. We set $\sigma_0 := \operatorname{pos}\{x_{r+1}, \dots, x_k\}$.

We claim that

$$(\operatorname{lin} \sigma_0) \cap \operatorname{lin}\{x_1, \dots, x_r\} = \{0\}. \tag{4}$$

Otherwise, for at least one $x_i \in \{x_{r+1}, \ldots, x_k\}$, x_{r+1} say, the projection parallel to $\text{lin}\{x_{r+2}, \ldots, x_k\}$ of x_{r+1} would be a point $x \neq 0$ in $\text{lin}\{x_1, \ldots, x_r\} = \text{pos}\{x_1, \ldots, x_r\}$, so that, for $x = \alpha_1 x_1 + \cdots + \alpha_r x_r$, $\alpha_1, \ldots, \alpha_r \geq 0$,

$$\begin{aligned} x_{r+1} &= x + \alpha_{r+2}x_{r+2} + \cdots + \alpha_k x_k, \qquad \alpha_{r+2}, \ldots, \alpha_k \geq 0, \\ &= \alpha_1 x_1 + \cdots + \alpha_r x_r + \alpha_{r+2}x_{r+2} + \cdots + \alpha_k x_k, \end{aligned}$$

a contradiction to the minimality of $\{x_{r+1}, \ldots, x_k\}$. (4) and $\{x_1, \ldots, x_r\} \subset U$ imply $\text{lin}\{x_1, \ldots, x_r\} = U$, so part (a) of the lemma follows.

(b) is a consequence of the way we proved (a).

□

It should be noted that U in 1.6 is uniquely determined, whereas, for $\dim U > 0$, we can choose σ_0 in many ways. If we write such a decomposition $\sigma = \sigma_0 + U$, we tacitly assume σ_0 to be of minimal dimension.

1.7 Definition. U in Lemma 1.6 is called the *cospan* of σ:

$$U = \text{cospan}\,\sigma.$$

The most elementary among the possible building bricks of a polytope is a simplex. Its counterpart for cones is the following.

1.8 Definition. A cone $\sigma = \text{pos}\{x_1, \ldots, x_k\}$ is called *simple* or a *simplex cone* if $x_1, \ldots, x_k$ are linearly independent. σ is said to be *simplicial* if each proper face of σ is a simplex cone.

Remark. We introduce the notion "simplex cone" to avoid a double meaning of "simplicial cone" occuring in the literature.

From linear algebra, we know Lemma 1.9.

1.9 Lemma.

(a) *Let* $\sigma = \text{pos}\{x_1, \ldots, x_k\}$ *be a simplex cone. Then,* $\text{conv}\{x_1, \ldots, x_k\}$ *is a* $(k-1)$*-simplex.*

(b) *Any simplex cone has* 0 *as an apex.*

1.10 Definition. We say a lattice vector is *primitive* or *simple* if its coordinates are relatively prime. If $\sigma = \text{pos}\{x_1, \ldots, x_k\}$, where $x_1, \ldots, x_k$ are primitive lattice vectors, then we call them *primitive generators* of σ.

A k-dimensional simplex cone $\sigma = \text{pos}\{x_1, \ldots, x_k\}$ in $\mathbb{R}^n$ is said to be *regular* if $x_1, \ldots, x_k$ are primitive, and if there exist primitive vectors $x_{k+1}, \ldots, x_n \in \mathbb{Z}^n$ such that

$$\det(x_1 \cdots x_n) = \pm 1.$$

If $\sigma = \sigma_0 + \operatorname{cospan} \sigma$ is an arbitrary lattice cone, we call σ *regular* if σ_0 is a simplex cone that is regular. $\{0\}$ is also called regular.

Some characterizations of regularity are readily shown:

1.11 Lemma. *The following conditions for an n-dimensional cone $\sigma =$ pos $\{x_1, \ldots, x_n\}$ with apex 0 and $x_1, \ldots, x_n$ primitive lattice vectors, are equivalent.*

(a) *σ is regular.*
(b) *Any lattice point of $\mathbb{Z}^n$ is an integral linear combination of $x_1, \ldots, x_n$.*
(c) *Any element of $\sigma \cap \mathbb{Z}^n$ is a nonnegative, integral, linear combination of $x_1, \ldots, x_n$.*
(d) *There exists a unimodular linear transformation that maps the canonical basis vectors $e_1, \ldots, e_n$ onto $x_1, \ldots, x_n$, respectively.*

In later sections we need to split a given cone σ into regular cones. The first step is cutting σ into simplex cones, which we achieve here:

1.12 Theorem. *Any cone $\sigma = \operatorname{pos}\{a_1, \ldots, a_r\}$ with apex 0 can be split into simplex cones $\sigma_1, \ldots, \sigma_k$ satisfying the following conditions*

(a) $\sigma = \sigma_1 \cup \cdots \cup \sigma_k$,
(b) *$\sigma_i \cap \sigma_j$ is a face of σ_i and of σ_j,* $\quad i, j = 1, \ldots, k$,
(c) $\sigma_i = \operatorname{pos}\{a_{i_1}, \ldots, a_{i_q}\}$ *for a subset* $\{a_{i_1}, \ldots, a_{i_q}\} \subset \{a_1, \ldots, a_r\}$, $i = 1, \ldots, k$.

PROOF. We aply III, Theorem 2.6 to the cell complex consisting of σ and the faces of σ. □

In the next section we shall need the following separation property (a consequence of a special case of the so-called Hahn–Banach theorem):

1.13 Lemma (Separation lemma). *Let $\sigma = \operatorname{pos}\{a_1, \ldots, a_r\}, \sigma' = \operatorname{pos}\{b_1, \ldots, b_s\}$ be cones with apex 0 such that $\dim \sigma' = n$ and $(\operatorname{relint} \sigma) \cap (\operatorname{int} \sigma') = \emptyset$. Then, there exists a hyperplane H such that $\sigma \subset H^+, \sigma' \subset H^-$ (closed half-spaces).*

PROOF. Equivalently to the lemma, we claim, that $A := \operatorname{conv}((-\sigma) \cup \sigma')$ has a supporting hyperplane passing through 0. Otherwise, since $\dim \sigma' = n$, we find that $A = \mathbb{R}^n$, and, hence,

$$0 = \alpha_1(-a_{i_1}) + \cdots + \alpha_p(-a_{i_p}) + \beta_1 b_{j_1} + \cdots + \beta_q b_{j_q}$$

for appropriate $\alpha_i, \beta_j \in \mathbb{R}_{>0}$, $p \geq 1$, and $q \geq 1$ such that among the vectors $a_{i_1}, \ldots, a_{i_p}, b_{j_1}, \ldots, b_{j_q}$ there are n linearly independent ones. Thus,

$$\alpha_1 a_{i_1} + \cdots + \alpha_p a_{i_p} = \beta_1 b_{j_1} + \cdots + \beta_q b_{j_q} \quad \in \quad \sigma \cap \sigma' =: \sigma_0.$$

Let σ_1, σ_1' be the smallest faces of σ, σ', respectively, which contain σ_0. Then, it is easy to see that $a_{i_1}, \ldots, a_{i_p} \in \sigma_1$, and $b_{j_1}, \ldots, b_{j_q} \in \sigma_1'$, so we obtain

(5) $$\dim \operatorname{conv}(\sigma_1 \cup \sigma_1') = n.$$

Let $x \in \operatorname{relint} \sigma_0$. Then, also, $x \in \operatorname{relint} \sigma_1$ and $x \in \operatorname{relint} \sigma_1'$, hence, $x \in \operatorname{relint} \operatorname{conv}(\sigma_1 \cup \sigma_1')$. If $k := \dim \sigma_1'$, choose $k+1$ affinely independent points $c_0, \ldots, c_k$ of σ_1' such that $x \in \operatorname{relint} \operatorname{conv}\{c_0, \ldots, c_k\}$. Since $\dim \sigma' = n$, we can choose points $c_{k+1}, \ldots, c_n$, such that $c_0, \ldots, c_n$ are vertices of an n-simplex $S \subset \sigma'$. We see that σ_1 intersects int S, and so,

$$(\operatorname{relint} \sigma_1) \cap (\operatorname{int} S) \neq \emptyset,$$

which contradicts the assumption of the lemma. □

As a particular application, we note a property for the dual cone $\breve{\sigma}$ of a cone σ with apex 0 (in 22 (c) we shall see that $\breve{\sigma}$ automatically is full-dimensional).

1.14 Corollary. *Let σ be a cone in $\mathbb{R}^n$ with apex 0 and $\breve{\sigma}$ be n-dimensional. Then,*

$$\operatorname{relint} \sigma \cap \operatorname{int} \breve{\sigma} \neq \emptyset.$$

PROOF. Otherwise, there would be a hyperplane H with $\sigma \subset H^+$ and $\breve{\sigma} \subset H^-$ by 1.13. For a normal vector v of H pointing into H^+, by definition of $\breve{\sigma}$, $\langle v, \sigma \rangle \geq 0$. Hence, $\breve{\sigma} \subset H^+ \cap H^- = H$, a contradiction to $\dim \breve{\sigma} = n$. □

Exercises

1. In $\mathbb{R}^2$, let σ be a two-dimensional lattice cone with apex 0. σ can be mapped by a unimodular transformation onto $\sigma' = \operatorname{pos}\{e_1, qe_1 + re_2\}$ for some $q, r \in \mathbb{Z}_{\geq 0}$, q, r relatively prime, $0 \leq q < r$.
2. A subset τ of a cone σ is a face of σ if and only if it satisfies the following conditions.

   a. τ is a cone.
   b. If $x \in \sigma \setminus \tau$ and $x' \in \sigma$, then, $x + x' \notin \tau$.

3. Carry out explicitly a subdivision of σ into simplex cones if $\sigma = \operatorname{pos} P$, P a 3-cube in $\{(\xi_1, \xi_2, \xi_3, 1)\} \subset \mathbb{R}^4$.
4. Extend Lemma 1.13 to the case where σ or σ' does not have 0 as an apex.

2. Dual cones and quotient cones

Now we turn to dual cones which are related to the original cones in the same way as polar polytopes are to the original polytopes. In fact, we shall see that there is a direct relationship between the two dualizations.

In I, 49 we have defined the dual cone of a cone $\sigma \in \mathbb{R}^n$ by

$$\check{\sigma} := \{x \in \mathbb{R}^n \mid \langle x, \sigma \rangle \geq 0\}.$$

For $m \in \mathbb{R}^n$, we set $\check{m} := (\mathbb{R}_{\geq 0}\, m)\check{}$. If a cone τ is included in σ, then, obviously, $\check{\sigma} \subset \check{\tau}$. Moreover, $\sigma^\perp := \{x \mid \langle x, \sigma \rangle = 0\}$ satisfies $\sigma^\perp = \check{\sigma} \cap (-\check{\sigma}) \subset \check{\sigma}$. Equality holds if σ is a linear subspace.

Examples. If σ is an angular region in $\mathbb{R}^2$, then so is $\check{\sigma}$ (see Figure 1a). If σ is a ray in $\mathbb{R}^n$, then $\check{\sigma}$ is a half-space $H^+ \supset \sigma$, where $H = \sigma^\perp$. If σ is a half-plane in $\mathbb{R}^3$, then, so is $\check{\sigma}$ (Figure 1b). For a quadrant σ in $\mathbb{R}^2$ or an octant in $\mathbb{R}^3$, $\sigma = \check{\sigma}$.

2.1 Theorem. *Let σ be an n-dimensional cone in $\mathbb{R}^n$ with apex 0, let $b_1, \ldots, b_r$ be inner normals of the facets of σ. Then*

$$\check{\sigma} = \operatorname{pos}\{b_1, \ldots, b_r\}.$$

PROOF. Each hyperplane $H_i := \{x \mid \langle x, b_i \rangle = 0\}$ supports σ; hence, $\langle \sigma, b_i \rangle \geq 0$. This implies $\check{\sigma} \supset \operatorname{pos}\{b_1, \ldots, b_r\}$. If x satisfies $\langle x, \sigma \rangle \geq 0$, but $x \notin \operatorname{pos}\{b_1, \ldots, b_r\} =: \tilde{\sigma}$, then, there exists a face $\operatorname{pos}\{b_{i_1}, \ldots, b_{i_s}\}$ of $\tilde{\sigma}$ and a b_{i_0}, such that $x = -\alpha b_{i_0} + \beta_1 b_{i_1} + \cdots + \beta_s b_{i_s}$ with positive $\alpha, \beta_1, \ldots, \beta_s$, and an a_k such that $\langle b_{i_j}, a_k \rangle = 0$ for $j = 1, \ldots, s$, but $\langle b_{i_0}, a_k \rangle > 0$. Now, $\langle x, a_k \rangle = -\alpha \langle b_{i_0}, a_k \rangle < 0$, a contradiction. □

For the cone σ in $\mathbb{R}^n$, set $V := \operatorname{aff} \sigma$, and let U be the orthogonal complement of V. Then, $\check{\sigma} = (\check{\sigma} \cap V) \oplus U$ and $\check{\sigma} \cap V$ is the dual of $\sigma \cap V$ in V.

2.2 Lemma. *Let σ and σ_i be cones in $\mathbb{R}^n$.*

(a) *If* $\dim \sigma = n$ *and σ has* 0 *as apex, then,* $\dim \check{\sigma} = n$, *and $\check{\sigma}$ has* 0 *as apex.*
(b) cospan $\check{\sigma} = \sigma^\perp$.
(c) *σ has* 0 *as apex, if and only if* $\dim \check{\sigma} = n$.
(d) $(\sigma_1 + \sigma_2)\check{} = \check{\sigma}_1 \cap \check{\sigma}_2$ *($\sigma_1 + \sigma_2$ Minkowski sum).*
(e) $(\sigma_1 \cap \sigma_2)\check{} = \check{\sigma}_1 + \check{\sigma}_2$.
(f) $\check{\check{\sigma}} = \sigma$.

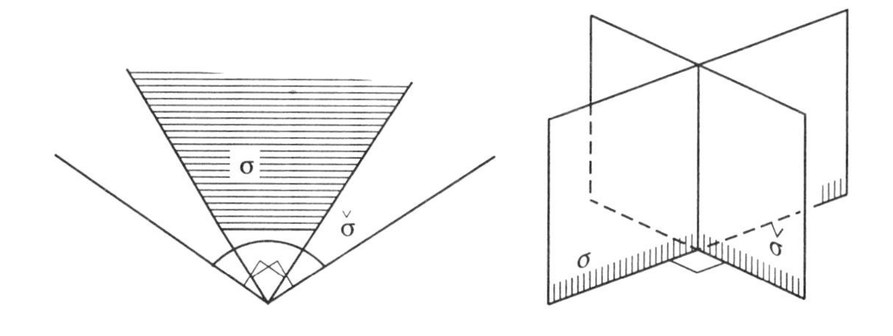

FIGURE 1a,b.

PROOF.

(d), (e), and (f) follow from the definition of dual cones.

(c) Let $\sigma \neq 0$ have an apex. Then, there is a hyperplane H with $H \cap \sigma = 0$, $H^+ \supset \sigma$. Let $b \in \sigma$ be a generator of the ray $\check{H}^+$. In a sufficiently small neighborhood of b, we find linearly independent vectors $b_1, \ldots, b_n$ such that, for $H_i := \{x \mid \langle x, b_i \rangle = 0\}$, we also have $\sigma \cap H_i = 0$ and $\sigma \subset H_i^+$, $i = 1, \ldots, n$ (otherwise $H \cap \sigma \neq 0$ would follow from σ being a closed set). Clearly, $b_i \in \check{\sigma}$, and, therefore, $\operatorname{pos}\{b_1, \ldots, b_n\} \subset \check{\sigma}$ implies $\check{\sigma}$ to be n-dimensional.

The converse is shown analogously.

(a) Let $u \in \operatorname{int} \sigma$. Then, for the facet normals $b_1, \ldots, b_r$ as in Theorem 2.1,

$$\langle u, b_i \rangle > 0, \qquad \text{for } i = 1, \ldots, r.$$

Hence, for $\tilde{H} := \{x \mid \langle u, x \rangle = 0\}$, by Theorem 2.1, we obtain,

$$\tilde{H} \cap \check{\sigma} = \{0\}.$$

(b) We obtain the following equivalent statements:

(b1)	$x \in \operatorname{cospan} \check{\sigma} = \check{\sigma} \cap (-\check{\sigma})$.
(b2)	$\langle x, \sigma \rangle \geq 0$ and $\langle -x, \sigma \rangle \geq 0$.
(b3)	$\langle x, \sigma \rangle = 0$.
(b4)	$x \in \sigma^\perp$.

□

2.3 Lemma. *Let τ be a face of σ, and let $0 \neq m \in \operatorname{relint}(\tau^\perp \cap \check{\sigma})$. Then,*

$$\check{\tau} = \check{\sigma} + \mathbb{R}_{\geq 0}(-m).$$

PROOF. If we can show that the obvious inclusions

$$\tau \subset \sigma \cap m^\perp \subset \sigma \cap (-m)\check{} \tag{1}$$

are equalities, then the claim follows immediately from 22 (e). For $x \in \sigma \cap (-m)\check{}$, we know that $\langle x, m \rangle \leq 0$, while $m \in \check{\sigma}$ means $\langle x, m \rangle \geq 0$; hence, $\sigma \cap m^\perp = \sigma \cap (-m)\check{}$. For $\tau \supset \sigma \cap m^\perp$ it suffices to verify the following:

$$\text{If } v \in \sigma \setminus \tau, \quad \text{then, } \langle m, v \rangle \neq 0.$$

Obviously we may assume that $v \neq -m$. Going over to the plane generated by 0, v, and m, we may assume that $n = 2$. Then, $\operatorname{lin} \tau = m^\perp$ and $v \notin \operatorname{lin} \tau$, hence, $\langle v, m \rangle \neq 0$. □

As we see from Lemma 2.2, the structure of $\check{\sigma}$ is largely determined if, in $\sigma = \sigma_0 + \operatorname{cospan} \sigma$, we know the structure of $\check{\sigma}_0$ relative to $\operatorname{lin} \sigma_0$. So, we are mainly interested in theorems about duals of full-dimensional cones with 0 as an apex.

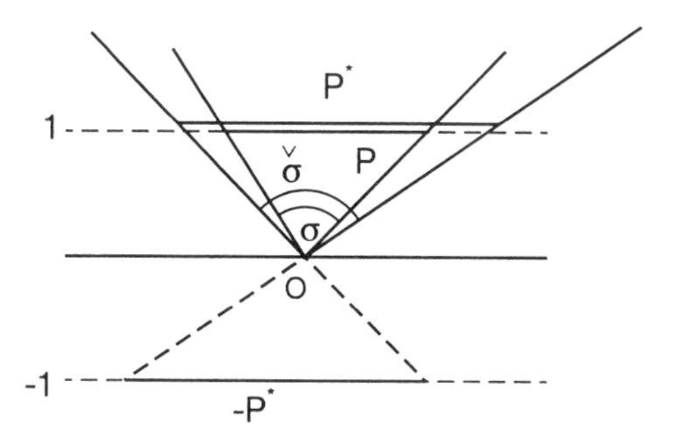

FIGURE 2.

It turns out that we only have to apply the results of II, 2 since there is a simple relationship between polar polytopes and dual cones:

2.4 Theorem. *Let σ be an n-dimensional cone in $\mathbb{R}^n$ with apex* 0. *Then, we can find an $(n-1)$-polytope P and a point $u \in \operatorname{relint} P$ such that*

(a) $\sigma = \operatorname{pos} P$,

(b) *If P^* denotes the polar of P in* aff P *with respect to the unit sphere centered in u, and $\tilde{P}^*$ is the reflection $2u - P^*$ of P^* in u (Figure 2), then,*

$$\check{\sigma} = \operatorname{pos} \tilde{P}^*.$$

PROOF. By 22 (c) and 114, there exists a unit vector $u \in (\operatorname{int} \sigma) \cap (\operatorname{int} \check{\sigma})$. We may assume that $u = (0, 1)$. Let us decompose every $x \in \mathbb{R}^n = \mathbb{R}^{n-1} \times \mathbb{R}$ in the form $x = (x', x_n)$. Since $u \in \operatorname{int} \check{\sigma}$ and $u^\perp = \mathbb{R}^{n-1} \times \{0\}$, we obtain $u^\perp \cap \sigma = \{0\}$. Then, $u^\perp + u = \mathbb{R}^{n-1} \times \{1\}$, and

$$P := (\mathbb{R}^{n-1} \times \{1\}) \cap \sigma =: P' \times \{1\}$$

is a polytope with $u \in \operatorname{relint} P$ and $0 \in \operatorname{relint} P'$. For the polar bodies P^* in $\mathbb{R}^{n-1} \times \{1\}$ and P'^* in $\mathbb{R}^{n-1}$, we, obviously, have

$$P^* = P'^* \times \{1\}.$$

But $\langle *, y' \rangle \leq 1$ on P' is equivalent to $\langle (*, 1), (y', -1) \rangle \leq 0$ on P, that is, to $(y', -1) \in -\check{\sigma}$, so that (since $y' \in P'^*$ is equivalent to $(y', 1) \in P^*$ or $(y', -1) \in P^* - 2u$)

$$\check{\sigma} = -\operatorname{pos}(P^* - 2u) = \operatorname{pos}(2u - P^*) = \operatorname{pos} \tilde{P}^*.$$

□

2.5 Definition. By the assignment $F \mapsto F^*$ of the faces of P respectively P^* in aff P according to II, 21, we obtain an assignment

$$\tau := \operatorname{pos} F \quad \longmapsto \quad \check{\tau} := \operatorname{pos}(-F^* + 2u)$$

between the faces of the cones $\sigma := \text{pos } P$ and $\check{\sigma} = \text{pos } \tilde{P}^*$. We call $\check{\tau}$ the *dual face* of τ.

From II, Theorem 2.5 we find Theorem 2.6.

2.6 Theorem. *Let σ be an n-dimensional cone with 0 as apex, and let $f_k(\sigma)$ denote the number of k-faces of σ. Then,*

$$f_k(\check{\sigma}) = f_{n-k}(\sigma) \qquad \textit{for } k = 0, \ldots, n-1.$$

2.7 Definition. Let $\sigma = \sigma_0 + \text{cospan } \sigma$ and $\sigma' = \sigma_0' + \text{cospan } \sigma'$ be two n-cones in $\mathbb{R}^n$. Assume there are polytopes P in σ_0 and P' in σ_0' of codimension 1 such that

$$\sigma_0 = \text{pos } P \text{ and } \sigma_0' = \text{pos } P'.$$

Then we call σ and σ' *combinatorially isomorphic*, $\sigma \approx \sigma'$, if $P \approx P'$ and $\dim \text{cospan } \sigma = \dim \text{cospan } \sigma'$. Equivalence classes of combinatorially isomorphic cones are said to be *types* of cones.

In II, Definition 2.7 we have introduced the quotient polytope P/F. It depends on the choice of the defining affine space U; its type, however, is uniquely determined. For cones, we introduce quotients as follows.

2.8 Definition. If σ is an n-dimensional cone with apex 0 and if τ is a face of σ, we define the *quotient cone σ/τ* as the type of

$$\text{pos}(P/F),$$

where $\sigma = \text{pos } P$, $\tau = \text{pos } F$, P an $(n-1)$-polytope, and P/F a quotient polytope.

If τ is $\{0\}$, then, $F = \emptyset$, and, thus, $\sigma/\{0\} = \sigma$.

Let us denote by $\mathcal{B}_1(\cdot)$ the set of proper, at least one-dimensional faces. By Theorem 2.4 and Theorem II, 2.9 we derive Theorem 2.9.

2.9 Theorem. *Let $[\tau, \sigma]$ be the set of proper faces of σ including τ. Then we have bijective, inclusion-preserving and, inclusion-reversing maps ψ and φ*

$$[\tau, \sigma] \underset{\psi}{\longrightarrow} \mathcal{B}_1(\sigma/\tau) \underset{\varphi}{\longrightarrow} \mathcal{B}_1(\check{\tau}).$$

The following result is of importance for the construction of toric varieties.

2.10 Theorem.

(a) *If σ is a lattice cone, then, so is $\check{\sigma}$.*
(b) *If σ is regular, then so is $\check{\sigma}$.*

PROOF. For the lattice cone σ, we may consider all generators over $\mathbb{Q}$ instead of $\mathbb{R}$: In the decomposition $\sigma = \sigma_0 + U$ for $U := \text{cospan } \sigma$, the space U and (by

16 (b)) the cone σ_0 have generators over $\mathbb{Q}_{\geq 0}$. Since $\check{\sigma} = \check{\sigma}_0 \cap \check{U}$ by 22 (d), we essentially have to analyze $\check{\sigma}_0$, and, up to a direct factor, we may do that in aff σ_0. Hence, we may assume that $\sigma = \text{pos}\{x_1, \ldots, x_t\}$ with primitive generators x_j is of full dimension n and has apex 0.

If $H = \text{lin}\,\tau$ for a facet τ of σ, then, $H = u^\perp$ for some $u \in \mathbb{Q}^n$. We may assume that $H = \text{lin}\{x_1, \ldots, x_{n-1}\}$; then, we may choose u as the solution of the system of equations $\langle x_j, u\rangle = \delta_{jn}$, for $j = 1, \ldots, n$. If $u_1, \ldots, u_l$ denote the vectors thus constructed for the different facets of σ, then, by 21, $\check{\sigma} = \text{pos}\{u_1, \ldots, u_l\}$, which proves (a).

For (b) we know, in addition, that $n = t = l$. Thus, the matrix $A := (u_1^t, \ldots, u_n^t)$ is inverse to the matrix $(x_1, \ldots, x_n)$. Since, by assumption, $\det A^{-1} = \pm 1$, Cramers rule implies that A has integral entries. □

For a linear map L between vector spaces, we denote by L^* the dual map; moreover, we identify the linear form $\langle a, \cdot\rangle$ with a.

2.11 Corollary. *Let $\sigma \subset \mathbb{R}^n$ and $\sigma' \subset \mathbb{R}^r$ be lattice cones, and let $L : \mathbb{R}^n \longrightarrow \mathbb{R}^r$ be a linear map such that $L(\mathbb{Z}^n) \subset \mathbb{Z}^r$ and $L(\sigma) \subset \sigma'$. Then, $L(\sigma)$ and $L^*(\check{\sigma}')$ are lattice cones, and*

$$L^*(\check{\sigma}') \subset \check{\sigma}.$$

PROOF. Using elementary rules from linear algebra, we find that

$$\begin{aligned} L^*(\check{\sigma}') &= \{L^*(x) \mid \langle x, \sigma'\rangle \geq 0\} \subset \{L^*(x) \mid \langle x, L(\sigma)\rangle \geq 0\} \\ &= \{L^*(x) \mid \langle L^*(x), \sigma\rangle \geq 0\} \subset \{v \mid \langle v, \sigma\rangle \geq 0\} = \check{\sigma}. \end{aligned}$$

With respect to appropriate bases for $\mathbb{Z}^n$ and $\mathbb{Z}^r$, L^* is represented by the transposed matrix of the matrix which represents L. Thus, $L(\mathbb{Z}^n) \subset \mathbb{Z}^r$ implies $L^*(\mathbb{Z}^r) \subset \mathbb{Z}^n$. Hence, by Lemma 2.10, $\check{\sigma}$, $\check{\sigma}'$, and, thus, $L(\sigma)$ and $L^*(\check{\sigma}')$ are lattice cones. □

Exercises

1. Let a regular octahedron P be embedded in the hyperplane $\{(\xi_1, \xi_2, \xi_3, 1)\}$ of $\mathbb{R}^4$. Find the dual cone of $\sigma := \text{pos}\,P$. (Use appropriate coordinates for the vertices of P).
2. In $\mathbb{R}^n$ a cone σ is self–dual, $\sigma = \check{\sigma}$, if and only if $\sigma = \text{pos}\{a_1, \ldots, a_n\}$ and $a_1, \ldots, a_n$ is an orthogonal basis of $\mathbb{R}^n$.
3. Let $y \in \sigma$. The following conditions are equivalent (each implies all others)
   a. $y \in \text{relint}\,\sigma$.
   b. $\langle y, u\rangle > 0$ for all $u \in \check{\sigma} \setminus \sigma^\perp$.
   c. $\check{\sigma} \cap (\text{pos}\{y\})^\perp = \sigma^\perp$.
   d. $\sigma + \mathbb{R}_{\geq 0}(-y) = \sigma + (-\sigma)$.
4. For cones, prove a statement analogous to that of II, 2, Exercise 4.

3. Monoids

3.1 Definition. A semi-group, that is a non-empty set S with an associative operation

$$+ : S \times S \longrightarrow S,$$

is called a *monoid* if it is commutative and has a zero, i.e., an element $0 \in S$ for which

$$s + 0 = s, \quad \text{for all } s \in S,$$

and it satisfies the cancellation law

$$s + x = t + x \text{ implies } s = t, \quad \text{for all } s, t, x \in S.$$

3.2 Lemma. *If σ is a cone in $\mathbb{R}^n$, then, $\sigma \cap \mathbb{Z}^n$ is a monoid.*

PROOF. From the definition of a cone σ, $x + y$ is in σ if $x, y \in \sigma$, in particular, $x + y \in \sigma \cap \mathbb{Z}^n$ if $x, y \in \sigma \cap \mathbb{Z}^n$. The zero vector is the zero of the monoid. □

3.3 Definition. A monoid S is said to be *finitely generated* if there exist $a_1, \ldots, a_r \in S$, called *generators*, such that

$$S = \mathbb{Z}_{\geq 0}\, a_1 + \cdots + \mathbb{Z}_{\geq 0}\, a_r.$$

A system of generators is called *minimal* if none of its elements is generated by the others.

3.4 Lemma (Gordan's Lemma). *If σ is a lattice cone in $\mathbb{R}^n$, then, the monoid $\sigma \cap \mathbb{Z}^n$ is finitely generated.*

PROOF. By Lemma 1.6 and Theorem 1.12, we may assume σ to be a simplex cone, i.e., $\sigma = \operatorname{pos}\{a_1, \ldots, a_k\}$ where $a_1, \ldots, a_k$ are simple and $k \leq n$.

Then the "fundamental parallelepiped"

$$F := \{\sum_{j=1}^{k} t_j a_j \mid 0 \leq t_j \leq 1\}$$

includes only finitely many lattice points. For each $x \in \sigma \cap \mathbb{Z}^n$, there exists a lattice point $y \in \sum_{j=1}^{k} \mathbb{Z}_{\geq 0}\, a_j$ such that $x - y \in F$. □

Example 1. For $\sigma = \operatorname{pos}\{(7, 2), (2, 5)\}$ in $\mathbb{R}^2$, it suffices to choose as additional generators $(1, 1)$, $(2, 1)$, $(3, 1)$, $(1, 2)$ (see Figure 3).

Remark. We can obtain an example of a monoid which is not finitely generated by considering a non-closed cone σ:

Let $\sigma' = \operatorname{pos}\{e_1, e_2\}$ in $\mathbb{R}^2$, $\sigma := \sigma' \setminus \operatorname{pos}\{e_2\}$. Then, $\sigma \cap \mathbb{Z}^2$ is not finitely generated (see also Exercise 2(a)).

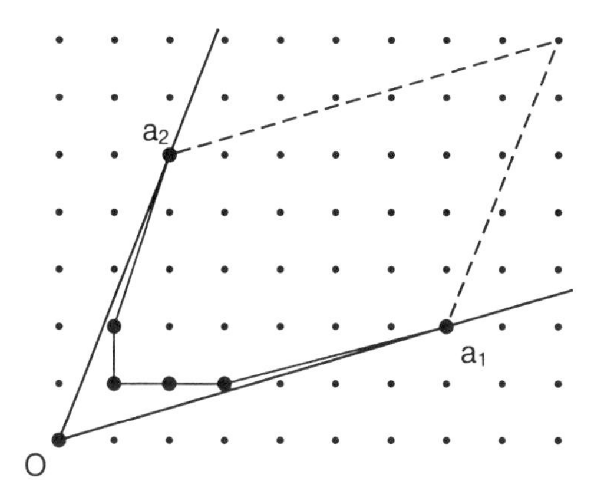

FIGURE 3.

3.5 Lemma. *If σ has an apex, then, the monoid $\sigma \cap \mathbb{Z}^n$ has (up to renumbering) precisely one minimal system of generators.*

PROOF. Let $a_1, \dots, a_t$ and $b_1, \dots, b_m$ be different minimal systems of generators and $b_1 \notin \{a_1, \dots, a_t\}$, say. There exist linear combinations, say

$$b_1 = \sum_{j=1}^{r} \lambda_j a_j \text{ for } \lambda_j \in \mathbb{Z}_{\geq 0}$$

and

$$a_i = \sum_{k=1}^{m} \mu_{ik} b_k, \quad \mu_{ik} \in \mathbb{Z}_{\geq 0} \text{ for } i = 1, \dots, t,$$

that yield a representation

$$b_1 = \sum_{k=1}^{m} \gamma_k b_k,$$

where $\gamma_1 = \sum_{j=1}^{r} \lambda_j \mu_{j1} > 0$, since the b_k's form a minimal system of generators as a monoid. On the other hand, $\gamma_1 \leq 1$; otherwise, $-b_1 \in \sigma$, though σ has an apex. That implies $\gamma_1 = 1$ and $\gamma_k = 0$ for $k \geq 2$, since $\sum_{k=2}^{m} \gamma_k b_k = 0$. As a consequence, $r = 1$ and, thus, $b_1 = a_1$, a contradiction. □

3.6 Lemma.

(a) *A monoid S with generators $a_1, \dots, a_k$ can be embedded as a subsemigroup into a group $G(S)$ which has $a_1, \dots, a_k$ as group generators (coefficients in $\mathbb{Z}$).*

(b) $$G(\sigma \cap \mathbb{Z}^n) = (\sigma - \sigma) \cap \mathbb{Z}^n .$$

PROOF.

(a) is shown by a standard procedure in linear algebra used, for example, when the semi-group of natural numbers is extended to the group of integers or $(\mathbb{Z}\setminus\{0\}, \cdot)$ is extended to $(\mathbb{Q}\setminus\{0\}, \cdot)$.

(b) This is clear from (a).

□

3.7 Definition. A monoid S is called *saturated* if $\alpha x \in S$ for $\alpha \in \mathbb{Z}_{\geq 0}$ and $x \in G(S)$ implies $x \in S$.

From the definition of $\sigma \cap \mathbb{Z}^n$, we have Lemma 3.8.

3.8 Lemma. *For every cone σ, the monoid $\sigma \cap \mathbb{Z}^n$ is saturated.*

We remark that non-saturated monoids can easily be constructed from $\sigma \cap \mathbb{Z}^n$ by omitting certain subsets of $\sigma \cap \mathbb{Z}^n$, as in the following examples:

Example 2. Let $x_1, \ldots, x_q \in \sigma \cap \mathbb{Z}^n$. Then, $S(x_1, \ldots, x_q) \cup \{0\}$, where

$$S(x_1, \ldots, x_q) := [(x_1 + \sigma) \cup \cdots \cup (x_q + \sigma)] \cap \mathbb{Z}^n$$

is a submonoid of $\sigma \cap \mathbb{Z}^n$ (Figure 4). Also, $[\mathbb{Z}^n \cap \operatorname{conv} S(x_1, \ldots, x_q)] \cup \{0\}$ is a submonoid.

Example 3. For every monoid $\sigma \cap \mathbb{Z}^n$, where the cone $\sigma = \operatorname{pos}\{a_1, \ldots, a_k\}$ is generated by primitive lattice vectors a_j,

$$S_0 := \{\alpha_1 a_1 + \cdots + \alpha_k a_k \mid \alpha_1, \ldots, \alpha_k \in \mathbb{Z}_{\geq 0}\}$$

is a submonoid. As is seen from Example 1, S_0 need not be equal to $\sigma \cap \mathbb{Z}^n$.

In general there are relationships between $a_1, \ldots, a_r$, which we discuss by using the following notion:

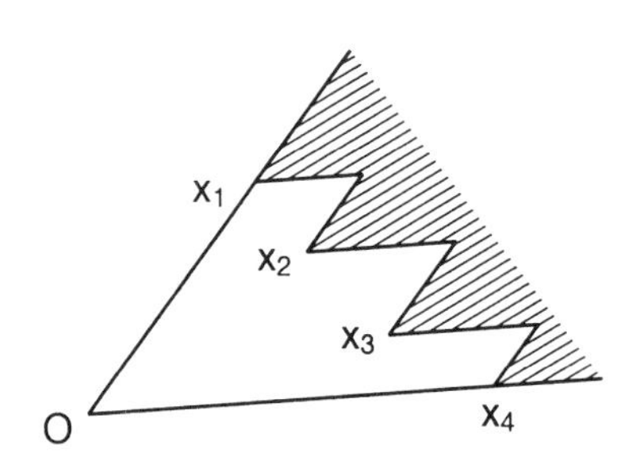

FIGURE 4.

3.9 Definition. Given a set $X = \{a_1, \ldots, a_r\} \subset \mathbb{Z}^n$, we call a vector $(\lambda, \mu) \in (\mathbb{R}_{\geq 0})^r \times (\mathbb{R}_{\geq 0})^r$ a *positive linear relation* of X if $\sum_{i=1}^r \lambda_i a_i = \sum_{i=1}^r \mu_i a_i$. We denote the set of all positive linear relations of X by poslin X.

3.10 Lemma.

(a) poslin X *is a lattice cone in* $\mathbb{R}^{2r}$.

(b) $\mathbb{Z}^{2r} \cap$ poslin X *is a finitely generated monoid in* $\mathbb{R}^{2r}$.

PROOF. If $\alpha, \beta \in$ poslin X, clearly $\alpha + \beta \in$ poslin X, also $t\alpha \in$ poslin X for every $t \in \mathbb{R}_{\geq 0}$. So, poslin X is a cone in $\mathbb{R}^{2r}$. Its linear hull is $V :=$ poslin $X -$ poslin X, and

$$\text{poslin } X = V \cap (\mathbb{R}_{\geq 0})^{2r}.$$

Since $(\mathbb{R}_{\geq 0})^{2r}$ is an intersection of half-spaces and, thus, a polyhedral cone in $\mathbb{R}^{2r}$, poslin X is a polyhedral cone by II 16. Clearly, it has rational generators. This proves (a).

(b) follows by Gordan's Lemma 3.4. □

We remark that poslin X also contains the trivial positive linear relations (μ, μ). These are generated by the relations

$$(0, \ldots, 0, \underset{i}{1}, 0, \ldots, 0, 0, \ldots, 0, \underset{i}{1}, 0, \ldots, 0), \qquad i = 1, \ldots, r.$$

Som even a nonempty set X of linearly independent vectors has a nonzero cone poslin X.

Exercises

1. Decompose $\sigma \subset \mathbb{R}^3$ into regular cones for $\sigma = \text{pos}\{(1, 0, 2), (1, 1, 3), (0, 2, 5)\}$.
2.
   a. A monoid $\sigma = (\mathbb{R}_{\geq 0}\, a_1 + \mathbb{R}_{\geq 0}\, a_2) \cap \mathbb{Z}^2$ in $\mathbb{R}^2$, a_1, a_2 primitive and linearly independent, has a_1, a_2 as generators if and only if the simplex

   $$\{x = \alpha_1 a_1 + \alpha_2 a_2 \mid \alpha_1 + \alpha_2 \leq 1, \quad 0 \leq \alpha_i \leq 1, \quad i = 1, 2\}$$

   does not contain a lattice point other than its vertices $0, a_1, a_2$.

   b. A statement analogous to (a) in $\mathbb{R}^3$ is false.
3. Let σ be a polyhedral cone with apex 0. If the monoid $\sigma \cap \mathbb{Z}^n$ is finitely generated, σ is a lattice cone.
4.
   a. Prove $\dim(\text{poslin } X) = r + \dim \mathcal{L}(X)$ (see II, 4), where X has r elements.

   b. Find a system of generators for poslin X if

   $$X = \{(1, 0, 0), (0, 1, 0), (1, 0, 1), (0, 1, 1), (1, 1, -1)\}.$$

   (Hint: First, apply a linear transformation which leaves the first two vectors fixed and maps $(1, 0, 1)$ onto $(0, 0, 1)$.)

4. Fans

In I, 4 we introduced the fan $\Sigma(K)$ of a convex body K as a set of outer normals

$$N(x) = -x + p_K^{-1}(x),$$

where an $x \in \operatorname{relint} F$ is assigned to each proper face F of K. If K is a polytope P and so possesses only finitely many faces, $\Sigma(K)$ consists of finitely many cones. Furthermore, if $\dim P = n$, all cones $\sigma \in \Sigma(P)$ have 0 as apex, and, to $x \in \operatorname{int} P$, there is assigned $-x + x = 0$ as a cone.

We now characterize those fans $\Sigma = \Sigma(P)$ which stem from the outer cones of n-polytopes P. In III, Definition 1.7 we have introduced fans, in particular, polyhedral, simplicial, and complete fans. If not stated otherwise, we shall always mean by "fan" a polyhedral fan.

As is illustrated by Figure 5, a cone of Σ need not be a face of a full-dimensional cone of Σ. Also Σ does not necessarily cover all of $\mathbb{R}^n$.

4.1 Definition. Let Σ, Σ' be fans in $\mathbb{R}^n$, and $L : \mathbb{R}^n \longrightarrow \mathbb{R}^n$ a linear map such that, for each $\sigma' \in \Sigma'$, there exists a $\sigma \in \Sigma$ satisfying $L(\sigma' \cap \mathbb{Z}^n) \subset \sigma \cap \mathbb{Z}^n$. Then, we say that L is also a *map of fans*

$$L : \Sigma' \longrightarrow \Sigma.$$

4.2 Theorem. *To every fan Σ, there exists a simplicial fan Σ' such that Σ and Σ' have the same 1-cones and the identity map* id *of $\mathbb{R}^n$ is a map of fans*

$$\mathrm{id} : \Sigma' \longrightarrow \Sigma.$$

PROOF. This follows from III, Theorem 2.6. □

The following notion will be fundamental for the "projectiveness" of the varieties to be introduced in Chapter VI. It specializes "polytopal".

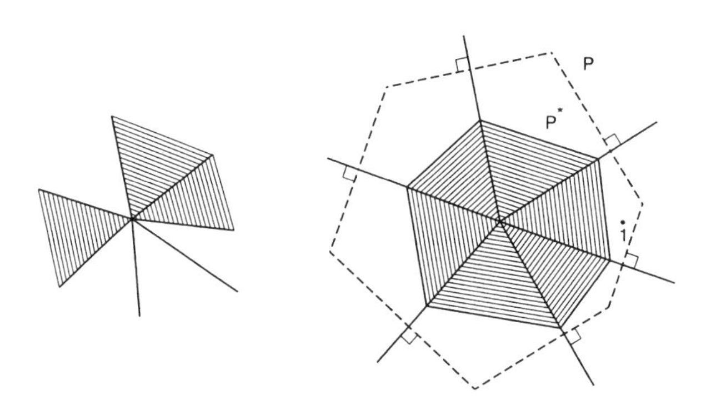

FIGURE 5.

4.3 Definition. A fan Σ is said to be *strongly polytopal* if there exists a polytope P^* such that $0 \in \operatorname{int} P^*$, and

$$\Sigma = \{\operatorname{pos} F \mid F \in \mathcal{B}(P^*)\}.$$

We call P^* a *spanning polytope* of Σ.

In particular, a strongly polytopal fan is complete. In analogy to polytopal spheres, we could call a fan polytopal if it is isomorphic to a strongly polytopal fan. We do, however, not need this notion. We use P^* in Definition 4.3 rather than P since the polar polytope of the spanning polytope of Σ will occur more often in later applications. Also, in accordance with I, Definition 4.14, we derive Theorem 4.4.

4.4 Theorem. *A fan Σ is strongly polytopal with spanning polytope P^* if and only if Σ is the fan $\Sigma = \Sigma(P)$ of (normal cones of) the polar polytope $P = P^{**}$ of P^*.*

PROOF. The polar polytope P of P^* is obtained as the intersection of all half-spaces $H^-_{v_i}$ (see I, Definition 6.1) where v_i is a vertex of P^*. So, if F is a facet of P^* with vertices $v_1, \ldots, v_k$, then, $\sigma = \operatorname{pos}\{v_1, \ldots, v_k\}$ is the corresponding cone, $v = H_{v_1} \cap \cdots \cap H_{v_k}$ is a vertex of P, and

$$N(v) = \sigma.$$

Since each face is an intersection of facets (II, Theorem 1.11), to an r-dimensional face of P^* there corresponds an $(n - r - 1)$-face of P, and, for a relative interior point v of that face, we find that $N(v) = \sigma$.

The converse is true by definition of $\Sigma(P)$ in I, 4.14. □

4.5 Theorem. *Given a complete fan Σ, there exists a strongly polytopal fan Σ' such that*

$$\mathrm{id} : \Sigma' \longrightarrow \Sigma$$

is a map of fans. $\Sigma' = \Sigma'(Z)$ can be chosen to be the fan of a zonotope Z.

PROOF. We extend each $(n - 1)$-cell of Σ to its linear hull and obtain, in this way, a hyperplane arrangement. By IV, Theorem 7.5, the theorem follows. □

From Theorem 4.4, it is readily deduced (compare IV, Definition 2.11):

4.6 Lemma. *P^*, Q^* are two spanning polytopes of the same fan Σ if and only if P and Q are strictly combinatorially isomorphic.*

Figure 6 illustrates Lemma 4.6. If a vertex v of P^* moves along the one-dimensional cone it spans, the polar hyperplane moves into a parallel position. In higher dimensions, v can always be "slightly moved" along $\mathbb{R}_{\geq 0}\, v$ without changing Σ if P^* is supposed to be simplicial. Dually, P must be simple, so that a

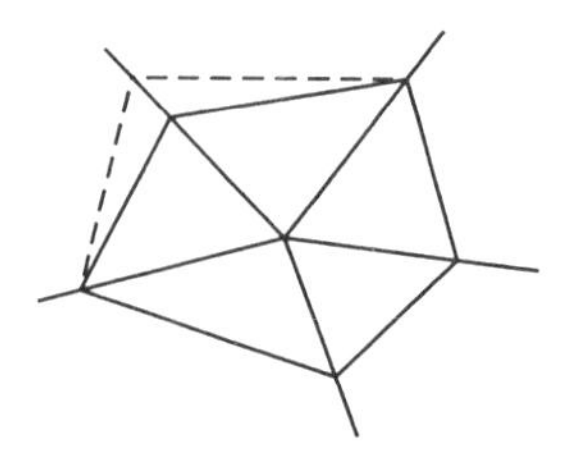
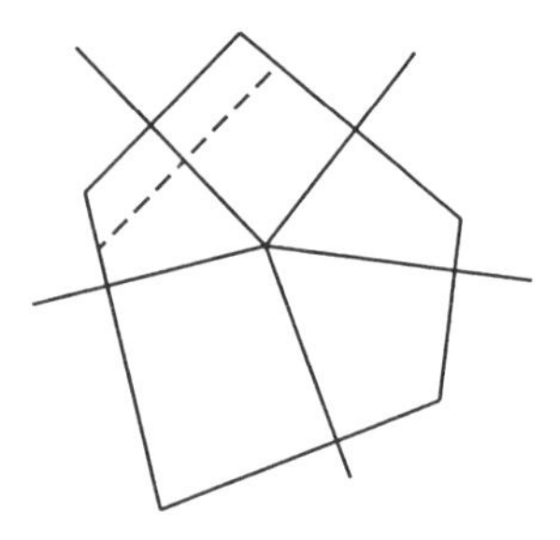

FIGURE 6.

little parallel displacement of the affine hull of a facet transforms P into a strictly combinatorially isomorphic polytope.

4.7 Theorem. *Every complete fan Σ in $\mathbb{R}^2$ is strongly polytopal.*

PROOF. We intersect the unit circle about 0 with the one-dimensional cones of Σ and choose as P^* the convex hull of the points thus obtained. □

A statement analogous to Theorem 4.7, for $n \geq 3$, does not hold. However, we can characterize the case of strong polytopality:

4.8 Theorem (Shephard's criterion). *Let $X := (a_1, \dots, a_k)$ be a finite sequence of lattice vectors in $\mathbb{Z}^n \subset U := \mathbb{R}^n$ that span the one-dimensional cones of a complete fan Σ, and let $\bar{X}_{\hat{U}}$ be a Gale transform of X. For each proper face $\sigma := \text{pos}\{a_{j_1}, \dots, a_{j_{k-r}}\}$ of Σ, we set $C(\sigma) := \text{conv}\,(\bar{X}_{\hat{U}} \setminus \{\hat{\bar{a}}_{j_1}, \dots, \hat{\bar{a}}_{j_{k-r}}\})$. Σ is strongly polytopal if and only if*

$$\bigcap_{\sigma \in \Sigma} \text{relint}\, C(\sigma) \neq \emptyset. \tag{1}$$

PROOF. We let $b_i := t_i a_i,\ t_i > 0,\ i = 1, \dots, k$. We wish to choose $t_1, \dots, t_k$ such that $P := \text{conv}\{b_1, \dots, b_k\}$ is a spanning polytope of Σ. From the Gale transform, $\bar{X}_{\hat{U}} = (\hat{\bar{a}}_1, \dots, \hat{\bar{a}}_k)$ of X (in $\mathbb{R}^{k-n-1}$; see II, Definition 4.16), we obtain a linear transform $\bar{X} = (\bar{a}_1, \dots, \bar{a}_k)$ of X (with U as linear space) by setting $\bar{a}_i := (\hat{\bar{a}}_i, 1) \in H_a \subset \mathbb{R}^{k-n},\ i = 1, \dots, k$.

Then, $\check{B} := (\check{b}_1, \dots, \check{b}_k) = (t_1^{-1}\bar{a}_1, \dots, t_k^{-1}\bar{a}_k)$ is a linear transform of $B := (b_1, \dots, b_k)$. Let $b := \check{b}_1 + \cdots + \check{b}_k$. Clearly, $b \neq 0$. Therefore, by II, Lemma 5.6, the perpendicular projection of $\check{B}$ onto a one-codimensional subspace H of $\mathbb{R}^{k-n}$ with normal b provides a Gale transform $\hat{\bar{B}} := (\hat{\bar{b}}_1, \dots, \hat{\bar{b}}_k)$ of B.

Let $c := (c', 1)$ be the point in which the ray $\mathbb{R}_{>0}\, b$ intersects H_a. Given a subsequence $Y = (a_{j_1}, \dots, a_{j_{k-r}})$ of X, we set $\widetilde{Y} = (\bar{a}_{i_1}, \dots, \bar{a}_{i_r})$. Then, the following statements are equivalent (Figure 7).

$$b \in \text{relint pos}\{\bar{a}_{i_1}, \dots, \bar{a}_{i_r}\}. \tag{2}$$

$$b \in \operatorname{relint} \operatorname{pos}\{\check{b}_{i_1}, \ldots, \check{b}_{i_r}\}. \tag{3}$$

$$c \in \operatorname{relint} \operatorname{pos}\{\bar{a}_{i_1}, \ldots, \bar{a}_{i_r}\}. \tag{4}$$

$$c \in \operatorname{relint} \operatorname{conv}\{\bar{a}_{i_1}, \ldots, \bar{a}_{i_r}\}. \tag{5}$$

$$c' \in \operatorname{relint} \operatorname{conv}\{\bar{\hat{a}}_{i_1}, \ldots, \bar{\hat{a}}_{i_r}\} = \operatorname{relint} C(\sigma). \tag{6}$$

A subsequence $(b_{j_1}, \ldots, b_{j_{k-r}})$ of B represents a face of P if and only if $\operatorname{pos}\{\check{b}_{j_1}, \ldots, \check{b}_{j_{k-r}}\}$ is a face of the cone spanned by P after the embedding into $\hat{U}$. By II, Theorem 4.14, this is equivalent to condition (3). Since (3) and (6) are equivalent, we conclude that a spanning polytope P exists if and only if c' can be found such that (6) is simultaneously satisfied for every $\sigma \in \Sigma$, that is, condition (1) is fulfilled. □

Examples of complete, nonstrongly polytopal fans:

Example 1. In $\mathbb{R}^3$ let $X = (a_1, \ldots, a_6)$ come from the vertices of a regular prism P with $0 \in \operatorname{int} P$, (see Figure 8).

Let the fan Σ be defined by the following facets:

$$\operatorname{pos}\{a_1, a_2, a_3\},\ \operatorname{pos}\{a_4, a_5, a_6\},\ \operatorname{pos}\{a_1, a_3, a_4\},\ \operatorname{pos}\{a_3, a_4, a_6\},$$
$$\operatorname{pos}\{a_1, a_2, a_5\},\ \operatorname{pos}\{a_1, a_4, a_5\},\ \operatorname{pos}\{a_2, a_3, a_6\},\ \operatorname{pos}\{a_2, a_5, a_6\}.$$

Then, the affine relationships

$$a_1 - a_3 - a_4 + a_6 = 0$$

and

$$a_1 - a_2 - a_4 + a_5 = 0$$

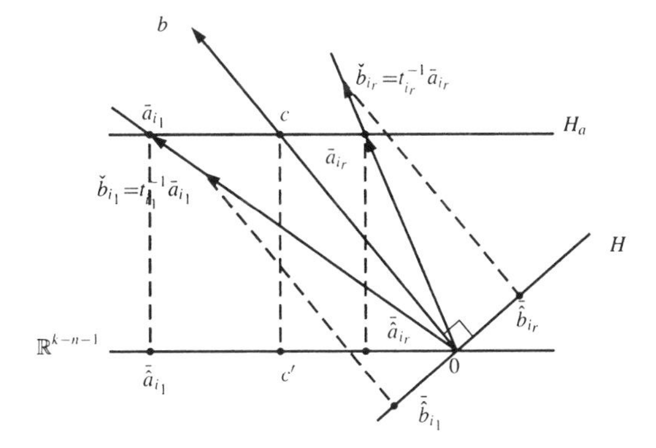

FIGURE 7.

provide a basis for the space of affine dependences of X. A Gale transform $\bar{X}_{\hat{U}}$ is formed by the rows of the transposed of the matrix

$$\begin{pmatrix} 1 & 0 & -1 & -1 & 0 & 1 \\ 1 & -1 & 0 & -1 & 1 & 0 \end{pmatrix},$$

(compare Figure 9).

It is readily seen that Shephard's condition is not satisfied: Look at the faces

$$\text{pos}\{a_2, a_6\}, \ \text{pos}\{a_3, a_4\}, \ \text{and} \ \text{pos}\{a_1, a_5\}.$$

From the combinatorial point of view, the subdivided prism is equivalent to an octahedron. However, its position in space can be varied only by moving the vertices along their positive hulls so that an octahedron is not obtained (see Exercise 1).

Example 2. In $\mathbb{R}^3$ choose a 3-simplex T, a vertex v of T, and a triangle $\Delta \subset (\mathbb{R}^3 \oplus \mathbb{R}) \setminus \text{aff}\, T$. $P := \text{conv}(T \cup \Delta)$ is then a polytope with 3 double-simplices and (in general) 4 simplices as facets. We denote the vertices of T by $1, 2, 3, 4 = v$ (Figure 10), the vertices of Δ (after projection) by 5, 6, 7. We consider a Schlegel diagram of P, namely a central projection into T.

We may place Δ such that we obtain facets $F_1, \ldots, F_7$, up to renumbering as follows:

$$\begin{aligned} F_1 &= [1,\ 2,\ 4,\ 5,\ 6] \\ F_2 &= [2,\ 3,\ 4,\ 6,\ 7] \\ F_3 &= [1,\ 3,\ 4,\ 5,\ 7] \\ F_4 &= [2,\ 5,\ 6,\ 7] \\ F_5 &= [4,\ 5,\ 6,\ 7] \\ F_6 &= [1,\ 2,\ 3,\ 5] \\ F_7 &= [2,\ 3,\ 5,\ 7]. \end{aligned}$$

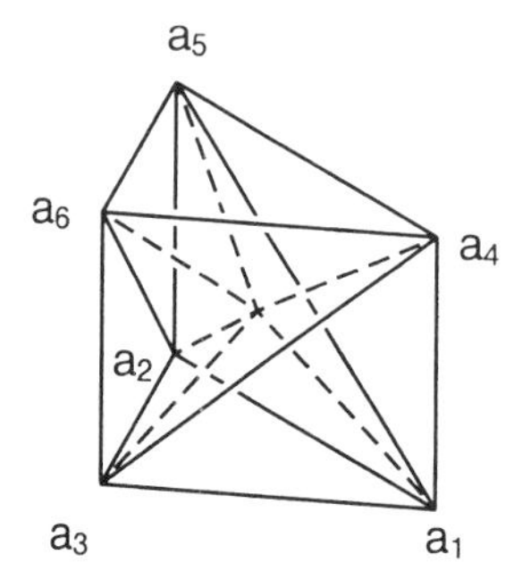

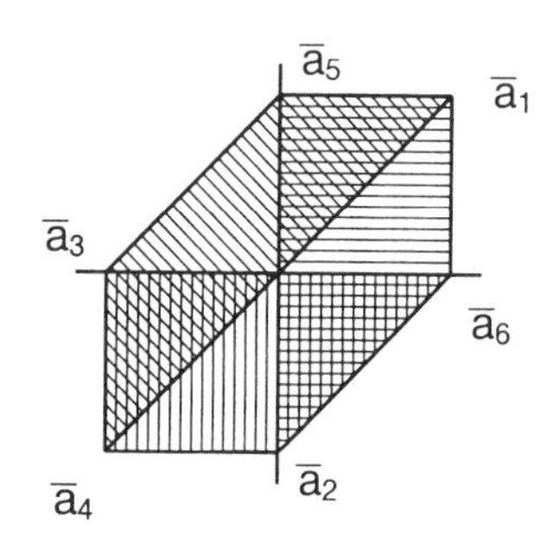

FIGURE 8, FIGURE 9.

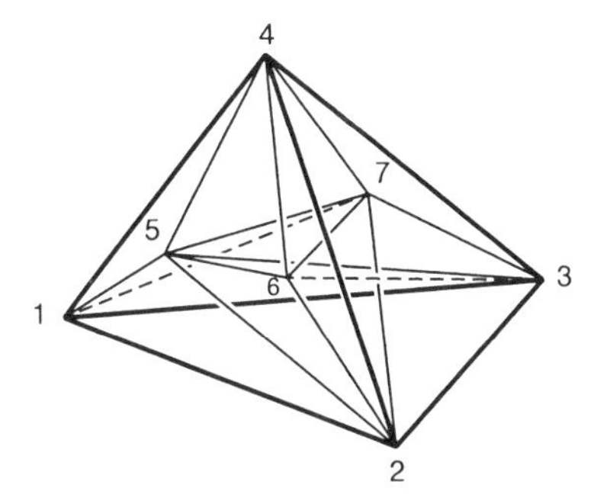

FIGURE 10.

We split the double simplices F_1, F_2, F_3 as follows:

Split F_1 along [2, 4, 5] into two simplices

Split F_2 along [3, 6] into three simplices

Split F_3 along [1, 7] into three simplices

In a Gale transform of the polytope P, we have, as cofaces, $[\bar{3}, \bar{7}]$, $[\bar{1}, \bar{5}]$, $[\bar{2}, \bar{6}]$ and four triangles. Without calculating the coordinates explicitly, we see that it is of the structure shown in Figure 11.

After the split we obtain additional cofaces $[\bar{2}, \bar{3}, \bar{4}, \bar{5}, \bar{6}]$, $[\bar{1}, \bar{2}, \bar{4}, \bar{5}, \bar{7}]$, $[\bar{1}, \bar{3}, \bar{6}, \bar{7}]$ which have no interior point in common. So the fan Σ obtained by projecting the splitted faces of P is not strongly polytopal.

The idea of Examples 1 and 2 underlies a general construction principle for nonstrongly polytopal fans:

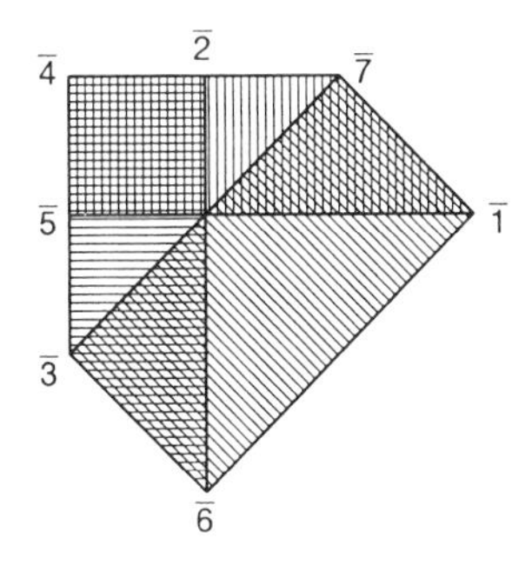

FIGURE 11.

4.9 Theorem.

(a) *There exist n-dimensional lattice polytopes, $n \geq 3$, with at least $f_0 - n$ facets (f_0 = number of vertices) which are simplicial but not simplices.*

(b) *If P is such a polytope, $0 \in \operatorname{int} P$, the facets that are not simplices can be dissected so that the fan obtained by projecting the new faces is not strongly polytopal.*

PROOF.

(a) Let C^n be an n-dimensional cube with center 0. In each edge s of C^n, consider that the hyperplane $H_s \supset s$ perpendicular to $\operatorname{lin} s$. We may assume that $0 \in H_s^-$. Then

$$P := \bigcap_{s \text{ edge of } C^n} H_s^-$$

is a polytope with $$f_0(P) = 2^n + 2n,$$

and $$f_{n-1}(P) = n \cdot 2^{n-1},$$

(compare Figure 12 for $n = 3$). Above each facet of C^n, a new vertex occurs. On each H_s, there are as many such vertices as the number of facets that the edge lies on, that is, $n - 1$ vertices $v_1, \ldots, v_{n-1}$. If v, v' are the end points of s, a facet of P is a bipyramid F over $\operatorname{conv}\{v_1, \ldots, v_{n-1}\}$. Since $s \setminus \{v, v'\} \subset \operatorname{relint} F$, F is a double simplex.

Since $f_0 - n = 2^n + n < n \cdot 2^{n-1}$ for $n \geq 3$, (a) holds.

(b) Let $F_1, \ldots, F_q, q \geq f_0 - n$, be the facets that are simplicial but not simplices. We apply Radon's theorem (I, Theorem 2.1) and find, for each F_i, a partition of vert F_i:

$$\operatorname{vert} F_i = V_i \cup V_i', \qquad V_i \cap V_i' = \emptyset,$$

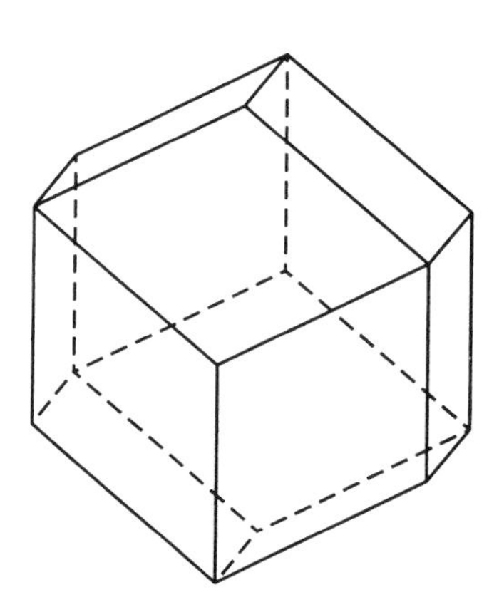

FIGURE 12.

such that

(1) $\quad (\operatorname{conv} V_i) \cap (\operatorname{conv} V_i')$ contains a relative interior point of F_i.

The Gale transform of vert P has dimension $f_0 - n - 1$, so the coface of each F_i is contained in a hyperplane H_i.

After the split, the cofaces are enlarged by an $\bar{x}_i$ for $x_i \in V_i \cup V_i'$, say $x_i \in V_i$. We may assume $\bar{x}_i \in H_i^+$. Then, $\bar{x}_j \in H_i^+$ for any other $x_j \in V_i$, since, otherwise,

$$0 \in \operatorname{relint} \operatorname{conv}(\{\bar{x} \mid x \in (\operatorname{vert} P) \setminus (\operatorname{vert} F_i)\} \cup \bar{V}_i)$$

would be a coface of P and conv V_i' would be a face of F_i, contrary to (1). Hence,

$$\bar{V}_i \subset H_i^+, \qquad \bar{V}_i' \subset H_i^-.$$

So, if we select, from each pair $(H_1^+, H_1^-), \ldots, (H_q^+, H_q^-)$, one half-space at random, the intersection of the interiors must be nonempty to fulfil Shephard's condition. Since $q \geq f_0 - n = 2 + \dim H_i$, the selection can be made such that the interiors have no point in common (also if some of the H_j are linearly dependent). So by Theorem 4.8, we can construct a fan which is not strongly polytopal.

□

Besides the face-splitting according to Theorem 4.9 there is still another way to find examples for nonstrongly polytopal fans. By definition of "nonpolytopal" (III, 4) we have the following theorem.

4.10 Theorem. *Let Σ_0 be an $(n-1)$-dimensional polyhedral sphere embedded into $\mathbb{R}^n$ as the boundary of a star-shaped set with 0 in its interior. If Σ_0 is nonpolytopal, the fan $\Sigma := \{\operatorname{pos} \sigma_0 \mid \sigma_0 \in \Sigma_0\}$ is complete but not strongly polytopal.*

As an example, consider the Barnette sphere (III, 4).

We now turn to a second fundamental property of fans:

4.11 Definition. A fan is called *regular* if all its cones are regular simplex cones.

Example 3. We start with a prism, as in Example 1, choosing $x_1 := e_1, x_2 := e_2, x_3 := e_3, x_4 := -e_2 - e_3, x_5 := -e_1 - e_3, x_6 := -e_1 - e_2$. Split the faces, as in Example 1, but also split $\operatorname{conv}\{x_4, x_5, x_6\}$ by a stellar subdivision (see III, 2) in direction $-e_1 - e_2 - e_3$.

The following characterization of regular fans is very useful.

4.12 Theorem (Oda's criterion). *A complete simplicial fan Σ is regular if and only if the following conditions are satisfied:*

(a) *There exists in Σ at least one regular n-cone.*

(b) *If* $\sigma = \operatorname{pos}\{x_1, x_2, \ldots, x_n\}$, $\sigma' = \operatorname{pos}\{x_1', x_2, \ldots, x_n\}$ *are two adjacent n-cones, there exist integers* $\alpha_2, \ldots, \alpha_n$ *such that*

$$x_1 + x_1' + \alpha_2 x_2 + \cdots + \alpha_n x_n = 0.$$

PROOF. Let Σ be regular, $\det \sigma = \det(x_1, x_2, \ldots, x_n) = 1$. Then, $\det(x_1', x_2, \ldots, x_n) = -1$, and, hence,

$$\det(x_1 + x_1', x_2, \ldots, x_n) = 0.$$

Since $x_2, \ldots, x_n$ are linearly independent, we can write

$$x_1 + x_1' = -\alpha_2 x_2 - \cdots - \alpha_n x_n.$$

From

$$\begin{aligned}\det(x_1, x_1', x_3, \ldots, x_n) &= \det(x_1, -\alpha_2 x_2, x_3, \ldots, x_n)\\ &= -\alpha_2 \det(x_1, \ldots, x_n) = -\alpha_2,\end{aligned}$$

we see that α_2 is integral. Similarly, $\alpha_3, \ldots, \alpha_n$ are shown to be integers. So (a) and (b) are true.

Conversely, let (a) and (b) be valid. For two adjacent n-cones $\sigma = \operatorname{pos}\{x_1, x_2, \ldots, x_n\}$, $\sigma' := \operatorname{pos}\{x_1', x_2, \ldots, x_n\}$, we deduce from (b) that

$$\det(x_1, x_2, \ldots, x_n) = -\det(x_1', x_2, \ldots, x_n).$$

Since Σ is complete, we shall see that all n-cones have the same absolute value of $\det \sigma$, that is $|\det \sigma| =: c$. Consider the set of all n-cones of Σ with the same absolute value of $|\det \sigma|$. Suppose it does not cover $\mathbb{R}^n$. Then, there is a gap which contains at least one n-cone σ_0, so the boundary of the gap contains at least one $(n-1)$-side of an n-cone σ' such that $|\det \sigma'| \neq |\det \sigma_0|$, contrary to the initial assumption.

By (a), therefore, all determinants of n-cones are ± 1. □

We remark that (a) in Theorem 4.12 cannot be left out, as is seen from the example illustrated in Figure 13 where all 3-cones have determinant ± 2.

Exercises

1. By direct arguments, show that, in Example 1 $x_1, \ldots, x_6$ cannot be moved on their positive hulls so that they become vertices of a polytope that spans Σ.
2. In Example 2, find all possible splits of F_1, F_2, F_3 according to Theorem 4.9 and investigate in which cases the resulting fan is polytopal and in which it is not.
3. Show that the fan, in Example 3, is regular but not strongly polytopal.
4. If a fan Σ is strongly polytopal, each fan obtained from Σ by a stellar subdivision is also strongly polytopal.

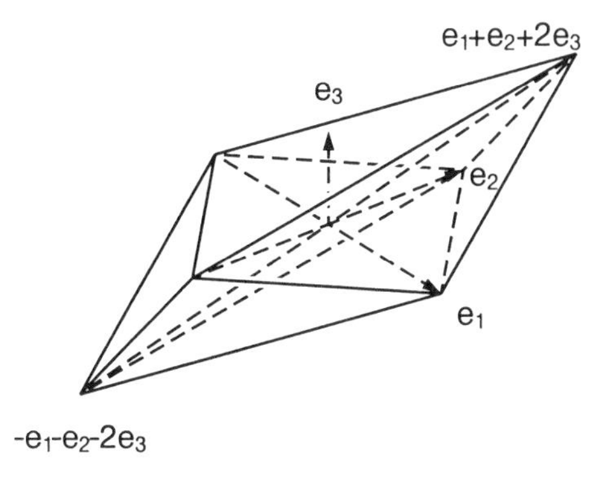

FIGURE 13.

5. The combinatorial Picard group

Given any fan Σ we denote the set of its dual cones by $\check{\Sigma}$,

$$\check{\Sigma} := \{\check{\sigma} \mid \sigma \in \Sigma\}.$$

Examples (Figure 14).

If Σ consists of the four quadrants of $\mathbb{R}^2$, the eight octands of $\mathbb{R}^3$, or, generally, the cones in which $\mathbb{R}^n$ is subdivided by the coordinate hyperplanes, all n-cones are self-dual. If we consider in Figure 14 the noncomplete fans consisting of $\{0\}$ and the one-dimensional cones, $\check{\Sigma}$ consists of $\mathbb{R}^2$ and the half-planes which occur in the illustrations of $\check{\Sigma}$.

5.1 Lemma. *If $\sigma = \sigma_1 + \cdots + \sigma_r$, where σ_i are the 1-dimensional faces of σ, then $\check{\sigma} = \check{\sigma}_1 \cap \cdots \cap \check{\sigma}_r$ is an intersection of half-spaces.*

PROOF. This is clear from the definitions and Lemma 2.2. □

If Σ is a fan and $\sigma \in \Sigma$, we assign to $\check{\sigma}$ the monoid

$$S := \check{\sigma} \cap \mathbb{Z}^n .$$

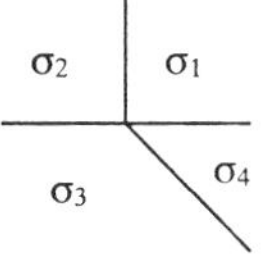

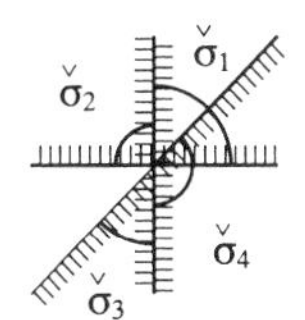

FIGURE 14.

We obtain a system

$$\mathcal{S} = \mathcal{S}(\Sigma) := \{S = \check{\sigma} \cap \mathbb{Z}^n \mid \sigma \in \Sigma\}$$

of monoids assigned to Σ. So, there are bijective relations between Σ, $\check{\Sigma}$, $\mathcal{S}(\Sigma)$

$$\Sigma \longleftrightarrow \check{\Sigma} \longleftrightarrow \mathcal{S}(\Sigma).$$

From any one of them, we can always reconstruct the others.

From the definitions and Lemma 2.2 we find that

5.2 Lemma. *For any two cones σ_1, σ_2 of a fan Σ,*
(a) $(\sigma_1 + \sigma_2)\check{}\cap \mathbb{Z}^n = (\check{\sigma}_1 \cap \mathbb{Z}^n) \cap (\check{\sigma}_2 \cap \mathbb{Z}^n)$, *and*
(b) $(\sigma_1 \cap \sigma_2)\check{}\cap \mathbb{Z}^n = (\check{\sigma}_1 \cap \mathbb{Z}^n) + (\check{\sigma}_2 \cap \mathbb{Z}^n)$.

The monoids of $\mathcal{S}(\Sigma)$ are all sub-semigroups of the additive group $\mathbb{Z}^n$. Now, we consider residue classes of the sub-semigroups in $\mathbb{Z}^n$.

To each $\check{\sigma} \in \check{\Sigma}$, we assign $m_\sigma \in \mathbb{Z}^n$ such that the following condition is satisfied:

(1) If σ_0 is a face of σ, $\quad m_\sigma - m_{\sigma_0} \in \operatorname{cospan} \check{\sigma}_0$.

Condition (1) guarantees that the inclusion of monoids in $\mathcal{S}(\Sigma)$ is preserved if we replace each monoid $\check{\sigma} \cap \mathbb{Z}^n$ by its residue class $m_\sigma + S = m_\sigma + (\check{\sigma} \cap \mathbb{Z}^n)$.

5.3 Definition. A system $\mathcal{P} := \{m_\sigma + \check{\sigma}\}_{\sigma \in \Sigma}$ of translated cones is called a *virtual polytope* (*with respect to the fan* Σ) provided $\{m_\sigma + \check{\sigma}\}_{\sigma \in \Sigma}$ satisfies (1).

If Σ is strongly polytopal, we may choose $\{m_\sigma\}_{\sigma \in \Sigma}$ such that

(2) $$\bigcap_{\sigma \in \Sigma} (m_\sigma + \check{\sigma}) =: P$$

is a lattice polytope, and $-P^*$ spans Σ. Then, $\mathcal{P}$ and P can be identified. More generally, (2) represents a Minkowski summand of the negative dual of the spanning polytope of Σ.

5.4 Lemma. *The virtual polytopes with respect to the same fan Σ are a commutative group $\tilde{\mathcal{G}}$ with respect to the following addition ($\mathcal{P}' := \{m'_\sigma + \check{\sigma}\}_{\sigma \in \Sigma}$)*

(3) $$\mathcal{P} + \mathcal{P}' := \{m_\sigma + m'_\sigma + \check{\sigma}\}_{\sigma \in \Sigma}.$$

The zero element is $\check{\Sigma}$.

Remark. In the case where $\mathcal{P}$ and $\mathcal{P}'$ can be identified with polytopes P and P', respectively, $\mathcal{P} + \mathcal{P}'$ corresponds to the Minkowski sum $P + P'$.

Proof of Lemma 5.4. Since $0 \in \check{\sigma}$, $\check{\sigma} + \check{\sigma} = \check{\sigma}$, hence,

$$(m_\sigma + \check{\sigma}) + (m'_\sigma + \check{\sigma}) = m_\sigma + m'_\sigma + \check{\sigma}.$$

Therefore, the addition of $\mathcal{P}$ and $\mathcal{P}'$ reduces to ordinary set addition. $\check{\Sigma}$ is the zero element since $(m_\sigma + \check{\sigma}) + \check{\sigma} = m_\sigma + \check{\sigma}$ for any $\sigma \in \Sigma$.

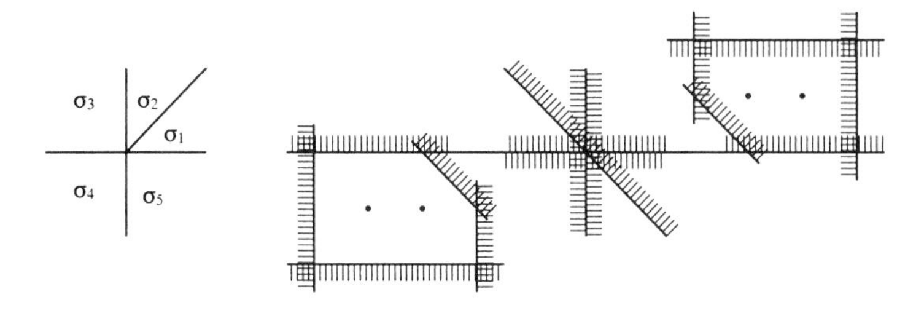

FIGURE 15.

The negative of $\mathcal{P} = \{m_\sigma + \check{\sigma}\}_{\sigma\in\Sigma}$ can always be obtained by choosing $-\mathcal{P} := \{-m_\sigma + \check{\sigma}\}_{\sigma\in\Sigma}$ as illustrated in Figure 15, 16, and 17 (with $\mathcal{M}$ as in Lemma 5.5).

We still have to show that (1) is preserved under the addition (3). Let σ_0 be a face of σ. Then,

$$m_\sigma - m_{\sigma_0} \in \operatorname{cospan} \check{\sigma}_0 \qquad \text{and} \qquad m'_\sigma - m'_{\sigma_0} \in \operatorname{cospan} \check{\sigma}_0$$

imply

$$(m_\sigma + m'_\sigma) - (m_{\sigma_0} + m'_{\sigma_0}) \in \operatorname{cospan} \check{\sigma}_0,$$

since cospan $\check{\sigma}_0$ is a linear space. Hence, (1) remains true. □

It should be noted that in $-\mathcal{P} = \{-m_\sigma + \check{\sigma}\}_{\sigma\in\Sigma}$ only the apexes $-m_\sigma$ are the negatives of the m_σ in $\mathcal{P}$, whereas the cones $\check{\sigma}$ remain unchanged.

We remark, further, that neither $\mathcal{P}$ nor $-\mathcal{P}$ needs to define a polytope, as Figure 17 illustrates (where Σ is the fan given by the quadrants of $\mathbb{R}^2$ and their faces).

The following lemma is immediate from Lemma 5.4 and the definition of monoids belonging to $\mathcal{S}(\Sigma)$:

5.5 Lemma.

(a) *The systems $\mathcal{M} := \{m_\sigma + \check{\sigma} \cap \mathbb{Z}^n\}_{\sigma\in\Sigma} = \{m_\sigma + S\}_{\sigma\in\Sigma}$ of residue classes assigned to the semi-groups of $\mathcal{S}(\Sigma)$ define a commutative group $\mathcal{G}$ with*

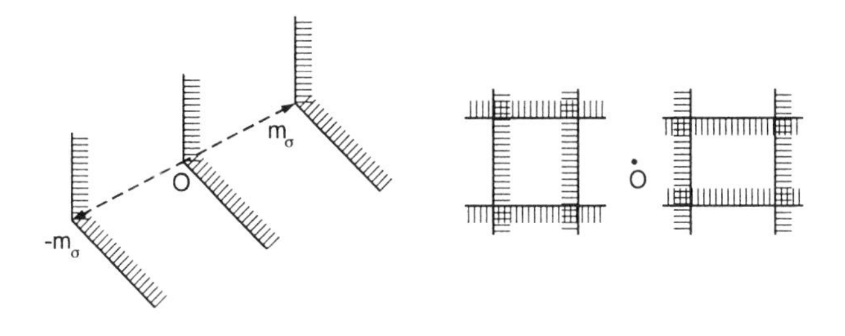

FIGURE 16, FIGURE 17.

respect to the addition ($\mathcal{M}' := \{m'_\sigma + S\}_{\sigma\in\Sigma}$)

(4) $$\mathcal{M} + \mathcal{M}' := \{m_\sigma + m'_\sigma + S\}_{\sigma\in\Sigma}.$$

The zero element is $\mathcal{S}(\Sigma)$.

(b) *The groups* $\mathcal{G}$ *and* $\tilde{\mathcal{G}}$ *are isomorphic.*

Many properties of virtual polytopes remain true under translations (applied simultaneously to all $m_\sigma + \check{\sigma}, \sigma \in \Sigma$) as is the case for polytopes. So, it is natural to assign to $\mathcal{G}$ or $\tilde{\mathcal{G}}$ the following group:

5.6 Definition. If $\mathcal{G}$ is the group of Lemma 5.5, we call $\mathcal{G}/\mathbb{Z}^n$ the *combinatorial Picard group* Pic Σ of Σ. We denote its elements by $\mathcal{P}$.

The Picard group Pic Σ is a finitely generated, commutative group, and, hence, by the fundamental theorem on commutative groups, is equivalent to a direct sum

$$\text{Pic}\,\Sigma \cong \mathbb{Z}^q \oplus \mathbb{Z}_{q_1} \oplus \cdots \oplus \mathbb{Z}_{q_p},$$

where $\mathbb{Z}_i$ denotes the finite cyclic group with i elements. $\mathbb{Z}_{q_1} \oplus \cdots \oplus \mathbb{Z}_{q_p}$ is called the *torsion* of the group, q its *Betti number*. Torsion can be nonzero, as is in the following example:

Example 1. In $\mathbb{R}^2$, let $\Sigma := \{\sigma_0, \sigma_1, \sigma_2\}$ where $\sigma_0 := \{0\}$, $\sigma_1 := \mathbb{R}_{\geq 0}\, e_1$, and

$$\sigma_2 := \mathbb{R}_{\geq 0}(e_1 + 2e_2).$$

We consider $m_{\sigma_0} = m_{\sigma_1} = 0, m_{\sigma_2} = e_1$. Then, $(m_{\sigma_1} + \check{\sigma}_1) \cap (m_{\sigma_2} + \check{\sigma}_2)$ is a cone whose apex $(0, \frac{1}{2})$ is not a lattice point—which it need not be since $\sigma_1 + \sigma_2 \notin \Sigma$ (Figure 18). The virtual polytope $\mathcal{P} := \{m_{\sigma_i} + \check{\sigma}_i\}_{i=0,1,2}$ is not obtained from $\check{\Sigma}$ by simultaneous translation via a lattice vector, so it does not represent the zero element of Pic Σ. However, $2\mathcal{P} = \{2m_{\sigma_i} + \check{\sigma}_i\}$ is obtained from $\check{\Sigma}$ by adding e_2 to each cone, and so does represent zero. It is readily seen that Pic $\Sigma \cong \mathbb{Z}_2$.

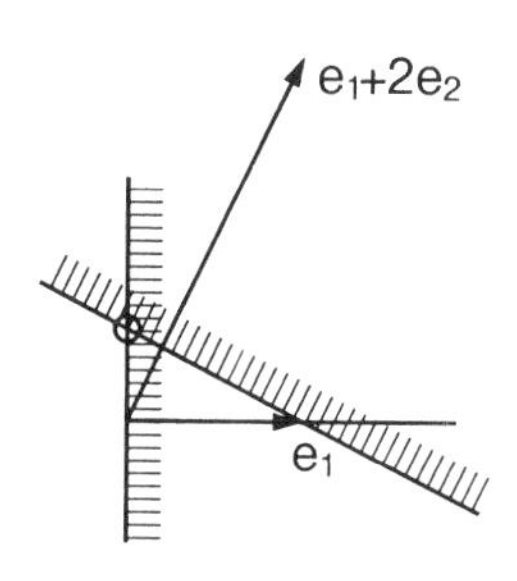

FIGURE 18.

A situation, as in Example 1, does not occur if at least one cone $\sigma \in \Sigma$ is n-dimensional.

5.7 Lemma. *For* Pic Σ *to be a torsion-free group, it is sufficient that* Σ *contains an* n*-cone* τ.

PROOF. Suppose Pic Σ contains an element of finite order. Then, there is a virtual polytope $\mathcal{P} = \{m_\sigma + \check{\sigma}\}_{\sigma \in \Sigma}$ and a natural number r such that $r\mathcal{P} = \{rm_\sigma + \check{\sigma}\}_{\sigma \in \Sigma}$ can be obtained from $\check{\Sigma}$ by adding a lattice vector c. Since τ is n-dimensional, $\{rm_\tau\} = \operatorname{cospan} \check{\tau}$ is a lattice point, and, hence, $rm_\tau = c$. Since m_τ is also a lattice point c_0, $\mathcal{P} = c_0 + \check{\Sigma}$, so that $\mathcal{P}$ represents the zero element of Pic Σ. □

If Σ contains an n-cone, we can calculate Pic Σ explicitly. First, we discuss the simplicial case:

5.8 Theorem. *Let* Σ *be a simplicial fan in* $\mathbb{R}^n$ *which contains at least one* n*-cone, and let* k *be the number of one-dimensional cones of* Σ*. Then,*

$$\operatorname{Pic} \Sigma \cong \mathbb{Z}^{k-n} .$$

PROOF. We assume, first, that Σ is not only simplicial but also regular. Since Σ is simplicial, a virtual polytope $\mathcal{P}$ is determined by an arbitrary choice of the m_ϱ for all one-dimensional $\varrho \in \Sigma$. For any such ϱ, we may replace m_ϱ by any other point of the hyperplane $m_\varrho + \varrho^\perp$. We may choose, for the sake of our proof, m_ϱ not necessarily as a lattice point. The hyperplane $m_\varrho + \varrho^\perp$ must, however, always contain at least one point of $\mathbb{Z}^n$. It is convenient to let m_ϱ be the foot of 0 on $m_\varrho + \varrho^\perp$. If $\varrho_1, \ldots, \varrho_k$ are all 1-cones of Σ, we write

$$m_{\varrho_i} = \alpha_i c_i, \qquad \alpha_i \geq 0, \quad \|c_i\| = 1, \quad i = 1, \ldots, k.$$

Then,

$$\langle c_i, x \rangle = \alpha_i \tag{3}$$

is an equation of $m_{\varrho_i} + \varrho_i^\perp$. Let d_i be a multiple of c_i such that the hyperplane $H_i = \{x \mid \langle d_i, x \rangle = 1\}$ has the following properties:

(a) There exists a lattice point on H_i.

(b) There does not exist a lattice point y such that $0 < \langle d_i, y \rangle < 1$.

Then, each hyperplane parallel to H_i and containing a lattice point is given by an equation

$$\langle d_i, x \rangle = r_i \tag{4}$$

where r_i is an arbitrary integer.

From the regularity of Σ, we deduce that (by Cramer's rule), for $d_{i_1}, \ldots, d_{i_m}$ representing $m \leq n$ simple vectors which span a face of Σ, the system (4) of equations ($i = i_1, \ldots, i_m$) has a lattice point as a solution. Therefore, the vectors $(r_1, \ldots, r_k) \in \mathbb{Z}^k$ can be chosen arbitrarily as representatives of virtual polytopes. Since we identify virtual polytopes which differ only by a translation vector $\in \mathbb{Z}^n$ and by considering Lemma 5.7, Pic $\Sigma \cong \mathbb{Z}^k / \mathbb{Z}^n \cong \mathbb{Z}^{k-n}$.

If Σ is simplicial but not regular, we proceed as above. However, the numbers $r_i \in \mathbb{Z}$ cannot be chosen arbitrarily since, in general, the systems (4) for $i = i_1, \ldots, i_m$ do not have integral solutions, only rational ones. Multiplying one r_i and all solutions of the respective equations (4) by an appropriate integral factor yields an integral solution. So the vectors $(r_1, \ldots, r_k)$ which represent defining elements of Pic Σ can be varied in each component. Since $\mathbb{Z}$ has only $\{0\}$ and groups isomorphic to $\mathbb{Z}$ as subgroups, it follows again that Pic $\Sigma \cong \mathbb{Z}^{k-n}$. □

5.9 Theorem. *Let Σ be a fan in $\mathbb{R}^n$ which contains at least one n-cone, and let $\varrho_1, \ldots, \varrho_k$ be the one-dimensional cones of Σ. We consider all maximal faces $\sigma_1, \ldots, \sigma_q$ of Σ which are not simplex cones, and set, for $\sigma_i = \varrho_{i_1} + \cdots + \varrho_{i_s}$, $i = 1, \ldots, q$,*

$$L_{\sigma_i} := \mathcal{L}(d_{i_1}, \ldots, d_{i_s}) \qquad \textit{(space of linear dependencies)}$$

and

$$L := L_{\sigma_1} + \cdots + L_{\sigma_q}, \qquad \lambda := \dim L.$$

Then,

$$\operatorname{Pic} \Sigma \cong \mathbb{Z}^{k-n-\lambda}.$$

PROOF. We consider the vectors $(r_1, \ldots, r_k)$ introduced in the proof of Theorem 5.8 as representatives of virtual polytopes. They can no longer be chosen as arbitrary lattice vectors. We must find the relationships which they satisfy.

Let σ_i be given as in the theorem, and let m_{σ_i} be the corresponding defining lattice vector of a virtual polytope. The one-faces of σ_i are spanned by vectors $d_{i_1}, \ldots, d_{i_s}$; they satisfy

$$\begin{aligned} \langle d_{i_1}, m_{\sigma_i} \rangle &= r_{i_1} \\ &\vdots \\ \langle d_{i_s}, m_{\sigma_i} \rangle &= r_{i_s} \end{aligned} \tag{5}$$

where the individual equations are introduced as in (4) (compare Figure 19).

It is useful to introduce the following matrix A. If L_{σ_1} is embedded in $\mathcal{L}(d_1, \ldots, d_k)$ canonically, we may write a basis of L_{σ_1} as row vectors in k components. We choose this basis as the first rows of A. The next rows are chosen as a basis of L_{σ_2}, analogously. Continuing in this way, the last rows of A are a basis of L_{σ_q}.

By II, Lemma 4.8, A may be considered to be composed of linear transforms (column vectors) of the sequences $(d_{i_1}, \ldots, d_{i_s})$ as follows

$$\begin{matrix} \text{basis of } L_{\sigma_1} \; \{ \\ \\ \\ \text{basis of } L_{\sigma_q} \; \{ \end{matrix} \begin{pmatrix} 0 \cdots\cdots 0\ \bar{d}_{1_1}\ 0 \cdots\cdots & & \cdots\cdots 0\ \bar{d}_{1_{s_1}}\ 0 \cdots\cdots \\ & \vdots & \\ \cdots & \bar{d}_{i_j} & \cdots \\ & \vdots & \\ 0 \cdots 0\ \bar{d}_{q_1}\ 0 \cdots\cdots & & \cdots\cdots 0\ \bar{d}_{q_{s_q}}\ 0 \cdots \end{pmatrix} =$$

Now, the conditions for $r = (r_1, \ldots, r_k) \in \mathbb{Z}^k$ to represent a virtual polytope are, by II, Lemma 4.8 (and the fact that $(d_{i_1}, \ldots, d_{i_s})$ is a linear transform of $(\bar{d}_{i_1}, \ldots, \bar{d}_{i_s})$)

$$Ar^t = 0.$$

Since A has rank λ, the vectors r span a subspace of $\mathbb{R}^k$ of dimension $k - \lambda$. So we obtain

$$\operatorname{Pic} \Sigma = \mathbb{Z}^{k-\lambda} / \mathbb{Z}^n \cong \mathbb{Z}^{k-n-\lambda}.$$

□

Remark. In the above proof, the vectors a in II, Lemma 4.8, attain a concrete meaning as points m_σ. In the case where the $\mathcal{P}_i$ are ordinary polytopes, it suffices to consider all vertices of the $\mathcal{P}_i$ as points m_σ.

Example 2. If Σ is simplicial, no nonzero space L_{σ_i} occurs, so that $\lambda = 0$ and Theorem 5.9 reduces to Theorem 5.8.

Example 3. Let Σ in $\mathbb{R}^3$ be spanned by the cube with vertices $\pm e_1 \pm e_2 \pm e_3$. We obtain L to be 4-dimensional and, hence, $\operatorname{Pic} \Sigma \cong \mathbb{Z}$.

Example 4. In Example 3, we replace the generator $e_1 + e_2 + e_3$ of a 1-cone by $e := 2e_1 + 2e_2 + 3e_3$, and change all faces containing $e_1 + e_2 + e_3$ by taking e as the generating vector instead of $e_1 + e_2 + e_3$. (Figure 20). Now, $\dim L = \lambda = 5$ so that $\operatorname{Pic} \Sigma = \{0\}$.

Remark. The fan of Example 4 cannot be spanned by the faces of a closed polyhedron (with planar faces). This is readily seen from the fact that those three spanning rectangles which meet on $\mathbb{R}_{\geq 0}(-e_1 - e_2 - e_3)$ determine the remaining three rectangles, the resulting polyhedron being a projective image of a cube. Then, however, e would have to be a multiple of $e_1 + e_2 + e_3$ which is not true.

Definition. We call $\mu(\Sigma) := k - n - \lambda$ the *combinatorial Picard number* of Σ.

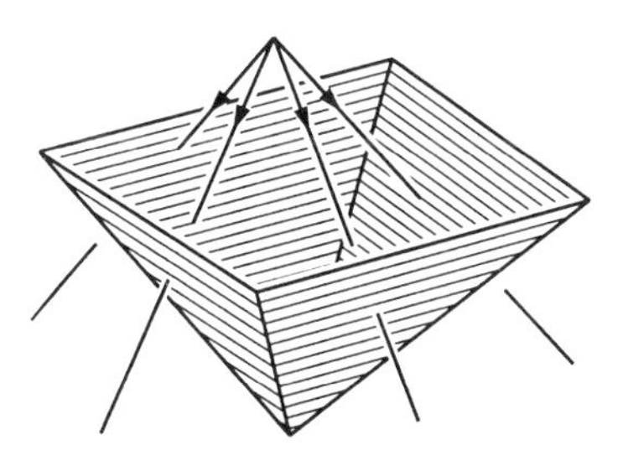

FIGURE 19.

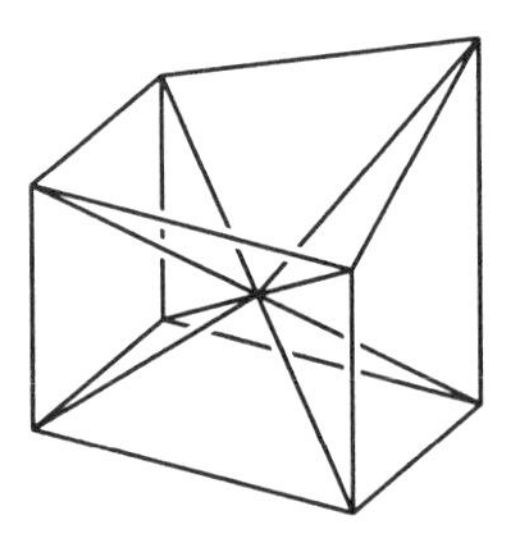

FIGURE 20.

Example 5. If Σ consists of an n-cone σ with k generators and all faces of σ, then $\lambda = k - n$ and $\mu(\Sigma) = 0$.

We investigate more thoroughly the case of complete fans Σ. In this case we may choose ($\Sigma^{(n)}$ the set of n-cones of Σ)

$$m_{\sigma_0} = m_\sigma, \qquad \text{if } \sigma_0 \text{ is a face of } \sigma \in \Sigma^{(n)},$$

where, of course, Equation (1) must be observed. So, if $\Sigma^{(n)} = \{\sigma_1, \ldots, \sigma_q\}$ and $a_i := m_{\sigma_i}, i = 1, \ldots, q$, the cones

$$\{a_1 + \check{\sigma}_1, \ldots, a_q + \check{\sigma}_q\}$$

determine an element $\mathcal{P}$ of Pic Σ. We write

$$\mathcal{P} := [a_1 + \check{\sigma}_1, \ldots, a_q + \check{\sigma}_q].$$

Definition. We call P an *associated* polytope of Σ, if $\Sigma = \Sigma(-P)$, that is, if Σ is spanned by $-P^*$ (compare Theorem 4.4), or, in other words, if Σ is the fan (of normal cones) of $-P$ (Figure 21).

Clearly,

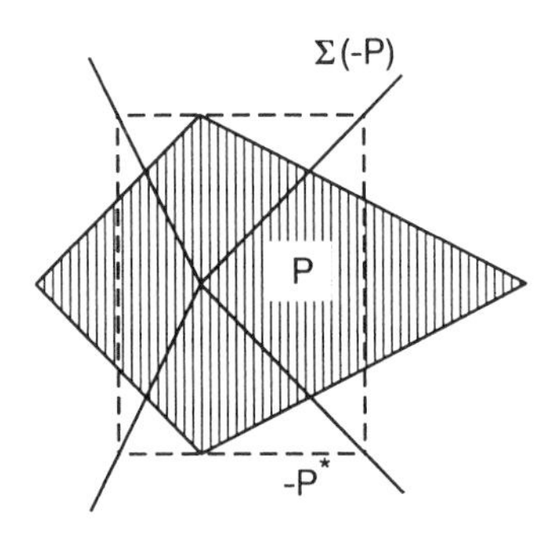

FIGURE 21.

5.12 Lemma. *Let Σ be complete and strongly polytopal, and, let $-P^*$ be a spanning polytope of Σ, so that P is an associated polytope of Σ. Then, for* vert $P = \{a_1, \dots, a_q\}$,

$$\mathcal{P} = \mathcal{P}(P) = [a_1 + \mathrm{pos}(P - a_1), \dots, a_q + \mathrm{pos}(P - a_q)]$$

is an element of Pic Σ *from which Σ can be reconstructed. Thus,*

$$\Sigma = \Sigma(-P) \quad \textit{for} \quad P = (a_1 + \mathrm{pos}(P - a_1)) \cap \cdots \cap (a_q + \mathrm{pos}(P - a_q)).$$

Any (Minkowski) summand P' of P can also be written in the form

$$P' = (a_1' + \mathrm{pos}(P - a_1)) \cap \cdots \cap (a_q' + \mathrm{pos}(P - a_q))$$

where an obvious assignment

$$a_i \longmapsto a_i'$$

provides a surjective map

$$\chi_{P'} : \mathrm{vert}\, P \longrightarrow \mathrm{vert}\, P'.$$

5.13 Definition. If P' is a lattice summand of an associated polytope P of the strongly polytopal fan Σ (possibly $P' = P$), we call (for $a_i' := \chi_{P'}(a_i)$, $a_i \in$ vert P)

$$\mathcal{P}(P') := [a_1' + \mathrm{pos}(P - a_1), \dots, a_q' + \mathrm{pos}(P - a_q)]$$

a *polytope element* of Pic Σ = Pic $\Sigma(-P)$.

5.14 Lemma. *Let $\Sigma = \Sigma(-P)$, and let P', P'' be lattice polytopes such that*

$$P = P' + P''.$$

Then,

$$\mathcal{P}(P) = \mathcal{P}(P') + \mathcal{P}(P'').$$

In particular, for any natural number r,

$$\mathcal{P}(rP) = r\mathcal{P}(P).$$

PROOF. This follows directly from the above definitions and Lemma 5.12. □

The following theorem will enable us to find a finite system of generators of Pic $\Sigma(-P)$, consisting of polytope elements.

5.15 Theorem. *For any $\mathcal{P} \in$ Pic $\Sigma(-P)$ there exists a lattice polytope P_0, strictly combinatorially isomorphic to P (hence also associated with Σ), and a natural number r such that*

$$\mathcal{P} = \mathcal{P}(P_0) - \mathcal{P}(rP).$$

PROOF. Let $P = (a_1 + \check{\sigma}_1) \cap \cdots \cap (a_q + \check{\sigma}_q)$. Since $\check{\sigma}_i \cap \mathbb{Z}^n = S_i$, we have a $1-1$-relationship between the $\check{\sigma}_i$ and the S_i. We can also represent P as an intersection of half-spaces,

$$P = H_1^- \cap \cdots \cap H_k^-, \qquad \text{where } 0 \in \operatorname{int} P \text{ is assumed,}$$

and $H_i \cap P$, $i = 1, \ldots, k$ are the facets of P.

Moreover, any representative of

$$\mathcal{P} = [b_1 + \check{\sigma}_1, \ldots, b_q + \check{\sigma}_q]$$

can be characterized by

$$\{\bar{H}_1^-, \ldots, \bar{H}_k^-\},$$

where each $\bar{H}_i^-$ is a translate of H_i^-, $i = 1, \ldots, k$. Further, by definition of $\mathcal{P}$, we obtain a natural assignment

$$H_{i_1} \cap \cdots \cap H_{i_p} \quad \longrightarrow \quad \bar{H}_{i_1} \cap \cdots \cap \bar{H}_{i_p}$$

for any subset $\{i_1, \ldots, i_p\} \subset \{1, \ldots, k\}$, as a result of the translations. In particular, the vertices a_i and b_i are intersections such that

$$\text{(1)} \quad H_{j_1} \cap \cdots \cap H_{j_s} = a_j \quad \longmapsto \quad b_j = \bar{H}_{j_1} \cap \cdots \cap \bar{H}_{j_s}, \qquad j = 1, \ldots, q.$$

For any positive integer r, we define

$$\begin{aligned} P_{(r)} : &= (rH_1^- + \bar{H}_1^-) \cap \cdots \cap (rH_k^- + \bar{H}_k^-) \\ &= H_1^{(r)-} \cap \cdots \cap H_k^{(r)-}, \end{aligned}$$

where $H_i^{(r)} := rH_i + \bar{H}_i$, $i = 1, \ldots, k$. Furthermore, we set

$$b_j^{(r)} := H_{j_1}^{(r)} \cap \cdots \cap H_{j_s}^{(r)}.$$

We claim that

(a) $b_j^{(r)} = ra_j + b_j$ is a point, and

(b) for sufficiently large r, $\{b_1^{(r)}, \ldots, b_q^{(r)}\} = \operatorname{vert} P_{(r)}$.

PROOF OF (a).

$$H_{j_\varrho}^{(r)} = rH_{j_\varrho} + b_{j_\varrho} \qquad \text{for } j_\varrho \in \{j_1, \ldots, j_s\},$$

hence,

$$\begin{aligned} b_j^{(r)} := H_{j_1}^{(r)} \cap \cdots \cap H_{j_s}^{(r)} &= (rH_{j_1} + b_j) \cap \cdots \cap (rH_{j_s} + b_j) \\ &= r(H_{j_1} \cap \cdots \cap H_{j_s}) + b_j \\ &= ra_j + b_j, \qquad \text{a point.} \end{aligned}$$

PROOF OF (b). For sufficiently large r, $ra_j + b_j \notin \operatorname{conv}\{ra_i + b_i \mid i \neq j\}$, since a_j is strictly separated from $(\operatorname{vert} P) \setminus \{a_j\}$ by a hyperplane. This proves $\{b_1^{(r)}, \ldots, b_q^{(r)}\} \subset \operatorname{vert} P_{(r)}$.

We show that $b_j^{(r)} \notin H_i^{(r)}$ for $i \notin \{j_1, \ldots, j_s\}$.

In fact, since a_j has positive distance from any H_i, with $i \notin \{j_1, \ldots, j_s\}$, we obtain $\frac{1}{r} b_j^{(r)} = a_j + \frac{1}{r} b_j \notin H_i + \frac{1}{r} b_i$ for sufficiently large r, and, hence, $b_j^{(r)} \notin r H_i + b_i = H_i^{(r)}$.

Therefore, if B, B' are sufficiently small balls with the same radius and centers a_j, $b_j^{(r)}$, respectively, the sets $B \cap P$ and $B' \cap P_{(r)}$ are translates of each other. Hence,

(2)

$$\mathrm{pos}(P - a_j) = \mathrm{pos}[(B \cap P) - a_j] = \mathrm{pos}[(B' \cap P_{(r)}) - b_j^{(r)}] = \mathrm{pos}(P_{(r)} - b_j^{(r)}).$$

In B, B', respectively,

$$\dim(H_{j_\varrho} \cap P) = \dim(H_{j_\varrho}^{(r)} \cap P_{(r)}) = n - 1,$$

so $H_{j_\varrho}^{(r)} \cap P_{(r)}$ is a facet $F_{j_\varrho}^{(r)}$ of $P_{(r)}$, $\varrho = 1, \ldots, s$.

Suppose $(\mathrm{vert}\, P_{(r)}) \setminus \{b_1^{(r)}, \ldots, b_q^{(r)}\} \neq \emptyset$. Then, the edge graph of any polytope being connected, we find an edge $[b_j^{(r)}, b]$ of $P_{(r)}$ for which $b \in \mathrm{vert}\, P_{(r)} \setminus \{b_1^{(r)}, \ldots, b_q^{(r)}\}$. (1) implies (up to renumbering the $H_{j_\varrho}^{(r)}$)

$$\mathrm{aff}[b_j^{(r)}, b] = H_{j_1}^{(r)} \cap \cdots \cap H_{j_p}^{(r)} =: g.$$

Since there is a $b_i^{(r)} \neq b_j^{(r)}$ on the line g (by Equation (2)), we find $b = b_j^{(r)}$; otherwise, there would be three vertices of $P_{(r)}$. This proves (b).

So, we have a bijection φ between vert P and vert $P_{(r)}$ which is inclusion-preserving for facets and maps each facet onto a facet with the same outer normal. Therefore, for sufficiently large r, we obtain a strict combinatorial isomorphism between P and $P_0 := P_{(r)}$. Since $r H_i + \bar{H}_i = H_i^{(r)}, i = 1, \ldots, k$, we conclude that

$$\mathcal{P} + \mathcal{P}(rP) = \mathcal{P}(P_0).$$

□

Example 6. Consider $\Sigma(P)$ to be the fan consisting of the quadrants of $\mathbb{R}^2$ and their sides, where P is the square with $a_1 = 0, a_2 = e_1, a_3 = e_2$, and $a_4 = e_1 + e_2$. Figure 22 illustrates an equation $\mathcal{P} + \mathcal{P}(3P) = \mathcal{P}(P_0)$.

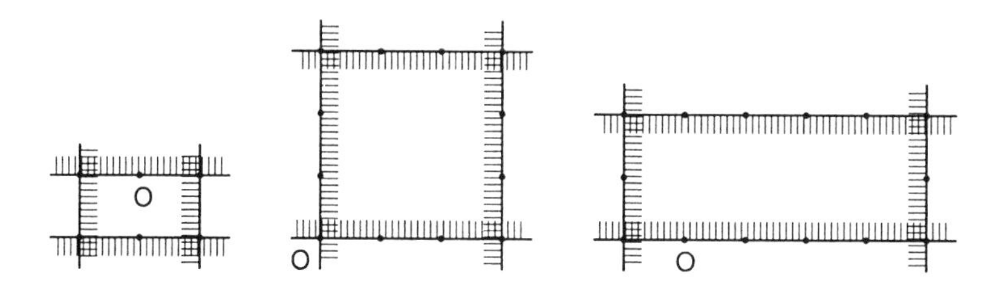

FIGURE 22.

5.16 Definition. If $\Sigma = \Sigma(P)$ is strongly polytopal, we call the group $\tilde{\mathcal{G}}$, in Lemma 5.4, the *polytope group* of Σ.

5.17 Theorem. *Let $\Sigma = \Sigma(P)$ be strongly polytopal. .*

(a) *The polytope group $\tilde{\mathcal{G}}$ is the smallest group into which the semi-group of all polytopes strictly combinatorially isomorphic to P can be embedded.*

(b) Pic Σ *can be generated by $f_{n-1}(P) - n - \lambda + 1$ polytope elements strictly isomorphic to P where λ is defined according to Theorem* 5.9.

PROOF. (a) is a consequence of Theorem 5.15; (b) follows from Theorems 5.9 and 5.15. □

As the following examples show, summands of P can also be chosen for generators of $\tilde{\mathcal{G}}$ or Pic Σ:

Example 7. If Σ consists of the cones into which $\mathbb{R}^n$ is split by the coordinate hyperplanes, we can set $P_i = [0, e_i]$ (line segments), $i = 1, \ldots, n$, and we obtain a system of n polytope elements generating

$$\operatorname{Pic}\Sigma \cong \mathbb{Z}^n .$$

Example 8. In the example of Figure 15, we may choose $P_1 := [0, e_1]$, $P_2 := [0, e_2]$, $P_3 := \operatorname{conv}\{e_1, e_2, e_1 + e_2\}$. Therefore,

$$\operatorname{Pic}\Sigma \cong \mathbb{Z}^3 .$$

Exercises

1. Find Pic Σ for Σ being spanned by the n-cube with vertices $\pm e_1 \pm \cdots \pm e_n$.
2. Let the facets of the cube C with vertices $0, e_1, 2e_2, 3e_3, e_1+2e_2, e_1+3e_3, e_2+3e_3, e_1+2e_2+3e_3$ be numbered $1, \ldots, 6$ such that opposite sides have sum equal to 7 (as in the case of a die). Let H_i be a half-space with face i on its boundary such that the outer normals p_i of H_i are outer normals of C, in the case $i = 2, 4, 5$, and point into the cube in the case $i = 1, 3, 6$. Then, $H_1^+, \ldots, H_6^+$ define a virtual polytope $\mathcal{P}$. Find the smallest k such that
$$\mathcal{P} + \mathcal{P}(kC) = \mathcal{P}(P_0)$$
for a three-dimensional polytope P_0.
3. Given any natural number r, find two polytopes P, P', $P \not\approx P'$, such that $r = \mu(\Sigma(P)) = \mu(\Sigma(P'))$.
4. If $P = P' \cdot P''$ is the join of two polytopes P', P'' (III, Definition 1.13), determine the combinatorial Picard number $\mu(\Sigma(P))$ from $\mu(\Sigma(P'))$ and $\mu(\Sigma(P''))$.

6. Regular stellar operations

In III, 1 and 2, we introduced the concepts "cell complex", "star", "closed star", "link", "join" and "stellar subdivision". The cell complexes considered here are fans, so all cells are cones. The join of two cones σ, σ', for which $\sigma \cap \sigma' = \{0\}$, (rather than equal to $\emptyset$ as in the case of polytopes) and also $(\text{lin}\,\sigma) \cap (\text{lin}\,\sigma') = \{0\}$, can be written as $\sigma + \sigma'$.

Of special interest are the following operations.

6.1 Definition. A stellar subdivision $s(\rho; \sigma)$ of a regular fan Σ (in direction p where $\rho = \mathbb{R}_{\geq 0}\, p$) is called *regular* if it preserves regularity. We also write $s(p; \sigma) = s(\rho; \sigma)$. Its inverse operation is, then, also called *regular.*

In the following, if we express a cone σ as $\sigma = \text{pos}\{q_1, \ldots, q_r\}$, we assume automatically $q_1, \ldots, q_r$ to be generators of $\sigma \cap \mathbb{Z}^n$ and, hence, simple vectors.

6.2 Theorem. *Let Σ be regular, $\sigma \in \Sigma, \sigma = \text{pos}\{x_1, \ldots, x_k\}, x_1, \ldots, x_k$ simple, and let $\rho = \mathbb{R}_{\geq 0}\, p$, p simple. $s(\rho; \sigma)$ is a regular stellar subdivision of Σ if and only if*

$$(1) \qquad p = x_1 + \cdots + x_k.$$

Proof. Let σ be a face of an n-dimensional regular cone $\bar{\sigma}$ (where $\bar{\sigma}$ need not be in Σ). We set

$$\bar{\sigma} = \text{pos}\{x_1, \ldots, x_k, x_{k+1}, \ldots, x_n\},$$

so that $\det \bar{\sigma} = \pm 1$. $s(\rho; \sigma)$ splits $\bar{\sigma}$ into n-dimensional cones,

$$\bar{\sigma}_1 := \text{pos}\{p, x_2, \ldots, x_k, x_{k+1}, \ldots, x_n\}, \ldots, \bar{\sigma}_k$$
$$:= \text{pos}\{x_1, \ldots, x_{k-1}, p, x_{k+1}, \ldots, x_n\}.$$

If (1) is true, we obtain

$$\det \bar{\sigma}_i = \det(x_1, \ldots, x_{i-1}, x_1 + \cdots + x_k, x_{i+1}, \ldots, x_n)$$
$$= \det \bar{\sigma} = \pm 1, \quad i = 1, \ldots, k.$$

So, all new cones are again regular.

Conversely, let $\bar{\sigma}_1, \ldots, \bar{\sigma}_k$ be regular, and let

$$p = \alpha_1 x_1 + \cdots + \alpha_k x_k, \qquad \alpha_1 > 0, \ldots, \alpha_k > 0.$$

Then,

$$\det(x_1, \ldots, x_{i-1}, \alpha_1 x_1 + \cdots + \alpha_k x_k, x_{i+1}, \ldots, x_n) = \alpha_i \det \bar{\sigma} = \pm \alpha_i = \pm 1.$$

Since $\alpha_i > 0$, this implies that $\alpha_i = 1, i = 1, \ldots, k$. $p = x_1 + \cdots + x_k$ is simple, since otherwise, for $p = rq, r > 1, q \in \mathbb{Z}^n$, we would have

$$\pm 1 = \det(rq - x_2 - \cdots - x_k, x_2, \ldots, x_n) = r \det(q, x_2, \ldots, x_n) = r \cdot s$$

for $s \in \mathbb{Z}$, a contradiction. Therefore, (1) is true. $\square$

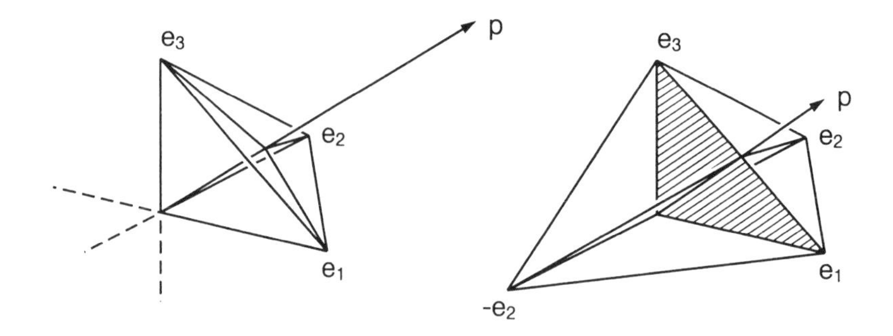

FIGURE 23a,b. (a) $\sigma = \bar{\sigma} = \operatorname{pos}\{e_1, e_2, e_3\}$, $p = e_1 + e_2 + e_3$ (b) $\sigma = \operatorname{pos}\{e_1, e_3\}$, $p = e_1 + e_3$.

Examples.

Under a unimodular transformation L, regularity is preserved, and stellar subdivisions are carried over as follows. Let a subdivision be applied, according to Theorem 6.2, in the direction $p = x_1 + \cdots + x_k$. Then, after applying L, we subdivide in the direction

$$L(p) = L(x_1) + \cdots + L(x_k).$$

6.3 Lemma. *Let $\sigma := \operatorname{pos}\{x_1, \ldots, x_k\}$, $k > 1$, be a regular cone, let σ be a face of the regular n-dimensional cone $\bar{\sigma} := \operatorname{pos}\{x_1, \ldots, x_k, x_{k+1}, \ldots, x_n\}$, and let $p := x_1 + \cdots + x_k$. Then, for $\sigma_i := \operatorname{pos}\{x_1, \ldots, x_{i-1}, p, x_{i+1}, \ldots, x_k, \ldots, x_n\}$, $i \in \{1, \ldots, k\}$, we can set $\check{\bar{\sigma}} = \operatorname{pos}\{y_1, \ldots, y_k, \ldots, y_n\}$ and $\check{\sigma}_i = \operatorname{pos}\{y'_1, \ldots, y'_{i-1}, y_i, y'_{i+1}, \ldots, y'_k, y_{k+1}, \ldots, y_n\}$ such that*

$$-y_i + y_j = y'_j, \qquad j = 1, \ldots, i-1, i+1, \ldots, k. \tag{2}$$

PROOF. Since (2) remains valid if a unimodular transformation is applied, we can assume $x_i = e_i$, $i = 1, \ldots, n$. Then, $\check{\sigma}_i$ has as generators $y'_1 = e_1 - e_i, \ldots, y'_{i-1} = e_{i-1} - e_i$, $y_i = e_i$, $y'_{i+1} = e_{i+1} - e_i, \ldots, y'_k = e_k - e_i$, $y_{k+1} = e_{k+1}, \ldots, y_n = e_n$. (See Figure 23b for the case $n = 3$, $k = 2$, $i = 1$). This proves the lemma. □

6.4 Lemma (Farey's lemma). *Let $\sigma = \operatorname{pos}\{x_1, x_2\}$ be a two-dimensional cone of a regular fan Σ in $\mathbb{R}^2$, x_1, x_2 simple, and let*

$$a = \alpha_1 x_1 + \alpha_2 x_2 \in (\operatorname{int} \sigma) \cap \mathbb{Z}^2$$

be simple. Then, by applying finitely many regular stellar subdivisions, Σ can be turned into a fan which contains $\mathbb{R}_{\geq 0} a$ as a one-dimensional cone.

PROOF. Up to a unimodular transformation, we may assume $x_1 = e_1$, $x_2 = e_2$. If $\alpha_1 = \alpha_2 = 1$, the lemma is proved after applying one subdivision. If $\alpha_1 > \alpha_2$, we apply $s(\mathbb{R}_{\geq 0}(e_1 + e_2); \sigma)$. Let L be the shear for which $L(e_1) = e_1$,

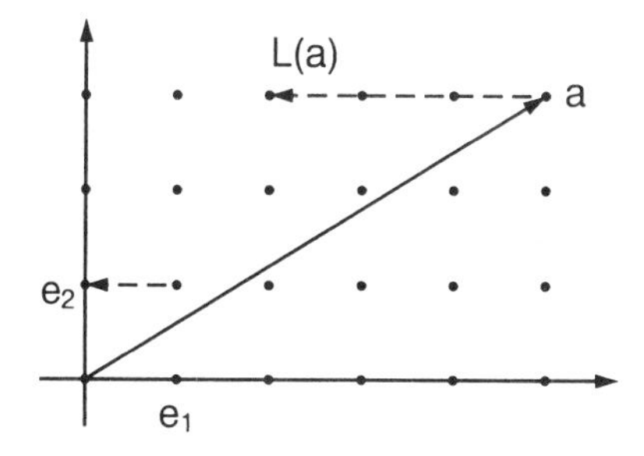

FIGURE 24.

$L(e_1 + e_2) = e_2$. We map $a = (\alpha_1, \alpha_2)$ onto $L(a) = (\alpha_1 - \alpha_2, \alpha_2) =: (\alpha_1', \alpha_2')$ (Figure 24).

If $\alpha_2 > \alpha_1$ we interchange the parts of e_1, e_2. So either $\alpha_1' < \alpha_1$ or $\alpha_2' < \alpha_2$. We divide α_1', α_2' by their greatest common divisor, and repeat the same procedure. After a finite number of steps, a will be transformed into $(1, 1)$ so that only one more subdivision is needed. □

Lemma 6.4 can be generalized to higher dimensions (see Exercise 4 below). However, there are no known analogs, for $n > 2$, to the following two strong theorems.

6.5 Theorem. *Let Σ, Σ' be regular and complete fans in $\mathbb{R}^2$. Then, there exist regular stellar subdivisions $s_1, \ldots, s_p, s_1', \ldots, s_q'$ such that*

$$s_p \circ \cdots \circ s_1 \Sigma = s_q' \circ \cdots \circ s_1' \Sigma' =: \Sigma'',$$

or, in other symbols,

$$\Sigma \xrightarrow{s_1} \cdots \xrightarrow{s_p} \Sigma'' \xleftarrow{s_q'} \cdots \xleftarrow{s_1'} \Sigma'.$$

PROOF. By applying Farey's lemma several times, we can arrange for each 1-cone of Σ to be a 1-cone of Σ' as well. So, let $\sigma = \text{pos}\{x_1, x_2\}$ be a 2-cone of Σ such that a 1-cone $\rho' \in \Sigma' \setminus \Sigma$ is contained in σ. Up to a unimodular transformation, we may assume $x_1 = e_1, x_2 = e_2$. Let $\rho' = \mathbb{R}_{\geq 0}\, a' = \mathbb{R}_{\geq 0}(\alpha_1', \alpha_2')$, a' a simple vector.

If there is no other 1-cone of $\Sigma' \setminus \Sigma$ contained in σ, we find $\alpha_2' = \det(e_1, a') = 1 = \det(a', e_2) = \alpha_1'$, so that, by applying $s(\mathbb{R}_{\geq 0}(e_1 + e_2); \sigma)$ we obtain ρ' as a cone of Σ (after subdivision).

Suppose a second 1-cone $\tilde{\rho} := \mathbb{R}_{\geq 0}\, \tilde{a} = \mathbb{R}_{\geq 0}(\tilde{\alpha}_1, \tilde{\alpha}_2)$, with simple $\tilde{a}$, in $\Sigma' \setminus \Sigma$ exists. If ρ' or $\tilde{\rho}$ equals $\rho := \mathbb{R}_{\geq 0}(e_1 + e_2)$, again we apply $s(\mathbb{R}_{\geq 0}(e_1 + e_2); \sigma)$. If, however, no cone of $\Sigma' \setminus \Sigma$ equals ρ, first, assume that $\rho \subset \rho' + \tilde{\rho} =: \sigma' \in \Sigma'$, $\det \sigma' = 1$. We can let $\tilde{\rho} \subset \rho + \mathbb{R}_{\geq 0}\, e_2$. Then, $\alpha_1' > \alpha_2' \geq 1, \tilde{\alpha}_2 > \tilde{\alpha}_1 \geq 1$. We obtain

$$1 = \alpha_1'\tilde{\alpha}_2 - \alpha_2'\tilde{\alpha}_1 > \alpha_2'(\tilde{\alpha}_2 - \tilde{\alpha}_1) \geq 1,$$

a contradiction.

If $\rho \not\subset \rho' + \tilde{\rho}$, we can assume that $\rho \subset \tilde{\rho} + \mathbb{R}_{\geq 0}\, e_2 \in \Sigma'$. Then, $\tilde{\alpha}_1 > \tilde{\alpha}_2 \geq 1$ and $1 = \tilde{\alpha}_1 \cdot 1 - \tilde{\alpha}_2 \cdot 0$, a contradiction. □

Given regular, complete fans Σ, Σ' in $\mathbb{R}^2$, does there always exist a fan Σ'' which, conversely to Theorem 6.5, can by successive regular stellar subdivisions, be transformed into Σ as well as into Σ'? The answer is no, as can be seen from the fans of Figure 25.

However, any regular, complete fan can be obtained by a chain of regular stellar subdivisions from one or the other fan of Figure 25.

6.6 Theorem. *Let Σ_0 be the fan spanned by $e_1, e_2, -e_1 - e_2$ in $\mathbb{R}^2$, and $\Sigma_{(k)}$ the fan spanned by $e_1, e_2, -e_1, -e_2 + ke_1$, $k \in \mathbb{Z}\setminus\{1, -1\}$ (according to Figure 25). Given any regular and complete fan Σ in $\mathbb{R}^2$, we can find regular stellar subdivisions $s_1, \dots, s_p$ or $s_1', \dots, s_q'$ such that*

$$s_p \circ \cdots \circ s_1 \Sigma_0 = \Sigma \quad or \quad s_q' \circ \cdots \circ s_1' \Sigma_{(k)} = \Sigma.$$

In other symbols,

$$\Sigma_0 \xrightarrow{s_1} \cdots \xrightarrow{s_p} \Sigma \quad or \quad \Sigma_{(k)} \xrightarrow{s_1'} \cdots \xrightarrow{s_q'} \Sigma$$

PROOF. Suppose in Σ there exists a convex quadrangle $A := \operatorname{conv}\{0, a_1, a_2, a_3\}$ whose vertices are $0, a_1, a_2, a_3$, such that the generator a_2 is adjacent to the generators a_1, a_3. We call A a reducible quadrangle. Up to a unimodular transformation, we may assume $a_1 = e_1, a_2 = e_1 + e_2$. Then, $\det(e_1 + e_2, a_3) = \alpha_{32} - \alpha_{31} = 1$, where $a_3 = (\alpha_{31}, \alpha_{32})$. Since $e_1 + e_2$ is a vertex of A, $\alpha_{31} < 1$. Similarly, since a_3 is a vertex of A, $\alpha_{32} > 0$. This readily implies $a_3 = (0, 1)$. Now, we apply $s^{-1}(\mathbb{R}_{\geq 0}\, a_2, \operatorname{pos}\{a_1, a_3\})$ and eliminate, in this way, a generator.

Doing this as often as possible, we obtain a fan Σ' without a reducible quadrangle. So, $\operatorname{conv}\{0, a_1, a_2, a_3\}$ is always a triangle A'.

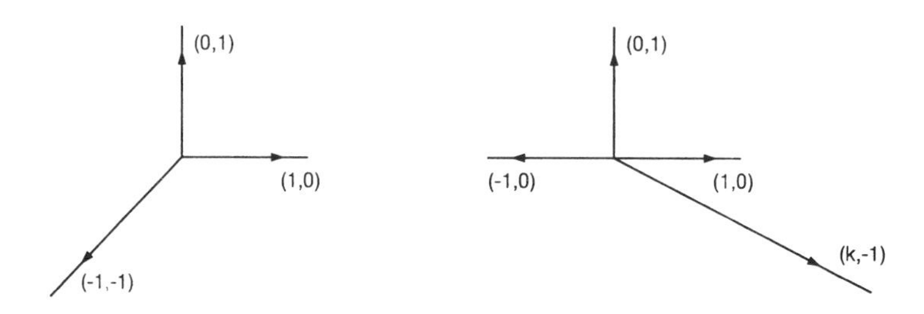

FIGURE 25.

In such a triangle A', we can assume $a_1 = e_1, a_2 = e_2$. Suppose, among the generators $a_3, a_4, \ldots$, following a_2 counterclockwise, there is at least one on or "above" the line g through a_1, a_2. Among these generators, let $a_i = (\alpha_{i1}, \alpha_{i2})$ be one with maximal α_{i2}. Then, $a_{i+1} = (\alpha_{i+1,1}, \alpha_{i+1,2})$ must lie "below" the line $\{ta_i \mid t \in \mathbb{R}\}$ (since Σ is complete). In the case $\alpha_{i+1,1} > 0$, either a_{i+1} is adjacent to a_1 and $\operatorname{conv}\{0, a_{i+1}, a_1, a_2\}$ is reducible, or the same contradiction is obtained for the generator $b \neq a_2$ adjacent to a_1.

In the case $\alpha_{i+1,1} < 0$, $\operatorname{conv}\{0, a_{i-1}, a_i, a_{i+1}\}$ is reducible. Therefore, $\alpha_{i+1,1} = 0$ and, hence, $a_{i+1} = -a_2$. Now $i = 3$, since, otherwise, $\operatorname{conv}\{0, a_{i-1}, a_i, a_{i+1}\}$ would again be reducible, and $a_4 = -e_2$ is readily seen to be adjacent to a_1. Now $1 = \det(a_2, a_3) = -\alpha_{32} = 1$, hence, $\alpha_{32} = -1$, and $\alpha_{31} \geq 2$. This proves Σ to be of type $\Sigma_{(k)}$.

However, if $a_3 = (\alpha_{31}, \alpha_{32})$ lies "below" g and "above" $h := \{te_1 \mid t \in \mathbb{R}\}$, we again obtain a contradiction to A' being a triangle. In the case $a_3 = -e_1$, "below" h there can be only one more generator a_4. Hence, Σ is of type $\Sigma_{(k)}$.

So let a_3 lie "below" h. Clearly $\alpha_{31} < 0$ (since Σ is complete). Suppose, for $a_4 = (\alpha_{41}, \alpha_{42}), \alpha_{41} > 0, \alpha_{42} < 0$. Then, $\operatorname{conv}\{0, a_4, a_1, a_2\}$ would be reducible. Therefore, either $\alpha_{42} = 0$ and $a_4 = a_1$, or $\alpha_{41} \leq 0$. In the former case, we find $\alpha_{31} = \alpha_{32} = -1$, hence Σ is of type Σ_0. In the latter case, $\alpha_{41} = 0$, since, otherwise, $0, e_2, a_3, a_4$ would be vertices of a convex quadrangle. So $a_4 = -e_2$, and Σ is again of type $\Sigma_{(k)}$. □

If we consider Theorems 6.5 and 6.6 and attempt to apply the case $n > 2$, there is no reasonable conjecture for $n > 2$ analogous to Theorem 6.6.

Oda's conjecture (strong version). . Theorem 6.5 is also true for three-dimensional fans.

As an example, we illustrate combinatorially how the fan of Example 1 in 4 and the fan with generators $e_1, e_2, e_3, -e_1 - e_2 - e_3$ can be succesively subdivided into a common regular fan. All stellar subdivisions can be chosen to be regular (Figure 26).

We remark that a "weak version" of Oda's conjecture meanwhile has been shown:

Any two, complete, regular, three-dimensional fans can be transformed into each other by a chain of finitely many operations which are either regular stellar subdivisions or inverses of such.

Regular stellar subdivisions can also be characterized by dual operations. As we have seen in III, 2, the dual combinatorial operations are "cutting off faces". How does regularity come in? Theorem 6.10 below will give an answer. First, we characterize regularity of a strongly polytopal fan $\Sigma = \Sigma(-P)$ with associated polytope P by properties of P.

6.7 Lemma. *A strongly polytopal fan $\Sigma = \Sigma(-P)$ (compare Definition* 4.3 *and* I, *Definition* 4.14*) is regular if and only if P possesses the following properties*

(a) *P is simple.*

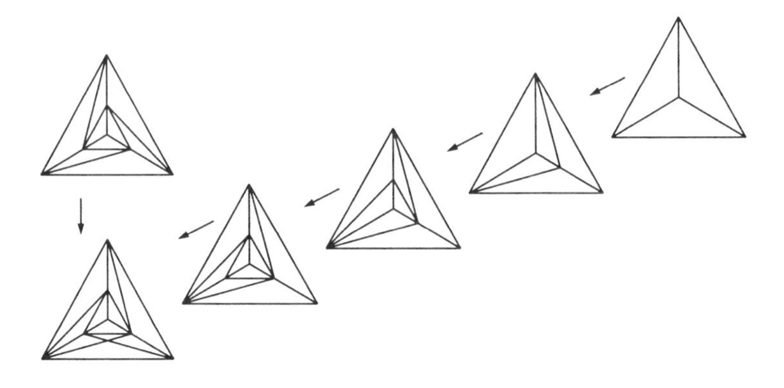

FIGURE 26.

(b) *For any vertex x_0 of P and the lattice points $x_1, \ldots, x_n$ adjacent to x_0 on edges of P, the vectors $x_1 - x_0, \ldots, x_n - x_0$ span $\mathbb{Z}^n$ integrally.*

PROOF. This is clear from Lemma 1.11, Theorem 4.4, and Theorem 2.10(b). □

6.8 Definition. We call a lattice polytope *lattice regular* if it satisfies conditions (a), (b) in Lemma 6.7.

Remark. The term "regular polytope" is defined by congruent edges and the existence of enough symmetries. To avoid confusion we use the words "lattice regular".

6.9 Lemma. *Let F be a proper face of a lattice regular polytope P. Then, the set $\mathcal{H}$ of all lattice points on edges of P, not on F but adjacent to vertices of F, lie in a hyperplane $H :=$ aff $\mathcal{H}$.*

PROOF. Let $a \in$ vert F, and let $a + b_1, \ldots, a + b_n$ be the adjacent lattice points of a on edges of P, where

$$a + b_1, \ldots, a + b_k \in F,$$
$$a + b_{k+1}, \ldots, a + b_n \notin F, \qquad k = \dim F.$$

Up to a translation, we can set $a = 0$. Since $\det(b_1, \ldots, b_n) = \pm 1$, we can, by a unimodular transformation, arrange $b_1 = e_1, \ldots, b_n = e_n$. Then, all points $b_i = e_i, i = k + 1, \ldots, n$, lie in the hyperplane

$$H = \{x = (\xi_1, \ldots, \xi_n) \mid \xi_{k+1} + \cdots + \xi_n = 1\}.$$

H remains invariant if we translate P, such that another vertex of F moves to 0, and then apply a unimodular transformation which leaves H (as a whole) fixed and maps the lattice points adjacent to 0 on edges onto $e_1, \ldots, e_n$. So, the lemma follows. □

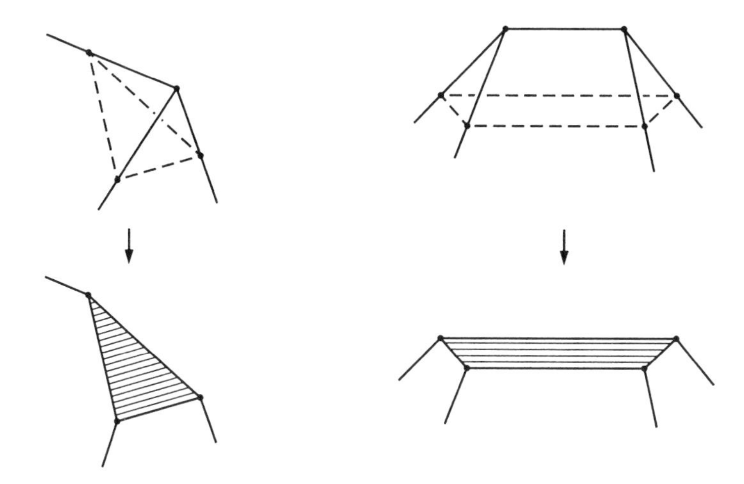

FIGURE 27.

6.10 Theorem. *To a regular stellar subdivision $s(\rho;\sigma)$ of a strongly polytopal regular fan $\Sigma = \Sigma(-P)$ there corresponds the following operation on P or a lattice polytope strictly combinatorially isomorphic to P.*

Let $\rho = \mathbb{R}_{\geq 0}\,u$, u simple, and let H be the supporting hyperplane of P with outer normal u. If the edges emanating from the face $F := P \cap H$, but not in F, do not all have at least three lattice points, we replace P by $2P$. Then, we cut off F from P by a hyperplane parallel to H that passes through the lattice points of P closest to those on F but not in F (Figure 27).

PROOF. This follows readily from Theorem 6.2 and Lemma 6.9. □

We remark that the spanning polytope P^* of Σ is not uniquely determined, but can be varied by moving the vertices on the 1-cones of Σ. Dually, the hyperplanes which carry facets of P can be translated, provided the combinatorial structure of P is not changed.

Exercises

1. Let the fan $\Sigma =: \Sigma_p$ in $\mathbb{R}^3$ have 3-cones $\sigma_1 := \operatorname{pos}\{e_1, e_2, e_3\}$, $\sigma_2 := \operatorname{pos}\{e_1, -e_1 - e_2, e_3\}, \sigma_3 := \operatorname{pos}\{e_2, -e_1 - e_2, e_3\}, \sigma_4 := \operatorname{pos}\{e_1, e_2, p\}, \sigma_5 := \operatorname{pos}\{e_1, -e_1 - e_2, p\}, \sigma_6 := \operatorname{pos}\{e_2, -e_1 - e_2, p\}$ where $p = -e_3 + re_1 + se_2$, $r, s \in \mathbb{Z}$. Find a sequence of regular stellar subdivisions and inverse regular stellar subdivisions which transforms Σ_p into Σ_{-e_3}.

2. Let Σ be the fan spanned by the simplex Δ with vertices $e_1, \ldots, e_n, -e_1 - \cdots - e_n$, and let Σ' be the fan into which $\mathbb{R}^n$ is split by the coordinate subspaces (spanned by the crosspolytope C). Find sequences $s_1, \ldots, s_p$ and $s'_1, \ldots, s'_q$ of regular stellar subdivisions such that

$$s_p \circ \cdots \circ s_1 \Sigma = s'_q \circ \cdots \circ s'_1 \Sigma'.$$

3. For $n = 3$, find polar polytopes P, P' (up to strict combinatorial isomorphism) of $P^* := \Delta$, $P'^* := C$ in Exercise 2 such that the construction of Theorem 6.10 can be carrried out explicitly for the dual operations of $s_1, \ldots, s_p, s'_1, \ldots, s'_q$, respectively.
4. Extend Farey's lemma to arbitrary dimension $n \geq 2$, and prove it.

7. Classification problems

We wish to classify fans under reasonable restrictions. From the point of view of applications in algebraic geometry, the main emphasis is placed on regular fans. For the sake of simplicity, we restrict ourselves to complete fans. Regularity is not an invariant under combinatorial isomorphisms of fans, not even invariant under all linear transformations. The appropriate equivalence relationship is given by unimodular transformations.

7.1 Definition. We call two fans Σ, Σ' *unimodular equivalent* if there exists a unimodular transformation $L : \mathbb{R}^n \longrightarrow \mathbb{R}^n$ which preserves $\mathbb{Z}^n$, such that L maps the cones of Σ bijectively onto the cones of Σ'.

So, on the one hand, by considering regular complete fans, we restrict the large variety of possible fans. On the other, combinatorially equivalent fans need not be unimodularly equivalent, which again enlarges the number of possible types. In fact, classification problems are solved only under strong limitations.

Complete fans are combinatorially isomorphic to polyhedral spheres. We note, first, that not all polyhedral spheres represent, conversely, complete fans (an example is given in III, Theorem 5.5). Already this requires sorting out polyhedral spheres if we want to have complete fans.

We now list several properties of fans and their logical dependencies.

7.2 Definition. We call a complete fan Σ *rational*, if all its cones are rational, and *polyhedral* if it is spanned by a (not necessarily convex) polyhedral sphere, that is, any $\sigma \in \Sigma$ has a representation $\sigma = \operatorname{pos} F_\sigma$, where $\dim \sigma = 1 + \dim F$ and $\{F_\sigma \mid \sigma \in \Sigma\}$ is a polyhedral sphere (see III, Definition 1.9).

The list below refers to properties of Σ which are not invariant under combinatorial isomorphisms (compare III, 5).

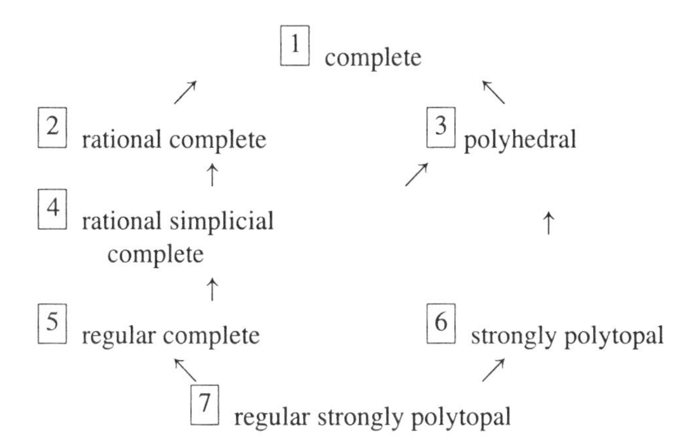

7.3 Lemma. *In the list above, no implication can be reversed.*

PROOF. $\boxed{1} \not\to \boxed{2}$: We choose a complete fan in $\mathbb{R}^2$, which contains $\operatorname{pos}\{e_1 + \sqrt{2}e_2\}$ as a 1-cone.

$\boxed{2} \not\to \boxed{4}$: See, for example, any nonsimplicial complete fan.

$\boxed{4} \not\to \boxed{5}$: Choose, for example, the complete fan in $\mathbb{R}^2$ with generators $e_1, e_2, -2e_1 - e_2$.

$\boxed{1} \not\to \boxed{3}$: See 5, Example 4 (Figure 20).

$\boxed{3} \not\to \boxed{4}$: Any nonsimplicial polytope provides a counterexample.

$\boxed{3} \not\to \boxed{6}$: The Barnette sphere (see III, 5) can be realized in $\mathbb{R}^4$ such that its cells span a fan Σ. Since the Barnette sphere is not polytopal, Σ is not polytopal, and, hence, not strongly polytopal. But, also, polytopal fans need not be strongly polytopal, see the Examples 1 and 2 in section 4.

$\boxed{5} \not\to \boxed{7}$: Consider the fan Σ illustrated in Figure 8 of 4. We choose $x_1 = e_1$, $x_2 = e_2, x_3 = e_3, x_4 = -e_2 - e_3, x_5 = -e_1 - e_3, x_6 = -e_1 - e_2$. We split the cone $\operatorname{pos}\{x_4, x_5, x_6\}$ by introducing the additional generator $x_7 = -e_1 - e_2 - e_3$. We leave the other cones unchanged. In this way, we obtain a regular complete fan which is not strongly polytopal.

$\boxed{6} \not\to \boxed{7}$: Consider the fan of Figure 13 in section 4. □

7.4 Definition. A regular complete fan is called *minimal* if it cannot be obtained from another regular complete fan by a regular stellar subdivision.

Theorem 6.6 can be looked at as a classification theorem:

7.5 Theorem. *Any two-dimensional minimal regular complete fan is unimodular equivalent to Σ_0 or $\Sigma_{(k)}$, $k \in \mathbb{Z} \setminus \{1, -1\}$ (see* Figure 25*).*

As we mentioned in section 6, no analog of Theorem 7.5 is known for dimension $n > 2$. Results have only been found for small numbers g of generators. We restrict ourselves to $n = 3$ and to $g \leq 6$.

7.6 Lemma. *Let Σ be a regular complete fan in $\mathbb{R}^3$, and suppose a_0, a_1, a_2, a_3 are generators such that $\overline{\text{st}}(\mathbb{R}_{\geq 0}\, a_0, \Sigma)$ consists of $\sigma_1 := \text{pos}\{a_0, a_1, a_2\}$, $\sigma_2 := \text{pos}\{a_0, a_2, a_3\}$, $\sigma_3 := \text{pos}\{a_0, a_3, a_1\}$ and their faces. If $0, a_0, a_1, a_2, a_3$ are the vertices of a polytope, either $2a_0 = a_1 + a_2 + a_3$ or a_0 can be eliminated by the inverse regular stellar subdivision $s^{-1}(\mathbb{R}_{\geq 0}\, a_0, \text{pos}\{a_1, a_2, a_3\})$.*

PROOF. Up to a unimodular transformation, we can let $a_1 = e_1$, $a_2 = e_2$, $a_0 = e_3$. For $a_3 = (\alpha, \beta, \gamma)$ we obtain from the regularity of Σ

$$\det(e_1, e_3, a_3) = -\beta = 1, \qquad \det(e_3, e_2, a_3) = -\alpha = 1.$$

a_3 is "above" $\text{lin}\{e_1, e_2\}$ and "below" $\text{aff}\{e_1, e_2, e_3\}$, hence, we obtain, from $\alpha = \beta = -1$, that $\gamma = 1$ or $\gamma = 2$. This implies $2a_0 = a_1 + a_2 + a_3$ or $a_0 = a_1 + a_2 + a_3$. In the latter case, we can apply Theorem 6.2. □

7.7 Theorem. *Any three-dimensional minimal regular complete fan with $g \leq 6$ generators is unimodular equivalent to one of the fans shown in* Figure 28 *where, in* b, $r \neq 1, -1$, *in* b′, $r \geq s$, $r \geq 0$, (r, s) *is different from* $(0, 0)$, $(1, 0)$, *and, in* c, (r, s) *is different from* $(1, 0)$, $(0, 1)$, $(-1, 0)$, *and* $(t, -1)$.

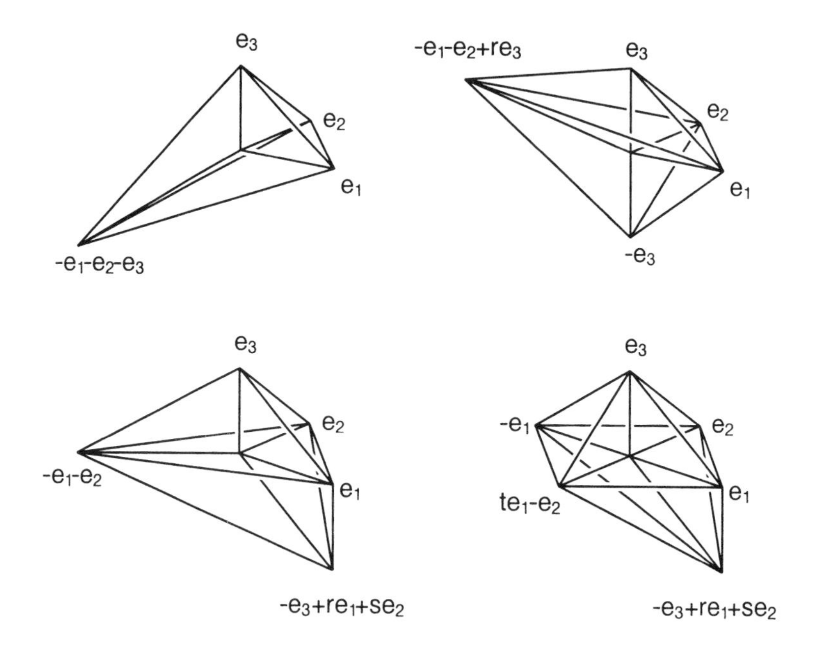

FIGURE 28.

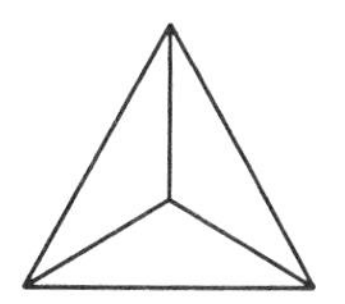
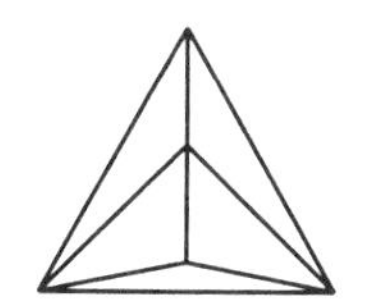
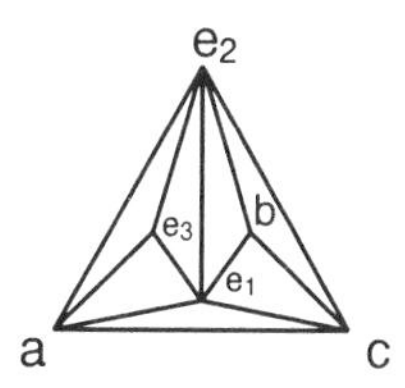

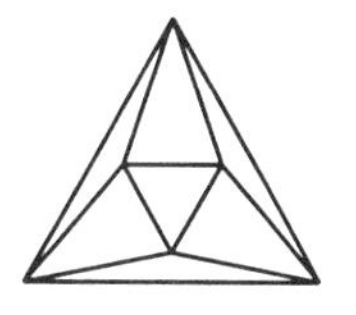

FIGURE 29a,b,c,d.

PROOF. First we find all polyhedral simplicial 2-spheres with at most six vertices. If there are only four vertices, we obtain Figure 29a, which is a Schlegel diagram.

Also for five vertices, there is, combinatorially only one type (Figure 29b). If there are six vertices, we know from II, Theorem 6.7 that the polytopal types are those illustrated in Figure 29c, d. By II, Theorem 6.7, any simplicial sphere with six vertices is isomorphic to one of them.

Now we look at the regular realizations. If we extend the arguments of Lemma 7.6, we see that, up to a unimodular transformation, only the upper left fan in Figure 28 is a regular realization in the case of four one-dimensional cones.

Let the given fan have five generators. Up to a unimodular transformation we assume e_1, e_2, e_3 to be three of them. Let $a = (\alpha_1, \alpha_2, \alpha_3)$ and $b = (\beta_1, \beta_2, \beta_3)$ be the remaining ones such that the following determinants of three-dimensional cones are to be considered:

$$\begin{aligned}
\det(e_2, a, e_3) &= -\alpha_1 = 1 \\
\det(a, e_1, e_3) &= -\alpha_2 = 1 \\
\det(e_2, a, b) &= -\alpha_1\beta_3 + \alpha_3\beta_1 = -1 \\
\det(a, e_1, b) &= -\alpha_2\beta_3 + \alpha_3\beta_2 = -1 \\
\det(e_1, e_2, b) &= \beta_3 = -1
\end{aligned}$$

We obtain $a = (-1, -1, \alpha_3)$ and $b = (\beta_1, \beta_2, -1)$ with $\alpha_3(\beta_1 - \beta_2) = 0$. If $\alpha_3 = 0$, we set $r := \beta_1$, $s := \beta_2$ and obtain the cases illustrated in the lower left fan of Figure 28 where, for reasons of symmetry, $r \geq s$ and $r \geq 0$ may be assumed. $(r, s) = (0, 0)$ has already been considered in case b. $(r, s) = (1, 0)$ yields $e_1 = b + e_3$. If $\alpha_3 \neq 0$, hence, $\beta_1 = \beta_2$, we obtain from $\det(a, e_1, b) = -1 + \alpha_3\beta_2 = -1$ that $\beta_1 = \beta_2 = 0$. For $r := \alpha_3$, we find the cases illustrated in the upper right of Figure 28, where $r \neq 1, -1$ because of Lemma 7.6.

So, let the fan have six generators. Again, we can assume that e_1, e_2, e_3 span a three-dimensional cone of the fan.

Claim The given fan does not have the combinatorial structure illustrated in Figure 29c.

PROOF. Let $a := (\alpha_1, \alpha_2, \alpha_3)$, $b := (\beta_1, \beta_2, \beta_3)$, $c := (\gamma_1, \gamma_2, \gamma_3)$ be the remaining generators such that the determinant equations for the 3-cones are
(1) $\det(e_1, e_3, a) = -\alpha_2 = 1$,
(2) $\det(e_2, e_3, a) = \alpha_1 = -1$,
(3) $\det(e_1, e_2, b) = \beta_3 = -1$,
(4) $\det(e_1, a, c) = \alpha_2\gamma_3 - \alpha_3\gamma_2 = -\gamma_3 - \alpha_3\gamma_2 = 1$,
(5) $\det(e_2, a, c) = \alpha_3\gamma_1 + \gamma_3 = -1$,
(6) $\det(e_1, b, c) = \beta_2\gamma_3 - \beta_3\gamma_2 = -1$, and
(7) $\det(e_2, b, c) = -\beta_1\gamma_3 - \gamma_1 = 1$.

The equations (4) and (5) imply

$$\alpha_3(\gamma_1 - \gamma_2) = 0. \tag{8}$$

From Lemma 7.6, we deduce that one of the following three cases must hold:

$$(9)\quad \alpha_3 = 0, \qquad (9')\quad \alpha_3 < 0 \quad \text{or} \quad (9'')\quad \alpha_3 \geq 2.$$

Case (9)

From (4), we find $\gamma_3 = -1$. Also, since $\beta_3 = -1$, we see that $0, e_1, e_2, b, c$ are vertices of a polytope, hence, by Lemma 7.6, $e_1 + e_2 + c = 2b$ and, therefore, $\gamma_3 = -1 = 2\beta_3 = -2$, a contradiction.

Case (9′)

From (8) we have $\gamma_1 = \gamma_2 =: \gamma$. From (6) and (7), we find that $(\beta_1 - \beta_2)\gamma_3 = 0$. Since $\alpha_3 < 0$ implies $\gamma_3 < 0$, we obtain $\beta_1 = \beta_2 =: \beta$. From (5) and (7), we obtain

$$\alpha_3(1 + \beta\gamma_3) = 1 + \gamma_3. \tag{10}$$

Since $\gamma_3 < 0$, we have $\alpha_3(1 + \beta\gamma_3) \leq 0$. Using (9′) this implies $1 + \beta\gamma_3 \geq 0$, and, hence,

$$(9'\text{A})\quad \beta \leq 0, \qquad \text{or} \quad (9'\text{B})\quad \beta = 1 \text{ and } \gamma_3 = -1$$

Case (9′A)

$\beta = 0$ implies $\gamma = -1$, and, hence, by (5), $\gamma_3 = \alpha_3 - 1$, so that $a = (-1, -1, \alpha_3)$, $b = (0, 0, -1)$, and $c = (-1, -1, \alpha_3 - 1)$. Now $a + b = c$, contrary to the minimality of the fan. So, let $\beta < 0$. We set $\alpha_3' := -\alpha_3$, $\beta' := -\beta$, $\gamma_3' := -\gamma_3$, so that $\alpha_3' > 0$, $\beta' > 0$, $\gamma_3' > 0$. Rewriting (10), we see that

$$1 + \alpha_3' = \gamma_3'(1 - \beta'\alpha_3') \tag{11}$$

with only positive parameters. The left side of (11) is positive, the right side nonpositive, a contradiction.

Case (9′B)

By (5), $\gamma = 0$, so that $b = (1, 1, -1)$, $c = (0, 0, -1)$, and, hence, $b = e_1 + e_2 + c$, contrary to the minimality of the fan.

Case (9″)

Geometrically, this means that the point e_3 lies "below or on" the affine plane H spanned by the points e_1, e_2, a. If b lies "above" the plane H' spanned by e_1, e_2, c, we interchange the roles of e_3 and b, so that (9) or (9′) occurs again.

Suppose, therefore, b lies below or on H'. Clearly, b and c also lie "below" H (otherwise the fan were not complete). Since $\beta_3 = -1$ and $\beta := \beta_1 = \beta_2$, we find $\beta \leq 0$. Furthermore, $\gamma_3 < 0$ and $\gamma := \gamma_1 = \gamma_2 < 0$. By (4), $1 = -\gamma_3 - \alpha_3\gamma > 3$, a contradiction.

This proves our claim.

In the octahedral case Figure 29d, we assume e_1, e_2, e_3 to be generators and to span a cone of the fan. We set $a = (\alpha_1, \alpha_2, \alpha_3)$, $b = (\beta_1, \beta_2, \beta_3)$, and $c = (\gamma_1, \gamma_2, \gamma_3)$, where a is adjacent to e_2, e_3 and b is adjacent to e_1, e_3. By calculating all determinants which involve a, b, or c, we obtain the following equations:

$$\text{(i)} \qquad \alpha_1 = -1, \qquad \beta_2 = -1, \qquad \gamma_3 = -1.$$

$$\alpha_2\beta_1 = 0, \qquad \alpha_3\gamma_1 = 0, \qquad \beta_3\gamma_2 = 0. \text{(ii)}$$

$$\text{(iii)} \qquad \alpha_2\beta_3\gamma_1 + \alpha_3\beta_1\gamma_2 = 0.$$

Up to change of notation, we can interchange e_1, e_2 and a, b simultaneously, also e_1, e_3 and a, c or e_2, e_3 and b, c. Therefore, there is no loss of generality if we let $\alpha_2 = 0$ to satisfy the first equation of (ii). For the second and third equations of (ii), we have solutions $\alpha_3 = \beta_3 = 0$ or $\alpha_3 = \gamma_2 = 0$ or $\gamma_1 = \beta_3 = 0$ or $\gamma_1 = \gamma_2 = 0$. In case $\gamma_1 = \beta_3 = 0$, it follows from (iii) that $\alpha_3 = 0$ or $\beta_1 = 0$ or $\gamma_2 = 0$. In all cases, one of the vectors a, b, c has only one nonzero coordinate and another one only two nonzero coordinates. We may assume $\alpha_2 = \alpha_3 = 0$ and $\beta_3 = 0$ so that

$$a = -e_1, \qquad b = \beta_1 e_1 - e_2, \qquad \text{and} \quad c = \gamma_1 e_1 + \gamma_2 e_2 - e_3.$$

We set $t := \beta_1, r := \gamma_1, s := \gamma_2$ and obtain the cases of the lower right of Figure 28 where $(r, s) \neq (1, 0), (0, 1), (-1, 0), (t, -1)$, since in these cases $c + e_3$ equals $e_1, e_2, -e_1, b$, respectively, so that the fan were not minimal. □

Remark. Instead of working directly with determinants, as in the last part of the proof of Theorem 7.7 one can use Oda's criterion (Theorem 4.12)

$$x_1 + x_1' + ux_2 + vx_3 = 0, \qquad u, v \in \mathbb{Z},$$

where x_1, x_1', x_2, x_3 are generators such that $\operatorname{pos}\{x_1, x_2, x_3\}$ and $\operatorname{pos}\{x_1', x_2, x_3\}$ are cones of the fan. u, v are then called weights (of the edge $[x_2, x_3]$). The classification proceeds by characterizing weighted edge graphs of simplicial spheres which belong to minimal complete regular fans.

Exercises

1. Any complete fan in $\mathbb{R}^2$ is combinatorially isomorphic to a regular complete fan.
2. The fan in $\mathbb{R}^3$, spanned by the faces of an icosahedron, is isomorphic to a regular complete fan.
3. Find out for which pairs of integral vectors (t, r, s) the fans in the lower right of Figure 28 are the same up to a unimodular transformation. [Hint: $(1, r, s)$, $(1, -r, r+s)$ is such a pair.]
4. Find (up to unimodular transformations) all minimal regular complete fans in $\mathbb{R}^4$ with $g \leq 6$ generators.

8. Fano polytopes

We consider a special class of fans; their cells are spanned by the faces of polytopes which are defined as follows.

8.1 Definition. Let $x_1, \ldots, x_r$ be simple lattice vectors which are the vertices of a polytope P with $0 \in \operatorname{int} P$. Then, P is called a *Fano polytope* provided the complete fan it spans is regular.

8.2 Theorem. *In* $\mathbb{R}^2$, *up to unimodular transformations, there exist five Fano polytopes, as illustrated in* Figure 30.

PROOF. We may suppose e_1, e_2 to be adjacent vertices of P. Let $a = (\alpha_1, \alpha_2) \neq e_2$ be adjacent to e_1. Then $\alpha_2 = \det(e_1, a) = -1$, and, since $e_1e_2 := \{(\xi_1, \xi_2) \mid \xi_1 + \xi_2 = 1\}$ is a supporting line of P, $\alpha_1 \leq 1$.

Similarly, for the vertex $b = (\beta_1, \beta_2) \neq e_1$ adjacent to e_2, $\beta_1 = -1$, $\beta_2 \leq 1$. We denote the vertices adjacent to a, b and different from e_1, e_2 by $a' = (\alpha'_1, \alpha'_2)$, $b' = (\beta'_1, \beta'_2)$, respectively.

First, suppose $\alpha_1 = \beta_2 = 1$. Then, $\alpha'_1 + \alpha'_2 = \det(a, a') = -1$, and $\beta'_1 + \beta'_2 = \det(b', b) = -1$, and, hence, $a' \in \{-e_1, -e_2\}$, $b' \in \{-e_1, -e_2\}$. It follows that either $a' \neq b'$ and $a' = -e_2$, $b' = -e_1$ or $a' = b'$ equal to $-e_1$ or $-e_2$. In both cases, no further vertex exists, and we obtain polytopes of type $\mathcal{F}_4$ or $\mathcal{F}_5$.

If $\alpha_1 = 1$ and $\beta_2 \leq 0$, then, again, $\alpha'_1 + \alpha'_2 = -1$. Furthermore, $\alpha'_2 \geq -1$. This implies $\beta'_2 > -1$, hence, $\beta_2 = 0$, and we obtain a polytope of type $\mathcal{F}_3$ or $\mathcal{F}_4$.

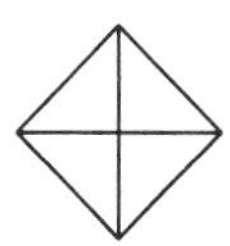
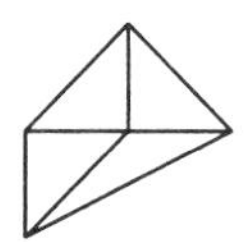
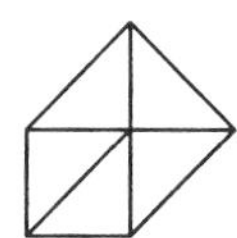
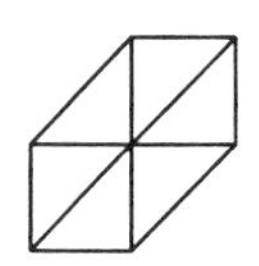

FIGURE 30. $\mathcal{F}_1, \mathcal{F}_2, \mathcal{F}_3, \mathcal{F}_4$, and $\mathcal{F}_5$

The same conclusion is drawn for $\beta_2 = 1$ and $\alpha_1 \leq 0$. So let $\alpha_1 \leq 0$ and $\beta_2 \leq 0$. Then, a, b belong to $\{-e_1, -e_2, -e_1 - e_2\}$, and we readily obtain one of the types $\mathcal{F}_1, \mathcal{F}_2, \mathcal{F}_3, \mathcal{F}_4$. □

Remark. Types $\mathcal{F}_3, \mathcal{F}_4, \mathcal{F}_5$ are not minimal in the sense of the preceding section. By inverse regular stellar subdivisions $\mathcal{F}_3$ can be reduced to $\mathcal{F}_1$, and $\mathcal{F}_4, \mathcal{F}_5$ just as well to $\mathcal{F}_1$ as to $\mathcal{F}_2$.

The direct analogs of $\mathcal{F}_1$ and $\mathcal{F}_2$ in $\mathbb{R}^3$ are the simplex $T = \text{conv}\{e_1, e_2, e_3, -e_1 - e_2 - e_3\}$ and the octahedron $\text{conv}\{e_1, -e_1, e_2, -e_2, e_3, -e_3\}$. We present a further example (Figure 31).

Remark. In dimensions three and four, all Fano polytopes (up to unimodular transformations) have been classified by Batyrev (there are 18 and 121 types, respectively). For higher dimensions, partial results are known. We prove one of them.

First, we remind ourselves of the split of polytopes (IV, 1). If P_1, P_2 are polytopes in complementary linear subspaces of $\mathbb{R}^n$, $0 \in \text{relint}\, P_1$,$0 \in \text{relint}\, P_2$, then, $P_1 \circ P_2 := \text{conv}(P_1 \cup P_2)$ is said to *split* into P_1, P_2. Dually, $P_1^* \oplus P_2^* = (P_1 \circ P_2)^*$.

Furthermore, we define the following polytopes that generalize $\mathcal{F}_5$:

8.3 Definition. A polytope $P_{(k)} := \text{conv}\{e_1, -e_1, \ldots, e_k, -e_k, e_1 + \cdots + e_k, -e_1 \ldots - e_k\}$, k even, or a unimodular copy of it, is called a *del Pezzo polytope*.

8.4 Theorem. *Any n-dimensional, centrally symmetric, Fano polytope P splits into line segments and del Pezzo polytopes,*

$$P = I_1 \circ \cdots \circ I_r \circ P_{(k_1)} \circ \cdots \circ P_{(k_s)}, \qquad r + k_1 + \cdots + k_s = n.$$

PROOF. The proof proceeds in several steps. First, we claim that

(1) Let $F := \text{conv}\{e_1, \ldots, e_n\}$, and $-F$ be facets of P. Then any further vertex $a = (\alpha_1, \ldots, \alpha_n)$ of P satisfies

$$-1 \leq \alpha_j \leq 1, \qquad j = 1, \ldots, n.$$

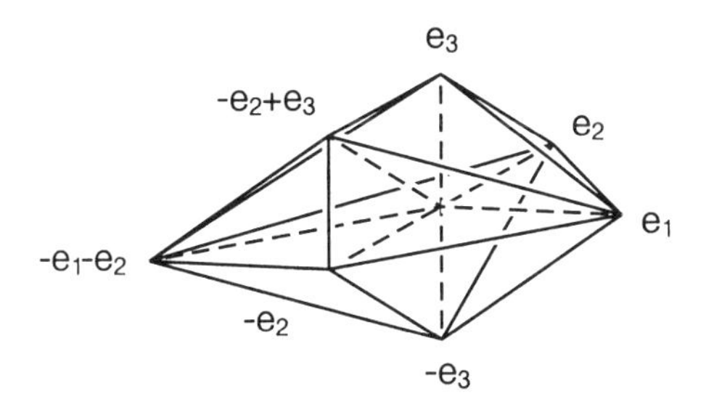

FIGURE 31.

Proof: We consider the unimodular transformation (for row vectors)

$$L = \begin{pmatrix} 1 & 0 & \cdots & 0 \\ 1 & 1 & & 0 \\ \vdots & \vdots & \ddots & \vdots \\ 1 & 0 & \cdots & 1 \end{pmatrix}.$$

Then, $L(F) = \operatorname{conv}\{e_1, e_1 + e_2, \ldots, e_1 + e_n\}$, $L(-F) = \operatorname{conv}\{-e_1, -e_1 - e_2, \ldots, -e_1 - e_n\}$. Let $F_j := \operatorname{conv}\{e_1, e_1 + e_2, \ldots, e_1 + e_{j-1}, e_1 + e_{j+1}, \ldots, e_1 + e_n\}$ be a facet of $L(F)$, and let $F_a := \operatorname{conv}(\{a\} \cup F_j)$ be a facet of $L(P)$ adjacent to $L(F)$. Since $H_1 := \{\xi_1 = 1\}$ and $-H_1$ are supporting hyperplanes of $L(P)$, we see that either $a = -e_1 - e_j$ or $\alpha_1 = 0$. In the latter case,

$$\alpha_j = \det(e_1, e_1 + e_2, \ldots, e_1 + e_{j-1}, a, e_1 + e_{j+1}, \ldots, e_1 + e_n) = -1,$$

since F_a spans a regular cone pos F_a.

The hyperplane aff F_a has (for $a \neq -e_1 - e_j$) an equation

$$\xi_j - \xi_1 = -1.$$

Its intersection with $H_0 := \{\xi_1 = 0\}$ supports the polytope $L(P) \cap H_0$ in H_0, and hence $H_j := \{\xi_j = -1\}$ supports

$$L(P) = \operatorname{conv}(L(F) \cup L(-F) \cup [L(P) \cap H_0]), \qquad j = 2, \ldots, n.$$

For reasons of symmetry, $-H_j$ also supports $L(P)$, $j = 2, \ldots, n$.

We obtain $L(P) \subset H_1^- \cap \cdots \cap H_n^- \cap (-H_1^-) \cap \cdots \cap (-H_n^-)$, which is also true for $a = -e_1 - e_j$. Hence, (1) holds for $L(P)$ instead of P. Since L^{-1} preserves $\xi_2, \ldots, \xi_n$, (1) also follows for P in the case $j = 2, \ldots, n$.

To verify (1) for $j = 1$, we consider the facet $F_b = \operatorname{conv}\{b, e_1 + e_2, \ldots, e_1 + e_n\}$, $b \neq e_1$, of $L(P)$. For $b = (\beta_1, \ldots, \beta_n)$, either $b = -e_1$ or $\beta_1 = 0$. In the latter case,

$$-\beta_2 - \cdots - \beta_n = \det(b, e_1 + e_2, \ldots, e_1 + e_n) = -1.$$

The supporting hyperplane aff F_b of $L(P)$ has an equation

$$\xi_2 + \cdots + \xi_n = 1$$

which remains invariant under L^{-1}. Therefore,

$$-1 \leq \xi_2 + \cdots + \xi_n \leq 1$$

for all points of P. Since $L^{-1}(0, \xi_2, \ldots, \xi_n) = (-\xi_2 - \cdots - \xi_n, \xi_2, \ldots, \xi_n)$, we obtain $|\xi_1| \leq 1$, and, hence, (1) for $j = 1$.

(2) Let $a = (\alpha_1, \ldots, \alpha_n)$, $a' = (\alpha'_1, \ldots, \alpha'_n)$ be vertices of P, both not contained in $F \cup (-F)$. Then, for $j = 1, \ldots, n$ neither $\alpha_j = \alpha'_j = 1$ nor $\alpha_j = \alpha'_j = -1$ is true.

Proof: Again, consider $L(P)$ so that, as we have seen above,

$$\alpha_1 = \alpha'_1 = 0.$$

Suppose, for a $j > 1, \alpha_j = \alpha'_j = -1$. Then, a, a', F_j would be contained in the supporting hyperplane $\{\xi_1 - \xi_j = 1\}$, and, hence, P would not be simplicial contradicting regularity. By symmetry, it follows that $\alpha_j = \alpha'_j = 1$ is not possible, either. Since α_j, α'_j remain invariant under L^{-1} for $j = 2, \ldots, n$, (2) follows for $j = 2, \ldots, n$. We may assume $n > 1$ (for $n = 1$, (2) is trivial). By interchanging the roles of α_1 and some $\alpha_j \neq \alpha_1$, (2) also follows for $j = 1$.

(3) Any vertex of P not in F or $-F$ can, up to renumbering of coordinates, be written as

$$a = (1, -1, \ldots, 1, -1, 0, \ldots, 0)$$

(there may be no zeros).

Proof: Since all vertices not in $L(F) \cup L(-F)$ of $L(P)$ lie in $\{\xi_1 = 0\}$, those of P which are not in F or $-F$ lie in

$$\xi_1 + \cdots + \xi_n = 0.$$

So, the claim follows from (1).

Proof of the theorem. According to (3), let $a = (1, -1, \ldots, 1, -1, 0, \ldots, 0)$ be a vertex of P not in F or $-F$. Since P is centrally symmetric, $-a = (-1, 1, \ldots, -1, 1, 0, \ldots, 0)$ is also a vertex. We write all vertices of P (except those in F and $-F$) as rows of a matrix:

$$\left(\begin{array}{ccccc|ccc} 1 & -1 & \cdots & 1 & -1 & 0 & \cdots & 0 \\ -1 & 1 & \cdots & -1 & 1 & 0 & \cdots & 0 \\ \hline 0 & 0 & \cdots & 0 & 0 & & & \\ \vdots & \vdots & \cdots & \vdots & \vdots & & * & \\ 0 & 0 & \cdots & 0 & 0 & & & \end{array}\right)$$
$$\underbrace{\hphantom{1 \quad -1 \quad \cdots \quad 1 \quad -1}}_{2k_1}$$

Because of (2), the lower left part of this matrix consists of zeros. The split of the matrix represents a split of the polytope into $P = P_{(k_1)} \circ P_0$, where $\dim P_0 = n - k_1$.

We repeat the procedure until there is no further vertex $a \notin F \cup (-F)$ and obtain a split

$$P = P_{(0)} \circ P_{(k_1)} \circ \cdots \circ P_{(k_s)}.$$

$P_{(0)}$ contains only vertices $e_i, -e_i, i = 1, \ldots, r = n - k_1 - \cdots - k_s$, and, hence, splits into r intervals. □

Exercises

1. Find three Fano polytopes with eight vertices in $\mathbb{R}^3$ each two of which are not related by a unimodular transformation.

2. Find (up to unimodular transformations) all Fano polytopes in $\mathbb{R}^3$ with at most six vertices.
3. Prove the converse of Theorem 8.4: All split polytopes presented there are Fano polytopes.
4. Prove that a Fano polytope in $\mathbb{R}^n$ with two facets F, $-F$ (symmetric with respect to 0) has, at most, $2n + 2$ vertices.

Part 2

Algebraic Geometry

VI

Toric Varieties

1. Ideals and affine algebraic sets

From its beginning, algebraic geometry is concerned with sets of zeros of finitely many polynomials. These affine algebraic sets form a basic part of the theory, usually as "charts" of which more general varieties are built up (by "gluing together"). The underlying field of coefficients may be general or restricted to one of the fields $\mathbb{Q}$, $\mathbb{R}$, $\mathbb{C}$, of rational, real, or complex numbers, depending on the topic discussed and the methods used.

We shall set up a framework which, on one side, fits the special type of varieties to be considered, and, on the other side, is general enough to include quite a number of basic algebraic geometric concepts. Though we stick to rather "classical" geometric objects, in many cases, we need modern terminology to formulate the equivalents of combinatorial geometric facts in algebraic geometry.

As a coordinate field, we choose the field $\mathbb{C}$ of complex numbers throughout. Many results could be extended to the cases of other fields, but we do not stress this possibility. There might be a question about "real" geometry which requires real coordinates. For our purposes, it is enough to give real illustrations by choosing polynomials "as real as possible" (for example, $\xi_1^2 + \xi_2^2 - 1$ representing a circle, rather than $\xi_1^2 + \xi_2^2 + 1$, which has no real zeros). We can do this because most facts considered do not depend on specific values of the coefficients of polynomials, avoiding exceptional cases. For the general setting, the coordinate field $\mathbb{C}$ is, in a way, more convenient, since any nonconstant polynomial has a zero in $\mathbb{C}$.

Polynomials are considered to be $\mathbb{C}$-valued functions of a complex vector space V. If $\xi := (\xi_1, \ldots, \xi_m)$ is a coordinate vector of V with respect to some basis, a polynomial f is given as a linear combination of monomials

$$f(\xi) = \sum \lambda_i \xi_1^{i_1} \cdots \xi_m^{i_m} =: \sum \lambda_i \xi^i, \qquad \text{for } i := (i_1, \ldots, i_m),$$

where $\lambda_i \in \mathbb{C}$, $i_j \in \mathbb{Z}_{\geq 0}$, and only finitely many λ_i are nonzero. The ring of all such polynomials (under ordinary addition and multiplication) is denoted by

$$\mathbb{C}[\xi_1, \ldots, \xi_m] =: \mathbb{C}[\xi].$$

Any (finite) set of polynomials determines a geometrical object in V, namely, the set of its common zeros. The sets thus obtained are called *affine algebraic sets* (see Definition 1.1); for a single nonconstant polynomial, they are called (affine algebraic) *curves*, if $\dim V = 2$, *surfaces*, if $\dim V = 3$, and *hypersurfaces* in general. On the other hand, for a given subset M of V, we consider the collection of all polynomials vanishing on M; it has the structure of an ideal in the ring of all polynomials. The relation between polynomial ideals and affine algebraic sets is our first object of study.

1.1 Definition. Let F be a subset of $\mathbb{C}[\xi] := \mathbb{C}[\xi_1, \ldots, \xi_m]$. The set

$$V(F) := \{\xi := (\xi_1, \ldots, \xi_m) \mid f(\xi) = 0, \qquad \text{for all } f \in F\},$$

is called the *(complex) affine algebraic set* $V(F) \hookrightarrow \mathbb{C}^m$ defined by F. If $F' \subset F$, we say $V(F)$ is an *(affine) algebraic subset* of $V(F')$. We set $V(f) := V(\{f\})$.

Evidently such an affine algebraic set is a closed subset of $\mathbb{C}^m$ (with respect to the usual topology).

First, we recall the definition and some elementary properties of ideals in a ring R (which is always assumed to be commutative with unit element 1).

1.2 Definition. An additive subgroup $\mathfrak{a}$ of R is called an *ideal* in R if

$$r \cdot a \in \mathfrak{a} \quad \text{holds for arbitrary} \quad a \in \mathfrak{a} \text{ and } r \in R,$$

or briefly

$$R \cdot \mathfrak{a} \subset \mathfrak{a}.$$

In other words, $\mathfrak{a}$ is closed with respect to multiplication by arbitrary ring elements. If $\mathfrak{a} \neq R$, we call $\mathfrak{a}$ *proper*. If F is a subset of R, then, the set of all finite R-linear combinations

$$R \cdot F := \{r_1 \cdot a_1 + \cdots + r_k \cdot a_k \mid r_j \in R,\ a_j \in F,\ k \in \mathbb{Z}_{\geq 0}\}$$

is an ideal, called the *ideal generated by* F. In particular, for $F = \{a\}$, the ideal $R \cdot a := R \cdot \{a\} =: (a)$ is called the *principal ideal generated by* a. As examples, we have the "zero ideal" $\mathfrak{o} := R \cdot (0) = (0)$ and the ideal $R = R \cdot 1 = (1)$.

If F is a subset of $\mathbb{C}[\xi]$ and if $\mathfrak{a}$ is the ideal generated by F, then, clearly, every polynomial in $\mathfrak{a}$ vanishes on $V(F)$, and we even have the following lemma:

1.3 Lemma. *If F is a subset of $\mathbb{C}[\xi]$ and $\mathfrak{a}$ the ideal generated by F, then $V(F) = V(\mathfrak{a})$.*

Thus, we see that every affine algebraic set can be defined by an ideal. On the other hand, given an arbitrary subset of $\mathbb{C}^m$, we associate with it an ideal as follows:

1.4 Definition. For a subset Z of $\mathbb{C}^m$, the set

$$\mathfrak{i}_Z := \{f \in \mathbb{C}[\xi] \mid f|_Z = 0\}$$

is an ideal, called the (*vanishing*) *ideal of* Z.

If V is an affine algebraic set $V(\mathfrak{a})$ in $\mathbb{C}^m$, then we also write $\mathfrak{a}_V$ instead of $\mathfrak{i}_{V(\mathfrak{a})}$.

Note that the inclusion $\mathfrak{a} \subset \mathfrak{a}_V$ in general is proper. Evidently, we have the following relationships:

1.5 Lemma.
(a) *If* $F \subset F'$, *then,* $V(F) \supset V(F')$.
(b) *If* $Z \subset Z'$, *then* $\mathfrak{i}_Z \supset \mathfrak{i}_{Z'}$.

To study the relationship between ideals and affine algebraic sets more closely, we, first define the sum and the product of two ideals (respectively cosets) of R so that they are again ideals (respectively, cosets). More generally, we set the following definition:

1.6 Definition. For an ideal $\mathfrak{a}$ of R and subsets S, S_j of R, define

$$\sum S_j := \{\sum_{\text{finite}} s_j \mid s_j \in S_j\},$$

$$S \cdot \mathfrak{a} := \{\textstyle\sum_{i=1}^m s_i a_i \mid m \in \mathbb{Z}_{\geq 1}, s_i \in S, a_i \in \mathfrak{a}, \},$$

$$(f + \mathfrak{a})(g + \mathfrak{a}) := fg + \mathfrak{a} \quad \text{for } f, g \in R.$$

Clearly, $S \cdot \mathfrak{a}$ is an ideal of R, included in $\mathfrak{a}$.

1.7 Lemma. *For ideals* $\mathfrak{a}$ *and* $\mathfrak{a}'$ *in* $\mathbb{C}[\xi]$, *the sets* $\mathfrak{a} + \mathfrak{a}'$ *and* $\mathfrak{a} \cdot \mathfrak{a}'$ *are also ideals, and*
(a) $V(\mathfrak{a} \cdot \mathfrak{a}') = V(\mathfrak{a} \cap \mathfrak{a}') = V(\mathfrak{a}) \cup V(\mathfrak{a}')$,
(b) $V(\mathfrak{a} + \mathfrak{a}') = V(\mathfrak{a}) \cap V(\mathfrak{a}')$, *and*
(c) $V(\mathbb{C}[\xi]) = \emptyset$, *and* $V(\mathfrak{o}) = \mathbb{C}^m$.

PROOF.
(a) The equation $f(a)f'(a) = 0$ implies $f(a) = 0$ or $f'(a) = 0$, since $\mathbb{C}$ has no zero divisors. Using Lemma 1.5, we obtain the equalities.
(b) For $a \in V(\mathfrak{a}) \cap V(\mathfrak{a}')$, $f \in \mathfrak{a}$, and $f' \in \mathfrak{a}'$, we obtain $(f + f')(a) = 0$, and, hence, $a \in V(\mathfrak{a} + \mathfrak{a}')$, which implies "$\supset$". The opposite inclusion follows from 15.
(c) is obvious, since $1 \in \mathbb{C}[\xi]$.

□

Remark. The properties above show that the collection of all affine algebraic subsets of $\mathbb{C}^m$ is the family of all closed sets of a topology on $\mathbb{C}^m$, which we shall call the *Zariski topology* (see Definitions 123 and 124). As we have seen above, affine algebraic sets are closed subsets of $\mathbb{C}^m$; hence, the Zariski topology is coarser than the usual topology on $\mathbb{C}^m$. It can be defined for affine algebraic varieties over an arbitrary field of coefficients. For $x = (x_1, \ldots, x_m) \in \mathbb{C}^m$, $\{x\} = V(\xi_1 - x_1, \ldots, \xi_m - x_m)$; hence, every one-point subset of $\mathbb{C}^m$ is affine algebraic and, thus, closed in the Zariski topology. We denote by

$$\mathfrak{m}_x := \mathbb{C}[\xi](\xi_1 - x_1) + \cdots + \mathbb{C}[\xi](\xi_m - x_m)$$

the ideal $\mathfrak{i}_{\{x\}}$.

Now, we consider some examples of ideals and affine algebraic sets:

Example 1. $\mathfrak{a} := R \cdot (\xi_1^2 + \xi_2^2 - 1)$ is an ideal in $R := \mathbb{C}[\xi_1, \xi_2]$.

Example 2. $\mathfrak{a} := R \cdot (\xi_1 - 1)(\xi_2 - 1)(\xi_3 - 1)$ is an ideal in $R = \mathbb{C}[\xi_1, \xi_2, \xi_3]$.

Example 3. $\mathfrak{a} := R \cdot (\xi_1^2 + \xi_2^2 - 1) + R \cdot (\xi_1 - \xi_2)$ is an ideal in $R = \mathbb{C}[\xi_1, \xi_2]$.

Example 4. $\mathfrak{a} : = R \cdot (\xi_1 - \xi_2) + R \cdot (\xi_1 + \xi_2) = R \cdot \xi_1 + R \cdot \xi_2$ is an ideal in $R = \mathbb{C}[\xi_1, \xi_2]$.

On the geometric side, $V(\mathfrak{a})$ is a (complex) circle in Example 1, the union of three planes in $\mathbb{C}^3$ in Example 2, the two intersection points of a complex line and a complex circle in Example 3, and one point in Example 4.

We are now going to introduce the notion of the (reduced) coordinate ring of an affine algebraic set $V(\mathfrak{a})$ in $\mathbb{C}^m$. Hilbert's famous "Nullstellensatz" shows that $V(\mathfrak{a})$, endowed with the Zariski topology, is entirely determined by its coordinate ring. First we recall the notion of a residue class ring of R: If $\mathfrak{a}$ is an ideal in R, then the set

$$R/\mathfrak{a} := \{f + \mathfrak{a} \mid f \in R\}$$

of all cosets, with sum and product of cosets defined as in 1.6 forms a ring.

1.8 Definition. If $\mathfrak{a}$ is an ideal in $R = \mathbb{C}[x]$ and $\mathfrak{a}_V$ is the ideal of all polynomials vanishing on $V(\mathfrak{a})$, we call $R/\mathfrak{a}_V =: R_V$, sometimes also denoted by $\bar{R}$, the *coordinate ring* or the *ring of regular functions* of the affine algebraic set $V(\mathfrak{a})$.

Let $f \in R_V$. Any two polynomials $g \in f + \mathfrak{a}_V$ and $g' \in f + \mathfrak{a}_V$ define the same map $g|_V : V \to \mathbb{C}$.

We remark that R_V has no nilpotent elements, that is, nonzero elements a power of which is zero. Note that the elements of R_V can be interpreted as the restrictions of polynomial functions to the affine algebraic set V from the ambient space $\mathbb{C}^m$ and conversely. In particular, the generators $\bar{\xi}_j := \xi_j + \mathfrak{a}_V$ of R_V are the restrictions of the coordinate functions, which explains the name "coordinate ring".

1.9 Lemma. *For every sequence $Y_1 \supset Y_2 \supset \cdots$ of algebraic sets, there is an integer r such that $Y_r = Y_{r+1} = \cdots$*

PROOF. Let $\mathfrak{i}_{Y_1} \subset \mathfrak{i}_{Y_2} \subset \cdots$ be the corresponding chain of ideals in $R := \mathbb{C}[\xi_1, \ldots, \xi_n]$, and let $\mathfrak{i} = \bigcup_{i=1}^{\infty} \mathfrak{i}_{Y_i}$. By Hilbert's Basissatz (compare, for example, van der Waerden [1967]) the ideal $\mathfrak{i}$ is finitely generated, that is, $\mathfrak{i} = R \cdot \eta_1 + \cdots + R \cdot \eta_k$ where $\eta_j \in R$. Since $\eta_j \in \mathfrak{i} = \bigcup_{i=1}^{\infty} \mathfrak{i}_{Y_i}$, $\eta_j \in \mathfrak{i}_{Y_{i_j}}$ for some $i_j \in \mathbb{Z}_{>0}$. Let $m := \max\{i_j \mid j = 1, \ldots, k\}$. Then, $\eta_j \in \mathfrak{i}_{Y_m}$ for $j = 1, \ldots, k$, and we get $\mathfrak{i}_{Y_m} = \mathfrak{i}_{Y_{m+1}} = \cdots$ which equals $\mathfrak{i}$. □

1.10 Definition. A proper ideal $\mathfrak{p}$ of R is called a *prime ideal* if $rs \in \mathfrak{p}$ implies $r \in \mathfrak{p}$ or $s \in \mathfrak{p}$. A proper ideal $\mathfrak{m}$ of R is called a *maximal ideal* if $\mathfrak{m} \subset \mathfrak{m}'$ for an arbitrary proper ideal $\mathfrak{m}'$ implies $\mathfrak{m}' = \mathfrak{m}$.

We note the following elementary result:

1.11 Lemma. *An ideal $\mathfrak{a}$ of R is*
(a) *prime if and only if $R/\mathfrak{a}$ is an integral domain,*
(b) *maximal if and only if $R/\mathfrak{a}$ is a field.*

As fields have no zero divisors, we evidently have

1.12 Corollary. *Every maximal ideal is prime.*

Affine algebraic sets whose vanishing ideal is prime play a special role:

1.13 Definition. An affine algebraic set $V \subset \mathbb{C}^m$ is called *irreducible* or an *affine algebraic variety* if it is not the union of two proper algebraic subsets. If $V_1 \subset V_2 \subset \mathbb{C}^m$ are two affine algebraic varieties, we say that V_1 is a *subvariety* of V_2. It determines an ideal denoted by $\mathfrak{i}_{V_1,V_2} := \{f \in R_{V_2} \mid f|_{V_1} = 0\} = \mathfrak{i}_{V_1}/\mathfrak{i}_{V_2} \subset \mathbb{C}[\xi_1, \ldots, \xi_m]/\mathfrak{i}_{V_2} = R_{V_2}$.

1.14 Lemma and Definition. *Each algebraic set X is a finite union of irreducible algebraic sets X_i. If we assume $X_i \not\subset X_j$ for $i \neq j$, then, the X_i are determined uniquely and are called the irreducible components of X.*

PROOF. Suppose X is not a finite union of irreducible algebraic sets. Then, in particular, X is not irreducible, so that $X = Y \cup Y'$ where $Y \subsetneq X$ and $Y' \subsetneq X$. If Y and Y' are both finite unions of irreducible sets, then, so is X. Therefore, at least one of them, Y say, is not a union of irreducible components. By repeating this reasoning and using induction, we can find an infinite sequence $X \supsetneq Y =: Y_1 \supsetneq Y_2 \supsetneq \cdots$ such that none of the Y_i is a finite union of irreducible algebraic sets. So, the sequence does not get stationary, contrary to Lemma 1.9.

Therefore $X = X_1 \cup \cdots \cup X_k$ is a union of finitely many irreducible components. We may assume $X_i \not\subset X_j$ for $i \neq j$ (otherwise we leave out X_i).

Suppose $X = X_1' \cup \cdots \cup X_m'$ is another such representation. For each $j \in \{1, \ldots, k\}$, $X_j = X_j \cap X = (X_j \cap X_1') \cup \cdots \cup (X_j \cap X_m')$. Since X_j is irreducible we find that $X_j \cap X_i' = X_j$, for some i, hence, $X_j \subset X_i'$. But, also,

$X'_i \subset X_q$ for some q, hence, $X_j \subset X'_i \subset X_q$, and, therefore, $X_j = X'_i = X_q$. So, each X_j is an X'_i, and, analogously, each X'_i is an $_p$. Therefore, the X_i are uniquely determined. □

1.15 Lemma. *An affine algebraic set X is irreducible if and only if $Y\mathfrak{i}_X$ is prime.*

PROOF. Assume X is irreducible, and let $f \cdot g \in \mathfrak{i}_X$. Then, $V(f \cdot g) = V(f) \cup V(g) \supset X$. Since X is irreducible, $V(f) \supset X$ or $V(g) \supset X$ which means $f \in \mathfrak{i}_X$ or $g \in \mathfrak{i}_X$. Hence, $\mathfrak{i}_X$ is prime.

If X is not irreducible, then, $X = X' \cup X''$, $X' \subsetneqq X$, $X'' \subsetneqq X$, hence, $\mathfrak{i}_X \subsetneqq \mathfrak{i}_{X'}$, $\mathfrak{i}_X \subsetneqq \mathfrak{i}_{X''}$. We may assume $f \in \mathfrak{i}_{X'} \setminus \mathfrak{i}_{X''}$, $g \in \mathfrak{i}_{X''} \setminus \mathfrak{i}_{X'}$. Then, $f \cdot g \in \mathfrak{i}_{X' \cup X''}$, so that $\mathfrak{i}_X$ is not prime. □

The "Nullstellensatz" is the higher dimensional analog of the "Fundamental Theorem of Algebra".

1.16 Theorem (Hilbert's Nullstellensatz). *Let $\mathfrak{a} \subset \mathbb{C}[\xi]$ be an ideal, and let $f \in \mathbb{C}[\xi]$. Then, $f \in \mathfrak{i}_{V(\mathfrak{a})}$ if and only if there exists a natural number k such that $f^k \in \mathfrak{a}$.*

1.17 Corollary (Weak version of Hilbert's Nullstellensatz). *Every maximal ideal of $\mathbb{C}[\xi]$ is of the form*

$$\mathfrak{m}_x = \mathbb{C}[\xi](\xi_1 - x_1) + \cdots + \mathbb{C}[\xi](\xi_m - x_m) \tag{1}$$

for a unique point $x = (x_1, \ldots, x_m)$, and, conversely, every such ideal is maximal. In particular, the maximal ideals of $\mathbb{C}[\xi]$ are finitely generated.

PROOF OF THEOREM 1.16. We need only show the "only if part" of the theorem. So, let $f^k \notin \mathfrak{a}$ for every k, and let $\mathfrak{m} \supset \mathfrak{a}$ be the maximal ideal of $R := \mathbb{C}[\xi]$ for which, also, $f^k \notin \mathfrak{m}$ for every k. First, we claim that, for each $i \in \{1, \ldots, m\}$, there exists an α_i such that $\xi_i - \alpha_i \in \mathfrak{m}$. We may assume $i = 1$. If $\xi_1 - \beta \notin \mathfrak{m}$ for each $\beta \in \mathbb{C}$, then, $f^{k_\beta} \in \mathfrak{m} + R(\xi_1 - \beta)$ for some natural number k_β, so there are polynomials $p_\beta \in \mathfrak{m}$ and $q_\beta \in R$ such that

$$p_\beta + q_\beta(\xi_1 - \beta) = f^{k_\beta}. \tag{2}$$

Since $\mathbb{C}$ is uncountable, we find nonnegative integers d and k such that $\{\beta \in \mathbb{C} \mid \text{degree of } q_\beta = d \text{ and } k_\beta = k\}$ is an infinite set. The polynomials $f \in R$ of degree $\leq d$ clearly form a vector space of finite dimension over $\mathbb{C}$. Therefore we can find distinct numbers $\beta_1, \ldots, \beta_r \in \mathbb{C}$ and numbers $\lambda_1, \ldots, \lambda_r \in \mathbb{C} \setminus \{0\}$ such that

$$\lambda_1 q_{\beta_1} + \cdots + \lambda_r q_{\beta_r} = 0. \tag{3}$$

For $q := \sum_{i=1}^{r} \lambda_i(\xi_1 - \beta_1)\cdots(\xi_1 - \beta_{i-1})(\xi_1 - \beta_{i+1})\cdots(\xi_1 - \beta_r) \in \mathbb{C}[\xi_1]$, we obtain, using (2) and (3),

$$\begin{aligned} f^k q &= \sum_{i=1}^{r} \lambda_i [p_{\beta_i} + q_{\beta_i}(\xi_1 - \beta_i)] \\ &\quad \times (\xi_1 - \beta_1)\cdots(\xi_1 - \beta_{i-1})(\xi_1 - \beta_{i+1})\cdots(\xi_1 - \beta_r) \\ &= \sum_{i=1}^{r} \lambda_i p_{\beta_i}(\xi_1 - \beta_1)\cdots(\xi_1 - \beta_{i-1})(\xi_1 - \beta_{i+1})\cdots(\xi_1 - \beta_r) \in \mathfrak{m}. \end{aligned}$$

On the other hand, we see, from $q \in \mathbb{C}[\xi_1]$ and from $\beta_1, \ldots, \beta_r$ being distinct, that $q(\beta_1) \neq 0$, so that q is not the zero polynomial. We write $q(\xi_1) = \gamma(\xi_1 - \gamma_1)\cdots(\xi_1 - \gamma_{r-1})$ (using the Fundamental Theorem of Algebra), where $\gamma, \gamma_1, \ldots, \gamma_{r-1} \in \mathbb{C}$ and $\gamma \neq 0$. Using (2), we obtain

$$\begin{aligned} f^k \cdot q \cdot q_{\gamma_1} \cdots q_{\gamma_{r-1}} &= \gamma \cdot f^k \cdot q_{\gamma_1}(\xi_1 - \gamma_1)\cdots q_{\gamma_{r-1}}(\xi_1 - \gamma_{r-1}) \\ &= \gamma \cdot f^k(f^{k_{\gamma_1}} - p_{\gamma_1})\cdots(f^{k_{\gamma_{r-1}}} - p_{\gamma_{r-1}}) \in \mathfrak{m}. \end{aligned}$$

Hence $f^{k+k_{\gamma_1}+\cdots+k_{\gamma_{r-1}}} \in \mathfrak{m}$, a contradiction.

Proceeding in the same way for $\xi_2, \ldots, \xi_m$, we find $\alpha_1, \ldots, \alpha_m$ such that

$$\mathfrak{m}_\alpha := R(\xi_1 - \alpha_1) + \cdots + R(\xi_m - \alpha_m) \subset \mathfrak{m}. \tag{4}$$

Let $h \in \mathfrak{m}$ be arbitrary. Then, we can write

$$h(\xi_1, \ldots, \xi_m) = h((\xi_1 - \alpha_1) + \alpha_1, \ldots, (\xi_m - \alpha_m) + \alpha_m) = g + c$$

for some $g = \sum_{i=1}^{m}(\xi_i - \alpha_i)g_i(\xi) \in \mathfrak{m}$ and $c \in \mathbb{C}$. From $c = h - g \in \mathfrak{m}$ we conclude $c = 0$. Therefore $\mathfrak{m} = \{h \in \mathbb{C}[\xi] \mid h(\alpha_1, \ldots, \alpha_m) = 0\} = \mathfrak{m}_\alpha$. Since $\mathfrak{m} \supset \mathfrak{a}$ we have $(\alpha_1, \ldots, \alpha_m) \in V(\mathfrak{a})$. On the other hand, $f \notin \mathfrak{m}$, hence $f(\alpha_1, \ldots, \alpha_m) \neq 0$ so that $f \notin \mathfrak{i}_{V(\mathfrak{a})}$. □

PROOF OF COROLLARY 1.17. Choose $f = 1$. Since it is shown that $\mathfrak{m} = \mathfrak{m}_\alpha$ in the proof of Theorem 1.16, the corollary follows from (4). □

We now list some useful consequences of Hilbert's Nullstellensatz. The first is essentially a restatement, where $\mathrm{Hom}_{\mathbb{C}\text{-alg}}(\mathbb{C}[\cdot], \mathbb{C})$ denotes the set of $\mathbb{C}$-algebra homomorphisms (which, by definition, send 1 to 1):

1.18 Corollary. *There is a one-to-one correspondence between the points x of $\mathbb{C}^m$, the maximal ideals $\mathfrak{m}_x$ of $\mathbb{C}[\xi_1, \ldots, \xi_m]$, and the evaluation homomorphisms $\mathbb{C}[\xi] \to \mathbb{C}$, $f \mapsto f(x)$:*

$$\mathbb{C}^m \longleftrightarrow \{\mathfrak{m} \subset \mathbb{C}[\xi] \mid \mathfrak{m} \text{ is a maximal ideal}\} \longleftrightarrow \mathrm{Hom}_{\mathbb{C}\text{-alg}}(\mathbb{C}[\xi], \mathbb{C}).$$

The proof of the last equivalence stems from the fact that those algebra homomorphisms are uniquely determined by their kernels, which are maximal ideals, (see Lemma 1.20).

The next result characterizes the points of an affine algebraic set in ideal-theoretic terms:

1.19 Corollary. *If $\mathfrak{a}$ is an ideal in $\mathbb{C}[\xi]$, then*

$$V(\mathfrak{a}) = \{x \in \mathbb{C}^m \mid \mathfrak{a} \subset \mathfrak{m}_x\};$$

in particular,

$$\mathfrak{i}_{V(\mathfrak{a})} = \{f \in \mathbb{C}[\xi] \mid f^k \in \mathfrak{a}\}, \quad \textit{and} \quad V(\mathfrak{a}) = V(\mathfrak{i}_{V(\mathfrak{a})}).$$

PROOF. We only show the last part. By Theorem 1.16, $V(\mathfrak{a}) = V(\{f \mid f^k \in \mathfrak{a} \text{ for some } k \in \mathbb{Z}_{>0}\}) = V(\mathfrak{i}_{V(\mathfrak{a})})$. □

By the following obvious lemma, the points of an affine algebraic set $V = V(\mathfrak{a})$ correspond precisely to the maximal ideals of the coordinate ring R_V (or of the residue class ring $R/\mathfrak{a}$):

1.20 Lemma. *Let $\mathfrak{a}$ be an ideal of a ring R. Then, the maximal ideals of the ring $\bar{R} := R/\mathfrak{a}$ are precisely those of the form $\bar{\mathfrak{m}} := \mathfrak{m}/\mathfrak{a}$ with $\mathfrak{m}$ maximal in R and $\mathfrak{a} \subset \mathfrak{m}$.*

That yields a generalization of Corollary 1.18:

1.21 Corollary. *If V is an affine algebraic set with coordinate ring R_V, then, there is a one-to-one correspondence between the sets*

$$V \longleftrightarrow \{\mathfrak{m} \subset R_V \textit{ maximal ideal}\} \longleftrightarrow \operatorname{Hom}_{\mathbb{C}\text{-alg}}(R_V, \mathbb{C}).$$

For the set of maximal ideals, there is the following standard notion in commutative ring theory:

1.22 Definition. The set of all maximal ideals of a ring R is called the (*maximal*) *spectrum* of R and is denoted by spec R.

Now, we introduce topological considerations into the investigation of affine algebraic sets. We recall that a topology on a set X is a collection of subsets of X, called open sets, such that arbitrary unions and finite intersections of open sets are open, and also $\emptyset$, X are open. X is then called a topological space. Bijective maps between topological spaces which preserve open sets, are said to be homeomorphisms.

On many occasions, we shall use "ordinary" topology which is induced in $\mathbb{C}^n$ by the ordinary open sets of $\mathbb{R}^{2n}$. But, for many purposes, another ("non–Hausdorff") topology, which we introduce now, is helpful.

1.23 Definition. A subset of an affine algebraic set X is called *Zariski open* if it is the complement of an algebraic subset which, in turn, is said to be *Zariski closed*.

1.24 Lemma and Definition. *Let $\{X_i\}_{i\in I}$ be a family of Zariski closed (open) subsets of an affine algebraic set X. Then, there exists a finite subset $I_0 \subset I$ such that $\bigcap_{i\in I} X_i = \bigcap_{i\in I_0} X_i$ ($\bigcup_{i\in I} X_i = \bigcup_{i\in I_0} X_i$). In particular, the Zariski open sets define a topology on X, called the Zariski topology.*

PROOF. For finite intersections, there is nothing to prove. Suppose it is false for infinite intersections. Then, for any finite subset $I_0 \subset I$, $\bigcap_{i\in I} X_i \subsetneq \bigcap_{i\in I_0} X_i$. By increasing I_0 succesively, we find a decreasing chain of algebraic sets which does not become stationary, contrary to Lemma 1.9.

Passing over to complements, we obtain the corresponding statement for unions. □

Remark. The finiteness properties of Zariski topology expressed in Lemma 1.24 have no analog in ordinary topology and make Zariski topology a powerful instrument.

Remark. If we associate with each ideal $\mathfrak{a}$ of R the subset $V(\mathfrak{a}) = \{\mathfrak{m} \in \operatorname{spec} R \mid \mathfrak{a} \subset \mathfrak{m}\}$, then, obviously, the collection of all these sets $V(\mathfrak{a})$ is the family of closed sets of a topology on spec R, which is also called Zariski topology. It is, then, a consequence of Corollary 1.21 that an affine algebraic set V, endowed with the Zariski topology, is homeomorphic to spec R_V of its coordinate ring R_V.

Remark. For an ideal $\mathfrak{a}$ of the polynomial ring $\mathbb{C}[\xi]$, the residue class ring $R := \mathbb{C}[\xi]/\mathfrak{a}$ is generated as a $\mathbb{C}$-algebra by the classes $\bar{\xi}_j := \xi_j + \mathfrak{a}$ for $j = 1, \dots, m$. Conversely, a finitely generated $\mathbb{C}$-algebra R can be written as a residue class ring $\mathbb{C}[\xi_1, \dots, \xi_m]/\mathfrak{a}$ for some m and a suitably chosen ideal $\mathfrak{a}$. Such a presentation and, hence, the affine algebraic set $V(\mathfrak{a})$ in $\mathbb{C}^m$ defined by these data, is not unique. The preceding Remark shows, however, that any two such "models" are homeomorphic. Thus, we can associate with such a finitely generated $\mathbb{C}$-algebra R a well-defined "abstract affine algebraic set," namely, its spectrum spec R, endowed with the Zariski topology. Moreover, if R is an integral domain, that is, without zero divisors, then every such ideal $\mathfrak{a}$ is prime, and we may consider spec R as an abstract affine algebraic variety.

Remark. In commutative ring theory, there is the more general notion of the prime spectrum spec R of a commutative ring R, that is the collection of all prime ideals $\mathfrak{p}$ of R. That notion is very useful in abstract algebraic geometry. If $R = R_V$ is the coordinate ring of an affine algebraic set, the elements of spec $\mathbb{R}$ are in one-to-one correspondence with the (irreducible) subvarieties of V. For our purposes, the more restricted notion of a maximal spectrum is sufficient.

From polynomials, we now proceed to rational and regular functions, thus, providing tools for the study of divisors in chapter VII.

Let Y be an affine algebraic variety, and let R_Y be its coordinate ring. Since R_Y is an integral domain, we can consider its *quotient field* K_Y:

1.25 Definition. By a *rational function*, we mean any element of a quotient field K_Y. We say that a rational function $f \in K_Y$ is *regular at* $p \in Y$ if in an open set U which contains p, we can express f as $f = \frac{g}{h}$ for $g, h \in R_Y, h(p) \neq 0$. If f is regular at each point of U, it is called *regular* on U. A rational function regular on Y is simply said to be a *regular function.*

1.26 Lemma. *A rational function* $f \in K_Y$ *is regular if and only if* $f \in R_Y$.

PROOF. If $f \in R_Y$, it is clearly regular. So, let $f \in K_Y$ be regular. For any p let $f = \frac{g_p}{h_p}$ where $g_p, h_p \in R_Y$ and $h_p(p) \neq 0$. We set $U_{p,f} := \{x \in Y \mid h_p(x) \neq 0\}$. By definition, $p \in U_{p,f}$, and so, $\{U_{p,f}\}_{p \in Y}$ is an open covering of Y. By Lemma 1.24 we can choose a finite subcovering $\{U_{p_i,f}\}_{p_i \in Y}$ for appropriate $p_i \in Y, i = 1, \ldots, k$. We set $g_i := g_{p_i}, h_i := h_{p_i}$. Consider the ideal $\mathfrak{i} := R_Y \cdot h_1 + \cdots + R_Y \cdot h_k \subset R_Y$, and suppose $\mathfrak{i} \neq R_Y$. Let $\mathfrak{m}$ be a maximal ideal which contains $\mathfrak{i}$. By Hilbert's Nullstellensatz, $\mathfrak{m} = \mathfrak{m}_x = \{v \in R_Y \mid v(x) = 0\}$ for some $x \in Y$. In particular, $h_i(x) = 0$ for $i = 1, \ldots, k$. But $x \in U_{p_i,f}$ for some i, so that $h_i(x) \neq 0$, a contradiction. Therefore $\mathfrak{i} = R_Y$, and $\mathfrak{i} \ni 1 = \sum f_i g_i$ for appropriate $f_i \in R_Y$, hence, $f = \sum f_i f g_i = \sum f_i h_i \in R_Y$. □

1.27 Lemma. *Let* U_f *be the set of all points of* Y *at which a given rational function* f *is regular. Then* U_f *is a nonempty Zariski open set.*

PROOF. Using the notation as in the proof of Lemma 1.26, we may, write $U_f = \bigcup_{p \in U_f} U_{p,f}$. Therefore, U_f is a union of Zariski open sets, and, hence, Zariski is open. □

Remark. Note that $f \in K_Y$ defines a map $f : U_f \to \mathbb{C}, p \mapsto f(p) = \frac{g_p(p)}{h_p(p)}$.

We extend the notions of Definition 1.8 and Definition 1.13 as follows.

1.28 Definition. A Zariski open subset U of an affine variety Y is called a *quasi-affine variety*. We define on it the *ring of regular functions*

$$R_U := \{f \in K_Y \mid f \text{ is regular on } U\}.$$

Let F be a subset of R_U. By an *algebraic subset* of U we mean the set

$$V = V(F, U) := \{x \in U \mid f(x) = 0 \text{ for all } f \in F\}$$

(compare Definition 1.1). For $F = \{f\}$, we set $V(f, U) := V(\{f\}, U)$. We say that an algebraic subset V of U is *irreducible* or a *subvariety* of U if it is not the union of two algebraic subsets of U, both different from V. As in the case of an affine variety each algebraic subset of U is a finite union of uniquely determined *irreducible components*. Each subvariety V of U determines an ideal

$$\mathfrak{i}_{V,U} := \{f \in R_U \mid f|_V = 0\}.$$

Remark. Let $U_1 \subset Y, U_2 \subset Y$ be two quasi-affine varieties. Then, $R_{U_1} = R_{U_2}$ does not imply $U_1 = U_2$. Example: $R_{\mathbb{C}^n} = R_{\mathbb{C}^n \setminus \{p\}}$ for $n \geq 2$ and some $p \in \mathbb{C}^n$.

1.29 Lemma. *Let U be a quasi-affine variety in Y. Then, the following is true.*

(a) *If $D \subset Y$ is irreducible, so is $D \cap U$. If $D \cap U \neq \emptyset$, then, $D \cap U$ is (in the Zariski topology) an open and dense subset of D.*

(b) *If $D \subset U$ is an irreducible algebraic subset of U, then, the Zariski closure $\bar{D}$ of D is irreducible in Y and satisfies $\bar{D} \cap U i D$. Moreover, $\mathfrak{i}_{\bar{D},Y} = \mathfrak{i}_{D,U} \cap R_Y$.*

(c) *The Zariski topology induces a topology on U (also called Zariski topology) whose closed sets are the algebraic subsets of U.*

PROOF.

(a) Let $D \subset Y$ be irreducible, $D \cap U \neq \emptyset$, and suppose $D \cap U$ is not irreducible. Then, $D \cap U = D_1 \cup D_2$ for D_1, D_2 Zariski closed in D, and we can find $f_1, f_2 \in R_U$ such that $f_1 \in \mathfrak{i}_{D_2,U} \setminus \mathfrak{i}_{D_1,U}$, $f_2 \in \mathfrak{i}_{D_1,U} \setminus \mathfrak{i}_{D_2,U}$. Clearly, $f_1 \cdot f_2 \in \mathfrak{i}_{D_1 \cup D_2,U}$. Choose a $p_1 \in D_1$ such that $f_1(p_1) \neq 0$. We have $f_1 = \frac{g_{p_1}}{h_{p_1}}$ where $g_{p_1}, h_{p_1} \in R_Y$ and $h_{p_1}(p_1) \neq 0$. Then $g_{p_1} \in (\mathfrak{i}_{D_2,U} \setminus \mathfrak{i}_{D_1,U}) \cap R_Y$. In an analogous way, we find $g_{p_2} \in (\mathfrak{i}_{D_1,U} \setminus \mathfrak{i}_{D_2,U}) \cap R_Y$. Since $g_{p_1} \cdot g_{p_2} \in \mathfrak{i}_{D\cap U,U}$, we obtain

$$D \supset (D \cap V(g_{p_1} \cdot g_{p_2}, Y)) \cup (D \cap (Y \setminus U)) \supset (D \cap U) \cup (D \cap (Y \setminus U)) = D.$$

Since D is irreducible and $D \cap U \neq \emptyset$, this implies $D = D \cap V(g_{p_1} \cdot g_{p_2}, Y)$, hence, $g_{p_1} \cdot g_{p_2} \in \mathfrak{i}_{D,Y}$. But $g_{p_1} \notin \mathfrak{i}_{D,Y}$, $g_{p_2} \notin \mathfrak{i}_{D,Y}$, contrary to $\mathfrak{i}_{D,Y}$ being prime (Lemma 1.15). Since $D \cap U = D \setminus [(Y \setminus U) \cap D]$ and since $(Y \setminus U) \cap D$ is a closed subset of the variety D we conclude that $D \cap U$ is a Zariski open and therefore dense subset of D.

(b) Let $D \subset U$ be irreducible. Then, $\mathfrak{i}_{D,U} \subset R_U$ is prime (Lemma 1.15). Also, $\mathfrak{i}_{D,U} \cap R_Y$ is obviously prime and defines the subvariety $\bar{D} \subset Y$. By definition, $\mathfrak{i}_{\bar{D},Y} = \mathfrak{i}_{D,U} \cap R_Y$ is the largest ideal in R_Y which is contained in $\mathfrak{i}_{D,U}$, so it corresponds to the smallest algebraic subset of Y containing D, and, hence, to $\bar{D}$. Note that $\bar{D} \cap U \supset D \cap U = D$.

Let $f \in \mathfrak{i}_{D,U} \subset R_U$. For $p \in D$, $f = \frac{g_p}{h_p}$ for $g_p, h_p \in R_Y, h_p(p) \neq 0$. In particular, $g_p = f \cdot h_p \in \mathfrak{i}_{\bar{D},Y}$. Since $g_p \in \mathfrak{i}_{\bar{D},Y}$, we obtain $g_p = f \cdot h_p \in \mathfrak{i}_{\bar{D}\cap U,U}$, and, since $\mathfrak{i}_{\bar{D}\cap U,U}$ is prime, we see that $f \in \mathfrak{i}_{\bar{D}\cap U,U}$. Finally, $\mathfrak{i}_{D,U} \subset \mathfrak{i}_{\bar{D}\cap U,U}$ and $D \supset \bar{D} \cap U \supset D$, hence, $D = \bar{D} \cap U$.

(c) By Lemma 1.24, $D = \bigcup_{i=1}^k D_i$ where the D_i are irreducible. Then, $\bar{D} = \bigcup_{i=1}^k \bar{D}_i$.

By (b), $D_i = \bar{D}_i \cap U$, hence, $\bar{D} \cap U = \bigcup_{i=1}^k (\bar{D}_i \cap U) = \bigcup_{i=1}^k D_i = D$. □

1.30 Lemma. *Let $U_1 \subset U_2 \subset Y$ be two quasi-affine varieties in the affine variety Y, and let $D \subset U_2$ be an irreducible subvariety of U_2 such that $\mathfrak{i}_{D,U_2} \subset R_{U_2}$ is generated by one element $g_D \in R_{U_2}$, that is, $\mathfrak{i}_{D,U_2} = g_D \cdot R_{U_2}$. Then, $\mathfrak{i}_{D\cap U_1,U_1} = g_D \cdot R_{U_1}$.*

PROOF. Let $f \in \mathfrak{i}_{D\cap U_1,U_1}$. For $p \in D \cap U_1$, we may write $f = \frac{g_p}{h_p}$ where $g_p, h_p \in R_Y$, $h_p(p) \neq 0$. g_p is regular on U_2 and $g_p|_{D\cap U_1} = 0$. Since D is irreducible, we see that $g_p|_D = 0$ which implies $g_p \in \mathfrak{i}_{D,U_2}$ and hence $g_p = g_D \cdot w_p$ where $w_p \in R_{U_2}$. Therefore $f = g_D \cdot \frac{w_p}{h_p}$ and $\frac{f}{g_D} = \frac{w_p}{h_p}$ follow which means that $\frac{f}{g_D}$ is regular in all points of $D \cap U_1$. Since $g_D(x) \neq 0$ in $U_1 \setminus D$, regularity of $\frac{f}{g_D}$ holds everywhere in U_1, and $f \in R_{U_1} \cdot g_D$ follows. □

1.31 Definition. Let X, Y be quasi-affine varieties, and let $\varphi : X \to Y$ be a map such that, for every $f \in R_Y$, $f \circ \varphi \in R_X$. Then, we call φ a *morphism.*

Any morphism φ defines a ring homomorphism $\varphi^* : R_Y \to R_X$ by $f \mapsto \varphi^*(f) := f \circ \varphi$.

Example 5. Let $U \subset Y$ be Zariski open in Y. Then the inclusion $U \hookrightarrow Y$ determines a morphism of quasi-affine varieties. $\varphi^* : R_Y \to R_U$ is the restriction of a regular function on Y to the subset U. We may, then, interpret $\varphi^* : R_Y \to R_U$ as the inclusion $R_Y \subset R_U$.

Example 6. Let U be an affine variety and $Z \subset U$ be a subvariety of U. Then, the inclusion $Z \hookrightarrow U$ determines a morphism $\varphi : Z \to U$. One easily proves that $\varphi^* : R_U \to R_Z$ is a surjection and $\ker \varphi^* = \mathfrak{i}_{Z,U}$. In particular, we get that $R_Z = R_U/\mathfrak{i}_{Z,U}$.

The following lemma is evident.

1.32 Lemma. *Let $\varphi : Z \to U$ be a morphism of affine varieties. Then, for $x \in Z$,*

$$\mathfrak{m}_{\varphi(x)} = \varphi^{*-1}(\mathfrak{m}_x).$$

1.33 Definition. A morphism φ is said to be an *isomorphism* if it is bijective and if φ^* is an isomorphism. By an *open inclusion*, we mean an isomorphism onto some Zariski open set. By a *closed embedding*, we mean an isomorphism onto a Zariski closed subset.

1.34 Lemma. *Let $\varphi : X \to Y$ be a morphism of affine varieties, and let $\varphi^* : R_Y \to R_X$ be the corresponding morphism of rings. Then,*

(a) *φ^* is an isomorphism if and only if φ is an isomorphism, and*
(b) *φ^* is surjective if and only if φ is a closed embedding.*

PROOF.

(a) We need only show the "only if" part. Let φ^* be an isomorphism and, hence, define a bijection between maximal ideals of R_X and $\varphi^{*-1}(R_X)$. Then, by Lemma 1.32, φ is a bijection.

(b) If φ is a closed embedding, it splits into an isomorphism $\varphi_0 : X \to \varphi(X)$ and the inclusion $\varphi(X) \subset Y$ where $\varphi(X)$ is closed in Y. Then, φ^* splits into a surjection $R_Y \to R_Y/\mathfrak{i}_{\varphi(X),Y} = R_{\varphi(X)}$ and an isomorphism $\varphi_0^* : R_{\varphi(X)} \to R_X$, so φ^* is a surjection.

Conversely, if φ^* is a surjection, then, its kernel is a prime ideal determining a subvariety Z of Y. The points of Z are in one-to-one correspondence with maximal ideals of R_Y which contain $\mathfrak{i}_Z$ and, hence, with ideals of the form $\varphi^{*-1}(\mathfrak{m})$, where

$\mathfrak{m}$ is maximal in R_X. Therefore, by Lemma 1.32, $Z = \varphi(X)$. By (a), we see that $\varphi : X \to \varphi(X)$ is an isomorphism. □

1.35 Definition. Let U be a quasi-affine variety. We say that U is an *affine variety* if and only if U is isomorphic to some affine variety $V \subset \mathbb{C}^m$.

Example 7. $\mathbb{C}\setminus\{0\} \longrightarrow Y = \{(x, y) \in \mathbb{C}^2 \mid xy = 1\}, x \longmapsto (x, x^{-1})$.

1.36 Definition. Let X be an affine algebraic variety. By the *dimension* dim X of X, we mean the supremum of all integers n such that there exists a chain $\emptyset \neq X_0 \subset X_1 \subset \cdots \subset X_n = X$ of distinct irreducible sets.

Example 8. The dimension of $\mathbb{C}^1$ is 1, since the only irreducible sets are the whole space and a single point.

1.37 Definition. Given a ring R, the *height* of a prime ideal $\mathfrak{p}$ is defined as the supremum of all integers such that there exists a chain $\mathfrak{p}_0 \subset \cdots \subset \mathfrak{p}_n = \mathfrak{p}$ of distinct prime ideals. The supremum of heights of all prime ideals of R is called the *(Krull) dimension* of R.

1.38 Lemma. *If Y is an affine algebraic variety, then, the dimension of Y is equal to the dimension of its coordinate ring R_Y.*

PROOF. The irreducible affine algebraic sets contained in $Y \subset \mathbb{C}^n$ correspond to those prime ideals in $R := \mathbb{C}[\xi_1, \ldots, \xi_n]$ which contain $\mathfrak{i}_Y$. These ideals are in one-to-one correspondence φ to prime ideals in $R_Y = \mathbb{C}[\xi_1, \ldots, \xi_n]/\mathfrak{i}_Y$:

$$\begin{array}{rcl} \varphi : R & \longrightarrow & R_Y = R/\mathfrak{i}_Y \\ \mathfrak{p} & \longmapsto & \varphi(\mathfrak{p}). \end{array}$$

So, the above definitions imply the lemma. □

We recall some elementary facts on rings and fields. Let $R_1 \supset R_2$ be two rings. We say that $a \in R_1$ is *integral* over R_2 if there exists a polynomial $w(x) = x^n + \alpha_{n-1}x^{n-1} + \cdots + \alpha_0$ in $R_2[x]$ such that $w(a) = 0$. The elements of R_1, which are integral over R_2, provide a ring $\bar{R}_2 \supset R_2$, and we have $R_1 \supset \bar{R}_2$.

If all elements of $R_1 \supset R_2$ are integral over R_2, we call R_1 *integral* over R_2. If $R_3 \supset R_2 \supset R_1$ are rings such that R_3 is integral over R_2 and R_2 is integral over R_1 then, R_3 is integral over R_1. Let $R_1 \supset R_2$ where R_2 is a domain with quotient field K. For an $a \in R_1$ which is integral over R_2, let $w(x) = x^k + a_{k-1}x^{k-1} + \cdots + a_0$, $a_i \in K$ for $i = 0, \ldots, k-1$ be a polynomial of minimal degree such that $w(a) = 0$. Then, it follows that $a_0, \ldots, a_{k-1} \in R_2$.

Let $K_1 \supset K_2$ be fields. We call elements $a_1, \ldots, a_l \in K_1$ *algebraically dependent* over K_2 if there exists a polynomial $w(x_1, \ldots, x_l) \in K_2[x_1, \ldots, x_l]$ such that $w(a_1, \ldots, a_l) = 0$, otherwise, *algebraically* independent. A maximal set of algebraically independent elements is called a *transcendency basis*. Its cardinality does not depend on the transcendency basis and is called the *transcendency degree* tr deg K_1 of K_1 over K_2.

A transcendency basis can be chosen from any set of generators of K_1 over K_2. If $\{a_1, \ldots, a_l\}$ is a transcendency basis, then, each element a of K_1 is algebraically dependent over the field $K_2(a_1, \ldots, a_l)$ which means that there exists a polynomial $w \in K_2(a_1, \ldots, a_l)[x]$ such that $w(a) = 0$ or, equivalently, there exists a polynomial $\widetilde{w} \in K_2[x_1, \ldots, x_l, x_{l+1}]$ such that $\widetilde{w}(a_1, \ldots, a_l, a) = 0$. (For more about transcendency degree compare, for example, Winter [1974], p. 41).

1.39 Lemma. *Let $X \subsetneqq Y$ be affine algebraic varieties in $\mathbb{C}^n$. Then,* tr deg $K_Y >$ tr deg K_X.

PROOF. We may choose the coordinate functions $\xi_1, \ldots, \xi_k$ (up to a permutation) as a transcendency basis of K_Y, $k =$ tr deg K_Y. So, we have $\mathfrak{i}_Y \subsetneqq \mathfrak{i}_X \subset \mathbb{C}[\xi_1, \ldots, \xi_n]$. If $f \in \mathfrak{i}_X \setminus \mathfrak{i}_Y$, then, f is algebraically dependent on the quotient field $\mathbb{C}(\xi_1, \ldots, \xi_k) \subset K_Y$. Thus, we find $f^m + a_{m-1} \cdot f^{m-1} + \cdots + a_0 = 0$ for $a_i \in \mathbb{C}(\xi_1, \ldots, \xi_k)$, that is, $a_i = g_i/h_i$, $g_i, h_i \in \mathbb{C}[\xi_1, \ldots, \xi_k]$. This implies $w_m \cdot f^m + w_{m-1} \cdot f^{m-1} + \cdots + w_0 = 0$ where $w_i = g_i \cdot h_0 \cdots h_{i-1} \cdot h_{i+1} \cdots h_{m-1} \in \mathbb{C}[\xi_1, \ldots, \xi_k]$, $i = 0, \ldots, m-1$, $w_m = h_0 \cdots h_{m-1}$. Since $f \in \mathfrak{i}_X$, we obtain $w_0 \in \mathfrak{i}_X$. But w_0 is not zero on Y, since, otherwise, we could divide by f. Thus, $w_0 = w_0(\xi_1, \ldots, \xi_k)$ is a nonzero polynomial determining the algebraic dependence of $\xi_1, \ldots, \xi_k$ on X. So we conclude that tr deg $K_X <$ tr deg K_Y. □

1.40 Lemma. *Let $X \subset Y$ be affine varieties in $\mathbb{C}^n$. If* tr deg $K_Y \geq$ tr deg $K_X + 2$ *then there is an affine variety X' such that $X \subsetneqq X' \subsetneqq Y$.*

PROOF. Let $\dim Y = k$. Since $Y \subset \mathbb{C}^n$ is an affine variety, the functions $\xi_1, \ldots, \xi_n$ generate the ring $R_Y = \mathbb{C}[\xi_1, \ldots, \xi_n]/\mathfrak{i}_Y$. We consider an element $w \in \mathfrak{i}_Y$ which is a nonzero polynomial represented as a sum $w = w_0 + \cdots + w_m$ of homogeneous polynomials w_i of degree i, where $i = 0, \ldots, m$. We may assume that $w_m \neq 0$. Substituting new coordinates $\xi'_n = \xi_n$, $\xi'_i = \xi_i - a_i \xi_n$ or $\xi_i = \xi'_i + a_i \xi_n$, $a_i \in \mathbb{C}$, $i = 1, \ldots, n-1$, we obtain

$$w(\xi'_1, \ldots, \xi'_n) = w_m(a_1, \ldots, a_{n-1}, 1) \cdot (\xi'_n)^m + \text{ terms in which } \xi'_n \text{ is of degree } < m.$$

Since w_m is a nonzero homogeneous polynomial, we find $a_1, \ldots, a_{n-1} \in \mathbb{C}$ such that $w_m(a_1, \ldots, a_{n-1}, 1) \neq 0$.

Then, we see that ξ'_n is integral over $\mathbb{C}[\xi'_1, \ldots, \xi'_{n-1}]/(\mathfrak{i}_Y \cap \mathbb{C}[\xi'_1, \ldots, \xi'_{n-1}]) \subset R_Y$. We write again ξ_i instead of ξ'_i, $i = 1, \ldots, n$. By repeating the same procedure, we find $\xi_1, \ldots, \xi_k, \xi_{k+1}, \ldots, \xi_n$ such that $\xi_1, \ldots, \xi_k$ provide a transcendency basis of K_Y, and $\xi_{k+1}, \ldots, \xi_n$ are integral over $\xi_1, \ldots, \xi_k$ which means that R_Y is integral over $\mathbb{C}[\xi_1, \ldots, \xi_k] \subset R_Y$.

Let $\xi_1, \ldots, \xi_l$ form a transcendency basis of R_X. Then, $\mathfrak{i}_{X,Y} \cap \mathbb{C}[\xi_1, \ldots, \xi_l] = 0$. In particular, tr deg $K_X = l$. By assumption, $l \leq k - 2$. Let $\mathfrak{p} \subset R_Y$

be a maximal ideal for which $\mathfrak{p} \cap (\mathbb{C}[\xi_1, \ldots, \xi_{l+1}] \cdot (R_Y \setminus \mathfrak{i}_{X,Y})) = 0$. By definition, $\mathfrak{p} \subset \mathfrak{i}_{X,Y}$. If $f, g \notin \mathfrak{p}$, then, by the maximality of the ideal $\mathfrak{p}$, we find $a \cdot f = p_1 + w_1(\xi_1, \ldots, \xi_{l+1}) \cdot c_1$ where $a \in R_Y$, $p_1 \in \mathfrak{p}$, $c_1 \notin \mathfrak{i}_{X,Y}$, and $b \cdot g = p_2 + w_2(\xi_1, \ldots, \xi_{l+1}) \cdot c_2$, where $b \in R_Y$, $p_2 \in \mathfrak{p}$, $c_2 \notin \mathfrak{i}_{X,Y}$. Hence, $a \cdot b \cdot f \cdot g = p + w_1(\xi_1, \ldots, \xi_{l+1}) \cdot w_2(\xi_1, \ldots, \xi_{l+1}) \cdot c_1 \cdot c_2$ with $p \in \mathfrak{p}$ and $c_1 \cdot c_2 \notin \mathfrak{i}_{X,Y}$, so that $f \cdot g \notin \mathfrak{p}$, and we conclude that $\mathfrak{p}$ is prime.

By definition, $\mathfrak{p} \subsetneq \mathfrak{i}_{X,Y} \subset R_Y$. Hence, $\mathfrak{p}$ defines a variety X' such that $X' \supsetneq X$. It remains to show that $Y \supsetneq X'$ or, equivalently, that $\mathfrak{p} \neq 0$ in R_Y.

In fact, suppose $\mathfrak{p} = 0$ in R_Y. Then, by the definition of $\mathfrak{p}$, for any nonzero ideal $\mathfrak{i} \supset \mathfrak{p} = 0$, $i \cap (\mathbb{C}[\xi_1, \ldots, \xi_{l+1}] \cdot (R_Y \setminus \mathfrak{i}_{X,Y})) \neq 0$. Since $l \leq k - 2$ and $\xi_1, \ldots, \xi_l$ form a transcendency basis of $R_X = R_Y/\mathfrak{i}_{X,Y}$, we can find an irreducible polynomial $v \in \mathbb{C}[\xi_1, \ldots, \xi_l, \xi_{l+2}] \setminus \mathbb{C}[\xi_1, \ldots, \xi_l]$ in $\mathfrak{i}_{X,Y}$. Then, the ideal $\mathfrak{i} := R_Y \cdot v$ satisfies $\mathfrak{i} \cap (\mathbb{C}[\xi_1, \ldots, \xi_{l+1}] \cdot (R_Y \setminus \mathfrak{i}_{X,Y})) \neq 0$.

Now, $f \cdot v = g \cdot h$ for some $f \in R_Y$, $g \in \mathbb{C}[\xi_1, \ldots, \xi_{l+1}]$, and $h \notin \mathfrak{i}_{X,Y}$. In particular, g and v are relatively prime polynomials in $\xi_1, \ldots, \xi_k$.

Let $w(x) := x^r + a_{r-1}x^{r-1} + \cdots + a_0, a_i \in \mathbb{C}[\xi_1, \ldots, \xi_k], i = 0, \ldots, r-1$, be a polynomial of minimal degree such that $w(h) = 0$. Then, $(\frac{g}{v})^r w(\frac{v}{g} \cdot f) = 0$ and

$$w'(x) := (\frac{g}{v})^r w(\frac{v}{g}x) = x^r + a_{r-1}\frac{g}{v}x^{r-1} + \cdots + a_0\frac{g^r}{v^r}$$

is a polynomial of minimal degree such that $w'(f) = 0$. Hence, $a_i \frac{g^{r-i}}{v^{r-i}} \in \mathbb{C}[\xi_1, \ldots, \xi_k]$ for $i = 0, \ldots, r-1$. In particular, v divides a_i for $i = 0, \ldots, r-1$. Hence, $a_i \in \mathfrak{i}_{X,Y}$ so that $h^r \in \mathfrak{i}_{X,Y}$, a contradiction. This proves that $\mathfrak{p} \neq 0$. □

From Lemmas 1.39 and 1.40, we conclude that each chain of varieties can be completed to a maximal chain, that is, $X_0 \subset X_1 \subset \cdots \subset X_l = X$ such that X_0 is a point and $\operatorname{tr deg} K_{X_i} = \operatorname{tr deg} K_{X_{i-1}} + 1 = (i-1) + 1 = i$. The other assertions being evident, we obtain the following theorem:

1.41 Theorem.

(a) *If Y is an affine variety, then* $\dim Y = \dim R_Y = \operatorname{tr deg} K_Y$.
(b) *Each chain of irreducible varieties can be completed to a chain of maximal length.*
(c) *If X is a variety in Y, then,* $\dim X$ *equals* $\dim Y$ *minus the height of* $\mathfrak{i}_X$ *in* R_Y.

Corollary. *Let $U \subset X$ be a quasi-affine variety in an affine variety X. Then,*

$$\dim U = \dim X = \operatorname{tr deg} K_X = \operatorname{tr deg} K_U.$$

PROOF. Let $Y_0 \subset Y_1 \subset \cdots \subset Y_l = U$ be a maximal chain in U. Then, $\bar{Y}_0 \subset \cdots \subset \bar{Y}_l = X$ is a maximal chain in X. Otherwise, by Lemma 1.29 and Theorem 1.40 we could complete it and intersect again with U obtaining a longer chain. □

Example 9. $\dim \mathbb{C}^n = n$.

Exercises

1. For each of the following algebraic sets A, find the ideal $\mathfrak{a}$ in $\mathbb{C}[\xi_1, \ldots, \xi_n]$ satisfying $A = V(\mathfrak{a})$:

   a. A is a finite set of points.
   b. A is the union of all coordinate hyperplanes of $\mathbb{C}^n$.
   c. A is an m-dimensional linear subspace of $\mathbb{C}^n$, $0 < m < n$.
   d. A is the union of a hyperplane H and a point not on H.

2. Prove that linear manifolds of $\mathbb{C}^n$ (translated linear subspaces) are algebraic varieties.
3. Find a subalgebra of $\mathbb{C}[\xi_1, \xi_2]$ which is not finitely generated.
4. Any morphism of quasiaffine varieties is continuous with respect to Zariski topology as well as with respect to ordinary topology.

2. Affine toric varieties

Let $R := \mathbb{C}[\xi_1, \ldots, \xi_{2n}]$ be the polynomial ring in $2n$ variables, $n \geq 1$. Then,

$$\mathfrak{a} := R(\xi_1\xi_{n+1} - 1) + \cdots + R(\xi_n\xi_{2n} - 1)$$

is an ideal in R. For $z_i := \xi_i + \mathfrak{a} \in R/\mathfrak{a}$, $i = 1, \ldots, 2n$, we, thus, have (writing simply 1 instead of $\bar{1} \in R/\mathfrak{a}$)

$$z_j z_{n+j} = 1,$$

and, hence $z_j^{-1} = z_{n+j}$ for $j = 1, \ldots, n$.

2.1 Definition. The elements of

$$\mathbb{C}[z, z^{-1}] := \mathbb{C}[z_1, \ldots, z_n, z_1^{-1}, \ldots, z_n^{-1}] = \mathbb{C}[\xi_1, \ldots, \xi_{2n}]/\mathfrak{a}$$

are called *Laurent polynomials*, whereas terms

$$\lambda \cdot z^a = \lambda z_1^{\alpha_1} \cdots z_n^{\alpha_n}, \text{ for } a = (\alpha_1, \ldots, \alpha_n) \in \mathbb{Z}^n, \quad \lambda \in \mathbb{C}^*,$$

are said to be *Laurent monomials*. .

The *monic* (i.e., $\lambda = 1$) Laurent monomials form a (multiplicative) group. The key for the construction of toric varieties is the fact that the mapping

$$\vartheta : \mathbb{Z}^n \to \mathbb{C}[z, z^{-1}], \ a \mapsto z^a \tag{1}$$

provides an *isomorphism* (again denoted by ϑ) between the (additive) group $\mathbb{Z}^n$ and the (multiplicative) group of monic Laurent monomials.

2.2 Definition. The *support* of a Laurent polynomial $f = \sum_{\text{finite}} \lambda_a z^a$, is defined as

$$\operatorname{supp}(f) := \{a \in \mathbb{Z}^n \mid \lambda_a \neq 0\}.$$

We, obviously, have

$$\operatorname{supp}(f \pm g) \subset \operatorname{supp}(f) \cup \operatorname{supp}(g) \quad \text{and} \quad \operatorname{supp}(fg) \subset \operatorname{supp}(f) + \operatorname{supp}(g)$$

for $f, g \in \mathbb{C}[z, z^{-1}]$, and $\operatorname{supp}(1) = \{0\}$. Hence, if a ring R of Laurent polynomials is "monomial" in the following sense, then the set $\bigcup_{f \in R} \operatorname{supp} f$ is a submonoid of $\mathbb{Z}^n$:

2.3 Definition. A ring R of Laurent polynomials is called a *monomial algebra* if it is a $\mathbb{C}$-algebra generated by Laurent monomials.

The set of all Laurent polynomials, with support in a given submonoid of $\mathbb{Z}^n$, clearly is a monomial algebra. For a lattice cone σ, we know that the set $\sigma \cap \mathbb{Z}^n$ is a submonoid, which, by Gordan's lemma, is even finitely generated. Thus, we have the following lemma

2.4 Lemma and Definition. *For a lattice cone σ, the ring*

$$R_\sigma := \{f \in \mathbb{C}[z, z^{-1}] \mid \operatorname{supp} f \subset \sigma\}$$

is a finitely generated, monomial algebra.

We note that R_σ, being a subring of $\mathbb{C}[z, z^{-1}]$, has no zero divisors. Moreover, we recall that every finitely generated $\mathbb{C}$-algebra without zero divisors defines an abstract affine algebraic variety, namely, its maximal spectrum. The varieties defined by the rings R_σ are basic for the algebro-geometric objects to be considered in the sequel:

2.5 Definition. For a lattice cone σ, the maximal spectrum $X_\sigma := \operatorname{spec} R_\sigma$ is called an (abstract) *affine toric variety* (or *torus embedding*).

As we always may consider such a cone σ as a lattice cone in the subspace $\operatorname{lin} \sigma$ with respect to the lattice $\operatorname{lin} \sigma \cap \mathbb{Z}^n$, we tacitly shall assume, in general, that σ is n-dimensional. In particular, that holds for the cones we have to consider in later sections, namely, those which are duals of strictly convex lattice cones in $\mathbb{R}^n$.

The reason to call the variety X_σ "toric" will be explained in 2.8 below.

Now, we now want to study the subvarieties that realize such an abstract variety X_σ in suitable affine spaces $\mathbb{C}^k$. We recall that these geometric realizations are obtained by introducing coordinates, which corresponds to a choice of generators of the monoid $\sigma \cap \mathbb{Z}^n$. In particular, we have to discuss transformations between different coordinate systems.

Example 1. The largest possible n-dimensional cone is $\sigma := \mathbb{R}^n$. Then, viewed as a monoid, $\sigma \cap \mathbb{Z}^n = \mathbb{Z}^n$ has generators $e_1, \dots, e_n, -e_1, \dots, -e_n$, so the associated algebra is $R_\sigma = \mathbb{C}[z_1, \dots, z_n, z_1^{-1}, \dots, z_n^{-1}]$. The corresponding toric variety X_σ can be described in $\mathbb{C}^{2n}$ with coordinates $\xi_1, \dots, \xi_{2n}$ as solution set of the equations

$$\xi_i \xi_{n+i} = 1,\ i = 1, \ldots, n.$$

Hence, $X_\sigma = V(\xi_1\xi_{n+1} - 1, \ldots, \xi_n\xi_{2n} - 1)$. For $n = 1$, we obtain a (complex) hyperbola with $\{\xi_1 = 0\}$ and $\{\xi_2 = 0\}$ as asymptotes (Figure 1); compare section 1, Example 7.

We may also realize X_σ as the set of points

$$T := \{(z_1, \ldots, z_n) \in \mathbb{C}^n \mid z_i \neq 0, i = 1, \ldots, n\} = (\mathbb{C} \setminus \{0\})^n,$$

which is isomorphic to $V(\xi_1\xi_{n+1} - 1, \ldots, \xi_n\xi_{2n} - 1)$ under the projection $\mathbb{C}^{2n} \to \mathbb{C}^n$. The inverse of the restricted projection $V(z_1z_{n+1} - 1, \ldots, z_nz_{2n} - 1) \to T$ is given by

$$(z_1, \ldots, z_n) \mapsto (z_1, \ldots, z_n, z_1^{-1}, \ldots, z_n^{-1}).$$

2.6 Definition. The set $T := (\mathbb{C} \setminus \{0\})^n =: (\mathbb{C}^*)^n$ is called a (*complex algebraic n-*)*torus*.

We note that T includes the *real* n-torus $(\mathbb{S}^1)^n$; in fact, T can be identified with $(\mathbb{S}^1)^n \times (\mathbb{R}_{>0})^n$ (see section 3). The name "algebraic torus" certainly reflects that relationship, a deeper relationship comes from the theory of algebraic groups. We remark that the notion of an algebraic torus should not be confused with that of a compact complex torus, which will, however, not be used in our text.

Remark. The realization of the torus T in $\mathbb{C}^{2n}$ provides a closed subset of $\mathbb{C}^{2n}$, whereas, as a subspace of $\mathbb{C}^n$, the torus T is not closed.

We mention that the choice of monomial generators $e_1, \ldots, e_n, -(e_1 + \ldots + e_n)$ corresponds to another realization of the n-torus, this time as an affine algebraic subvariety in $\mathbb{C}^{n+1}$. In the coordinates $z_1, \ldots, z_n, z_{n+1}$, the variety is defined by the single equation $z_1 \cdots z_n \cdot z_{n+1} = 1$ (see Exercise 4).

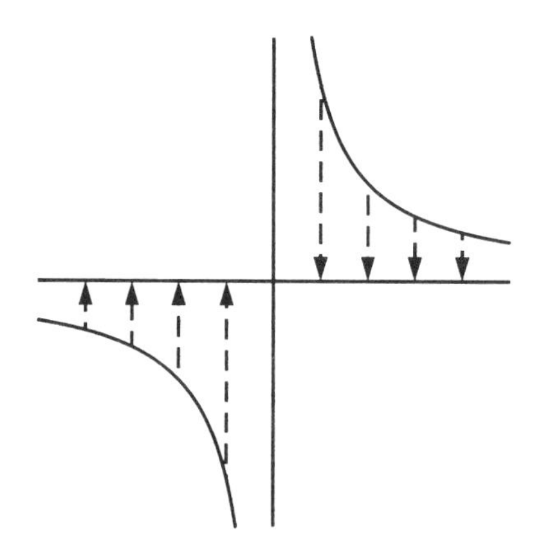

FIGURE 1.

As in the special case of the torus T, we are going to look for suitable coordinates in some affine space $\mathbb{C}^k$ in which an abstract affine toric variety X_σ can be realized as an affine algebraic variety given by a finite set of algebraic equations.

Let $A = (a_1, \ldots, a_k)$ be a system of generators of the monoid $\sigma \cap \mathbb{Z}^n$ (which exist by Gordan's lemma). If we set

$$u_i := z^{a_i} \in \mathbb{C}[z, z^{-1}], \quad \text{for } i = 1, \ldots, k,$$

then, $R_\sigma = \mathbb{C}[u_1, \ldots, u_k]$. With the indeterminates $\xi_1, \ldots, \xi_k$, there is an algebra homomorphism

$$\varphi_A : \mathbb{C}[\xi_1, \ldots, \xi_k] \to \mathbb{C}[u_1, \ldots, u_k], \quad \xi_i \mapsto u_i \text{ for } i = 1, \ldots, k.$$

Our aim is to determine the ideal $\mathfrak{a} := \mathfrak{a}_A := \ker \varphi_A$ (which depends on the choice of A), so that

$$\mathbb{C}[\xi_1, \ldots, \xi_k]/\mathfrak{a}_A \cong \mathbb{C}[u_1, \ldots, u_k] = R_\sigma.$$

The definition of $\mathfrak{a}_A$ can be rephrased as follows. For every $f \in \mathbb{C}[\xi]$, $f \in \mathfrak{a}_A$ if and only if $\varphi_A(f) = 0$, which, in turn, is equivalent to $f(u) = 0$ in $\mathbb{C}[u]$.

We consider integral positive linear relations (ν, μ) (see V 39) for the system A, i.e., equations of the form

$$\nu \cdot A = \mu \cdot A, \quad \text{where } \mu, \nu \in \mathbb{Z}^k_{\geq 0}, \tag{2}$$

and $\nu \cdot A := \sum_{j=1}^k \nu_j a_j$. Such a relationship provides a monomial equation

$$z^{\nu \cdot A} = (z^{a_1})^{\nu_1} \cdots (z^{a_k})^{\nu_k} = (z^{a_1})^{\mu_1} \cdots (z^{a_k})^{\mu_k} = z^{\mu \cdot A},$$

or, in terms of the generators $u_j = z^{a_j}$ of R_σ, a binomial relationship

$$u_1^{\nu_1} \cdots u_k^{\nu_k} - u_1^{\mu_1} \cdots u_k^{\mu_k} = 0. \tag{3}$$

Thus, it is clear that for such a relationship (ν, μ) the corresponding binomial $\xi^\nu - \xi^\mu$ is an element of the ideal $\mathfrak{a}_A$. In fact, we want to show that $\mathfrak{a}_A$ is generated by those binomials:

2.7 Theorem. *For every lattice cone σ, the corresponding affine toric variety X_σ is realized by the affine algebraic variety $V(\mathfrak{a}_A)$ in $\mathbb{C}^k$, where $A = (a_1, \ldots, a_k)$ is a system of generators of the monoid $\sigma \cap \mathbb{Z}^n$ and the ideal $\mathfrak{a}_A$ of $\mathbb{C}[\xi_1, \ldots, \xi_k]$ is generated by finitely many binomials of the form $\xi^\nu - \xi^\mu$.*

PROOF. By Lemma V 3.10, the monoid of all integral, positive, linear relationships (2) is finitely generated. Hence, it suffices to show that every element of $\mathfrak{a}_A$ is a sum of binomials as above. For a polynomial $f = \sum \lambda_\nu \xi^\nu$, $f(u) = \sum \lambda_\nu u^\nu = \sum \lambda_\nu z^{\nu \cdot A}$, so for every $a \in \sigma \cap \mathbb{Z}^n$, the coefficient c_a of z^a is $\sum_{\nu; \nu \cdot A = a} \lambda_\nu$. $f \in \mathfrak{a}_A$ if and only if all these c_a's vanish. Hence, in that case, if $\lambda_\nu \neq 0$ for some multi–index $\nu \in \mathbb{Z}^k_{\geq 0}$, there is another one, say $\mu \neq \nu$, which satisfies $\nu \cdot A = \mu \cdot A$ and $\lambda_\mu \neq 0$. The corresponding binomial $\lambda_\nu(\xi^\nu - \xi^\mu)$ is in $\mathfrak{a}_A$; subtracting it from f yields a polynomial in $\mathfrak{a}_A$ with strictly fewer terms than f. The proof is thus obtained by an obvious induction. □

We have seen in Corollary 1.17 that the points of $V := V(\mathfrak{a}_A)$ correspond bijectively to the maximal ideals $\mathfrak{m}$ of the polynomial ring $\mathbb{C}[\xi]$, which contain $\mathfrak{a}_A$, and also to the ideals $\bar{\mathfrak{m}} := \mathfrak{m}/\mathfrak{a}_A$ of the ring R_σ, which is just the coordinate ring R_V. Hence, the generators $u_1, \ldots, u_k$ of R_σ are the *coordinate functions* on X_σ that realize X_σ in $\mathbb{C}^k$. A point with coordinate vector $x = (x_1, \ldots, x_k) \in \mathbb{C}^k$ represents a point of X_σ if and only if the relationships $x^\nu = x^\mu$ hold for all (ν, μ) that satisfy (2).

In 2.7, the description of the affine variety $V(\mathfrak{a}_A)$ in $\mathbb{C}^k$ representing an abstract affine toric variety X_σ, allows some immediate conclusions which will be used in the sequel:

2.8 Theorem. *Fix a system $A = (a_1, \ldots, a_k)$ of generators for the monoid $\sigma \cap \mathbb{Z}^n$ and set $V := V(\mathfrak{a}_A)$.*

(a) *The mapping $\gamma : T \to V$ given by $t := (t_1, \ldots, t_n) \mapsto (t^{a_1}, \ldots, t^{a_k})$ sends T bijectively onto the open subset $V \cap \mathbb{C}^{*k}$ of V.*

(b) *For arbitrarily chosen $x \in V$ and $t \in T$, the point $(t^{a_1}x_1, \ldots, t^{a_k}x_k)$ also belongs to V.*

PROOF.

(a) It is clear that each $\gamma(t)$ satisfies the defining binomial relationship from above and that all points in $\gamma(T)$ have nonzero coordinates. Choose a lattice point a in σ such that all translated points $a + e_i$ also lie in σ, where $e_1, \ldots, e_n$ denotes the canonical basis of $\mathbb{R}^n$. The Laurent monomials $z^a =: f_0(u)$ and $z^{a+e_i} =: f_i(u)$ all belong to the coordinate ring $R_\sigma = \mathbb{C}[u] \subset \mathbb{C}[z, z^{-1}]$. For a point $\gamma(t)$, $f_i(\gamma(t)) = t_i t^a = t_i f_0(\gamma(t))$, hence we recover $t_i = f_i(\gamma(t))/f_0(\gamma(t))$; in particular, γ is injective. On the other hand, each point x of $V \cap \mathbb{C}^{*k}$ lies in $\operatorname{im}(\gamma)$: As the monomials f_i take nonzero values on x,

$$x = \gamma(f_1(x)/f_0(x), \ldots, f_n(x)/f_0(x)).$$

(b) is readily verified.

□

Thus, we have seen that each n-dimensional, affine toric variety includes the n-torus as an open subset. Moreover, (b) says that the torus T, looked at as an abelian group, operates in a natural way on such a variety (see section 5 for more details on the torus action).

Example 2. For $\sigma := \operatorname{pos}(\{e_1, e_2\})$, the monoid $\sigma \cap \mathbb{Z}^2$ has linearly independent generators e_1, e_2. Therefore, $\mathfrak{a} = \mathfrak{o}$, the zero ideal, and

$$X_\sigma = \mathbb{C}^2.$$

The same is true for each cone

$$\sigma = \operatorname{pos}(\{e_1 + \nu e_2, e_2\}), \quad \text{for } \nu \in \mathbb{Z}.$$

More generally, if $\sigma = \operatorname{pos}(\{a_1, \ldots, a_n\})$ is a regular lattice cone in $\mathbb{R}^n$, then, again, $\mathfrak{a} = \mathfrak{o}$; hence, X_σ can by identified with the affine space $\mathbb{C}^n$.

Example 3. For $\sigma := \operatorname{pos}(\{e_1, e_1 + 2e_2\})$, the monoid $\sigma \cap \mathbb{Z}^2$ is generated by $a_1 = e_1, a_2 = e_1 + 2e_2, a_3 = e_1 + e_2$. There is a linear relationship

$$a_1 + a_2 = 2a_3,$$

and, hence, a monomial equation

$$u_1 u_2 = u_3^2.$$

X_σ is a quadratic cone with "singularity" in (0, 0, 0) (Figure 2) (more on singularities in section 8).

We are going to discuss in more detail the transition from one such system of coordinate functions on X_σ to another one. There are two cases. Here, we can profit from the decomposition $\sigma = \sigma_0 + \operatorname{cospan} \sigma$ of Lemma V 1.6. By Theorem 2.12 below, we essentially have to consider the following cases:

(1) The cone σ has an apex.

Then, by V, Lemma 3.5, the monoid $\sigma \cap \mathbb{Z}^n$ has a minimal system $A = (a_1, \ldots, a_k)$ of generators, which is unique up to renumbering. Therefore, we have distinguished minimal systems of coordinates $u_1 := z^{a_1}, \ldots, u_k := z^{a_k}$. We may, however, introduce additional generators and, thus, additional coordinates. As an example, consider the an affine plane $\mathbb{C}^2$ as affine toric variety X_σ for $\sigma := \operatorname{pos}(\{e_1, e_2\}) \subset \mathbb{R}^2$ (see Example 2 above). Choosing the additional generator $a_3 := e_1 + e_2$ for the monoid $\sigma \cap \mathbb{Z}^2$, we obtain the representation $\mathbb{C}[\xi_1, \xi_2, \xi_3]/(\xi_1\xi_2 - \xi_3) = \mathbb{C}[u_1, u_2, u_3]$ for the abstract coordinate ring R_σ. The corresponding model for X_σ is the quadric surface in $\mathbb{C}^3$ given by the equation

$$u_3 = u_1 \cdot u_2,$$

which is identified with $\mathbb{C}^2$ through the parametrization

$$(u_1, u_2) \mapsto (u_1, u_2, u_1 \cdot u_2),$$

with the inverse given by the projection

$$(u_1, u_2, u_3) \mapsto (u_1, u_2).$$

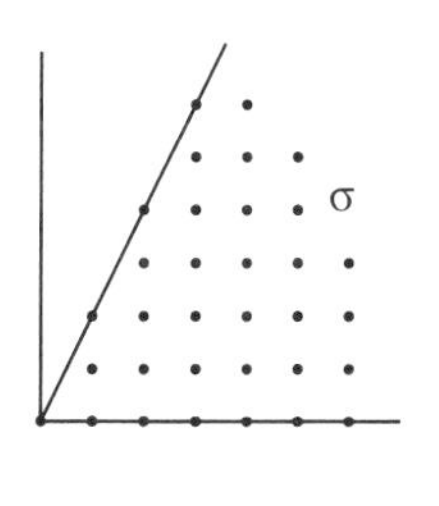

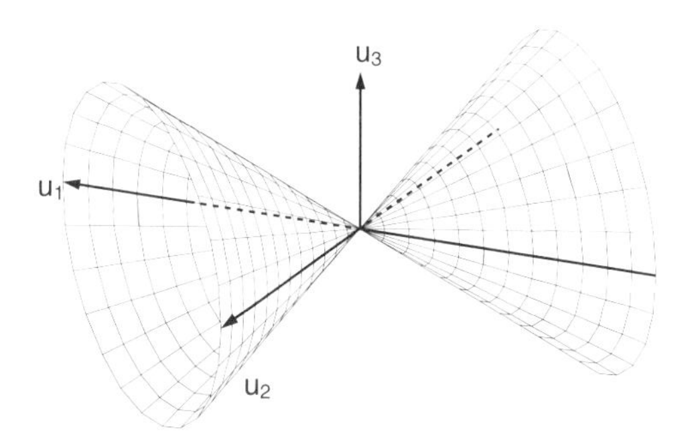

FIGURE 2.

This can be easily generalized to the case of an arbitrary cone σ with an apex. The additional generators a_j for $j = k+1, \ldots, k'$ are positive integral combinations $a_j = \sum_{i=1}^{k} \nu_{ji} a_i$, so we have the corresponding relationships $u_j = u_1^{\nu_{j1}} \cdots u_k^{\nu_{jk}}$. The model $V \subset \mathbb{C}^k$ for X_σ that corresponds to the minimal system of generators, with coordinates $(u_1, \ldots, u_k)$, is mapped to the model $V' \subset \mathbb{C}^{k'}$ via

$$(u_1, \ldots, u_k) \mapsto (u_1, \ldots, u_k, u_{k+1}, \ldots, u_{k'}) ,$$

where the extra coordinates u_j are given by the relations from above. The inverse map is again given by the projection onto the first k coordinates $(u_1, \ldots, u_k)$.
(2) cospan $\sigma \neq \{0\}$.

Here, in any case (except $\sigma = \mathbb{R}$), there exist different minimal systems of generators of $\sigma \cap \mathbb{Z}^n$. We consider the following:

Example 4. For the half-plane $\sigma = \operatorname{pos}(\{e_1, -e_1, e_2\})$ and fixed $\nu \in \mathbb{Z}_{\geq 0}$ the monoid $\sigma \cap \mathbb{Z}^2$ has $e_1, -e_1, b = \nu e_1 + e_2$ as generators, (see Figure 3 for $\nu = 2$). In particular, $u_1 := z^{e_1} = z_1, u_2 := z^{-e_1} = z_1^{-1}, u_3 := z^{e_2} = z_2$ and $v_1 := u_1$, $v_2 := u_2$, $v_3 := z^{\nu e_1 + e_2} = z_1^{\nu} z_2$ are two coordinate systems, and

$$\begin{aligned} v_1 &= u_1 & u_1 &= v_1 \\ v_2 &= u_2 & u_2 &= v_2 \\ v_3 &= u_1^{\nu} u_3, \text{ and} & u_3 &= v_2^{\nu} v_3 \end{aligned}$$

are transformation formulae for the coordinates.

We now intend to introduce a notion of isomorphy under which X_σ and $\mathbb{C}^* \times \mathbb{C}$ are isomorphic:

2.9 Definition. Let $\Phi : \mathbb{C}^k \to \mathbb{C}^m$ be a monomial mapping (i.e. every nonzero component of Φ is a monomial in the coordinates of $\mathbb{C}^k$), and let $X_\sigma \hookrightarrow \mathbb{C}^k$ and $X_{\sigma'} \hookrightarrow \mathbb{C}^m$ be (realizations of) affine toric varieties. If $\Phi(X_\sigma) \subset X_{\sigma'}$, then $\varphi := \Phi|_{X_\sigma}$ is a morphism called an *(affine) toric morphism* from X_σ to $X_{\sigma'}$. If φ is bijective and the inverse mapping $\varphi^{-1} : X_{\sigma'} \to X_\sigma$ is again a toric morphism, then, we call φ a *(toric) isomorphism*, we say that X_σ and $X_{\sigma'}$ are isomorphic, and write $X_\sigma \underset{\text{toric}}{\cong} X_{\sigma'}$ or, briefly, $X_\sigma \cong X_{\sigma'}$.

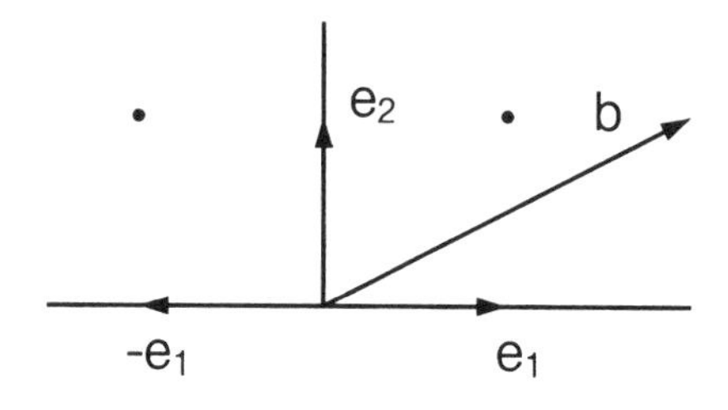

FIGURE 3.

Moreover, if $\sigma = \sigma'$, then, we call φ a *coordinate transformation.*

Because of Lemma 2.10 below, we do not have to really distinguish between the abstract affine toric varieties and their geometric realization. We also want to justify this for toric morphisms. Such morphisms $X_\sigma \to X_{\sigma'}$ are closely related to monomial homomorphisms $\vartheta : R_{\sigma'} \to R_\sigma$ of the coordinate algebras, where the algebra homomorphism ϑ is called *monomial* if it is given by monomials in the monomial generators:

Remark. Every toric morphism $\varphi : X_\sigma \to X_{\sigma'}$, uniquely determines a monomial homomorphism $\varphi^* : R_{\sigma'} \to R_\sigma$ and vice versa.

PROOF. Let φ be given. Then, by definition, there is a commutative diagram

$$\begin{array}{ccccc} X_\sigma & \cong & V(\mathfrak{a}) & \hookrightarrow & \mathbb{C}^k \\ \downarrow{\scriptstyle\varphi} & & \downarrow & & \downarrow{\scriptstyle\Phi} \\ X_{\sigma'} & \cong & V(\mathfrak{a}') & \hookrightarrow & \mathbb{C}^m \end{array}$$

where Φ is monomial, i.e., of the form $\Phi(x) = (\lambda_1 x^{c_1}, \ldots, \lambda_m x^{c_m})$. The corresponding algebra homomorphism

$$\Phi^* : \mathbb{C}[y_1, \ldots, y_m] \longrightarrow \mathbb{C}[x_1, \ldots, x_k], \quad g \mapsto g \circ \Phi$$

maps y_j to $\lambda_j x^{c_j}$ and, thus, is monomial. It is readily seen that

$$\Phi(V(\mathfrak{a})) \subset V(\mathfrak{a}') \quad \text{if and only if} \quad \Phi^*(\mathfrak{a}') \subset \mathfrak{a}. \tag{4}$$

Hence, Φ^* induces a monomial homomorphism

$$\begin{array}{ccccc} \varphi^* : R_{\sigma'} = & \mathbb{C}[y]/\mathfrak{a}' & \longrightarrow & \mathbb{C}[x]/\mathfrak{a} = & R_\sigma \\ & \| & & \| & \\ & \mathbb{C}[v] & \longrightarrow & \mathbb{C}[u] & \end{array}$$

Conversely, a monomial homomorphism $\vartheta : R_{\sigma'} \to R_\sigma$ can be lifted to a monomial homomorphism $\Theta : \mathbb{C}[y] \to \mathbb{C}[x]$ with $\Theta(\mathfrak{a}') \subset \mathfrak{a}$. That corresponds to a monomial morphism

$$\Phi : \mathbb{C}^k \longrightarrow \mathbb{C}^m, \quad x \mapsto (\Theta(y_1), \ldots, \Theta(y_m))$$

with $\Phi^* = \Theta$. We have $\Theta(\mathfrak{a}') \subset \mathfrak{a}$, hence, $\Phi(V(\mathfrak{a})) \subset V(\mathfrak{a}')$ by (4), so Φ induces a toric morphism $\varphi : X_\sigma \to X_{\sigma'}$. □

We note that toric morphisms are continuous with respect to the induced metric topology, since they are given by polynomials. In particular, if two affine toric varieties X_σ and $X_{\sigma'}$ are isomorphic, then, they are homeomorphic. Hence, on an abstract toric variety, there is a well-defined *"complex" topology*, which is finer than the Zariski topology.

Remark. For the "gluing maps" in section 3, we need, in particular, the toric morphisms that come from an inclusion $\sigma' \subset \sigma$ of n-dimensional cones. For systems of monoid generators A of $\sigma \cap \mathbb{Z}^n$ and A' of $\sigma' \cap \mathbb{Z}^n$, there are positive integral linear combinations $a_i' = \nu^{(i)} \cdot A$, which yield monomial identities $z^{a_i'} = z^{\nu^{(i)} \cdot A}$, i.e., $v_i = u^{\nu^{(i)}}$. It is clear that the corresponding toric morphism induces an isomorphism between the copies of T which are included in the affine toric varieties X_σ and $X_{\sigma'}$.

Up to a toric isomorphism, the geometric realization of an abstract toric variety does not depend on the choice of coordinates:

2.10 Lemma. *If $R_\sigma \cong \mathbb{C}[u_1, \dots, u_k] = \mathbb{C}[\xi]/\mathfrak{a}$ and $R_\sigma \cong \mathbb{C}[v_1, \dots, v_m] = \mathbb{C}[\eta]/\mathfrak{a}'$ are two representations of R_σ by coordinate rings, then, there exists a coordinate transformation between $(u_1, \dots, u_k)$ and $(v_1, \dots, v_m)$.*

PROOF. Let $A = (a_1, \dots, a_k)$ and $B = (b_1, \dots, b_m)$ be systems of generators of the monoid $\sigma \cap \mathbb{Z}^n$, and let $u_i := z^{a_i}$ and $v_j := z^{b_j}$ be the corresponding coordinates. We can represent each b_j as a positive integral linear combination of the a_i's, and conversely. This provides the coordinate transformation. □

With an appropriate notion of isomorphisms, on each of the levels

lattice cones — coordinate algebras — affine toric varieties

one object determines the other two. So, for lattice cones, we set $\sigma \subset \mathbb{R}^n = \operatorname{lin} \sigma$ and $\sigma' \subset \mathbb{R}^m = \operatorname{lin} \sigma'$, we call σ and σ' *isomorphic* and write $\sigma \cong \sigma'$, if $m = n$ and there is a unimodular transformation $L : \mathbb{R}^n \to \mathbb{R}^n$ with $L(\sigma) = \sigma'$. Then, the monoids $\sigma \cap \mathbb{Z}^n$ and $\sigma' \cap \mathbb{Z}^n$ are isomorphic as well.

We call the coordinate algebras R_σ and $R_{\sigma'}$ *monomially isomorphic*, and we write $R_\sigma \underset{\text{mon}}{\cong} R_{\sigma'}$ or, briefly, $R_\sigma \cong R_{\sigma'}$, if there exist mutually inverse monomial homomorphisms $R_\sigma \leftrightarrow R_{\sigma'}$.

Then, we obtain the following theorem:

2.11 Theorem. *For lattice cones $\sigma \subset \mathbb{R}^n = \operatorname{lin} \sigma$ and $\sigma' \subset \mathbb{R}^m = \operatorname{lin} \sigma'$, the following conditions are equivalent:*

$$(a)\ \sigma \cong \sigma' \qquad (b)\ R_\sigma \underset{\text{mon}}{\cong} R_{\sigma'} \qquad (c)\ X_\sigma \underset{\text{toric}}{\cong} X_{\sigma'}.$$

PROOF. The implications "a) ⇒ b) ⇒ c)" are clear, so we are left with "c) ⇒ a)". We have seen in the Remark above that every toric morphism $\varphi : X_\sigma \to X_{\sigma'}$ induces a (monomial) homomorphism of monomial $\mathbb{C}$-algebras $\varphi^* : R_{\sigma'} \to R_\sigma$. First, we show that φ^* extends uniquely to a monomial homomorphism

$$\Theta : \mathbb{C}[w_1, w_1^{-1}, \dots, w_m, w_m^{-1}] \longrightarrow \mathbb{C}[z_1, z_1^{-1}, \dots, z_n, z_n^{-1}]$$

of the Laurent polynomial algebras with $R_{\sigma'} \subset \mathbb{C}[w, w^{-1}]$ and $R_\sigma \subset \mathbb{C}[z, z^{-1}]$: Choose a lattice point $b \in \sigma'$ such that all lattice points $b + e_j$ for $j = 1, \dots, m$ also belong to σ' (this is possible since σ' is of full dimension). With $g_0 := w^b$ and $g_j := w^{b+e_j} = w_j g_0$ we set $\Theta(w_j) := \varphi^*(g_j)/\varphi^*(g_0)$, which is a well-defined Laurent monomial of $\mathbb{C}[z, z^{-1}]$ since $\varphi^*(g_0)$ and $\varphi^*(g_j)$ are Laurent monomials.

Thus, we have $\Theta(w_j) = \lambda_j z^{c_j}$ with $\lambda_j \in \mathbb{C}^*$ and $c_j = (c_{j1}, \ldots, c_{jn}) \in \mathbb{Z}^n$ for $j = 1, \ldots, m$.

As $\varphi : X_\sigma \to X_{\sigma'}$ is a toric isomorphism by assumption, we immediately find that Θ is a Laurent monomial $\mathbb{C}$-algebra isomorphism, so its inverse maps z_i to $\mu_i w^{d_i}$ with $\mu_i \in \mathbb{C}^*$ and $d_i = (d_{i1}, \ldots, d_{im}) \in \mathbb{Z}^m$ for $i = 1, \ldots, n$. Thus, we have

$$\begin{aligned} w_j = \Theta^{-1}(\lambda_j z^{c_j}) &= \lambda_j \cdot \prod_{i=1}^n \Theta^{-1}(z_i^{c_{ji}}) \\ &= \lambda_j \cdot \prod_{i=1}^n \mu_i^{c_{ji}} w^{c_{ji} d_i} \qquad \text{for } j = 1, \ldots, m, \end{aligned}$$

so we obtain (using an analogous calculation for $z_i = \Theta(\mu_i w^{d_i})$) the equations

$$\sum c_{ji} d_{ik} = \delta_{jk}, \quad \lambda_j \prod_{i=1}^n \mu_i^{c_{ji}} = 1, \quad \sum d_{ij} c_{jk} = \delta_{ik},$$

and

$$\mu_i \prod_{j=1}^m \lambda_j^{d_{ij}} = 1.$$

Consequently, the integral matrices

$$(c_{ji})_{j=1,\ldots,m;i=1,\ldots,n} \quad \text{and} \quad (d_{ij})_{i=1,\ldots,n;j=1,\ldots,m}$$

are inverse to each other; in particular, $m = n$.

By the construction of Θ, $\Theta(R_{\sigma'}) \subset R_\sigma$. Let $A = (a_1, \ldots, a_k)$ and $B = (b_1, \ldots, b_l)$ denote systems of generators of the monoids $\sigma \cap \mathbb{Z}^n$ and $\sigma' \cap \mathbb{Z}^n$, respectively. Then, Θ maps each generator $v_s = w^{b_s}$ of $R_{\sigma'}$ to a monomial $\kappa_s u^{p_s} = \kappa_s z^{A \cdot p_s}$ in the generators $u_i = z^{a_i}$ of R_σ, where $\kappa_s \in \mathbb{C}^*$ and $p_s \in \mathbb{Z}^k_{>0}$ for $s = 1, \ldots, l$. On the other hand, $\Theta(w^{b_s}) = \prod_{j=1}^n (\lambda_j z^{c_j})^{b_{sj}}$, so $\sum_{j=1}^n c_j b_{sj} = \sum_{i=1}^k a_i p_{si}$ is a positive integral linear combination of the a_i's. Together with the analogous statement for Θ^{-1}, this implies that σ and σ' are isomorphic lattice cones. □

Let us discuss the case of a lattice cone σ with nontrivial $U := \operatorname{cospan} \sigma$. Via the decomposition $\sigma = \sigma_0 + U$ of Lemma V 1.6, it can be reduced to the case of cones with apex 0:

2.12 Theorem. *For an n-dimensional lattice cone σ in $\mathbb{R}^n$, set $d := n - \dim \sigma_0$. Then, the affine toric variety X_σ has the structure of a Cartesian product*

$$X_\sigma \cong X_{\sigma_0} \times \mathbb{C}^{*d}.$$

In particular, if the cone σ is regular, then,

$$X_\sigma \cong \mathbb{C}^{n-d} \times \mathbb{C}^{*d}.$$

PROOF. We choose sets of generators $a_1, \dots, a_{k'}$ for $\sigma_0 \cap \mathbb{Z}^n$ and $b_1, \dots, b_{k''}$ for $U \cap \mathbb{Z}^n$, respectively. Their union generates the monoid $\sigma \cap \mathbb{Z}^n$. Every linear relation of the form $\sum \nu_i a_i = \sum \mu_j b_j$ is trivial, so any linear relation involving both sets of generators splits uniquely as the sum of a relationship involving only $a_1, \dots, a_{k'}$ and another one involving the only $b_1, \dots, b_{k''}$. Let $u_1, \dots, u_{k'}$ and $v_1, \dots, v_{k''}$ be the corresponding coordinate functions for representations V' of X_{σ_0} in $\mathbb{C}^{k'}$ and V'' of $X_U \cong \mathbb{C}^{*d}$ in $\mathbb{C}^{k''}$, respectively. Then, $u_1, \dots, u_{k'}, v_1, \dots, v_{k''}$ are the coordinates for the affine subvariety V of $\mathbb{C}^{k'+k''}$ representing X_σ. There is no algebraic relationship of the form $f(u) = g(v)$ with nonconstant polynomials f and g, so a point $(x, y) \in \mathbb{C}^{k'} \times \mathbb{C}^{k''}$ lies in V if and only if x lies in V' and y in $V'' \cong \mathbb{C}^{*d}$, i.e., $V = V' \times V''$, which proves the claim. □

Exercises

1. For $k \in \mathbb{Z}_{>0}$, find coordinates and equations for X_σ, where $\sigma := \text{pos}(\{e_1, e_1 + ke_2\})$.
2. Find coordinates and an equation for X_σ, where $\sigma := \text{pos}(\{e_1, e_2, e_1 + e_2 + 2e_3\})$.
3. Determine X_σ for $\sigma = \mathbb{R}^n$.
4. The n-torus T has the two representations by the points assigned to

$$\text{spec}\, \mathbb{C}[u_1, \dots, u_n, u_1^{-1}, \dots, u_n^{-1}] \cong \text{spec}\, \mathbb{C}[u_1, \dots, u_n, (u_1 \cdots u_n)^{-1}].$$

   Find the change of coordinate functions.

3. Toric varieties

We are now going to construct general toric varieties by "gluing" affine ones. The gluing information is encoded in a fan Σ in $\mathbb{R}^n$, but not in the most straightforward way: For the present purpose, it turns out to be essential to pass to the duals $\check{\sigma}$ (see I, 49 and V, 2) of the cones $\sigma \in \Sigma$ and to consider the associated affine toric varieties $X_{\check{\sigma}}$. The condition that the cones of Σ have 0 as apex guarantees that their duals are all of the same dimension n. If τ is a face of a cone σ, then, we have an inclusion $\check{\sigma} \subset \check{\tau}$. By the Remark preceding Lemma 2.10, the induced toric morphism $\psi_{\tau,\sigma} : X_{\check{\tau}} \to X_{\check{\sigma}}$ is an isomorphism given by a coordinate transformation on the copies of the complex algebraic n-torus T included in both varieties as an open subset. We want to show that $\psi_{\tau,\sigma}$ is an open inclusion. That will allow glueing any two such affine toric varieties $X_{\check{\sigma}}, X_{\check{\sigma}'}$ along the open subsets corresponding to the common face $\tau := \sigma \cap \sigma'$.

To prove that claim, we choose a simple lattice vector $m \in \text{relint}(\tau^\perp \cap \check{\sigma})$. Its existence follows from the proof of V, Lemma 2.3, and it satisfies

$$\check{\tau} = \check{\sigma} + \mathbb{R}_{\geq 0}(-m)$$

and, hence,

$$\check{\tau} \cap \mathbb{Z}^n = (\check{\sigma} \cap \mathbb{Z}^n) + \mathbb{Z}_{\geq 0}(-m).$$

The monoid $\check{\tau} \cap \mathbb{Z}^n$ is, thus, obtained from $\check{\sigma} \cap \mathbb{Z}^n$ by introducing the additional generator $-m$ that lies in cospan $\check{\tau}$. We may assume that $a_k := m$ occurs as the last member in a system of generators $a_1, \ldots, a_k$ of $\check{\sigma} \cap \mathbb{Z}^n$. The monoid $\check{\tau} \cap \mathbb{Z}^n$ is, thus, generated by $a_1, \ldots, a_k, a_{k+1} := -m$. It is now readily seen that all nontrivial relationships between these generators are sums of relationships involving only $a_1, \ldots, a_k$, and the obvious relationship $a_k + a_{k+1} = 0$. In fact, given a relationship $\sum_{i=1}^{k+1} \nu_i a_i = \sum_{j=1}^{k+1} \mu_j a_j$ with $\nu_i, \mu_j \in \mathbb{Z}_{\geq 0}$ and $\nu_{k+1} \geq \mu_{k+1}$, say, we may subtract $\mu_{k+1} a_{k+1}$ on both sides and, thus, reduce it to $\mu_{k+1} = 0$. Adding $\nu_{k+1} a_k$ on both sides and using $a_k + a_{k+1} = 0$, then, leaves us with a relationship involving only $a_1, \ldots, a_k$.

On the side of the monomial algebras $R_{\check{\sigma}}$ and $R_{\check{\tau}}$ with the corresponding generators $u_i := z^{a_i}$, the additive relationship $a_k + a_{k+1} = 0$ corresponds to the multiplicative relationship $u_k u_{k+1} = 1$ in $R_{\check{\tau}}$. As that is the only "new" defining relationship when passing from $R_{\check{\sigma}}$ to $R_{\check{\tau}}$, and as the generators u_i are just the coordinate functions on the toric varieties $X_{\check{\sigma}}$ and $X_{\check{\tau}}$, we, thus, see that the projection $(x_1, \ldots, x_k, x_{k+1}) \mapsto (x_1, \ldots, x_k)$ identifies $X_{\check{\tau}}$ with the open subset of $X_{\check{\sigma}}$ given by $(x_k \neq 0)$ (called a "principal" Zariski open subset). We have, thus, proved the following result:

3.1 Lemma. *With the above notation, we have a natural identification*

$$X_{\check{\tau}} \cong X_{\check{\sigma}} \setminus \{u_k = 0\}.$$

Remark. In general, the "hyperplane section" $X_{\check{\sigma}} \cap \{u_k = 0\}$ is not a linear subspace, as is seen in the following example. For $\sigma := \mathrm{pos}\{e_1, e_2, e_3\} \subset \mathbb{R}^3$ and the face $\tau := \mathrm{pos}\{e_1\}$, we may choose $m := a_4 := e_2 + e_3$. Then, $X_{\check{\sigma}}$ is the affine 3-space $\mathbb{C}^3$, embedded in $\mathbb{C}^4$ by the coordinates $u_i := z_i$ for $i \leq 3$ and $u_4 = z_2 z_3$. Therefore, the "hyperplane section" $X_{\check{\sigma}} \cap (u_4 = 0)$ in $\mathbb{C}^4$ consists of the two coordinate planes $(u_2 = u_4 = 0)$ and $(u_3 = u_4 = 0)$ that meet in a line, namely, the u_1-axis.

For two cones $\sigma, \sigma' \in \Sigma$, let $\tau := \sigma \cap \sigma'$ be the common face. Choosing an appropriate coordinate system $v_1, \ldots, v_l$ for $X_{\check{\sigma}'}$, according to the lemma above, we have isomorphisms

$$X_{\check{\sigma}} \setminus \{u_k = 0\} \cong X_{\check{\tau}} \cong X_{\check{\sigma}'} \setminus \{v_l = 0\}.$$

Their composition yields a toric isomorphism (see 2.9)

$$\psi_{\sigma,\sigma'} : X_{\check{\sigma}} \setminus \{u_k = 0\} \xrightarrow{\cong} X_{\check{\sigma}'} \setminus \{v_l = 0\}$$

corresponding to the coordinate transformation

$$(u_1, \ldots, u_k, u_{k+1}) \mapsto (v_1, \ldots, v_l, v_{l+1})$$

of $X_{\check{\tau}}$.

3.2 Definition. The above isomorphism $\psi_{\sigma,\sigma'}$ is called the *gluing map* that *glues together* $X_{\check{\sigma}}$ and $X_{\check{\sigma}'}$ along $X_{\check{\tau}}$.

Using these gluing data, we can now construct a general toric variety as follows:

3.3 Definition. Let Σ be a fan in $\mathbb{R}^n$. In the disjoint union $\bigcup_{\sigma\in\Sigma} X_{\check{\sigma}}$, we identify two points $x \in X_{\check{\sigma}}$ and $x' \in X_{\check{\sigma}'}$ that are mapped to each other by the gluing map $\psi_{\sigma,\sigma'}$. The set of points, thus, obtained is called the *toric variety* (or *torus embedding*) X_Σ determined by Σ.

We still have to check some properties in order to justify the notation *toric variety* for X_Σ. For our purpose, it suffices to see that X_Σ is a topological space endowed with an open covering by affine complex varieties that intersect (Zariski) open subvarieties. Such a covering is given by the (copies of the) affine toric varieties $X_{\check{\sigma}}$, for $\sigma \in \Sigma$, that are naturally included in X_Σ, thus, defining the topology. So, these data are provided by construction. In fact, we consider two different topologies on X_Σ, corresponding to the two topologies on affine toric varieties, namely, the "ordinary" or "complex" topology and the Zariski topology (Definition 1.24). Next, we note the following result, essentially already stated in 2.8:

3.4 Lemma. *Each affine toric variety $X_{\check{\sigma}}$ contains the torus T as a Zariski open, dense subset.*

PROOF. The zero cone $o := \{0\}$ is a face of every cone $\sigma \in \Sigma$, and its dual $\check{o} = \mathbb{R}^n$ yields $X_{\check{o}} = T$. □

Moreover, all these embedded tori $\psi_{o,\sigma}(T)$ are identified under the gluing maps (see also the Remark preceding 2.10), thus, proving the following lemma:

3.5 Lemma. *There is a natural, open, dense embedding of the torus T into X_Σ.*

Finally, we note the following result:

3.6 Lemma. *With respect to the "ordinary" topology defined by the ordinary topologies on the affine toric varieties $X_{\check{\sigma}}$, the space X_Σ is Hausdorff.*

PROOF. If two different points x, x' lie in the same affine open subvariety $X_{\check{\sigma}}$, they have disjoint, open neighborhoods. Assume that we have $x \in X_{\check{\sigma}} \setminus X_{\check{\sigma}'}$ and $x' \in X_{\check{\sigma}'} \setminus X_{\check{\sigma}}$ for different cones $\sigma, \sigma' \in \Sigma$. As these cones intersect in a common face τ, we find a lattice point $m \in \tau^\perp$ satisfying $m \in \operatorname{relint}(\tau^\perp \cap \check{\sigma})$ and $-m \in \operatorname{relint}(\tau^\perp \cap \check{\sigma}')$. Then, $u_k := z^m$ and $v_l := z^{-m}$ are coordinate functions on $X_{\check{\sigma}}$ and $X_{\check{\sigma}'}$, respectively. It follows from 31 that they satisfy $u_k(x) = 0$ and $v_l(x') = 0$ and that we have $u_k v_l = 1$ on the intersection $X_{\check{\tau}} = X_{\check{\sigma}} \cap X_{\check{\sigma}'}$. It is, thus, clear that the sets $\{|u_k| < 1\} \subset X_{\check{\sigma}}$ and $\{|v_l| < 1\} \subset X_{\check{\sigma}'}$ have the desired property. □

We remark that only a weaker "separation property" holds for the Zariski topology.

We can also express the gluing construction on the abstract level (see 2.5), without the explicit use of coordinates. The natural inclusion of a face in a cone of Σ induces natural inclusion maps (denoted "$\hookrightarrow$" and "$\hookleftarrow$", respectively):

$$\begin{array}{ccc} \check{\tau} & \hookleftarrow & \check{\sigma}, \\ R_{\check{\tau}} & \hookleftarrow & R_{\check{\sigma}}, \end{array}$$

and

$$X_{\check{\tau}} = \operatorname{spec} R_{\check{\tau}} \quad \hookrightarrow \quad \operatorname{spec} R_{\check{\sigma}} = X_{\check{\sigma}},$$

thus, providing the gluing maps.

For the following examples, we recall the definition of the complex projective n-space $\mathbb{P}^n$ as the space of lines (i.e., one-dimensional complex linear subspaces) in $\mathbb{C}^{n+1}$. Any nonzero vector $v := (\eta_0, \eta_1, \ldots, \eta_n)$ defines a line $\mathbb{C} \cdot v$, and two such vectors $v, v' \in \mathbb{C}^{n+1} \setminus \{0\}$ define the same line if and only if one is a (nonzero) scalar multiple of the other. We may, thus, associate with any element $\mathbb{C} \cdot v$ of $\mathbb{P}^n$ its *homogeneous coordinates* $[\eta_0, \eta_1, \ldots, \eta_n]$ where at least one component η_i is nonzero, and all components are determined only up to a common (nonzero) scalar factor. The subset $U_j := \{[\eta_0, \eta_1, \ldots, \eta_n] \in \mathbb{P}^n \mid \eta_j \neq 0\}$ can be identified with the affine n-space $\mathbb{C}^n$ by means of the bijective mapping

$$\begin{array}{ccc} U_j & \longrightarrow & \mathbb{C}^n \\ [\eta_0, \eta_1, \ldots, \eta_n] & \longmapsto & (\eta_0/\eta_j, \ldots, \eta_{j-1}/\eta_j, \eta_{j+1}/\eta_j, \ldots, \eta_n/\eta_j), \end{array}$$

that defines the jth system of *inhomogeneous coordinates* (also called *affine coordinates*) $(\zeta_{j,i})_{i=1,\ldots,n}$ on $\mathbb{P}^n$. The projective n-space is, thus, endowed with a covering by $n+1$ copies of the affine n-space. For $0 \leq j < k \leq n$, the transition from the coordinates $(\zeta_{j,i})$ to $(\zeta_{k,i})$ that provides the gluing of U_j and U_k is readily seen to be given by a monomial transformation. To see that $\mathbb{P}^n$ is a toric variety, we note that the intersection $\bigcap_{i=0}^n U_i$ is immediately identified with $\mathbb{C}^{*n}$, that is, the torus T is embedded in $\mathbb{P}^n$. Moreover, the natural torus action (see section 5) on $\mathbb{C}^n = U_0$ by componentwise multiplication extends to $\mathbb{P}^n$ in the obvious way

$$((t_1, \ldots, t_n), [\eta_0, \eta_1, \ldots, \eta_n]) \longmapsto [\eta_0, t_1\eta_1, \ldots, t_n\eta_n].$$

In Example 1, we show that the complex projective plane $\mathbb{P}^2$ is a toric variety; the general case of $\mathbb{P}^n$ is discussed in Example 5.

Example 1. Let the projective plane $\mathbb{P}^2 = \{[\eta_0, \eta_1, \eta_2] \mid \eta_i \in \mathbb{C},$ not all $\eta_i = 0\}$ be given, the homogeneous coordinates η_0, η_1, η_2 being determined only up to a common multiple.

It is covered by three affine planes $A_0 := \{(1, \eta_1\eta_0^{-1}, \eta_2\eta_0^{-1}) \mid \eta_0 \neq 0\}$,
$A_1 := \{(\eta_0\eta_1^{-1}, 1, \eta_2\eta_1^{-1}) \mid \eta_1 \neq 0\}$ and $A_2 := \{(\eta_0\eta_2^{-1}, \eta_1\eta_2^{-1}, 1) \mid \eta_2 \neq 0\}$.
Setting $z_1 := \eta_1\eta_0^{-1}$, $z_2 := \eta_2\eta_0^{-1}$, we obtain
$A_0 = \{(z_1, z_2)\}$, $A_1 = \{(z_1^{-1}, z_2z_1^{-1})\}$, and $A_2 = \{(z_2^{-1}, z_1z_2^{-1})\}$.

We find isomorphic coordinate rings, each representing an affine plane (compare Figure 4):

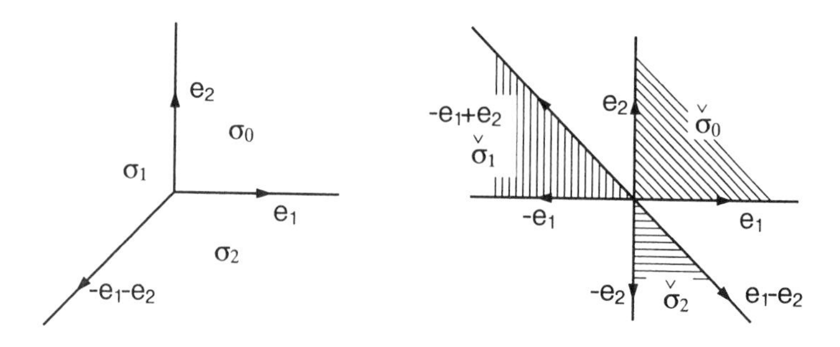

FIGURE 4.

$$\begin{aligned}
\sigma_0 &:= \mathbb{R}_{\geq 0}\, e_1 + \mathbb{R}_{\geq 0}\, e_2, & \sigma_1 &:= \mathbb{R}_{\geq 0}\, e_2 + \mathbb{R}_{\geq 0}(-e_1 - e_2),\\
R_{\check{\sigma}_0} &= \mathbb{C}[z_1, z_2], & R_{\check{\sigma}_1} &= \mathbb{C}[z_1^{-1}, z_1^{-1} z_2],\\
X_{\check{\sigma}_0} &= A_0, & X_{\check{\sigma}_1} &= A_1,\\
\sigma_2 &:= \mathbb{R}_{\geq 0}\, e_1 + \mathbb{R}_{\geq 0}(-e_1 - e_2) & &\\
R_{\check{\sigma}_2} &= \mathbb{C}[z_1 z_2^{-1}, z_2^{-1}], & &\\
X_{\check{\sigma}_2} &= A_2. & &
\end{aligned}$$

Example 2. Given $\mathbb{P}^1 \times \mathbb{P}^1 = \{([\eta_0, \eta_1], [\zeta_0, \zeta_1]) \mid (\eta_0, \eta_1) \neq (0, 0), (\zeta_0, \zeta_1) \neq (0, 0)\}$.

We may cover $\mathbb{P}^1 \times \mathbb{P}^1$ by four "charts", that is, affine planes. We set

$$z_1 := \eta_1 \eta_0^{-1}, \qquad z_2 := \zeta_1 \zeta_0^{-1}.$$

$A_0 = \{(z_1, z_2)\} = X_{\check{\sigma}_0}$, $A_1 = \{(z_1^{-1}, z_2)\} = X_{\check{\sigma}_1}$, $A_2 = \{(z_1, z_2^{-1})\} = X_{\check{\sigma}_2}$, $A_3 = \{(z_1^{-1}, z_2^{-1})\} = X_{\check{\sigma}_3}$, where $\sigma_0, \ldots, \sigma_3$ are given as is seen from Figure 5a.

Figure 5b illustrates the restriction of $\mathbb{P}^1 \times \mathbb{P}^1$ to its real points.

Example 3 Hirzebruch surfaces $\mathcal{H}_k$. We consider a hypersurface in $\mathbb{P}^1 \times \mathbb{P}^2 = \{([\eta_0, \eta_1], [\zeta_0, \zeta_1, \zeta_2]) \mid (\eta_0, \eta_1) \neq (0, 0), (\zeta_0, \zeta_1, \zeta_2) \neq (0, 0, 0)\}$ given by an equation

$$\eta_0^k \zeta_0 = \eta_1^k \zeta_1, \qquad k \in \mathbb{Z}\,.$$

It is called a Hirzebruch surface $\mathcal{H}_k$. By a modification of the arguments in Example 2, we find four affine planes as charts whose gluing together depends on

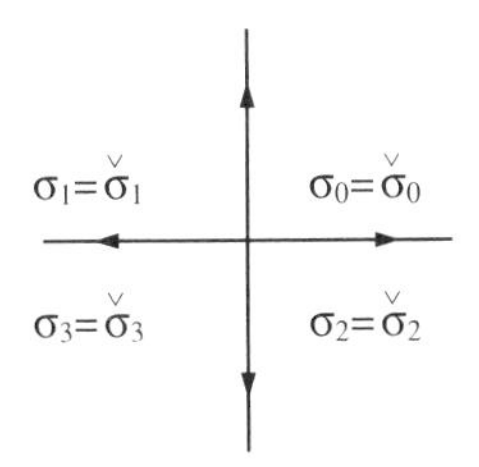

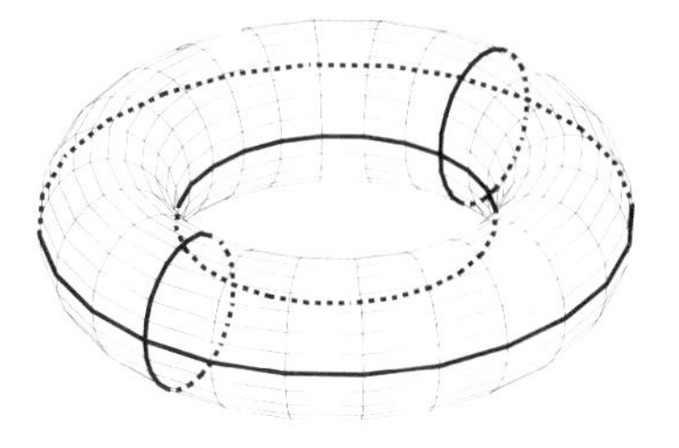

FIGURE 5a,b.

k. They are given by the spectra of the following rings (compare Figure 6):

$$\begin{aligned}
R_{\check{\sigma}_0} &= \mathbb{C}[z^{e_1}, z^{e_2}] = \mathbb{C}[z_1, z_2],\\
R_{\check{\sigma}_2} &= \mathbb{C}[z^{-e_2}, z^{e_1+ke_2}] = \mathbb{C}[z_2^{-1}, z_1 z_2^k],\\
R_{\check{\sigma}_1} &= \mathbb{C}[z^{-e_1}, z^{e_2}] = \mathbb{C}[z_1^{-1}, z_2],\\
R_{\check{\sigma}_3} &= \mathbb{C}[z^{-e_1-ke_2}, z^{-e_2}] = \mathbb{C}[z_1^{-1} z_2^{-k}, z_2^{-1}].
\end{aligned}$$

We return to these examples later on. We now discuss a method to get a simplified "real" picture of what a toric variety looks like. We write the complex parameter $z \in \mathbb{C}^*$ using polar coordinates (r, ϑ): $z = re^{i\vartheta} = r(\cos\vartheta + i\sin\vartheta)$ with $r := |z| > 0$ and $0 \le \vartheta := \arg(z) < 2\pi$ (Figure 7).

The absolute value map

$$\mathbb{C}^* \longrightarrow \mathbb{R}_{>0}, \quad z \longmapsto |z|$$

is continuous (with respect to the usual metric topology) surjective, a group homomorphism; for every $r > 0$, the "fiber" $\{z \in \mathbb{C}^* \mid |z| = r\}$ is a circle of radius r.

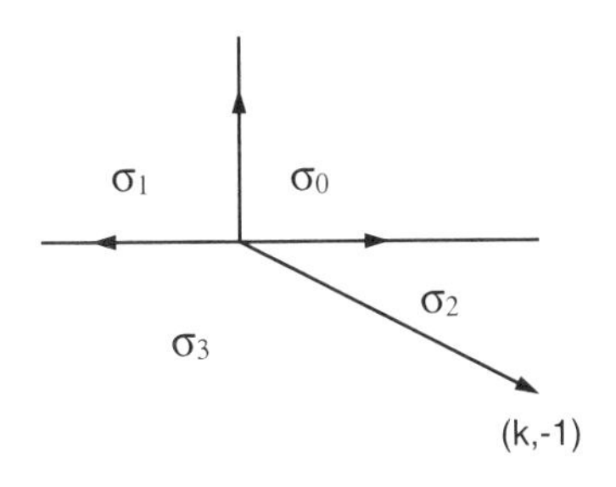

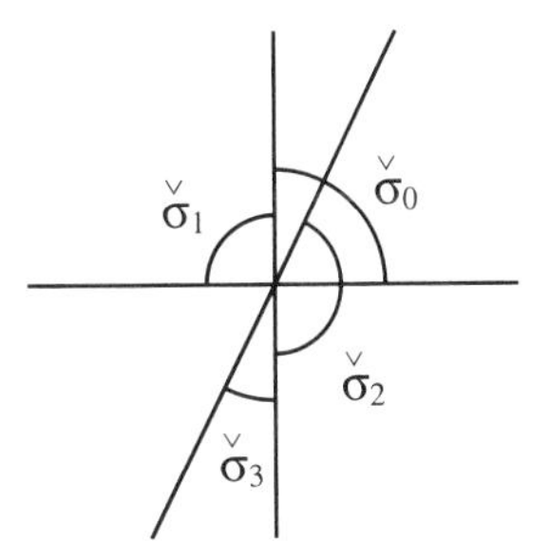

FIGURE 6.

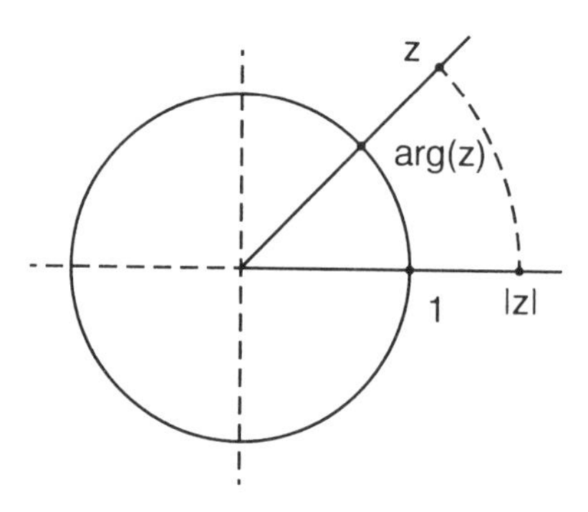

FIGURE 7.

Applying the absolute value to each component of $(z_1, \ldots, z_n) \in T$, we obtain the map

$$\Psi : T = (\mathbb{C}^*)^n \longrightarrow (\mathbb{R}_{>0})^n, \quad (z_1, \ldots, z_n) \longmapsto (|z_1|, \ldots, |z_n|)$$

that is again a continuous surjective homomorphism. Each fibre $\Psi^{-1}(r_1, \ldots, r_n)$ is the cartesian product of n circles, so it is a compact real torus of real dimension n. We note that Ψ admits an obvious continuous surjective extension to a map $\bar{\Psi} : \mathbb{C}^n \to (\mathbb{R}_{\geq 0})^n$.

On the other hand, applying the "argument" mapping

$$\vartheta : \mathbb{C}^* \longrightarrow \mathbb{S}^1 := \{z \in \mathbb{C}^* \mid |z| = 1\}, \quad z \longmapsto z/|z|$$

to each component of $(z_1, \ldots, z_n) \in T$, we obtain the map

$$\Theta : T = (\mathbb{C}^*)^n \longrightarrow (\mathbb{S}^1)^n, \quad (z_1, \ldots, z_n) \longmapsto (z_1/|z_1|, \ldots, z_n/|z_n|).$$

That is again a continuous surjective homomorphism. On the compact real n-torus $(\mathbb{S}^1)^n$ embedded in T, that map is the identity; in topology, such a map is called a "retraction". Each fibre $\Theta^{-1}(x)$, for some point $x \in (\mathbb{S}^1)^n$, is isomorphic to $(\mathbb{R}_{>0})^n$. It is obvious to see that the product mapping

$$\begin{array}{ccc} (\Psi, \Theta) : T & \longrightarrow & (\mathbb{R}_{>0})^n \times (\mathbb{S}^1)^n, \\ (z_1, \ldots, z_n) & \longmapsto & ((|z_1|, \ldots, |z_n|), (z_1/|z_1|, \ldots, z_n/|z_n|)) \end{array}$$

actually is a homeomorphism (with respect to the product topology) and a group isomorphism, thus, exhibiting T as a product of two factors:

3.7 Definition. We call the subgroup $(\mathbb{S}^1)^n \subset (\mathbb{C}^*)^n$ the *compact factor* T_{cp}, and $(\mathbb{R}_{>0})^n \subset (\mathbb{C}^*)^n$ the *radial factor* T_{rad} of the torus $T = (\mathbb{C}^*)^n$.

For an affine toric variety $V = X_{\check{\sigma}}$ that is realized in $\mathbb{C}^k$ by the choice of a system of generators $(a_1, \ldots, a_k)$ of the monoid $\check{\sigma} \cap \mathbb{Z}^n$, the embedded torus inherits this product structure. Moreover, if two affine toric varieties $X_{\check{\sigma}}$ and $X_{\check{\sigma}'}$ are glued, it is immediately clear from the description that the gluing map $\psi_{\sigma,\sigma'}$ respects this product structure and, thus, identifies the compact factors and the radial factors of the embedded tori, respectively.

We have already noted that the projection $\Psi : T \to T_{\text{rad}}$ onto the radial factor can be extended to a map $\bar{\Psi}$ of the toric variety $\mathbb{C}^n$ onto $(R_{\geq 0})^n$, i.e., onto the closure of the radial factor. This fact can be generalized to any affine toric variety $X_{\check{\sigma}}$: For any geometric realization $V = V(\mathfrak{a}_A)$ in some $\mathbb{C}^k$ given by a choice of generators as mentioned above, the equality $\bar{\Psi}(V) = V \cap (\mathbb{R}_{\geq 0})^k$ holds. That follows easily from the structure of the binomial relationship (3) preceding Theorem 2.7. Moreover, it is clear that this set is just the closure (in the "ordinary" topology) of the radial factor of the embedded torus in V. We denote with $X_{\check{\sigma}}^{\text{rad}}$ the corresponding subset of $X_{\check{\sigma}}$.

3.8 Definition. The subset $\bigcup_{\sigma\in\Sigma} X_{\check{\sigma}}^{\text{rad}}$ of a toric variety X_Σ is called the *manifold with corners* $\text{Mc}(X_\Sigma)$ associated to X_Σ.

Figure 8 shows $\text{Mc}(X_\Sigma)$ for Examples 1 to 3.

Remark. As the binomial relationships and the gluing maps for affine toric varieties are given by normalized monomial functions (that is with coefficient 1), toric varieties can be defined over any field. In particular, using the field $\mathbb{R}$ of real numbers instead of $\mathbb{C}$, we, thus, obtain the "real part" of a (complex) toric variety X_Σ. Then, the manifold with corners associated with X_Σ is also contained in the real part (see Figure 8a where $\text{Mc}(X_\Sigma)$ is the dotted area). However, the theory of toric varieties does not wholly work for $\mathbb{R}$ instead of $\mathbb{C}$ since, often, Hilbert's Nullstellensatz is used, which is no longer true if we replace $\mathbb{C}$ by $\mathbb{R}$.

Example 4. We obtain the toric variety $\mathbb{P}^2 \times \mathbb{P}^1$ by the fan of Figure 9a (spanned, for example, by the bitetrahedron $\text{conv}\{e_1, e_2, -e_1 - e_2, e_3, -e_3\}$). Figure 9b shows $\text{Mc}(\mathbb{P}^2 \times \mathbb{P}^1)$; Figure 9c illustrates $\mathbb{P}^2 \times \mathbb{P}^1$ and a homeomorphic image

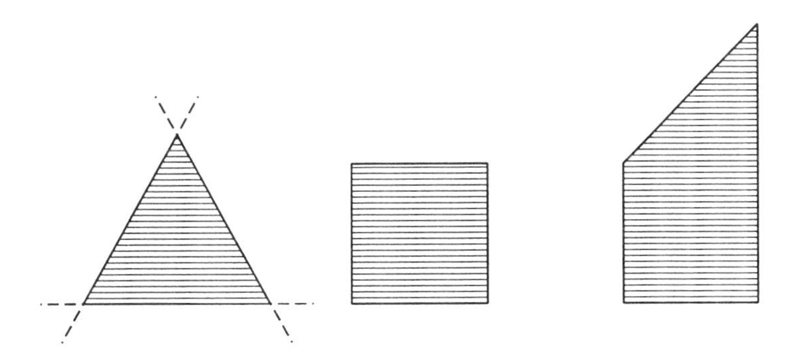

FIGURE 8a,b,c.

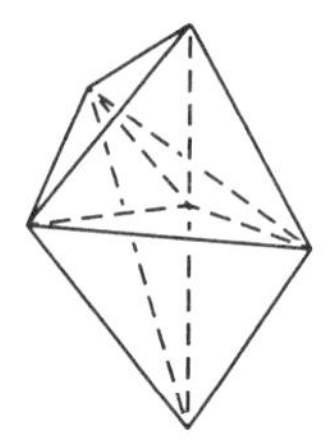

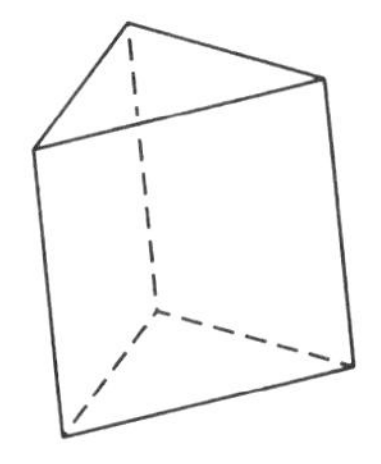

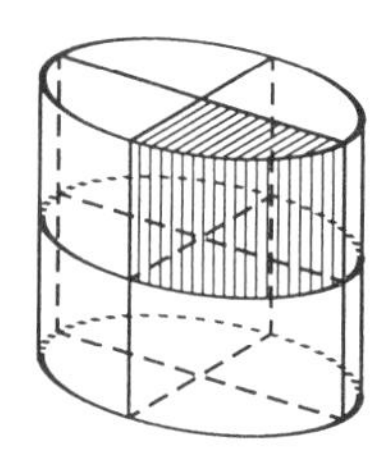

FIGURE 9a,b,c.

of $\mathrm{Mc}(\mathbb{P}^2 \times \mathbb{P}^1)$ (using a circular disc with diametrically opposed points on the boundary, identified as a model for the real part of $\mathbb{P}^2$, a line segment with end points identified for the real part of $\mathbb{P}^1$).

Remark. In our Examples 1 to 4, the defining fan Σ is always spanned by a convex polytope P. In all these cases, the associated manifold, with corners $\mathrm{Mc}(X_\Sigma)$, is seen to be (isomorphic to) the polar polytope P^* (see I, 6) of P.

Example 5. The n-dimensional complex projective space $\mathbb{P}^n$: Generalizing Example 1, we see that $\mathbb{P}^n$ is the toric variety determined by the fan Σ with the n-dimensional cones

$$\sigma_0 := \mathrm{pos}(e_1, \ldots, e_n) \quad \text{and}$$

$$\sigma_i := \mathrm{pos}(e_1, \ldots, e_{i-1}, e_{i+1}, \ldots, e_n, -(e_1 + \cdots + e_n)) \quad \text{for } i = 1, \ldots, n.$$

The $n+1$ toric varieties $X_{\check{\sigma}_i}$ for $i = 0, \ldots, n$, covering X_Σ, are just the copies $U_i \cong \mathbb{C}^n$ of the affine n-space that correspond to the system of inhomogeneous coordinates.

Example 6. So far, all fans have been complete, that is, they have covered $\mathbb{R}^n$. We shall give a characteristic property for toric varieties X_Σ with complete Σ (compactness) in section 9. As an example, for a noncomplete fan, let Σ consist of the four one-dimensional cones in Example 2 together with $\{0\}$. Then, X_Σ is obtained from $\mathbb{P}^1 \times \mathbb{P}^1$ by deleting four points, namely, the origins of the four charts $A_0, \ldots, A_3$ in Example 2.

3.9 Definition. An affine toric variety X_σ is said to be *regular* or *smooth* or *nonsingular* if σ is regular (see V, section 1), *quasi-smooth* or *$\mathbb{Q}$-factorial* if σ is a simplex cone. A toric variety X_Σ is called *regular* or *smooth* or *nonsingular* if, for any $\sigma \in \Sigma$, $X_{\check{\sigma}}$ is regular. We say X_Σ is *quasi-smooth* if, for any $\sigma \in \Sigma$, $X_{\check{\sigma}}$ is quasi-smooth.

Examples 1 to 5 are regular toric varieties. Example 7 is nonregular.

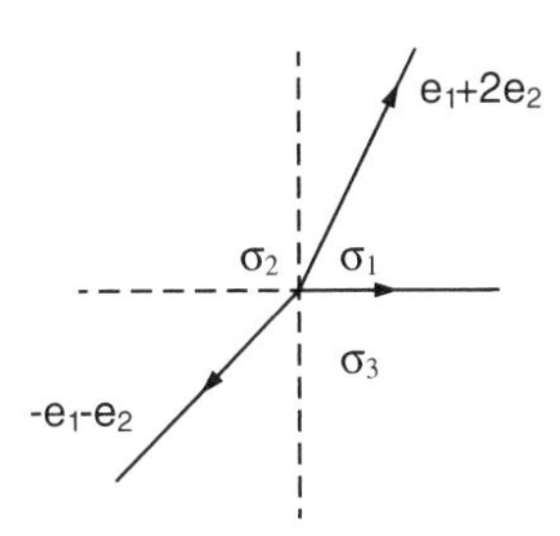

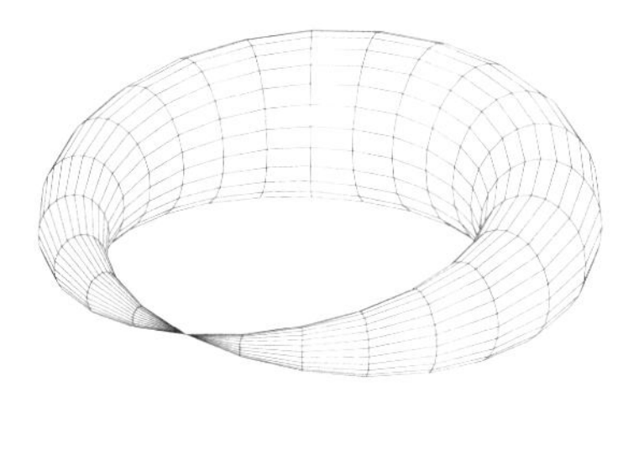

FIGURE 10a,b.

Remark. For our purposes, we do not need to discuss more refined notions like tangent spaces and differentials. Of course, these concepts are very important in the general theory of complex varieties; thus, one has to use them for a more thorough study of toric varieties.

Example 7. *The projective quadratic cone*: For the complete fan Σ in $\mathbb{R}^2$, defined by $e_1, e_1 + 2e_2$, and $-e_1 - e_2$ (Figure 10a), we know from section 2, Example 3, that $X_{\check{\sigma}_1}$ is the affine quadratic cone, whereas $X_{\check{\sigma}_2}$ and $X_{\check{\sigma}_3}$ are affine planes as they are defined by regular cones. We illustrate the real part of X_Σ in Figure 10b: the affine quadratic cone $X_{\check{\sigma}_1}$ is completed by the "circle at infinity" that represents a complex projective line l. That line l, in turn, is the union of a coordinate axis of $X_{\check{\sigma}_2}$ and another one of $X_{\check{\sigma}_3}$. The real picture is that of a "pinched (real) torus"—the inner tube of a tire with a "meridian" squeezed to a point.

Example. Let $a_0, \ldots, a_n$ be simple lattice vectors which span $\mathbb{R}^n$ nonnegatively and, hence, define a (uniquely determined) complete fan Σ. We call X_Σ a *weighted projective space*.

Exercises

1. Let Σ be the fan obtained from the regular octahedron $\operatorname{conv}\{e_1, e_2, e_3, -e_1, -e_2, -e_3\}$ by projecting its faces from 0. Show that $X_\Sigma = \mathbb{P}^1 \times \mathbb{P}^1 \times \mathbb{P}^1$.
2. In $\mathbb{R}^2$, let Σ be obtained by projecting from 0 the faces of the square $Q := \operatorname{conv}\{e_1 + e_2, e_1 - e_2, -e_1 + e_2, -e_1 - e_2\}$ and adding $\{0\}$. Find the affine toric varieties which define X_Σ.
3. Let $\Sigma := \{\mathbb{R}_{\geq 0}\, e_1 + \mathbb{R}_{\geq 0}\, e_2, \mathbb{R}_{\geq 0}\, e_3 + \mathbb{R}_{\geq 0}(-e_1 - e_2 - e_3), \text{Faces}\}$. Describe X_Σ.
4. Find analogs of the Hirzebruch surfaces in dimension three which are hypersurfaces in $\mathbb{P}^1 \times \mathbb{P}^3$. Find their fans and their defining affine toric varieties.

4. Invariant toric subvarieties

In the present section, we want to associate with each cone σ in a fan Σ a toric variety $X_{\Sigma/\sigma}$ that is embedded into X_Σ as a closed subvariety. In the systematic discussion of the torus action on a toric variety, we shall see that $X_{\Sigma/\sigma}$ is an orbit closure.

To begin with, let σ be a cone with apex 0 in $\mathbb{R}^n$, and let τ be a proper face of the dual cone $\check{\sigma}$. We note that, for dimensional reasons, τ is never the dual of a cone with apex 0 in $\mathbb{R}^n$, (see V, Lemma 2.2(c)).

Let $a_1, \ldots, a_k$ be a system of generators of the monoid $\tau \cap \mathbb{Z}^n$. We extend it to a system $a_1, \ldots, a_k, a_{k+1}, \ldots, a_q$ of generators of $\check{\sigma} \cap \mathbb{Z}^n$. These systems provide coordinates

$$(u_1, \ldots, u_k), \qquad (u_1, \ldots, u_k, u_{k+1}, \ldots, u_q)$$

(with $u_i := z^{a_i}$ for $i = 1, \ldots, q$ as in section 2) for the affine toric varieties X_τ and $X_{\check{\sigma}}$. We may assume that $a_{k+1}, \ldots, a_q \notin \tau \cap \mathbb{Z}^n$. In that case, it is easy to see that none of the coordinate functions $u_{k+1}, \ldots, u_q$ is invertible on $X_{\check{\sigma}}$, so there is a natural mapping

$$\begin{array}{rcl} \varphi : X_\tau & \longrightarrow & X_{\check{\sigma}} \\ (u_1, \ldots, u_k) & \longmapsto & (u_1, \ldots, u_k, 0, \ldots, 0). \end{array}$$

In fact, in all monomial relationships $\prod_{i=1}^q u_i^{\alpha_i} = \prod_{j=1}^q u_j^{\beta_j}$, as in section 2, formula (3) (following 2.6), with $\alpha_i > 0$ for some $i > k$, $\beta_j > 0$ for some $j > k$.

In the special case $\tau = \{0\}$, the cone $\check{\sigma}$ has an apex, so the origin $0 \in \mathbb{C}^q$ is a point of $X_{\check{\sigma}}$, and we obtain $\varphi(X_\tau) = \{0\} \subset X_{\check{\sigma}}$.

Since every linear relationship between $a_1, \ldots, a_k$ in $\check{\sigma}$ also holds in τ (and vice versa), the equations that characterize X_τ, according to Theorem 2.7, remain unchanged under φ. Therefore,

4.1 Lemma. *φ is an injective affine toric morphism.*

We may, thus, identify X_τ and the closed subvariety $\varphi(X_\tau) = X_{\check{\sigma}} \cap (u_{k+1} = \cdots = u_q = 0)$ of $X_{\check{\sigma}}$.

4.2 Definition. A closed subvariety of the affine toric variety $X_{\check{\sigma}}$ is called an *invariant affine (closed) toric subvariety* if it is of the form $\varphi(X_\tau) = X_{\check{\sigma}} \cap (u_{k+1} = \cdots = u_q = 0)$ as above.

We note that, in general, if a closed subvariety Y of an affine toric variety X is an affine toric variety itself, it is not invariant under the torus action introduced in the next section. As an example, consider the affine quadratic cone $Y \hookrightarrow \mathbb{C}^3$ or the affine plane $\mathbb{C}^2$ realized in $\mathbb{C}^3$ by the coordinates $(u_1, u_2, u_1 + u_2)$. For that reason, we call subvarieties, satisfying the conditions of the definition, "invariant".

Example 1. If $\sigma = \check{\sigma} = \operatorname{pos}(e_1, e_2)$ is the first quadrant in $\mathbb{R}^2$, and $\tau = \operatorname{pos}(e_1)$, then $\varphi(X_\tau)$ is the ξ_1 axis of the affine plane $X_{\check{\sigma}} = \mathbb{C}^2$.

Example 2. In $\mathbb{R}^2$, we consider the cone $\sigma = \operatorname{pos}(e_1, e_1 + 2e_2)$, its dual $\check{\sigma} = \operatorname{pos}(2e_1 - e_2, e_2)$, and the face $\tau = \operatorname{pos}(2e_1 - e_2)$ of $\check{\sigma}$. With respect to the generators $a_1 = 2e_1 - e_2, a_2 = e_1, a_3 = e_2$ of the monoid $\check{\sigma} \cap \mathbb{Z}^2$, we have three coordinates $u_i = z^{a_i}$ for $X_{\check{\sigma}}$, satisfying the equation $u_1 u_3 = u_2^2$. With respect to these coordinates, $X_\tau = X_{\check{\sigma}} \cap (u_2 = u_3 = 0)$, i.e., the ξ_1-axis. For $\tau' = \operatorname{pos}(a_3)$, we get $X_{\tau'} = X_{\check{\sigma}} \cap (u_1 = u_2 = 0)$, i.e., the ξ_3-axis.

Remark. We can also give an alternative description of the invariant affine toric subvarieties without using coordinates: A linear endomorphism of $\mathbb{R}^n$ that maps the monoid $\check{\sigma} \cap \mathbb{Z}^n$ surjectively onto $\tau \cap \mathbb{Z}^n$ induces a surjective homomorphism

$$\begin{array}{rcll} R_{\check{\sigma}} & \longrightarrow & R_\tau, & \text{hence an injective map} \\ X_\tau = \operatorname{spec} R_\tau & \longrightarrow & \operatorname{spec} R_{\check{\sigma}} = X_{\check{\sigma}} & \end{array}$$

(maximal ideals of R_τ are lifted to maximal ideals of $R_{\check{\sigma}}$).

Now let a fan Σ and a fixed cone $\sigma \in \Sigma$ be given. We consider the star $\operatorname{st}(\sigma, \Sigma)$ of σ in Σ, i.e., the set of all cones $\sigma' \in \Sigma$ that contain σ as a face (see III, 1). Using the orthogonal projection $\pi : \mathbb{R}^n \to \sigma^\perp$ onto the linear subspace $\sigma^\perp$ of $\mathbb{R}^n$, we obtain the collection of cones (Figure 11)

$$\Sigma/\sigma := \{\pi(\sigma') \mid \sigma' \in \operatorname{st}(\sigma, \Sigma)\}$$

that is easily seen to be a fan in $\sigma^\perp$, called a quotient fan of Σ (compare III, Definition 3.3).

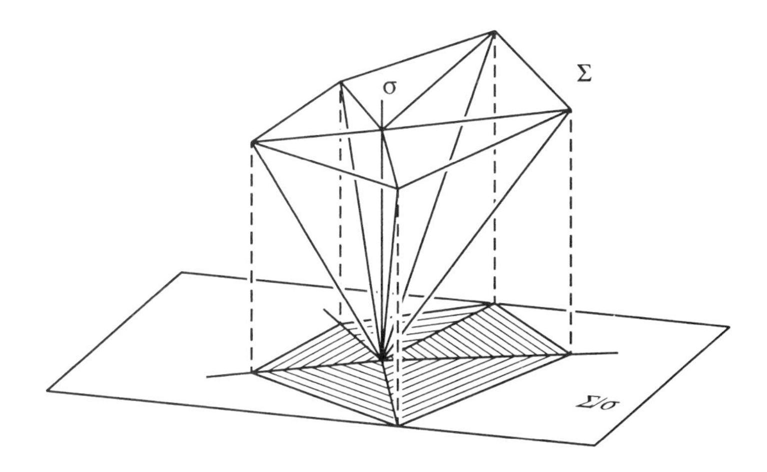

FIGURE 11.

Obviously, the elements $\pi(\sigma') \in \Sigma/\sigma$ are lattice cones with respect to the lattice $\pi(\mathbb{Z}^n)$. We now consider the dual $(\Sigma/\sigma)\check{}$ of the fan in $\sigma^\perp$.

4.3 Lemma.

(a) *We have the equality*

$$(\Sigma/\sigma)\check{} = \{\check{\sigma}' \cap \sigma^\perp \mid \sigma' \in \mathrm{st}(\sigma, \Sigma)\},$$

and each cone $\tau = \check{\sigma}' \cap \sigma^\perp \in (\Sigma/\sigma)\check{}$ *is a face of the corresponding cone* $\check{\sigma}'$.

(b) *The following diagram is commutative* ($\check{\mathrm{st}}\,(\cdot, \cdot)$ *denoting the set of dual cones)*

$$\begin{array}{ccc} \mathrm{st}\,(\sigma, \Sigma) & \xrightarrow{\sigma' \mapsto \check{\sigma}'} & \check{\mathrm{st}}(\sigma, \Sigma) \\ \downarrow \pi & & \downarrow \cap \sigma^\perp \\ \Sigma/\sigma & \xrightarrow{\pi(\sigma') \mapsto \pi(\sigma')\check{}} & (\Sigma/\sigma)\check{} \end{array}$$

PROOF. This follows easily from the definition of dual cones. □

We now show that we can embed the toric variety $X_{\Sigma/\sigma}$, given by the fan Σ/σ in $\sigma^\perp$, into X_Σ.

4.4 Lemma.

(a) *To each cone* $\tau = \check{\sigma}' \cap \sigma^\perp$ *in* $(\Sigma/\sigma)\check{}$ *there corresponds an invariant toric embedding* $X_\tau \hookrightarrow X_{\check{\sigma}'}$.

(b) *Each such embedding* $X_\tau \hookrightarrow X_{\check{\sigma}'}$ *is compatible with the gluing of the affine toric varieties in* $X_{\Sigma/\sigma}$ *and in* X_Σ*, respectively. As a consequence, the choice of* σ *determines an embedding of* $X_{\Sigma/\sigma}$ *into* X_Σ *as an invariant closed toric subvariety.*

PROOF.

(a) τ is an element of $(\Sigma/\sigma)\check{}$ and a face of $\check{\sigma}'$. So, we can apply Lemma 4.1.

(b) is a consequence of (a).

□

Note that, under π, the lattice $\mathbb{Z}^n$ need not be projected onto $\sigma^\perp \cap \mathbb{Z}^n$. Furthermore, Σ need not be simplicial.

Example 3. In the fan Σ defining the projective plane $\mathbb{P}^2$ (see section 3, Example 1), we choose $\sigma := \mathrm{pos}(e_2)$. Then,

$$\begin{aligned} \mathrm{st}(\sigma, \Sigma) &= \{\sigma_0, \sigma_1, \sigma\}, \quad \check{\mathrm{st}}(\sigma, \Sigma) = \{\check{\sigma}_0, \check{\sigma}_1, \mathbb{R}\,e_1 + \mathbb{R}_{\geq 0}\,e_2\} \quad \text{and} \\ \sigma^\perp &= \mathbb{R}\,e_1, \\ \Sigma/\sigma &= \{\mathrm{pos}(e_1), \mathrm{pos}(-e_1), \{0\}\}, \quad (\Sigma/\sigma)\check{} = \{\mathrm{pos}(e_1), \mathrm{pos}(-e_1), \mathbb{R}\,e_1\}. \end{aligned}$$

Thus, we see that $X_{\Sigma/\sigma}$ is a projective line, namely, a coordinate line of $\mathbb{P}^2$.

Example 4. For $\mathbb{P}^1 \times \mathbb{P}^1$ (see section 3, Example 2), we obtain the following invariant toric subvarieties: the four projective lines

$$\mathbb{P}^1 \times \{0\}, \ \mathbb{P}^1 \times \{\infty\}, \ \{0\} \times \mathbb{P}^1, \text{ and } \{\infty\} \times \mathbb{P}^1$$

(their real points form the four circles in Figure 5b), and the four points in which these lines intersect. The defining cones for these invariant toric subvarieties can readily be found.

Example 5. For the Hirzebruch surface $\mathcal{H}_k$ (see section 3, Example 3), generalizing the previous example, which is just the special case $k = 0$, we obtain four projective lines and their four points of intersection as proper toric subvarieties. An illustration of $\mathcal{H}_1$ will be given in section 6, Example 3.

Example 6. For $\mathbb{P}^2 \times \mathbb{P}^1$ (see section 3, Example 4), the fan Σ is spanned by the bitetrahedron $\operatorname{conv}\{e_1, e_2, -e_1 - e_2, e_3, -e_3\}$ in $\mathbb{R}^3$ (see Figure 9a in section 3). It contains five one-dimensional cones (spanned by the vertices), nine two-dimensional cones (spanned by the edges), and six three-dimensional cones (spanned by the facets). Correspondingly, there are five two-dimensional, nine one-dimensional, and six zero-dimensional, invariant, (affine) toric subvarieties. In the illustration of the real part of X_Σ given in Figure 9c, the two-dimensional toric subvarieties are represented by the horizontal plane in the middle ($\mathbb{P}^2 \times \{0\}$) and the horizontal planes on top and at the bottom that have to be identified (both represent $\mathbb{P}^2 \times \{\infty\}$), the two vertical planes through the center, and the cylindrical boundary (each of them represents a product $l_i \times \mathbb{P}^1$, where l_i is a coordinate line of $\mathbb{P}^2$). The one-dimensional toric subvarieties are projective lines. Two of them are represented by the two diametrically arranged pairs of parallel lines on the cylindrical boundary and another two by similar pairs on the top and the bottom plane; these pairs are to be identified. One is given by the pair consisting of the "upper" and the "lower" boundary circle, which are to be identified, and the identifying diametrically opposite points. Another one is represented by the middle circle with the same identification. The remaining three are given by inner line segments. The nonnegative part of each projective line is contained in the hatched part of Figure 9c or is represented by an edge of the prism of Figure 9b.

Exercises

1. Find all invariant toric subvarieties of $\mathbb{P}^n$ (considered as a toric variety; see Example 5 in section 3).
2. Find all invariant toric subvarieties of X_Σ in Exercise 1 of section 3.
3. Any invariant toric subvariety of a toric variety X_Σ determined by an $(n-1)$-face σ of Σ, is either a torus, an affine line, or a projective line.
4. Any invariant toric subvariety of a regular toric variety is again regular.

5. The torus action

Let $T := (\mathbb{C}^*)^n$ be the n-dimensional algebraic torus. It is clear that T is a group with respect to coordinatewise multiplication. As such, it "operates" on itself. In this section, we discuss the extension of that operation to a natural torus action on a toric variety.

We recall that an *action* (or *operation*) of a group G (with the composition written multiplicatively) on a set X is a mapping $G \times X \to X$, $(g, x) \mapsto gx$, that satisfies the two conditions $g(hx) = (gh)x$ and $ex = x$ for arbitrary $g, h \in G$ and $x \in X$, where $e \in G$ denotes the neutral element of the group. In particular, we always have $g^{-1}(gx) = x$, so for fixed $g \in G$, the mapping $X \to X, x \mapsto gx$ is a bijection. We also say that G is a transformation group acting (or operating) on X. For a fixed element $x \in X$, the subset $Gx := \{gx \mid g \in G\}$ of X is called the *orbit* of (or through) x. It is clear that each point of X lies on a unique orbit. If there is only one orbit, the action is called *transitive*. In particular, a group acts on itself transitively by multiplication. If H is a subgroup of G, then, an action of G on X obviously induces an action of H on X.

Now let σ be a cone in $\mathbb{R}^n$ with apex 0, and let $A = (a_1, \ldots, a_k)$ be a system of generators of the monoid $\check{\sigma} \cap \mathbb{Z}^n$. As we know from Theorem 2.7, the choice of A determines a geometric realization of the affine toric variety $X_{\check{\sigma}}$ as an affine algebraic subvariety V of $\mathbb{C}^k$. Moreover, we know from Theorem 2.8(a) that the mapping $T \to \mathbb{C}^k$ given by $t \mapsto (t^{a_1}, \ldots, t^{a_k})$ maps T bijectively onto the open dense subset $V \cap (\mathbb{C}^*)^k$ of V, the embedded torus (cf. Lemma 3.4). Next, we recall from 2.8(b) that, for $x = (x_1, \ldots, x_k) \in V$ and $t \in T$, we have $(t^{a_1}x_1, \ldots, t^{a_k}x_k) \in V$. The mapping $T \times V \to V$, thus defined, is a torus action on V that extends the natural action of the torus on itself. The embedded torus is an orbit of that action, called the *"big" orbit.*

To see that the action is independent of the choice of A, we recall from 2.10 that, for every two systems of coordinates $(u_1, \ldots, u_k)$ and $(v_1, \ldots, v_m)$ on an abstract affine toric variety $X_{\check{\sigma}}$, corresponding to different choices A and A' of monoid generators and thus to different geometric realizations $V = V(\mathfrak{a})$ in $\mathbb{C}^k$ and $V' = V(\mathfrak{a}')$ in $\mathbb{C}^m$, there is a coordinate transformation $\varphi : V \to V'$. In the remark preceding 210, we have already noted that φ induces a bijection—in fact, a coordinate transformation—between the embedded copies of the torus in V and V', respectively. Then, it is easy to see that, for every given point $x = (x_1, \ldots, x_k) \in V$, its image $\varphi(x) = (y_1, \ldots, y_m) \in V'$, and for every $t \in T$, $\varphi(t^{a_1}x_1, \ldots, t^{a_k}x_k) = (t^{a'_1}y_1, \ldots, t^{a'_m}y_m)$. The identification of the abstract affine toric variety $X_{\check{\sigma}}$ with its geometric realization $V \subset \mathbb{C}^k$ is, thus, compatible with the torus action on V, so,

5.1 Theorem. *With the above notation, the map*

$$T \times X_{\check{\sigma}} \to X_{\check{\sigma}}\,, \quad (t, x) \mapsto tx := (t^{a_1}x_1, \ldots, t^{a_k}x_k)$$

is an action of the torus on the affine toric variety $X_{\check{\sigma}}$, called the natural torus action. It extends the natural action of the torus on itself. The embedded torus is an orbit of the action, called the "big" orbit.

Now, let Σ be a fan in $\mathbb{R}^n$. For a cone $\sigma \in \Sigma$ and a face τ of σ, it is obvious that the identification $X_{\check{\tau}} \cong X_{\check{\sigma}} \setminus (u_k = 0)$ of Lemma 3.1 is compatible with the natural torus action on these affine toric varieties. This immediately implies that all gluing maps $\psi_{\sigma,\sigma'}$ respect the torus action, too. That proves the following result:

5.2 Theorem. *For a fan Σ in $\mathbb{R}^n$, the natural torus actions on the affine toric varieties $X_{\check{\sigma}}$, for $\sigma \in \Sigma$, are compatible with the gluing maps and, thus, yield a natural torus action on the toric variety X_Σ. The statements of Theorem* 5.1 *carry over to this more general case.*

As the torus action on an affine toric variety $X_{\check{\sigma}}$ is given by monomials, the following result is immediately clear:

5.3 Theorem. *With respect to the "complex" and the Zariski topology, the torus action $T \times X_\Sigma \to X_\Sigma$ is given by a continuous map. The big orbit is an open, dense subset, so every other orbit is contained in its closure.*

Thus, the torus acts on $X = X_\Sigma$ as a group of continuous transformations; in particular, for each element $t \in T$, the map $X \to X$, $x \mapsto tx$ is a homeomorphism.

Without entering further into the discussion, we mention that toric varieties are algebraic varieties, the torus is an algebraic group, and it acts as an algebraic transformation group.

Now, we want to discuss the behavior of invariant toric subvarieties with respect to the torus action. As at the beginning of the previous section, we first consider the affine case. Using the same notation, we, thus, look at a cone σ with apex 0 in $\mathbb{R}^n$ and a face τ of the dual cone $\check{\sigma}$. With respect to a suitable "toric" coordinate system $u_1, \ldots, u_k, u_{k+1}, \ldots, u_q$ on $X_{\check{\sigma}}$, $X_\tau \cong X_{\check{\sigma}} \cap (u_{k+1} = \cdots = u_q = 0)$. From the description of the natural torus action on $X_{\check{\sigma}}$ given in Theorem 5.1, we get the following justification of the name "invariant, affine, toric subvariety":

5.4 Lemma. *Each invariant, affine, toric subvariety X_τ of an affine toric variety $X_{\check{\sigma}}$ is invariant under the natural torus action on $X_{\check{\sigma}}$.*

The smallest invariant, affine, toric subvariety of $X_{\check{\sigma}}$ which can be thus obtained, corresponds to the smallest face of $\check{\sigma}$, namely, to cospan $\check{\sigma}$. We want to show that cospan $\check{\sigma}$ it is a distinguished orbit of the torus action on $X_{\check{\sigma}}$.

5.5 Lemma. *The invariant, affine, toric subvariety of $X_{\check{\sigma}}$ corresponding to* cospan $\check{\sigma}$ *is an embedded torus of dimension* dim cospan $\check{\sigma} = n - \dim \sigma$. *The* *subvariety is the only closed orbit of the natural torus action on $X_{\check{\sigma}}$, and it lies in the closure of every other such orbit.*

PROOF. To simplify the notation, we write $\gamma := \operatorname{cospan} \check{\sigma}$. We can choose a system of generators $a_1, \ldots, a_k$ of the monoid $\check{\sigma} \cap \mathbb{Z}^n$ with the following properties: The elements $a_1, \ldots, a_p, a_{p+1} := -(a_1 + \cdots + a_p)$ with $0 \leq p := \dim \gamma$ are a minimal system of generators for the monoid $\gamma \cap \mathbb{Z}^n$, and $a_i \notin \gamma$ for $i = p+2, \ldots, k$. Without loss of generality, we may assume that $a_i = e_i$ for $i = 1, \ldots, p$ holds. The corresponding coordinates $u_1, \ldots, u_p, u_{p+1} := (u_1 \cdots u_p)^{-1}$ are invertible functions on $X_{\check{\sigma}}$, whereas $u_{p+2}, \ldots, u_k$ are not invertible. To see that the torus acts transitively on X_γ, we choose an arbitrary point $x = (x_1, \ldots, x_p, x_{p+1}, 0, \ldots, 0) \in X_\gamma$ (with $x_i \neq 0$ and $x_1 \cdots x_p \cdot x_{p+1} = 1$). Then, $t = (x_1^{-1}, \ldots, x_p^{-1}, 1, \ldots, 1)$ lies in the torus T, and satisfies $tx = (1, \ldots, 1, 1, 0, \ldots, 0)$. This proves that the torus T acts transitively on X_γ, so X_γ is an orbit. Moreover, we can immediately identify X_γ with the subtorus $T' := \{t \in T \mid t_{p+1} = \cdots = t_n = 1\}$ of dimension p and identify the induced action of T' on X_γ with the natural action of T' on itself.

Next, we want to prove that X_γ is the only closed T-orbit in $X_{\check{\sigma}}$. For every point $x \in X_{\check{\sigma}}$, we know that the first $p+1$ coordinates are always nonzero. Hence, for a point $x \notin X_\gamma$, there is at least one nonzero coordinate x_i with $i \geq p+2$, and it is clear that this coordinate is nonzero for all points on the orbit Tx. We observe that, by suitably choosing the basis elements $e_{p+1}, \ldots, e_n$ of $\mathbb{Z}^n$, we may assume, for every generator a_i with $i \geq p+2$, the last $n-p$ components to be nonnegative, and at least one of them to be strictly positive.

This is a consequence of the general fact that each cone σ with apex 0 is contained in a regular cone, which we prove as follows. Let H be a hyperplane for which $H \cap \sigma = \{0\}$ (compare V, 1). Up to a unimodular transformation, we may assume $H = \operatorname{pos}\{e_1, \ldots, e_{n-1}\}$, and $\sigma \subset H + \mathbb{R}_{\geq 0}\, e_n$, but not $\sigma \subset H$. For an appropriate natural number r, the hyperplane $H + re_n$ intersects σ in a lattice polytope P. Any translation of P by a translation vector $c \in H \cap \mathbb{Z}^n$ parallel to H may be extended to a unimodular shear of $\mathbb{R}^n$ with axis H. We can choose c such that the shear maps P and, hence, σ into $\operatorname{pos}\{e_1, \ldots, e_n\}$.

Now consider the subtorus $T'' := \{t \in T \mid t_1 = \cdots = t_p = 1\}$ that is complementary to T'. For every $\lambda \in \mathbb{C}^*$, the element $t''(\lambda) := (1, \ldots, 1, \lambda, \ldots, \lambda)$ lies in T'', and $t''(\lambda)\, x = (x_1, \ldots, x_p, x_{p+1}, \lambda^{\beta_{p+2}} x_{p+2}, \ldots, \lambda^{\beta_k} x_k)$ with all exponents $\beta_{p+2}, \ldots, \beta_k$ positive. All these points $t''(\lambda)\, x$ are in the orbit Tx. On the other hand, the limit point $\lim_{\lambda \to 0} t''(\lambda)\, x = (x_1, \ldots, x_p, x_{p+1}, 0, \ldots, 0)$ clearly lies in X_γ, so the orbit Tx is not closed. As X_γ is an orbit itself, it lies in the closure of Tx, as follows easily from the continuity of the action. □

5.6 Lemma. *Every invariant, affine, toric subvariety X_τ of $X_{\check{\sigma}}$ contains a unique orbit O_τ that is relatively open and dense in X_τ, that is, X_τ is an orbit closure. Conversely, every orbit closure in $X_{\check{\sigma}}$ is such an invariant toric subvariety X_τ for a unique face τ of $\check{\sigma}$.* □

PROOF. Using Theorem 5.3, the first part follows easily from the next result. For the second, we use a decomposition of the cone $\check{\sigma}$ as given in V, Lemma 1.6. We can find a basis $\{e_1, \ldots, e_p, e_{p+1}, \ldots, e_n\}$ of the lattice $\mathbb{Z}^n$ such that $\{e_1, \ldots, e_p\}$ is a basis for $\gamma \cap \mathbb{Z}^n$ (with $\gamma := \operatorname{cospan} \check{\sigma}$ as before).Consider the linear subspace $U' := \operatorname{lin}\{e_{p+1}, \ldots, e_n\}$. Then, $\upsilon := \check{\sigma} \cap U'$ is a (lattice) cone with apex 0 satisfying $\check{\sigma} = \gamma + \upsilon$ and $\dim \gamma + \dim \upsilon = n$. By suitably modifying the basis $\{e_{p+1}, \ldots, e_n\}$ of $U' \cap \mathbb{Z}^n$ if necessary, we may assume that $\upsilon \subset \operatorname{pos}\{e_{p+1}, \ldots, e_n\}$ holds. Every face of $\check{\sigma}$ can be uniquely written in the form $\tau = \gamma + \tau_0$, where $\tau_0 = \tau \cap \upsilon$ is a face of υ. Let $A(\upsilon) := \{a_{p+2}, \ldots, a_k\}$ be the unique minimal set of generators of the monoid $\upsilon \cap \mathbb{Z}^n$ (see V, Lemma 3.5 applied to $\operatorname{lin}(\upsilon)$). Then, $A(\tau_0) := A(\upsilon) \cap \tau_0$ is the minimal set of generators of $\tau_0 \cap \mathbb{Z}^n$.

Now consider a point $x \in X_{\check{\sigma}}$ that is not in the closed orbit. Hence, with respect to the coordinate system $(u_1, \ldots, u_p, u_{p+1}, u_{p+2}, \ldots, u_k)$ corresponding to the minimal set of generators $A := A(\check{\sigma}) := \{a_1, \ldots, a_p, a_{p+1}, a_{p+2}, \ldots, a_k\}$ with $a_i := e_i$ for $i \leq p$ and $a_{p+1} = -(a_1 + \ldots + a_p)$ (see the proof of Lemma 5.5), there are coordinates $x_i \neq 0$ for some $i \geq p+2$, so $A(x) := \{a_i \in A \mid x_i \neq 0,\ i \geq p+2\}$ is not empty. Then, $\tau_0 := \upsilon \cap \operatorname{lin} A(x)$ obviously is the smallest face of υ containing $A(x)$. From the structure of the relationships defining $X_{\check{\sigma}}$, it is not difficult to check that $A(x) = A(\tau_0)$ holds. Then, $\tau := U' + \tau_0$ has the property that the point x lies in the open, dense orbit of X_τ, so X_τ is the closure of the orbit Tx. □

5.7 Theorem. *The natural torus action on an invariant toric subvariety X_τ of $X_{\check{\sigma}}$ is given by the induced action of a subtorus T' of T.*

PROOF. We can choose a basis $(e_1, \ldots, e_n)$ of the lattice $\mathbb{Z}^n$ such that $(e_1, \ldots, e_r)$ is a basis of $\operatorname{lin} \tau$, and $\tau \subset \operatorname{cospan} \check{\sigma} + \operatorname{pos}\{e_1, \ldots, e_r\}$. Accordingly, we have a system of monoid generators $a_1, \ldots, a_m, a_{m+1}, \ldots, a_k$ for $\check{\sigma} \cap \mathbb{Z}^n$ with $\tau = \operatorname{pos}\{a_1, \ldots, a_m\}$, and $a_i \notin \tau$ for $i \geq m+1$. Let $u_1, \ldots, u_m$ and $u_1, \ldots, u_m, u_{m+1}, \ldots, u_k$ be the corresponding coordinates on X_τ and $X_{\check{\sigma}}$, respectively. Then, the natural torus action of $(\mathbb{C}^*)^r$ on X_τ is given by $s(x_1, \ldots, x_m) = (s^{a_1}x_1, \ldots, s^{a_m}x_m)$ for $s = (s_1, \ldots, s_r) \in (\mathbb{C}^*)^r$. It is immediately clear that this is compatible with the induced action of $T' := \{t \in T \mid t_{r+1} = \cdots = t_n = 1\}$ on the invariant affine toric subvariety $X_\tau \cong X_{\check{\sigma}} \cap (u_{m+1} = \cdots = u_k = 0)$. We remark that the complementary torus $T'' := \{t \in T \mid t_1 = \cdots = t_r = 1\}$ acts trivially on X_τ. □

For a fixed cone σ in the fan Σ, we now consider the invariant toric subvariety $X_{\Sigma/\sigma}$ of X_Σ. It is endowed with its own natural torus action. By Lemma 5.6, we know that, in the affine case, an invariant toric subvariety is an orbit closure and the natural torus action is induced from the natural torus action on the ambient variety by restricting it to a suitable subtorus. It is now easy to see that this holds in the general case, too.

5.8 Theorem.

(a) *Any orbit is the embedding T_k of a torus of some dimension k between 0 and n.*

(b) *X_Σ is the disjoint union of all its orbits; their number is finite.*

PROOF.

(a) follows from Lemma 5.6.

The first part of (b) is a consequence of the definition of an orbit. The second part follows inductively, since any toric subvariety $X_{\Sigma/\sigma}$ contains a torus T_k of dimension k as a dense subset and $X_{\Sigma/\sigma} \setminus T_k$ is covered by lower dimensional tori whose closures are again toric subvarieties associated with the proper faces of Σ/σ. □

5.9 Definition. We call each T_k an *embedded torus*. $T_n = T$ is said to be the *big torus* in X_Σ (compare Lemma 3.5).

Example 1. $\mathbb{P}^1$ is covered by the one-dimensional torus $\mathbb{C}^*$ and two zero-dimensional tori, one consisting of $0 \in \mathbb{C}$, the other of ∞ (on the Riemann sphere of complex numbers).

Example 2. $\mathbb{P}^1 \times \mathbb{P}^1$ (see Example 2 in 3). $T = \mathbb{C}^{*2}$ is the two-dimensional orbit. There are four one-dimensional orbits and four zero-dimensional ones.

Remark. Torus actions can be used to characterize toric varieties. In fact, the original definition and the name of toric varieties stem from the theory of algebraic groups (Demazure 1970, Mumford et al. 1973, T. Oda 1973).

Exercises

1. Find all orbits on a Hirzebruch surface (see Example 5 in section 4).
2. Find all orbits of $X_\Sigma \cong \mathbb{P}^1 \times \mathbb{P}^1 \times \mathbb{P}^1$ as introduced in Exercise 1 of section 3.
3. Prove that any torus action on $\mathbb{P}^n$ (Example 4 in section 3) is a projective linear transformation of $\mathbb{P}^n$.
4. Consider the subgroup $\mathcal{T}_0$ of $\mathcal{T}$ consisting of all matrices C with $|c_1| = \cdots = |c_n| = 1$. Find all orbits of $\mathcal{T}_0$ for $\mathbb{P}^1 \times \mathbb{P}^1$ (see Example 2 above).

6. Toric morphisms and fibrations

In section 2 we introduced toric morphisms for affine toric varieties. We will extend them, now, to general toric varieties and characterize the extensions combinatorially.

6.1 Theorem. *Let $L_0 : \mathbb{R}^n \longrightarrow \mathbb{R}^r$, $L_0(\mathbb{Z}^n) \subset \mathbb{Z}^r$ be a linear map which induces a map of fans*

$$L : \Sigma \longrightarrow \Sigma'$$

(see V. 4.1*). Then, L gives rise, in a natural way, to a map*

$$\Psi : X_\Sigma \longrightarrow X_{\Sigma'}$$

whose restriction $\Psi_\sigma := \Psi|_{X_{\check{\sigma}}}$ to any affine piece $X_{\check{\sigma}}$ of X_Σ is an affine toric morphism

$$\Psi_\sigma : X_{\check{\sigma}} \longrightarrow X_{\check{\sigma}'}.$$

In particular, this map is continuous with respect to the complex and to the Zariski topology on X_Σ and $X_{\Sigma'}$.

6.2 Definition. We call $\Psi =: \bar{L}$ a *toric morphism.*

PROOF OF THEOREM 6.1. Let $\sigma \in \Sigma$. By V, Corollary 2.11, the dual map L^* yields $L^*(\check{\sigma}') \subset \check{\sigma}$ if $L(\sigma) \subset \sigma'$. Furthermore, $\check{\sigma}$, $\check{\sigma}'$, $L^*(\check{\sigma}')$ are lattice cones. Let $a_1, \ldots, a_k$ be generators of $\check{\sigma} \cap \mathbb{Z}^n$ and $b_1, \ldots, b_m$ be generators of $\check{\sigma}' \cap \mathbb{Z}^r$. We write

$$\begin{aligned} L^*(b_i) &:= \alpha_{i1}a_1 + \cdots + \alpha_{ik}a_k, \\ &\alpha_{ij} \in \mathbb{Z}_{\geq 0}, \quad i = 1, \ldots, m, \quad j = 1, \ldots, k. \end{aligned} \tag{1}$$

Because

$$\check{\sigma}' \cap \mathbb{Z}^r \xrightarrow{\;L^*\;} \check{\sigma} \cap \mathbb{Z}^n \tag{2}$$

is a homomorphism of monoids, we obtain a ring homomorphism

$$R_{\check{\sigma}'} \longrightarrow R_{\check{\sigma}}$$

and the induced morphism

$$\operatorname{spec} R_{\check{\sigma}} =: X_{\check{\sigma}} \xrightarrow{\;\bar{L}\;} X_{\check{\sigma}'} := \operatorname{spec} R_{\check{\sigma}'}$$

of the affine varieties defined by $R_{\check{\sigma}}$ and $R_{\check{\sigma}'}$. In the coordinates $u_1 := z^{a_1}, \ldots, u_k := z^{a_k}$ of $X_{\check{\sigma}}$ corresponding to $a_1, \ldots, a_k$ and the coordinates $(w_1, \ldots, w_m)$ of $X_{\check{\sigma}'}$ corresponding to $b_1, \ldots, b_m$,

$$(u_1, \ldots, u_k) \longmapsto (u_1^{\alpha_{11}} \cdots u_k^{\alpha_{1k}}, \ldots, u_1^{\alpha_{m1}} \cdots u_k^{\alpha_{mk}}) = (w_1, \ldots, w_m), \tag{3}$$

so that we obtain an affine toric morphism.

If $(u_1, \ldots, u_k)$ is represented in another affine chart $X_{\check{\tau}}$, $\tau \in \Sigma$, by $(v_1, \ldots, v_q)$, the change of coordinates is obtained by a linear transformation of the lattice vectors which define the coordinates (see proof of Lemma 2.10 and the definition of gluing map in section 3). Since L is the same linear map for all generators, $\bar{L}$ is readily seen to be compatible with the gluing maps. As affine toric morphisms are continuous with respect to both topologies (see the discussion following Definition 2.9), it is clear that the maps obtained by gluing are again continuous. □

6.3 Definition. Let X_Σ, $X_{\Sigma'}$ be toric varieties with embedded tori $T \subset X_\Sigma$, $T' \subset X_{\Sigma'}$.

Let $\Psi : X_\Sigma \longrightarrow X_{\Sigma'}$ be a map, and $\alpha : T \longrightarrow T'$ a homomorphism such that

$$\Psi(c \cdot x) = \alpha(c) \cdot \Psi(x) \qquad \text{for all } c \in T. \tag{1}$$

Then, we call Ψ *equivariant* (with respect to α).

6.4 Theorem. *Every toric morphism Ψ is equivariant with respect to a suitable homomorphism of the embedded tori.*

PROOF. Since T is dense in X_Σ, it suffices to prove (1) for all $x \in T$. Let $\Psi|_T$ be given by

$$v_1 = u_1^{\alpha_{11}} \cdots u_q^{\alpha_{1q}},$$
$$\vdots$$
$$v_p = u_1^{\alpha_{p1}} \cdots u_q^{\alpha_{pq}}.$$

Then, for $c = (c_1, \ldots, c_q)$, $a_i := (\alpha_{i1}, \ldots \alpha_{iq})$,

$$(c_1 u_1)^{\alpha_{i1}} \cdots (c_q u_q)^{\alpha_{iq}} = c_1^{\alpha_{i1}} \cdots c_q^{\alpha_{iq}} u_1^{\alpha_{i1}} \cdots u_q^{\alpha_{iq}} = c^{a_i} v_i$$

so that $\alpha(c) := (c^{a_1}, \ldots, c^{a_q})$ defines a homomorphism $T \longrightarrow T'$ with respect to which Ψ is equivariant. □

Example 1. Let $\sigma := \text{pos}\{e_1, e_1 + 2e_2\}$, $\sigma' := \text{pos}\{e_1, e_1 + 4e_2\}$, and let L_0 be defined by $L_0(e_1) = e_1$, $L_0(e_1 + 2e_2) = e_1 + 4e_2$, so $L_0(\sigma) \subset \sigma'$ holds. Then, L_0 can be represented by the symmetric matrix $A := \begin{pmatrix} 1 & 0 \\ 0 & 2 \end{pmatrix}$, and L_0^* by the transposed matrix $A^t = A$.

The vectors $a_1 = 2e_1 - e_2$, $a_2 = e_2$ and $a_3 = e_1$ generate $\check{\sigma} \cap \mathbb{Z}^2$, and $b_1 = 4e_1 - e_2$, $b_2 = e_2$ and $b_3 = e_1$ generate $\check{\sigma}' \cap \mathbb{Z}^2$ (Figure 12). Then,

$$L_0^*(b_1) = 2a_1, \qquad L_0^*(b_2) = 2a_2, \qquad L_0^*(b_3) = a_3,$$

so that in the corresponding affine coordinates $u_i = z^{a_i}$ of $X_{\check{\sigma}}$ and $v_i = z^{b_i}$ of $X_{\check{\sigma}'}$, for $i = 1, 2, 3$ (with $u_1 u_2 = u_3^2$ and $v_1 v_2 = v_3^4$),

$$(u_1, u_2, u_3) \longmapsto (u_1^2, u_2^2, u_3).$$

The map is surjective but not injective.

Example 2. Let Σ be the complete fan in $\mathbb{R}^2$ with generators $e_1, e_2, -2e_1 - e_2$, $-e_1 - e_2$, and let Σ' be the fan of $\mathbb{P}^2$ with generators $e_1, e_2, -e_1 - e_2$. We consider $L_0 = \text{id} = L_0^*$.

The toric variety X_Σ is covered by three affine planes $\mathbb{C}^2$, together with $X_{\check{\sigma}}$, defined by the nonregular cone $\sigma := \text{pos}\{e_2, -2e_1 - e_2\}$ (Figure 13). We note

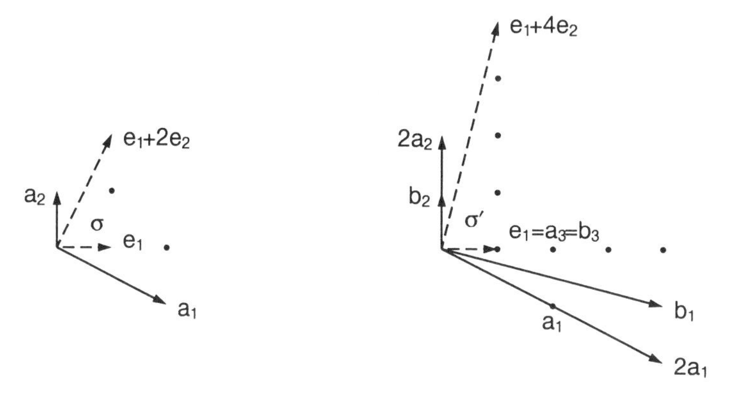

FIGURE 12.

that $X_{\check{\sigma}}$ is the affine quadratic cone, see Example 3 of section 2. We study the restriction of $\Psi = \bar{L}$ onto $X_{\check{\sigma}}$. In addition to $a_1 = -e_1 + 2e_2$, $a_2 = -e_1$, we need a third generator $a_3 = -e_1 + e_2$ for $\check{\sigma} \cap \mathbb{Z}^2$, whereas $b_1 = -e_1 + e_2$ and $b_2 = -e_1$ generate $\check{\sigma}' \cap \mathbb{Z}^2$. Now, $b_1 = a_3$ and $b_2 = a_2$ so that, for $u_i := z^{a_i}$, $i = 1, 2, 3$, the affine toric morphism $\Psi_\sigma := \Psi|_{X_{\check{\sigma}}}$ is given by

$$(u_1, u_2, u_3) \quad \longmapsto \quad (u_3, u_2)$$

and is a toric morphism of the quadratic cone $X_{\check{\sigma}}$ defined by $u_1 u_2 = u_3^2$ into the affine coordinate plane $X_{\check{\sigma}'} \cong \mathbb{C}^2$ of $\mathbb{P}^2$. The line $\{(u_1, 0, 0)\}$ is mapped onto one point $(0, 0)$, and the points $(u_3, 0)$ for which $u_3 \neq 0$ have no inverse image. Therefore, Ψ_σ is neither injective nor surjective. (Ψ, however, is surjective.)

6.5 Definition. Let $\Sigma = \Sigma' \cdot \Sigma''$ be the join of two fans Σ', Σ'' (see III, 1.12 to 1.14) such that

(a) Σ' is contained in a k-dimensional subspace U of $\mathbb{R}^n$, $0 < k < n$,

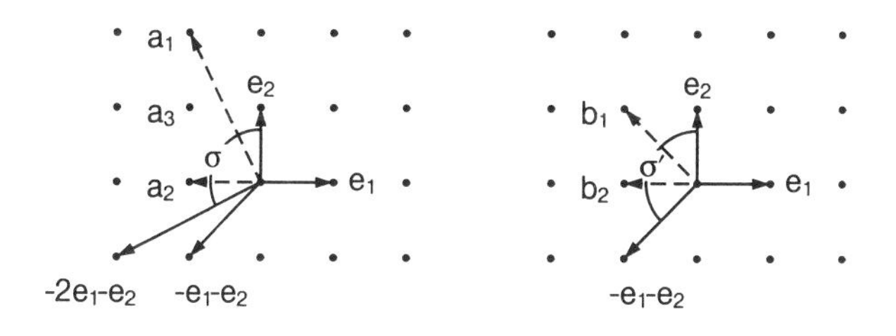

FIGURE 13.

(b) Σ'' can be projected bijectively onto a fan Σ_0 lying in the orthogonal complement $U^\perp$ of U.

Then, we call Σ_0 a *projection fan* of Σ *perpendicular* to Σ', and say that Σ has a projection fan (with respect to Σ', Σ'').

Remark. The following example shows that $\Sigma = \Sigma' \cdot \Sigma''$ need not have a projection fan with respect to a given decomposition.

Let $\Sigma' := \{\mathbb{R}_{\geq 0}\, e_1, \mathbb{R}_{\geq 0}\, e_2, \{0\}\}$, $\Sigma'' := \{\mathbb{R}_{\geq 0}\, e_3, \mathbb{R}_{\geq 0}(-e_1 - e_2 + e_3), \{0\}\}$. Then, the join $\Sigma = \Sigma' \cdot \Sigma''$ exists, but the projection of Σ'' onto $\mathbb{R}\, e_3$ is not injective. However,

6.6 Lemma. *If* $\Sigma = \Sigma' \cdot \Sigma''$ *and* Σ' *is complete (relative to* $U = \operatorname{lin} |\Sigma'|$*), then,* Σ *has a projection fan* Σ_0 *(in* $U^\perp$*).*

PROOF. Let π be the perpendicular projection onto $U^\perp$. Suppose, for $\sigma'', \tau'' \in \Sigma''$, $\sigma'' \cap \tau'' = \{0\}$, we have $\pi(\sigma'') \cap \pi(\tau'') \ni a \neq 0$. Let $b \in \pi^{-1}(a) \cap \sigma''$ and $c \in \pi^{-1}(a) \cap \tau''$. Then, the line g through b, c is parallel to U, and the linear hull of g is a plane, which intersects U in a line g'. If ϱ_1, ϱ_2 are the two rays of g' emanating from 0, $\varrho_1 + \mathbb{R}_{\geq 0}\, b \subset \sigma' \cdot \sigma''$, $\varrho_1 + \mathbb{R}_{\geq 0}\, c \subset \sigma' \cdot \tau''$ for some $\sigma' \in \Sigma'$ and $\varrho_2 + \mathbb{R}_{\geq 0}\, b \subset \tau' \cdot \sigma''$, $\varrho_2 + \mathbb{R}_{\geq 0}\, c \subset \tau' \cdot \tau''$ for some $\tau' \in \Sigma'$ (σ', τ' exist because of the completeness of Σ'). Since $\varrho_1 + \mathbb{R}_{\geq 0}\, b$, $\varrho_1 + \mathbb{R}_{\geq 0}\, c$, $\varrho_2 + \mathbb{R}_{\geq 0}\, b$, $\varrho_2 + \mathbb{R}_{\geq 0}\, c$ all lie in a half-plane, one of the first two sets intersects one of the second two outside U, so that $[(\sigma' \cdot \sigma'') \cap (\tau' \cdot \tau'')] \setminus U \neq \emptyset$ or $[(\sigma' \cdot \tau'') \cap (\tau' \cdot \sigma'')] \setminus U \neq \emptyset$, contrary to the definition of "join". □

6.7 Theorem. *Let* $\Sigma, \Sigma', \Sigma''$ *be regular fans in* $\mathbb{R}^n$ *such that* $\Sigma = \Sigma' \cdot \Sigma''$*, and let* Σ_0 *be the projection fan of* Σ *perpendicular to* $\operatorname{lin} |\Sigma'|$*. Then, the projection* $\pi : \Sigma \longrightarrow \Sigma_0$ *induces a map of fans such that, for any* $\sigma_0 \in \Sigma_0$*, we have an isomorphism*

$$\bar{\pi}^{-1}(X_{\check{\sigma}_0}) \cong X_{\Sigma'} \times X_{\check{\sigma}_0}. \tag{5}$$

(Here $\check{\sigma}_0$ *is the dual cone of* σ_0 *relative to* $\operatorname{lin} \sigma_0$*, and* $\bar{\pi}$ *is defined according to Theorem 6.1 and Definition 6.2). If* $\Sigma_0 = \Sigma''$ *(up to a unimodular transformation) we obtain, in particular, an isomorphism*

$$X_\Sigma \cong X_{\Sigma'} \times X_{\Sigma''}.$$

6.8 Definition. We call $\bar{\pi}$ a *fibration* of X_Σ with *base space* X_{Σ_0} and *fibers* $\bar{\pi}^{-1}(p_0) \cong \{p\} \times X_{\Sigma'} \cong X_{\Sigma'}$ for any $p_0 \in X_{\Sigma_0}$ and a $p \in \bar{\pi}^{-1}(p_0)$. Then, $X_{\Sigma'}$ is said to be the *typical fiber* of the fibration. In short, we say that X_Σ is an $X_{\Sigma'}$*-fiber bundle* over X_{Σ_0}. In the case $X_\Sigma \cong X_{\Sigma'} \times X_{\Sigma''}$, we call the fibration *trivial.*

PROOF OF THEOREM 6.7. By Definition 6.5, each $\sigma'' \in \Sigma''$ is injectively mapped by π onto a $\sigma_0 \in \Sigma_0$, whereas $\pi(\sigma') = \{0\}$ for every $\sigma' \in \Sigma'$. So we

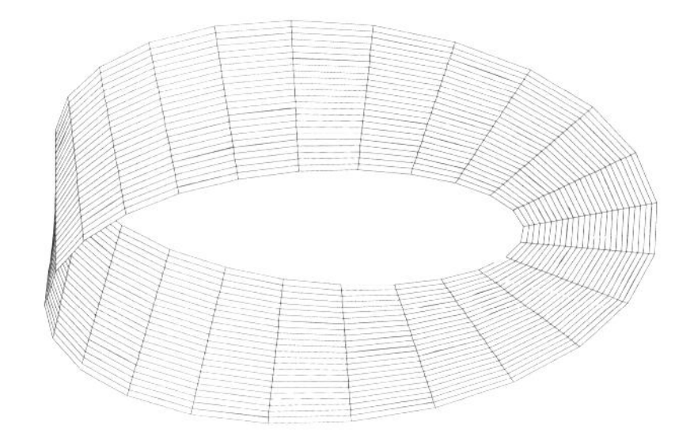

FIGURE 14.

find $\pi(\sigma' \cdot \sigma'') = \sigma_0$. Since $\pi(\mathbb{Z}^n) \subset \mathbb{Z}^n$ the linear map π induces a map of fans which we denote also by π:

$$\pi : \Sigma \longrightarrow \Sigma_0.$$

We consider the induced toric morphism

$$\bar{\pi} : X_\Sigma \longrightarrow X_{\Sigma_0}.$$

From $X_{(\sigma' \cdot \sigma'')^\vee} \cong X_{\check{\sigma}'} \times X_{\check{\sigma}''}$ and $X_{\check{\sigma}''} \cong X_{\check{\sigma}_0}$ (dual cones $\check{\sigma}'$, $\check{\sigma}''$, $\check{\sigma}_0$, relative to appropriate lower dimensional spaces), we find that

$$\bar{\pi}|_{X_{(\sigma' \cdot \sigma'')^\vee}} : X_{(\sigma' \cdot \sigma'')^\vee} \longrightarrow X_{\check{\sigma}_0}.$$

For a fixed σ'', this is true for each $\sigma' \in \Sigma'$, so that

$$\bar{\pi}^{-1}(X_{\check{\sigma}_0}) \cong X_{\Sigma'} \times X_{\check{\sigma}_0}.$$

This readily implies the theorem. □

Example 3. Let Σ be the fan of a Hirzebruch surface $\mathcal{H}_k$ (Example 3 in section 3). Then, $\Sigma = \Sigma' \cdot \Sigma''$, where Σ' covers a line, so that $X_{\Sigma'}$ is a projective line. Therefore all fibers are projective lines, "twisted" in X_Σ, and $\mathcal{H}_k$ is a $\mathbb{P}^1$-fiber bundle over $\mathbb{P}^1$.

We illustrate the real points of $\mathcal{H}_1$ as a closed Möbius strip (Figure 14).

We denote the real projective plane by $\mathbb{P}^2_{(r)}$, represent it as a circular disc with diametrically opposite points identified, and identify $\mathbb{P}^1_{(r)}$ (real projective line) with a circle so that $\mathbb{P}^2_{(r)} \times \mathbb{P}^1_{(r)}$ is represented by a full torus in $\mathbb{R}^3$. By the equation

$$\eta_0\zeta_0 = \eta_1\zeta_1,$$

a twisted band in $\mathbb{P}^2_{(r)} \times \mathbb{P}^1_{(r)}$ is given which is a (projective) Möbius strip. The fibers are the lines (represented by line segments) of which the Möbius strip is composed. The real part of $X_{\Sigma'} \times X_{\check{\sigma}''}$ is obtained from the Möbius strip by leaving out one of the fibers, that is, "cutting the Möbius strip" so that it becomes an ordinary strip, the cartesian product of an affine line and a projective line.

Exercises

1. Let $\Sigma := \{\text{pos}\{e_1, e_1 + 3e_2\}, \text{pos}\{e_1 + 3e_2, e_2\}, \text{pos}\{e_2, -e_1\}, \text{pos}\{-e_1, -e_2\}, \text{pos}\{-e_2, e_1\}, \text{faces}\}$ and let L be the perpendicular projection onto $\mathbb{R}\, e_2$. Find $\bar{L}$ explicitly.
2. Given $\mathbb{P}^n$ as toric variety and $L = 2 \cdot I$, I the identity map, describe $\bar{L}$ in projective coordinates.
3. If a point p is deleted from a projective plane $\mathbb{P}^2$, a fibration of $\mathbb{P}^2 \setminus \{p\}$, with a projective line as base space exists. Find a fan for $\mathbb{P}^2 \setminus \{p\}$ and describe the fibration according to Theorem 6.11.
4. Find the three-dimensional analog of Example 3 (Σ' two-dimensional).

7. Blowups and blowdowns

We are now going to study a special case of a toric morphism and its inverse operation.

7.1 Definition. Let Σ be a regular fan, and let $s(p;\sigma)$ be a regular stellar subdivision of Σ (V, Definition 6.1). Then, the toric morphism

$$\Psi_\sigma : X_{s(p;\sigma)\Sigma} \longrightarrow X_\Sigma$$

induced by the identity map I, $\Psi_\sigma = \bar{I}$ (see Theorem 6.1), is called an *equivariant blowdown* or in short a *blowdown* of $X_{s(p;\sigma)\Sigma}$. The inverse operation Ψ_σ^{-1} is said to be an *(equivariant) blowup* of X_Σ.

Example 1. Let $\sigma := \text{pos}\{e_1 - e_2, e_2\}$ be a cone of some regular fan Σ in $\mathbb{R}^2$. By $s(e_1;\sigma)$, we decompose σ into $\sigma_1 = \text{pos}\{e_1 - e_2, e_1\}$ and $\sigma_2 = \text{pos}\{e_1, e_2\}$. Then, $\check{\sigma} = \text{pos}\{e_1, e_1 + e_2\}$, $\check{\sigma}_1 = \text{pos}\{e_1 + e_2, -e_2\}$, and $\check{\sigma}_2 = \text{pos}\{e_1, e_2\}$ (Figure 15).

We set $u_1 := z^{e_1} = z_1$, $u_2 := z^{e_1+e_2} = z_1 z_2$, $u_1' := z^{-e_2} = z_2^{-1}$, $u_2' := z^{e_1+e_2} = z_1 z_2 (= u_2)$, $u_1'' := z^{e_1} = z_1 (= u_1)$, $u_2'' := z^{e_2} = z_2$.

The two affine planes $X_{\check{\sigma}_1} = \{(u_1', u_2')\}$, $X_{\check{\sigma}_2} = \{(u_1'', u_2'')\}$ are mapped under the blowdown Ψ into the affine plane $X_{\check{\sigma}} = \{(u_1, u_2)\}$:

$$\begin{array}{ccccccc} X_{\check{\sigma}_1} & \longrightarrow & X_{\check{\sigma}} & \qquad & X_{\check{\sigma}_2} & \longrightarrow & X_{\check{\sigma}} \\ (u_1', u_2') & \longmapsto & (u_1' u_2', u_2') & & (u_1'', u_2'') & \longmapsto & (u_1'', u_1'' u_2''). \end{array}$$

The projective line with charts $\{(u_1', 0)\} = \{(z_2^{-1}, 0)\}$, $\{(0, u_2'')\} = \{(0, z_2)\}$ is mapped onto the point $(0, 0)$. For $(u_1, u_2) \neq (0, 0)$, the blowdown is bijective. We illustrate the blowdown for $X_{\check{\sigma}_1}$ in Figure 16. The line $u_2' = 0$ which is projected onto $(0, 0)$ lies on a hyperbolic paraboloid.

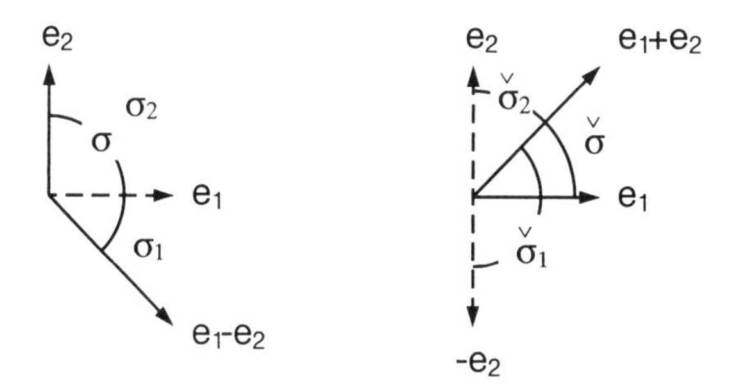

FIGURE 15.

7.2 Theorem. *Let X_Σ be a regular toric variety, and let X_{Σ_0} be an invariant toric subvariety defined by the star* $\mathrm{st}(\sigma;\,\Sigma) \approx \Sigma_0$ *of σ in Σ; $1 < k := \dim \sigma \leq n$.*

(a) *Under the blowup Ψ_σ^{-1}, any point x of X_{Σ_0} is replaced by a $(k-1)$-dimensional projective space.*

(b) *The blowdown Ψ_σ is a toric morphism which is bijective outside $\Psi_\sigma^{-1}(X_{\Sigma_0})$.*

PROOF. Let $\sigma = \mathrm{pos}\{a_1, \ldots, a_k\}$, and let $p := a_1 + \cdots + a_k$. First, we assume that $\dim \bar{\sigma} = n$ for a $\bar{\sigma} \in \mathrm{st}(\sigma;\,\Sigma)$, $\bar{\sigma} = \mathrm{pos}\{a_1, \ldots, a_k, a_{k+1}, \ldots, a_n\}$. Then, by the regular stellar subdivision in direction p, $\bar{\sigma}$ is split into n-dimensional cones $\sigma_i := \mathrm{pos}\{a_1, \ldots, a_{i-1}, p, a_{i+1}, \ldots, a_k, \ldots, a_n\}$, $i = 1, \ldots, k$. By V, Lemma

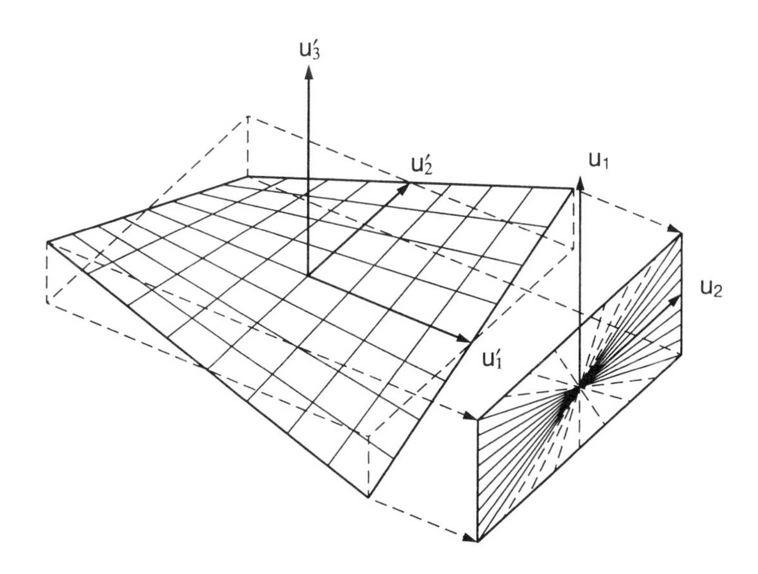

FIGURE 16.

6.3, we can set

$$\check{\bar{\sigma}} = \text{pos}\{b_1, \dots, b_k, \dots, b_n\},$$

and

$$\check{\sigma}_i = \text{pos}\{b_1 - b_i, \dots, b_{i-1} - b_i, b_i, b_{i+1} - b_i, \dots, b_k - b_i, b_{k+1}, \dots, b_n\},$$

where $i = 1, \dots, k$. Therefore, $X_{\check{\bar{\sigma}}}$ has coordinates $(u_1, \dots, u_n) = (z^{b_1}, \dots, z^{b_n}\}$ and $X_{\check{\sigma}_i}$ has coordinates $(v_1, \dots, v_n) = (z^{b_1 - b_i}, \dots, z^{b_{i-1}-b_i}, z^{b_i}, z^{b_{i+1}-b_i}, \dots, z^{b_k - b_i}, z^{b_{k+1}}, \dots, z^{b_n})$. $\Psi_\sigma|X_{\check{\sigma}_i}$ is given by the equations

$$\begin{aligned} u_1 &= v_1 v_i, \dots, u_{i-1} = v_{i-1} v_i, u_i = v_i, \\ u_{i+1} &= v_{i+1} v_i, \dots, u_k = v_k v_i, u_{k+1} = v_{k+1}, \dots, u_n = v_n. \end{aligned}$$

Its inverse is defined for $v_i \neq 0$. If $v_i = 0$, the images of all points $(v_1, \dots, v_{i-1}, 0, v_{i+1}, \dots, v_k, \mathring{v}_{k+1}, \dots, \mathring{v}_n)$ are $(0, \dots, 0, \mathring{v}_{k+1}, \dots, \mathring{v}_n)$, for any fixed set $\mathring{v} := \{\mathring{v}_{k+1}, \dots, \mathring{v}_n\}$ of coordinates, that is, the fiber $X_0^{(i)}$ of $\Psi_\sigma|X_{\check{\sigma}_i}$ above $(0, \dots, 0, \mathring{v}_{k+1}, \dots, \mathring{v}_n)$ is an affine $(k-1)$-space $X_{\mathring{v}}^{(i)}$. Any two such $X_{\mathring{v}}^{(i)}$ are isomorphic to each other and to $X_0^{(i)}$. But the $X_0^{(i)}$, $i = 1, \dots, k$, can be considered as the affine charts of the projective $(k-1)$-space defined by generators $b_1 - b_2, b_2 - b_3, \dots, b_{k-1} - b_k, b_k - b_1$ they all lie in $(\text{pos}\{p\})^\perp$ and have sum 0.

If the maximal dimension of a cone $\bar{\sigma} \in \text{st}(\sigma; \Sigma)$ is less than n, say $r = \dim \bar{\sigma} < n$, $\check{\bar{\sigma}}$ has a cospan $\neq \{0\}$. We define the σ_i as above, replacing n by r, $i = 1, \dots, k$. Then,

$$\text{cospan}\, \check{\bar{\sigma}} = \text{cospan}\, \check{\sigma}_1 = \cdots = \text{cospan}\, \check{\sigma}_k,$$

so that $X_{\check{\bar{\sigma}}}, X_{\check{\sigma}_1}, \dots, X_{\check{\sigma}_k}$ have the same coordinates u_j for $j > k$. Hence, all arguments above can be applied. □

In Example 1, $n = k = 2$, hence, $\bar{\sigma} = \sigma$, and the σ_i are those introduced there.

Example 2. Suppose $\sigma = \bar{\sigma} = \text{pos}\{e_1, e_2, e_3\}$ in $\mathbb{R}^3$ (see V, 6, Figure 23a). Independent of a fan to which σ belongs, we obtain a blowup in $(0, 0, 0)$ of $X_{\check{\bar{\sigma}}} = \mathbb{C}^3$ in which $(0, 0, 0)$ is replaced by a projective plane. The fan of the projective plane has generators $e_1 - e_2, e_2 - e_3, e_3 - e_1$. The affine charts of the projective plane have coordinates $(z_1 z_2^{-1}, z_2 z_3^{-1})$, $(z_2 z_3^{-1}, z_3 z_1^{-1})$, and $(z_3 z_1^{-1}, z_1 z_2^{-1})$.

Example 3. Let $\sigma := \text{pos}\{e_1, e_3\}$ in $\mathbb{R}^3$ (see V, 6, Figure 23b). If for some fan Σ, $\bar{\sigma} = \text{pos}\{e_1, e_3, e_2\} \in \text{st}(\sigma, \Sigma)$, we set $\sigma_1 := \text{pos}\{e_1 + e_3, e_3, e_2\}$, $\sigma_2 := \text{pos}\{e_1, e_1 + e_3, e_2\}$. Then $\check{\bar{\sigma}} = \text{pos}\{e_1, e_3, e_2\}$, $\check{\sigma}_1 = \text{pos}\{e_1, e_3 - e_1, e_2\}$, $\check{\sigma}_2 = \text{pos}\{e_1 - e_3, e_3, e_2\}$. As coordinates for $X_{\check{\sigma}_1}$, we obtain $(v_1, v_2, v_3) = (z^{e_1}, z^{e_3 - e_1}, z^{e_2})$, and, for $X_{\check{\sigma}_2}$, $(v_1', v_2', v_3') = (z^{e_1 - e_3}, z^{e_3}, z^{e_2})$. If $(u_1, u_2, u_3) = (z^{e_1}, z^{e_3}, z^{e_2})$ are the coordinates of $X_{\check{\bar{\sigma}}}$, the blowdown Ψ_σ induces in $X_{\check{\sigma}_1}$ the map $(v_1, v_2, v_3) \mapsto (v_1, v_1 v_2, v_3)$, which is bijective for $v_1 \neq 0$ and maps $\{(0, v_2, v_3)\}$ onto $\{(0, 0, v_3)\}$. Similarly, Ψ_σ induces in $X_{\check{\sigma}_2}$ the map $(v_1', v_2', v_3') \mapsto$

$(v_1'v_2', v_2', v_3')$, which again is bijective for $v_2' \neq 0$ and maps $\{(v_1', 0, v_3')\}$ onto $(0, 0, v_3')\}$. Since $v_2 = v_1'^{-1}$ and $v_3 = v_3'$ for any fixed $v_3 = v_3' = p$, the sets $\{(0, v_2, p)\}$ and $\{(v_1', 0, p)\}$ can be considered charts of a projective line which is mapped under Ψ_σ onto $(0, 0, p)$.

In the above arguments, if e_2 is replaced by $-e_2$, we see that, for p^{-1} instead of p, the inverse image of $(0, 0, p^{-1})$ is also a projective line. Since $\{(0, 0, p)\}$, $\{(0, 0, p^{-1})\}$ are affine charts of the projective line X_{Σ_0}, we obtain an illustration of case (b) of Theorem 7.2.

If σ bounds a second 3-cone $\bar{\bar{\sigma}} = \text{pos}\{e_1, e_3, b\}$, then, clearly, $\langle b, e_2 \rangle < 0$, so that, in the above arguments, e_2 is replaced by $-e_2$. As a result, we find that, under the blowup Ψ_σ^{-1}, any point of the projective line with charts $\{(0, 0, v_3)\}$, $\{(0, 0, v_3^{-1})\}$ is replaced by a projective line.

Now, we will translate the results on fans, which we collected in V, 6, into the language of toric varieties.

7.3 Definition. We call X_Σ *complete* if Σ is complete, that is, covers all of $\mathbb{R}^n$.

Remark. In section 9, we shall see that "complete" is equivalent to "compact".

From V, Theorem 6.5,

7.4 Theorem. *Let X_Σ, $X_{\Sigma'}$ be regular, complete, two-dimensional toric varieties. Then there exist equivariant blowups $\Psi_1^{-1}, \dots, \Psi_p^{-1}, \Psi_1'^{-1}, \dots, \Psi_q'^{-1}$, such that*

$$X_\Sigma \underset{\Psi_1^{-1}}{\longrightarrow} \cdots \underset{\Psi_p^{-1}}{\longrightarrow} X_{\Sigma''} \underset{\Psi_q'^{-1}}{\longleftarrow} \cdots \underset{\Psi_1'^{-1}}{\longleftarrow} X_{\Sigma'}$$

where $X_{\Sigma''}$ is again a regular, complete, two-dimensional toric variety.

V, Theorem 6.6 implies

7.5 Theorem. *Any regular, complete, two-dimensional toric variety can be successively blowndown into either a Hirzebruch surface $\mathcal{H}_k$, $k \neq \pm 1$, or into $\mathbb{P}^2$. Hence, there exist equivariant blowdowns $\Psi_1, \dots, \Psi_r, \Psi_1', \dots, \Psi_s'$ such that either*

$$X_\Sigma \underset{\Psi_1}{\longrightarrow} \cdots \underset{\Psi_r}{\longrightarrow} \mathcal{H}_k, \quad k \neq \pm 1, \qquad \text{or} \quad X_\Sigma \underset{\Psi_1'}{\longrightarrow} \cdots \underset{\Psi_s'}{\longrightarrow} \mathbb{P}^2.$$

Remark. In the case $k = \pm 1$, $\mathcal{H}_k$ can be further blowndown to $\mathbb{P}^2$.

Oda's conjecture (strong version). Theorem 7.4 is also true for regular, complete, three-dimensional toric varieties.

Remark. A weak version of Oda's conjecture has been shown for arbitrary dimension. Given any two regular, complete, n-dimensional toric varieties X_Σ, $X_{\Sigma'}$, there exists a sequence of operations, that are either blowups or blowdowns and transform X_Σ into $X_{\Sigma'}$ (see Appendix to this section).

Exercises

1. Transform any Hirzebruch surface into $\mathbb{P}^2$ by an alternating sequence of equivariant blowups and blowdowns.
2. Translate the stellar subdivisions, in the example illustrated in Figure 26 of V, 6, into equivariant blowups, where the left upper fan is given by V, 4, Example 2, and the right upper fan is one of a projective 3-space. Present equations for all blowdowns.
3. Let Σ consist of one regular n-cone σ and the faces of σ. Any equivariant blowup Ψ^{-1} gives rise to a fibration $\Psi^{-1}(X_{\Sigma_0}) \longrightarrow X_{\Sigma_0}$ where X_{Σ_0} is a toric subvariety of X_Σ.
4. By using Farey's lemma generalized to higher dimensions (V, 6, Exercise 4), prove *De Concini–Procesi's Theorem*: Given regular fans Σ, Σ' which have the same point sets $|\Sigma| = |\Sigma'|$, there exists a fan Σ'' obtained from Σ by a finite sequence of regular stellar subdivisions such that each cone of Σ'' is contained in a cone of Σ'. In terms of toric varieties, there exist equivariant blowups $\Psi_1^{-1}, \ldots, \Psi_q^{-1}$ and a toric morphism ϕ such that

$$X_\Sigma \xrightarrow[\Psi_1^{-1}]{} \cdots \xrightarrow[\Psi_q^{-1}]{} X_{\Sigma''} \xleftarrow[\phi]{} X_{\Sigma'}.$$

8. Resolution of singularities

If an n-dimensional toric variety X_Σ is regular, it is composed of pieces $\mathbb{C}^k \times \mathbb{C}^{*n-k}$, and hence, in whichever sense of the word, nonsingular. In the case $n = 1$, X_Σ is either an affine line or a projective line or $\mathbb{C}^*$. Therefore,

8.1 Lemma. *Any one-dimensional toric variety is regular.*

If $n = 2$, the simplest case of a "singularity" is that of the apex of a quadratic conical surface (equation $u_1u_2 = u_3^2$; see section 2). In its fan, a cone $\sigma = \operatorname{pos}\{e_1, e_1 + 2e_2\}$ occurs which has determinant equal neither to 1 nor to -1. The size of $|\det \sigma|$ is, in a way, a measure of how bad a singularity is. In higher dimensions, a similar observation can be made for n-dimensional simplex cones σ (which may have arbitrarily large $|\det \sigma|$).

We are interested in resolving singularities. Before this we will give a pragmatic definition of singularity.

8.2 Definition. A point $x \in X_\Sigma$ is called *singular* or a *singularity* of X_Σ if some affine chart $X_{\check\sigma}$, $\sigma \in \Sigma$, to which it belongs, is not of the type $\mathbb{C}^k \times \mathbb{C}^{*n-k}$ according to Theorem 2.12.

8.3 Lemma. *If a singularity x of X_Σ lies in an orbit T_k of X_Σ, according to Theorem 5.8, then, all points of the orbit T_k are singularities. The big torus does not contain singularities.*

PROOF. If one point of T_k lies in $X_{\check{\sigma}}$, so do all points. Therefore, the lemma is true by the definition of a singularity. Since $T = \mathbb{C}^{*n}$, there is no singularity in T. □

8.4 Definition. Let X_Σ be a toric variety which has singularities, and let $X_{\Sigma'}$ be a regular toric variety for which $|\Sigma| = |\Sigma'|$, and a toric morphism

$$\Psi : X_{\Sigma'} \longrightarrow X_\Sigma$$

exists isomorphic on the tori. Then, we call Ψ a *resolution* of (the singularities of) X_Σ.

8.5 Theorem. *Any toric variety X_Σ with singularities possesses a resolution Ψ. We may choose Ψ as composed of morphisms $\Psi_1, \ldots, \Psi_q$ which stem from stellar subdivisions $s_1, \ldots, s_q$:*

$$\begin{array}{ccccc} \Sigma' & \longleftarrow_{s_q} & \cdots & \longleftarrow_{s_1} & \Sigma \\ \vdots & & & & \uparrow \\ \downarrow & & & & \vdots \\ X_{\Sigma'} & \longrightarrow_{\Psi_q} & \cdots & \longrightarrow_{\Psi_1} & X_\Sigma. \end{array}$$

PROOF. First, by stellar subdivisions, we turn Σ into a simplicial fan Σ' (compare V, Theorem 4.2). Let $\sigma \in \Sigma'$ be a nonregular maximal cone, $\dim \sigma = k$. We apply induction on k and assume that a $(k-1)$-face σ_0 of σ has been made regular by stellar subdivisions of Σ. Thus, σ is split into k-simplices. Let τ be one of them, $\tau = \tau_0 + \varrho$, where $\tau_0 \subset \sigma_0$ and τ_0 is regular. Up to a unimodular transformation, we can assume $\tau_0 = \operatorname{pos}\{e_1, \ldots, e_{k-1}\}$. Then, $\varrho = \mathbb{R}_{\geq 0}\, a$, a simple, $a = \alpha_1 e_1 + \cdots + \alpha_{k-1} e_{k-1} + \alpha e_k$ (in $\mathbb{R}^k$ spanned by τ), $\alpha_1, \ldots, \alpha_{k-1}, \alpha \in \mathbb{Z}_{\geq 0}$.

If $\alpha = 1$, τ is regular. If $\alpha > 1$, there exists a lattice point $b = \beta_1 e_1 + \cdots + \beta_{k-1} e_{k-1} + \frac{1}{\alpha} a$, $0 \leq \beta_i < 1$, $i = 1, \ldots, k-1$. We obtain

$$\alpha' := |\det(e_1, \ldots, e_{i-1}, b, e_{i+1}, \ldots, e_{k-1}, a)| < \alpha = \det(e_1, \ldots, e_{k-1}, a).$$

We apply to Σ the stellar subdivision $s(b; \Sigma)$ in direction b. All k-dimensional cones affected by $s(b; \Sigma)$ split into cones with smaller determinant of generators. Hence, after a finite number of steps, we end up with only regular cones. □

Example 1. Let $\sigma := \operatorname{pos}\{e_1, e_1 + 2e_2\}$ belong to any fan Σ; $\check{\sigma} = \operatorname{pos}\{e_2, 2e_1 - e_2\}$. The coordinates of $X_{\check{\sigma}}$ are $(u_1, u_2, u_3) = (z^{e_2}, z^{2e_1 - e_2}, z^{e_1})$, so that $u_1 u_2 = u_3^2$ is the equation of the conical surface. We subdivide σ in direction $\mathbb{R}_{\geq 0}(e_1 + e_2)$. Let $\sigma_1 := \operatorname{pos}\{e_1, e_1 + e_2\}$, $\check{\sigma}_1 := \operatorname{pos}\{e_2, e_1 - e_2\}$. Then, we may set $X_{\check{\sigma}_1} =$

$\{(v_1, v_2)\} = \{(z^{e_2}, z^{e_1 - e_2})\}$, and Ψ_1 is given by

$$(u_1, u_2, u_3) = (v_1, v_1 v_2^2, v_1 v_2).$$

So, first, we map each point (v_1, v_2) onto the point $(v_1, v_2, v_1 v_2)$ of a regular quadratic surface, and then project onto the conical surface $\{(v_1, v_1 v_2^2, v_1 v_2)\}$. Hereby, the line $\{(0, v_2)\}$ is mapped, first, onto the line $\{(0, v_2, 0)\}$ and then onto the point $(0, 0, 0)$, whereas the lines $\{(c, v_2)\}$ for $c \neq 0$ are mapped onto the lines $\{(c, v_2, cv_2)\}$ and, then, onto the parabolas $\{(c, cv_2^2, cv_2)\}$ (Figure 17).

The mapping Ψ_1 in Example 1 is based on a stellar subdivision and looks much like a blowdown (which it is not, since $X_{\check{\sigma}}$ is not regular). In fact, we can look at Ψ_1 as induced by a blowdown, if we consider the affine space $\mathbb{C}^3$, in which $X_{\check{\sigma}}$ is embedded, as an affine toric variety X_τ, $\tau = \check{\tau} = \text{pos}\{e_1, e_2, e_3\}$. By the regular stellar subdivision in direction $e_1 + e_2$, we define a blowup Ψ^{-1} of $X_{\check{\tau}}$. One of the charts of $\Psi^{-1}(X_{\check{\tau}})$ is given by $X_{\check{\sigma}_0}$ where $\sigma_0 = \text{pos}\{e_1, e_1 + e_2, e_3\}$, and, hence, $\check{\sigma}_0 = \text{pos}\{e_2, e_1 - e_2, e_3\}$. The coordinates of $X_{\check{\sigma}_0}$ are, then, $(v_1, v_2, v_3) = (z^{e_2}, z^{e_1 - e_2}, z^{e_3})$, so that $(u_1, u_2, u_3) = (v_1, v_1 v_2^2, v_3)$. Now, $v_3 = v_1 v_2$ again represents the above surface, this time obtained from the quadratic cone by blowing up $\mathbb{C}^3$ along $\{(0, u_2, 0)\}$.

What we have seen in Example 1 refers to a general idea of how to resolve singularities. In the example, the conical surface, we started with, and the embedding space were both toric varieties. This need not be so. In many cases, it is useful to choose the embedding space as a toric variety and embed singular nontoric varieties in it. We do not develop the general theory but illustrate it only in a further example.

Example 2. Let $\mathbb{P}^2$ be given as a toric variety (see section 3, Example 1) by the fan with 1-cones generated by $e_1, e_2, -e_1 - e_2$. The homogeneous coordinates are

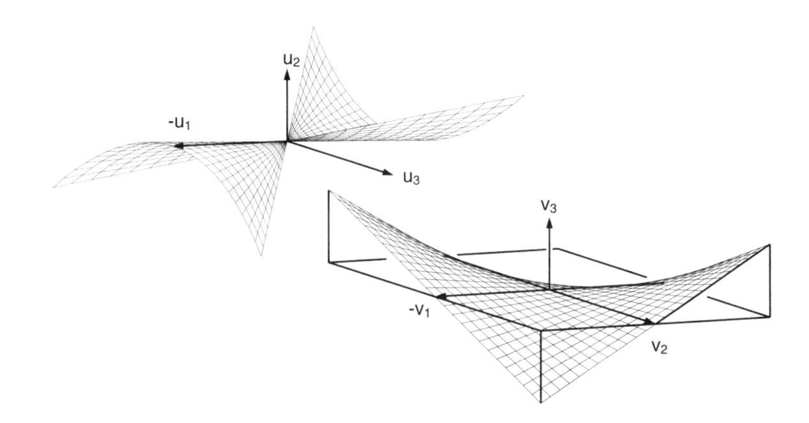

FIGURE 17.

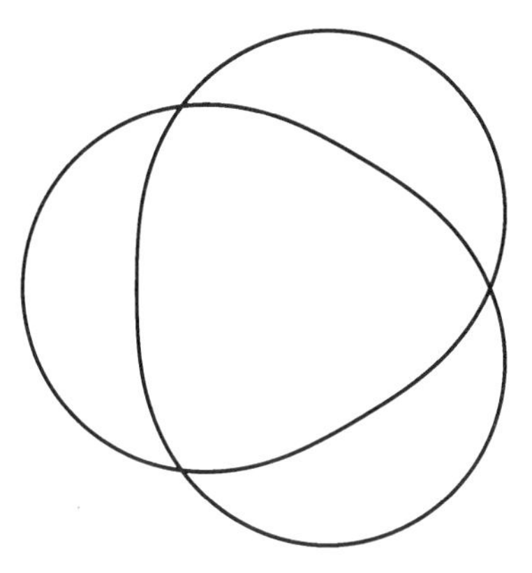

FIGURE 18.

$[\xi_1 : \xi_2 : \xi_3]$ (related to the coordinates of the affine toric charts as explained in section 3). We consider a curve given by the equation

$$\xi_1^2\xi_2^2 - \xi_1^2\xi_3^2 + \xi_2^2\xi_3^2 = 0. \tag{1}$$

It has singularities in $\mathbb{C}(1, 0, 0)$, $\mathbb{C}(0, 1, 0)$, and $\mathbb{C}(0, 0, 1)$, that is, in the zeros of the three affine charts of $\mathbb{P}^2$ (Figure 18 provides a qualitative picture). Here "singularity" can be understood as in the above definition although the curve is not a toric variety (by Lemma 8.1).

As in Example 1 of section 3, we write the affine charts as $A_0 = \{(z^{e_1}, z^{e_2})\} = \{(z_1, z_2)\}$, $A_1 = \{(z^{-e_1}, z^{-e_1+e_2})\} = \{(z_1^{-1}, z_1^{-1}z_2)\}$, $A_2 = \{(z^{e_1-e_2}, z^{-e_2})\} = \{(z_1z_2^{-1}, z_2^{-1})\}$.

We blow up 0 in each of the affine planes A_1, A_2, A_3, replacing 0 by a projective line $\mathbb{P}^1$. The resolution of singularities can be geometrically understood as follows.

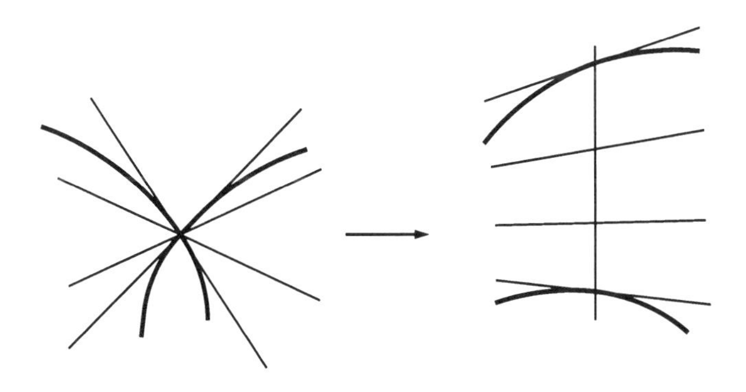

FIGURE 19.

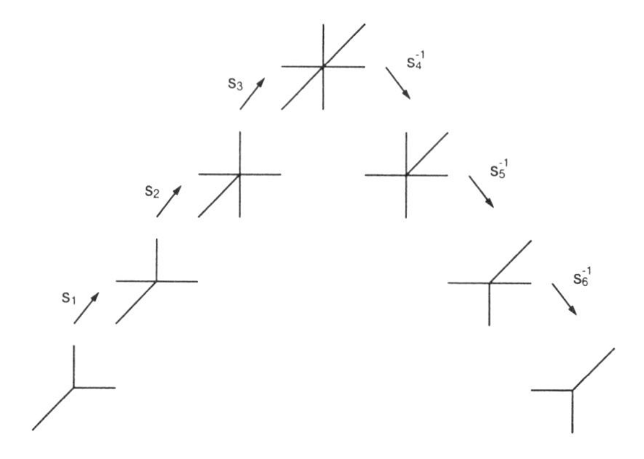

FIGURE 20.

As Figure 19 illustrates, each line of A_i passing through 0 of A_i is "lifted" under the blowup to a line of the ruled surface. So, the two tangents of the curve in 0 "take along" the point of tangency to different points of $\mathbb{P}^1$ (Figure 19). Doing this for all three charts, we obtain, from (1) a curve without self-intersection. It is contained in the surface obtained from $\mathbb{P}^2$ by the three blowups, called a *del Pezzo surface*.

In the present example, we may even go a step further and transform the new curve back into $\mathbb{P}^2$ without creating new singularities. We choose a different system of charts of $\mathbb{P}^2$, those obtained from the generators $-e_1$, $-e_2$, $e_1 + e_2$. We apply blowdowns given by the inverse stellar subdivisions $s_4^{-1}, s_5^{-1}, s_6^{-1}$, as illustrated in Figure 20 (s_1, s_2, s_3 the stellar subdivisions giving rise to the above blowups). It is readily checked that the curve obtained has (in coordinates $w_1 = z^{-e_1}$, $w_2 = z^{-e_2}$) an equation

$$1 - w_1^2 + w_2^2 = 0,$$

so that its extension in $\mathbb{P}^2$ is a projective ellipse.

Exercises

1. Consider an arbitrary complete (rational) fan Σ in $\mathbb{R}^2$ with 3 generators. Find a resolution of X_Σ.
2. Given the fan Σ as in Figure 13 of V, 4. Resolve the singularities of X_Σ.
3. In Example 2 we have described the resulting curve after the blowups and blowdowns in w_1, w_2-coordinates. Find the equations of the curve in coordinates given by the other two charts of the projective plane.

4. Consider the real part of the curve (1) in A_1 after choosing $\xi_3 = \sqrt{-1}$. Show explicitly that the curve consists of two graphs of functions $\xi_2 = f(\xi_1)$, $\xi_2 = g(\xi_1)$. Find the tangents in (0,0).

9. Completeness and compactness

A fan Σ is called complete if its cones cover $\mathbb{R}^n$. We have also called a toric variety X_Σ complete if Σ is complete (section 7). We now wish to characterize the completeness of X_Σ by topological means.

9.1 Theorem. *A toric variety X_Σ is compact if and only if Σ is complete.*

PROOF.

I. Let X_Σ be compact. By induction, we show that Σ is complete. For $n = 1$, the projective line is the only compact toric variety; its fan is complete. For $n > 1$, suppose Σ is not complete. Then there is a one-dimensional cone $\rho \in \Sigma$ on the boundary of $|\Sigma|$; so $\Sigma_0 := \pi(\mathrm{st}(\rho; \Sigma))$ is also not complete. But the subvariety X_{Σ_0} is closed in X_Σ and, hence, also compact. Therefore, by induction, Σ_0 would have to be complete, a contradiction.

II. If a sequence of points has no accumulation point, either it has a subsequence in a subvariety which also has no accumulation point or we find a subsequence $(x_i)_{i\in\mathbb{Z}_{>0}}$ in the big torus such that, for each $m \in \mathbb{Z}^n_{>0}$, either $(z^m(x_i))_{i\in\mathbb{Z}_{>0}}$ or $(z^{-m}(x_i))_{i\in\mathbb{Z}_{>0}}$ is bounded. In the former case, we assume the assertion to be true by induction. In the latter case, we let $M := \{m \mid (z^m(x_i))_{i\in\mathbb{Z}_{>0}} \text{ is bounded}\}$. Then, pos M is a cone for which $M = (\mathrm{pos}\, M) \cap \mathbb{Z}^n$ and pos $M \cup \mathrm{pos}(-M) = \mathbb{R}^n$. Hence, pos M contains a half-space, and we find a $\sigma \in \Sigma$ such that $\check{\sigma} \subset \mathrm{pos}\, M$. In $X_{\check{\sigma}}$, all coordinates $u^i_1, \ldots, u^i_k$ are bounded, hence, $(x_i)_{i\in\mathbb{Z}_{>0}}$ has an accumulation point, a contradiction. □

Since the embedded torus T is dense in X_Σ, from Theorem 9.1,

9.2 Theorem. *If Σ is complete, X_Σ is a compactification of the torus T.*

9.3 Theorem.
(a) *Each toric variety X_Σ possesses a toric compactification $X_{\Sigma'}$ ($\Sigma \subset \Sigma'$ and $X_{\Sigma'}$ compact).*
(b) *If X_Σ is smooth, $X_{\Sigma'}$ can also be chosen smooth.*

PROOF.
(a) follows from III, Theorem 2.8.
(b) We apply Theorem 8.5 to $X_{\Sigma'}$. From the construction of Σ' in the proof of III, Theorem 2.8, it is readily seen that all cones of Σ can be left unchanged when Σ' is made regular.

□

Example. Given $\Sigma = \{\mathbb{R}_{\geq 0}\, e_1, \mathbb{R}_{\geq 0}\, e_2, \mathbb{R}_{\geq 0}(-e_1 - e_2), \{0\}\}$ we obtain as X_Σ a noncompact variety $\mathbb{P}^2 \setminus \{p, q, r\}$ where p, q, r are noncollinear points. Filling in the three 2-cones, we compactify by adding p, q, and r.

Exercises

1. Given $\Sigma := \{\mathbb{R}_{\geq 0}\, e_1, \mathbb{R}_{\geq 0}(-e_1), \mathbb{R}_{\geq 0}(-e_2), \mathbb{R}_{\geq 0}(e_1 + 2e_2), \mathbb{R}_{\geq 0}\, e_3, \mathbb{R}_{\geq 0}(e_1 + e_2 - e_3), \mathbb{R}_{\geq 0}(e_2 - e_3), \{0\}\}$, find a regular compactification $X_{\Sigma'}$ of X_Σ such that Σ and Σ' have the same 1-cones.
2. Let Σ consist only of $\{0\}$ and finitely many 1-cones. Describe X_Σ.
3. Let Σ be the complete fan in $\mathbb{R}^3$ with generators $e_1, -e_1, e_2, -e_2, e_3, e_1 + e_2 - e_3$ such that the four octands of $\mathbb{R}^3$, which contain $\mathbb{R}_{\geq 0}\, e_3$, are in Σ. Consider Σ', as introduced in part II of the proof of Theorem 9.3, and describe $X_{\Sigma'}$.
4. Let $\mathbb{P}^2$ be given as toric variety by the fan Σ with generators $e_1, e_2, -e_1 - e_2$. Let $\Psi_1^{-1}, \ldots, \Psi_k^{-1}, \ldots$ be the blowups obtained successively by introducing the new generators $e_1 + e_2, \ldots, e_1 + ke_2, \ldots$ If we apply the infinite sequence of blowups, we obtain a generalized toric variety X in an infinite-dimensional space. Collect information about X; show that X is not compact.

VII

Sheaves and projective toric varieties

1. Sheaves and divisors

In VI, Lemma 1.27, we introduced rational functions as functions whose restriction on an appropriate Zariski open set U_0 is regular, that is, represented by a quotient $f = g/h$ of polynomials g, h with h nowhere 0 on U_0. Even more concretely, we may choose U_0 to be a Zariski open subset of the torus T so that the rational functions on X_Σ are all given by rational functions on T.

For further investigation of a toric variety X_Σ, it is useful to study systems of rational functions on X_Σ. This may be done in several ways; one is that of sheaves; another is that of divisors. We first introduce the idea of sheaves. We do not need sheaves in full generality. Dealing with rational functions makes things easier than dealing with general objects of algebraic geometry.

1.1 Definition. Let R be a (commutative) ring with 1, and let M be a commutative group (written additively) together with a multiplication

$$\begin{array}{ccc} R \times M & \longrightarrow & M \\ (a, x) & \longmapsto & ax \end{array}$$

such that the following rules are satisfied,

$$a(x + y) = ax + ay,$$

$$(a + b)x = ax + bx,$$

$$a(bx) = (ab)x, \qquad \text{for all} \quad a, b \in R, \quad x, y \in M,$$

and $$1 \cdot x = x.$$

Then, we call M an R-*module* or, briefly, a *module*.

Remark. A module is a generalized vector space, the field of scalars being replaced by a ring. Of course, any vector space over a field K is a K-module.

1.2 Definition. A *module homomorphism* $\varphi : M \longrightarrow M'$ of an R-module M into an R'-module M', R a subring of R', is a map for which

$$\varphi(ax + by) = a\varphi(x) + b\varphi(y), \qquad \text{for all} \quad a, b \in R, \quad x, y \in M.$$

If $R = R'$ and φ is bijective, we call it a *module isomorphism.*

Example 1. Consider the set M of all polynomials in one variable with integral coefficients. Then M is a $\mathbb{Z}$-module. It has an extension to a $\mathbb{C}$-vector space of polynomials with complex coefficients. If a homomorphism of this vector space into another $\mathbb{C}$-vector space is restricted to M, we obtain a module homomorphism.

Clearly,

1.3 Lemma. *Any commutative group G can be considered a $\mathbb{Z}$-module by setting $1 \cdot x = x$, $(-1) \cdot x = -x$, and $k \cdot x = x + (k-1) \cdot x$, $k \in \mathbb{Z}$.*

Therefore, if we discuss modules, we include commutative groups by considering them as $\mathbb{Z}$-modules.

1.4 Definition. Let X be a toric variety with Zariski topology. To each open subset U of X, let a ring $\mathcal{F}(U)$ of rational $\mathbb{C}$-valued functions on U be given such that the following is true:

(a) For any pair $V \subset U$ of open subsets of X and any $f \in \mathcal{F}(U)$, the restriction $f|_V$ belongs to $\mathcal{F}(V)$.

(b) Let $U = \bigcup U_\alpha$ be a union of open sets U_α, α in some index set, and, for each α, let an element $f_\alpha \in \mathcal{F}(U_\alpha)$ be given such that

$$f_\alpha|_{U_\alpha \cap U_\beta} = f_\beta|_{U_\alpha \cap U_\beta}$$

for any such pair f_α, f_β. Then, there exists an element $f \in \mathcal{F}(U)$ such that

$$f|_{U_\alpha} = f_\alpha, \qquad \text{for all} \quad \alpha.$$

Then, we call the collection $\mathcal{F} := \mathcal{F}(U)_{U \subset X}$ a *sheaf of rational functions* on X, in short, a *sheaf* (in this book).

A rational function f is determined by its restriction $f|_U$ to any Zariski open set U. Therefore, for the sake of a simplified notation, we identify f and $f|_U$. Restriction from U to V determines the inclusion $\mathcal{F}(U) \hookrightarrow \mathcal{F}(V)$.

1.5 Lemma. *Let $\mathcal{F}$ and $\mathcal{F}'$ be sheaves and $\{U_\alpha\}_{\alpha \in I}$ a covering of X by Zariski open sets. If $\mathcal{F}(U) = \mathcal{F}'(U)$ for each U contained in some U_α, then, $\mathcal{F} = \mathcal{F}'$.*

PROOF. Let V be an arbitrary Zariski open set of X. Then, we deduce from (b) that $\mathcal{F}(V) = \bigcap_{\alpha \in I} \mathcal{F}(V \cap U_\alpha) = \bigcap_{\alpha \in I} \mathcal{F}'(V \cap U_\alpha) = \mathcal{F}'(V)$; hence, $\mathcal{F} = \mathcal{F}'$. $\square$

Remark. (b) is a "gluing property" by which collections of "local" functions are pasted together to "global" functions on X (as an open set).

1.6 Theorem. *To each open subset U of a toric variety X_Σ, we assign the ring of regular functions on U, and denote it by $\mathcal{O}(U)$. Then, $\mathcal{O}$ is a sheaf of rational functions on X_Σ.*

1.7 Definition. $\mathcal{O}$ is called the *structure sheaf* $\mathcal{O}_{X_\Sigma}$ of X_Σ.

PROOF OF THEOREM 1.6. (a) and (b) are immediate from the definitions. □

1.8 Theorem. *For the sheaf $\mathcal{O} = \mathcal{O}_{X_\Sigma}$ of a toric variety,*
(a) $\mathcal{O}(X_{\check{\sigma}}) = R_{\check{\sigma}}$ *is a ring of Laurent polynomials for any* $\sigma \in \Sigma$.
(b) $\mathcal{O}(X_{\{0\}^\vee}) \cong \mathbb{C}[z, z^{-1}] = \mathbb{C}[z_1, \ldots, z_n, z_1^{-1}, \ldots, z_n^{-1}]$, *and*
(c) $\mathcal{O}(X_\Sigma) = \mathbb{C}$ *if* Σ *is complete.*

PROOF.
(a) This has been shown in the proof of VI, Lemma 1.26.
(b) is a special case of (a).
(c) Let $f \in \mathcal{O}(X_\Sigma)$. Then, f is a Laurent polynomial, and $f \in \bigcap_{\sigma\in\Sigma} R_{X_{\check{\sigma}}}$ implies supp $f \in \bigcap_{\sigma\in\Sigma} \check{\sigma} = \{0\}$ so that f is a constant function.

□

Remark. From Theorem 1.8, we see that the structure sheaf defines on each affine piece $X_{\check{\sigma}}$ of a toric variety (as a special open subset) a module of Laurent polynomials. For other open subsets this is not true, in general. As an example, consider, in $X_{\check{\sigma}_1}$ of Example 2 below, an open subset $U := X_{\check{\sigma}_1} \setminus \{(z_1, z_2) \mid z_1 - z_2 = 0\}$. Then, $\frac{1}{z_1 - z_2} \in \mathcal{O}(U)$, but $\frac{1}{z_1 - z_2}$ is not a Laurent polynomial.

1.9 Definition. Let $\mathcal{F}$ be a sheaf of rational functions on X_Σ which assigns to each open set U of X_Σ an $\mathcal{O}(U)$-module $\mathcal{F}(U)$ of rational functions on U. Then, $\mathcal{F}$ is called a *sheaf of $\mathcal{O}_{X_\Sigma}$-modules.*

The next theorem presents the sheaves we are mainly concerned with in the present chapter.

1.10 Theorem. *Let X_Σ be a toric variety. To each cone $\tau \in \Sigma$, let a vector $m_\tau \in \mathbb{Z}^n$ be given such that the following is true:*
(a) *If τ_0 is a face of τ, then,* $m_\tau - m_{\tau_0} \in$ cospan $\check{\tau}_0$.
Then, we obtain a sheaf $\mathcal{F}$ of $\mathcal{O}$-modules of rational functions by setting

$$\mathcal{F}(\emptyset) := \{0\}, \tag{1}$$

$$\mathcal{F}(U) := z^{m_\tau}\mathcal{O}(U) \quad \textit{for any nonempty, open set } U \subset X_{\check{\tau}}, \textit{ and} \tag{2}$$

$$\mathcal{F}(U_1 \cup \cdots \cup U_s) := z^{m_{\tau_1}}\mathcal{O}(U_1) \cap \cdots \cap z^{m_{\tau_s}}\mathcal{O}(U_s) \tag{3}$$

$$\textit{for open sets } U_i \subset X_{\check{\tau}_i}, \quad i = 1, \ldots, s.$$

PROOF. First, we show that $\mathcal{F}(U)$ is well defined. An open set U may be contained in different affine charts $X_{\check{\tau}}, X_{\check{\tau}'}$, which are also open in X_Σ. Then, we must show that

$$(I) \qquad z^{m_\tau}\mathcal{O}(U) = z^{m_{\tau'}}\mathcal{O}(U).$$

In fact, $\tau_0 := \tau \cap \tau'$ is a common face of τ and τ'. Therefore, by (a), $m_\tau - m_{\tau_0} \in$ cospan $\check{\tau}_0$ and $m_{\tau'} - m_{\tau_0} \in$ cospan $\check{\tau}_0$, hence, $m_\tau - m_{\tau'} \in$ cospan $\check{\tau}_0$. From $\tau_0 \subset \tau$ and $\tau_0 \subset \tau'$, we obtain $R_{\check{\tau}_0} \supset R_{\check{\tau}}$ and $R_{\check{\tau}_0} \supset R_{\check{\tau}'}$. It follows that $z^{m_\tau - m_{\tau'}} \in R_{\check{\tau}_0}$, and, hence, by (2), $z^{m_\tau} R_{\check{\tau}_0} = z^{m_{\tau'}} R_{\check{\tau}_0}$. So, the claim (I) is true.

If U is not contained in any $X_{\check{\tau}}$ but intersects $X_{\check{\tau}_1}, \dots, X_{\check{\tau}_s}$, we set

$$U = (U \cap X_{\check{\tau}_1}) \cup \cdots \cup (U \cap X_{\check{\tau}_s})$$

and, by applying (3), obtain

$$\mathcal{F}(U) = z^{m_{\tau_1}}\mathcal{O}(U \cap X_{\check{\tau}_1}) \cap \cdots \cap z^{m_{\tau_s}}\mathcal{O}(U \cap X_{\check{\tau}_s}).$$

So, using (I), we see that $\mathcal{F}(U)$ is again well defined. The sheaf properties (a) and (b) are evident. □

Example 2. Consider $\mathbb{P}^2$ as represented by the fan in Example 1 of VI, 3. We have $\mathcal{O}(X_{\check{\sigma}_0}) = \mathbb{C}[z_1, z_2]$, $\mathcal{O}(X_{\check{\sigma}_1}) = \mathbb{C}[z_1^{-1}, z_1^{-1}z_2]$, $\mathcal{O}(X_{\check{\sigma}_2}) = \mathbb{C}[z_1 z_2^{-1}, z_2^{-1}]$.

If we choose $\mathcal{F}(X_{\check{\sigma}_0}) := \mathcal{O}(X_{\check{\sigma}_0}), \mathcal{F}(X_{\check{\sigma}_1}) := z_1\mathcal{O}(X_{\check{\sigma}_1}), \mathcal{F}(X_{\check{\sigma}_2}) := z_2\mathcal{O}(X_{\check{\sigma}_2})$, all other elements of $\mathcal{F}$ are determined. We find $m_{\sigma_0} = 0$, $m_{\sigma_1} = e_1$, $m_{\sigma_2} = e_2$ and, for $\sigma_{ij} := \sigma_i \cap \sigma_j$, $i, j = 0, 1, 2$, $i < j$, choose $m_{\sigma_{01}} = 0$, $m_{\sigma_{02}} = 0$, $m_{\sigma_{12}} = e_1$ (see Figure 1a).

Another sheaf is obtained by setting $\mathcal{F}'(X_{\check{\sigma}_0}) := \mathcal{O}(X_{\check{\sigma}_0})$, $\mathcal{F}'(X_{\check{\sigma}_1}) := z_1^{-1}\mathcal{O}(X_{\check{\sigma}_1})$, $\mathcal{F}'(X_{\check{\sigma}_2}) := z_2^{-1}\mathcal{O}(X_{\check{\sigma}_2})$, $m'_{\sigma_0} := 0$, $m'_{\sigma_1} := -e_1$, $m'_{\sigma_2} := -e_2$. Figure 1b illustrates that the compatibility condition (a) can be satisfied: For the cones $\sigma_{ij} := \sigma_i \cap \sigma_j$, $i, j = 0, 1, 2$, $i < j$, we choose $m_{\sigma_{01}} := 0$, $m_{\sigma_{02}} := 0$, $m_{\sigma_{12}} := -e_1$.

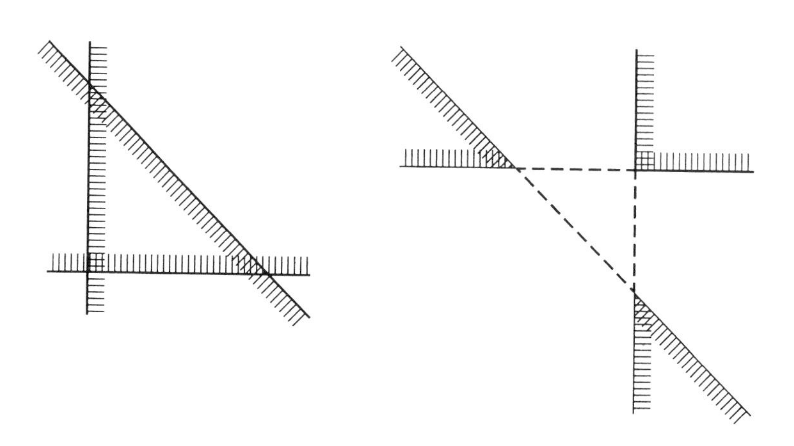

FIGURE 1a,b.

There is a difference between $\mathcal{F}$ and $\mathcal{F}'$ in Example 2 which we shall consider later. $\mathcal{F}$ has "global sections", namely, $\mathcal{F}(X_{\check{\sigma}_0}) \cap \mathcal{F}(X_{\check{\sigma}_1}) \cap \mathcal{F}(X_{\check{\sigma}_2}) = \{\alpha z_1 + \beta z_2 + \gamma \mid \alpha, \beta, \gamma \in \mathbb{C}\}$, that is, all linear functions in z_1, z_2. However, $\mathcal{F}'$ has no such elements except 0.

1.11 Definition. Let $\mathcal{F}$ be a sheaf of $\mathcal{O}$-modules on a toric variety X_Σ. Each $f \in \mathcal{F}(U)$ is called a *section* of $\mathcal{F}$, and we say that the section z^{m_τ} in Theorem 1.10(2) *generates* $\mathcal{F}$ on the open piece $X_{\check{\tau}}$. If $f \in \mathcal{F}(X_\Sigma)$, we call f a *global section* of $\mathcal{F}$ and $\mathcal{F}(X_\Sigma)$ the $\mathcal{O}(X_\Sigma)$-module of *global sections*, of $\mathcal{F}$.

We now turn to objects which are based on the zeros and poles of rational functions and which are in close relationship to sheaves of such functions.

1.12 Definition. A subset D of a Zariski open subset Y of X_Σ is called a *prime divisor* on Y if, for each $\sigma \in \Sigma$, the intersection $D \cap X_{\check{\sigma}} \cap Y$ is an $(n-1)$-dimensional subvariety of $X_{\check{\sigma}} \cap Y$. In particular, in the case $Y = X_\Sigma$, we say that D is an *invariant prime divisor* if $D \cap X_{\check{\sigma}}$ is always an invariant affine toric subvariety of $X_{\check{\sigma}}$. Formal linear combinations

$$D = n_1 D_1 + \cdots + n_r D_r, \quad n_i \in \mathbb{Z}, \quad i = 1, \ldots, r,$$

are called (*Weil*) *divisors* on Y, in particular, *invariant divisors* on $Y = X_\Sigma$ if $D_1, \ldots, D_r$ are invariant prime divisors. By the *sum* of two divisors $D = \sum_i n_i D_i$, $D' = \sum_i n_i' D_i$, we mean $D + D' := \sum_i (n_i + n_i') D_i$. Let $Y_1 \subset Y_2 \subset X_\Sigma$ be Zariski open sets in X_Σ. If $D = \sum_i n_i D_i$ is a divisor on Y_2, we call $D|_{Y_1} := \sum_{D_i \cap Y_1 \neq \emptyset} n_i (D_i \cap Y_1)$ the *restriction* of D to Y_1. If $n_i \geq 0$ for $i = 1, \ldots, r$, we say that D is *effective* and write

$$D \geq 0.$$

1.13 Lemma. *Let X be a quasi-affine variety such that R_X is a unique factorization domain (as is always true if $X = X_{\check{\sigma}}$ for a regular $\sigma \in \Sigma$, compare the next lemma), and let D be a prime divisor on X. Then, $\mathfrak{i}_{D,X}$ is generated by one function.*

PROOF. $\mathfrak{i}_{D,X}$ is prime since D is prime. Each $f \in \mathfrak{i}_{D,X}$ may be written as $f = f_1 \cdots f_k x$ where $f_1, \ldots, f_k$ are prime elements and x is a unit of R_X. Then, for at least one $i \in \{1, \ldots, k\}$ the prime ideal $R_X f_i$ is contained in $\mathfrak{i}_{D,X}$. But $\mathfrak{i}_{D,X}$ has height 1, so that $R_X f_i = \mathfrak{i}_{D,X}$. □

1.14 Lemma. *Let σ be a regular cone, and let D be a prime divisor on $X_{\check{\sigma}}$. Then, $\mathfrak{i}_{D,X_{\check{\sigma}}} \subset R_{X_{\check{\sigma}}}$ is generated by one function.*

PROOF. By VI, 2.12, $X_{\check{\sigma}} \cong \mathbb{C}^{*k} \times \mathbb{C}^{n-k} \subset \mathbb{C}^n$ for some $k \in \{0, \ldots, n\}$. By Definition 1.12, $D' := D \cap X_{\check{\sigma}}$ is prime on $X_{\check{\sigma}}$. By VI, 1,29 b), $\bar{D}' \subset \mathbb{C}^n$ is also prime. Since the ring $\mathbb{C}[\xi_1, \ldots, \xi_n]$ is a unique factorization domain, we see, from Lemma 1.13 that $\mathfrak{i}_{\bar{D}',\mathbb{C}^n} \subset \mathbb{C}[\xi_1, \ldots, \xi_n]$ is generated by one element. Therefore, by VI, Lemma 1.30, $\mathfrak{i}_{D,X_{\check{\sigma}}}$ is also generated by one element. □

Let f be a rational function on X_Σ, and let D be a prime divisor on Y. We assume that Σ, has at least, one 1-cone. We can find an open set $X_{\check{\tau}}$ with $\dim \tau = 1$ such that $D \cap X_{\check{\tau}} \neq \emptyset$. Since $X_{\check{\tau}} \cong \mathbb{C} \times \mathbb{C}^{*n-1}$, we see that $X_{\check{\tau}}$ is smooth.

By Lemmas 1.14 and VI, Lemma 1.30, $\mathfrak{i}_{D\cap X_{\check{\tau}}\cap Y, X_{\check{\tau}}\cap Y} \subset R_{X_{\check{\tau}}\cap Y}$ is generated by one function g_D, so that $\mathfrak{i}_{D\cap X_{\check{\tau}}\cap Y, X_{\check{\tau}}\cap Y} = R_{X_{\check{\tau}}\cap Y} \cdot g_D$. Since $\mathfrak{i}_{D\cap X_{\check{\tau}}\cap Y}$ is prime, $g_D \in R_{X_{\check{\tau}}\cap Y}$ is a prime element. We represent f as a quotient $f = \frac{g}{h}$ for $g, h \in R_{X_{\check{\tau}}\cap Y}$. We set

$$n_{D,g} := \max\{r \mid g_D^r \text{ divides } g,\ r \geq 0\},$$

$$n_{D,h} := \max\{s \mid g_D^s \text{ divides } h,\ s \geq 0\},$$

and

$$n_{D,f} := n_{D,g} - n_{D,h}.$$

1.15 Lemma.

(a) *$n_{D,f}$ does not depend on the choice of the representation $f = \frac{g}{h}$, or on the choice of the generator g_D of $\mathfrak{i}_{D\cap X_{\check{\tau}}\cap Y, X_{\check{\tau}}\cap Y}$.*

(b) *$n_{D,f}$ remains unchanged if we replace $X_{\check{\tau}} \cap Y$ by another quasi-affine open set U for which $D \cap U \neq \emptyset$ and $\mathfrak{i}_{D\cap U, U} \subset R_U$ is generated by one element.*

(c) *$n_{D,f} \neq 0$ only for a finite number of prime divisors.*

PROOF.

(a) Suppose $f = \frac{g}{h} = \frac{g'}{h'}$. Then $gh' = g'h$, and hence, $n_{D,g} + n_{D,h'} = n_{D,g'} + n_{D,h}$. This implies $n_{D,f} = n_{D,g} - n_{D,h} = n_{D,g'} - n_{D,h'}$. If g'_D is a further generator of $\mathfrak{i}_{D\cap X_{\check{\tau}}\cap Y, X_{\check{\tau}}\cap Y}$, then, $g_D = g'_D \cdot x$ for some unit x of $R_{X_{\check{\tau}}\cap Y}$, so that g_D^r divides g if and only if g'^r_D divides g, analogously for h.

(b) We set $U' := X_{\check{\tau}} \cap Y$. Suppose $g_{D\cap U}$ generates $\mathfrak{i}_{D\cap U, U} \subset R_U$ and $g_{D\cap U'}$ generates $\mathfrak{i}_{D\cap U', U'} \subset R_{U'}$. We have $D \not\subset (X_\Sigma \setminus U) \cup (X_\Sigma \setminus U') = X_\Sigma \setminus (U \cap U')$ so that $D \cap U \cap U' \neq \emptyset$. Now, by VI, Lemma 1.30, we obtain

$$R_{U\cap U'} \supset \mathfrak{i}_{D\cap U\cap U', U\cap U'} = R_{U\cap U'} \cdot g_{D\cap U} = R_{U\cap U'} \cdot g_{D\cap U'}.$$

Therefore, $g_{D\cap U}$ and $g_{D\cap U'}$ differ only by a unit on $R_{U\cap U'}$.

(c) Given a prime divisor D, we know there exists an $X_{\check{\tau}}$, $\dim \tau = 1$, such that $D \cap X_{\check{\tau}} \cap Y \neq \emptyset$. Then, for any representation $f = \frac{g}{h}$, where $g, h \in R_{X_{\check{\tau}}}$ we consider all prime divisors $D_{i,\tau}$, $i = 1, \ldots, l$, which are irreducible components of $V(g \cdot h, X_{\check{\tau}}) \subset X_{\check{\tau}}$. But $n_{D,f} \neq 0$ implies $D \subset V(g \cdot h, X_{\check{\tau}})$ and $\bar{D} \cap X_{\check{\tau}} \subset V(g \cdot h, X_{\check{\tau}})$. From the irreducibility of D and by VI, Lemma 1.29 b), we get $\bar{D} \cap X_{\check{\tau}} = D_{i,\tau}$, for some i, and $D = \bar{D} \cap X_{\check{\tau}} \cap Y = D_{i,\tau} \cap X_{\check{\tau}} \cap Y$. Since there are only finitely many $X_{\check{\tau}}$ and finitely many $D_{i,\tau}$, for each τ we conclude that only finitely many divisors D satisfy $n_{D,f} \neq 0$. □

1.16 Definition. A divisor is called *principal* on Y if it is of the form

$$(f) := \sum_{D \text{ prime on } Y} n_{D,f} D.$$

We say a divisor is *locally principal* or a *Cartier divisor* if, for each $x \in Y$, there exists a Zariski open subset U of Y such that $x \in U$ and the divisor is principal on

U. Two Cartier divisors D, D' are called *linearly equivalent* if $D - D'$ is principal. If $D - D' \geq 0$, we write $D \geq D'$. We say a Cartier divisor is T-*invariant* or, in short, *invariant* if it remains the same under each torus action.

Example 3. Let $X_\Sigma = \mathbb{C}^n$, and let $f_i \in \mathbb{C}[\xi_1, \ldots, \xi_n]$ be prime polynomials, $i = 1, \ldots, k$. For $f = f_1^{a_1} \cdots f_k^{a_k}$, $a_i \in \mathbb{Z}$, and setting $D_i := \{f_i = 0\}$, $i = 1, \ldots, k$, $(f) = \sum_{i=1}^k a_i D_i$.

From Lemma 1.14 we obtain Lemmas 1.17 and 1.18.

1.17 Lemma. *Let σ be a regular cone, and let D be an arbitrary Weil divisor on X_σ. Then D is principal on X_σ.*

1.18 Lemma. *Let Y be a Zariski open subset of a toric variety X_Σ, and let $D_0 = \sum_i n_i D_i$ be a Weil divisor on Y. Then, $\{f \in K_Y \mid (f) + D_0 \geq 0\}$ is a vector space (of rational functions) over $\mathbb{C}$.*

PROOF.
Let $(f) + D_0 \geq 0$, $(f') + D_0 \geq 0$, and let D be a prime divisor on Y. We find a regular cone τ such that $D \cap X_\tau \cap Y \neq \emptyset$. Then by Lemma 1.14, $\mathfrak{i}_{D \cap X_\tau \cap Y, X_\tau \cap Y} \subset R_{X_\tau \cap Y}$ is generated by one element g_D. We set $f = \frac{g}{h}$, $f' = \frac{g'}{h'}$, where $g, h, g'h' \in R_{X_\tau \cap Y}$. Then,

$$f + f' = \frac{gh' + g'h}{hh'}.$$

We set $n_D := n_i$ for $D = D_i$, and $n_D = 0$ otherwise. By assumption, $n_{D,f} + n_D \geq 0$. From $f = \frac{gh'}{hh'}$, we obtain $n_{D,gh'} + n_D \geq n_{D,hh'}$ and from $f' = \frac{g'h}{hh'}$ we find $n_{D,g'h} + n_D \geq n_{D,hh'}$. Hence, by definition of $n_{D,gh'}$ and $n_{D,g'h}$, $n_{D,gh'+g'h} + n_D \geq n_{D,hh'}$, so that

$$n_{D,f+f'} + n_D \geq 0.$$

Since this is true for any D, we conclude that $(f + f') + D_0 \geq 0$. □

1.19 Lemma. *Let $f = \sum_{i=1}^k a_i z^{m_i}$ be a Laurent polynomial on X_Σ where $a_i \neq 0$, $i = 1, \ldots, k$. Define $f^t(z) := f(t \cdot z)$, t in the torus T. Then, the following vector spaces (of polynomials) over $\mathbb{C}$ coincide:*

$$\operatorname{lin}\{f^t \mid t \in T\} = \operatorname{lin}\{z^{m_i} \mid i = 1, \ldots, k\}.$$

PROOF. We note that

$$f^t(z) = f(t \cdot z) = \sum_{i=1}^k a_i (tz)^{m_i} = \sum_{i=1}^k a_i t^{m_i} z^{m_i}.$$

We wish to choose $t^{(i)} = (t_1^{(i)}, \ldots, t_k^{(i)})$, $i = 1, \ldots, k$, such that the system

$$\sum_{i=1}^k a_i t^{(1)m_i} z^{m_i} = f^{t^{(1)}}(z), \ \ldots, \ \sum_{i=1}^k a_i t^{(k)m_i} z^{m_i} = f^{t^{(k)}}(z),$$

looked at as a system of linear equations in the variables $z^{m_1}, \ldots, z^{m_k}$, may be resolved. The determinant of the coefficient matrix is readily seen to be

$$\Delta := a_1 \cdots a_k \begin{vmatrix} t^{(1)m_1} & \ldots & t^{(1)m_k} \\ \vdots & & \vdots \\ t^{(k)m_1} & \ldots & t^{(k)m_k} \end{vmatrix}.$$

Let $p = (p_1, \ldots, p_n)$ be a sequence of different prime numbers, and let $t^{(i)} := (p_1^i, \ldots, p_n^i)$, $i = 1, \ldots, k$. A little calculation shows that, up to a nonzero constant, Δ is a non-vanishing Vandermonde determinant (as used in the proof of II, Theorem 3.11). So, the z^{m_j} occuring in f may be expressed as linear combinations of $f^{t^{(1)}}, \ldots, f^{t^{(k)}}$. □

1.20 Lemma. *Let f be a Laurent polynomial on $X_{\check{\sigma}}$ such that $(f) \geq 0$. Then, f is regular on $X_{\check{\sigma}}$.*

PROOF. From $(f) \geq 0$, we obtain $(f^t) \geq 0$, and, hence, by Lemma 1.19, $(z^{m_i}) \geq 0$, which implies $m_i \in \check{\sigma}$, and z^{m_i} is regular on $X_{\check{\sigma}}$, $i = 1, \ldots, k$. Then, $f = \sum_{i=1}^k a_i z^{m_i}$ is also regular. □

1.21 Lemma. *Let f be a rational function on $X_{\check{\sigma}}$. If $(f|_U) \geq 0$ for some open $U \subset X_{\check{\sigma}}$, then, f is regular on U.*

PROOF. We write $(f) = \sum_{D \text{ prime}} n_{D.f} D$. The coefficients $n_{D.f}$ can be negative only in the case $D \cap U = \emptyset$ or, equivalently, $D \subset X_{\check{\sigma}} \setminus U$. We set $Y := \bigcup_{n_{D.f}<0} D \subset X_{\check{\sigma}} \setminus U$. Fix a point $p \in U$, and choose any $g \in \mathfrak{i}_{Y.X_{\check{\sigma}}} \setminus \mathfrak{i}_{p.X_{\check{\sigma}}}$. For any prime divisor $D \subset Y$, $g \in \mathfrak{i}_D$ and $n_{D.g} > 0$, which implies $(f \cdot g^k) > 0$ for sufficiently large k. By Lemma 1.20, $f \cdot g^k \in R_{X_{\check{\sigma}}}$. Finally, $f = \frac{f \cdot g^k}{g^k}$ where $g^k(p) \neq 0$, and $f \cdot g^k, g^k \in R_{X_{\check{\sigma}}}$. Hence, f is regular at each $p \in U$. □

We are now able to build up sheaves by the aid of divisors.

1.22 Theorem. *Each Cartier divisor D on $X = X_\Sigma$ determines a sheaf $\mathcal{L}_D$ as follows:*

$$\mathcal{L}_D(U) = \{f \in K_X \mid (f) + D \geq 0 \text{on } U\}.$$

PROOF. The sheaf properties are readily verified by using the preceding lemmas. □

We shall prove special properties of these sheaves in the following section.

Exercises

1. Let $\mathbb{P}^1 \times \mathbb{P}^1$ be given as a toric variety X_Σ (see VI, 3, Example 2). Find sheaves $\mathcal{F}_1, \mathcal{F}_2, \mathcal{F}_3$ such that the global sections are as follows:

$$\mathcal{F}_1(X_\Sigma) = \{\alpha_0 + \alpha_1 z_1 + \alpha_2 z_2 + \alpha_3 z_1 z_2 \mid \alpha_0, \ldots, \alpha_3 \in \mathbb{C}\}$$
$$\mathcal{F}_2(X_\Sigma) = \{\alpha_0 + \alpha_1 z_1 \mid \alpha_0, \alpha_1 \in \mathbb{C}\}$$
$$\mathcal{F}_3(X_\Sigma) = \{0\}.$$

2. Consider $\mathbb{P}^2 \times \mathbb{P}^1$ as a toric variety X_Σ (VI, 4, Example 6). Does there exist a sheaf according to Theorem 1.11 satisfying $\mathcal{F}_1(X_\Sigma) = \{\alpha_0 + \alpha_1 z_1 + \alpha_2 z_2 + \alpha_3 z_3 \mid \alpha_0, \ldots, \alpha_3 \in \mathbb{C}\}$?
3. Consider $\mathbb{P}^n$ as a toric variety. Find the structure sheaf and a sheaf whose global sections are all linear functions in n complex variables.
4. Define isomorphisms between sheaves, and show that, if $\mathcal{F}$, $\mathcal{F}'$ are sheaves according to Theorem 1.10 for which $m_{\sigma'} = m_\sigma + a$ for all $\sigma \in \Sigma$ and a fixed lattice vector a, then, $\mathcal{F}$, $\mathcal{F}'$ are isomorphic.

2. Invertible sheaves and the Picard group

Now we will now investigate further the sheaves introduced in section 1. We define tensor products of them and introduce the so-called Picard group.

2.1 Definition. We call two sheaves $\mathcal{F}, \mathcal{F}'$ of $\mathcal{O}_{X_\Sigma}$-modules *isomorphic* , $\mathcal{F} \cong \mathcal{F}'$, if there exists, for any open set U of X_Σ, an isomorphism φ_U between the $\mathcal{O}(U)$-modules $\mathcal{F}(U), \mathcal{F}'(U)$ such that $\varphi_U|_V = \varphi_V$ for each open subset $V \subset U$.

2.2 Definition. A sheaf $\mathcal{F}$ of $\mathcal{O}_{X_\Sigma}$-modules is said to be *invertible* if there exists a covering $\{U_\alpha\}$ of X_Σ by Zariski open sets such that

$$\mathcal{F}(U_\alpha) \cong \mathcal{O}_{X_\Sigma}(U_\alpha)$$

for all $U_\alpha \in \{U_\alpha\}$.

2.3 Lemma. *The sheaves introduced in Theorem 1.10 are invertible.*

PROOF. We choose $\{U_\alpha\} = \{X_{\check{\sigma}}\}_{\sigma \in \Sigma}$. By definition $\mathcal{F}(X_{\check{\sigma}}) = z^{m_\sigma}\mathcal{O}(X_{\check{\sigma}})$. Multiplication by z^{-m_σ} clearly provides a module isomorphism

$$\mathcal{F}(X_{\check{\sigma}}) \xrightarrow{\cdot z^{-m_\sigma}} \mathcal{O}(X_{\check{\sigma}}).$$

□

For $\mathcal{F}(X_{\check{\sigma}})$, the meaning of "invertible" can be made concrete.

If we set $\mathcal{F}'(X_{\check\sigma}) := z^{-m_\sigma}\mathcal{O}(X_{\check\sigma})$, then, by set multiplication, $\mathcal{F}(X_{\check\sigma})\cdot\mathcal{F}'(X_{\check\sigma}) = \mathcal{O}(X_{\check\sigma})$. In combinatorial terms (see Figure 2) it means that

$$(m_\sigma + \check\sigma) + (-m_\sigma + \check\sigma) = \check\sigma.$$

So, $\mathcal{O}(X_{\check\sigma})$ attains the meaning of a unit element. This observation will be extended to the sheaves themselves. Before we do so, we will achieve two things. In Theorem 2.13, it will be shown that all invertible sheaves on a toric variety X_Σ are of the type introduced in Theorem 1.10. Here, we will introduce a multiplication for invertible sheaves which allows us to define unit elements and inverse elements in the set of invertible sheaves on X_Σ.

2.4 Lemma. *The sheaf $\mathcal{L}_D$ introduced in Theorem 1.22 is invertible.*

PROOF. We can cover X_Σ by Zariski open sets U contained in $X_{\check\sigma}$ for some $\sigma \in \Sigma$ such that $D = (g)$ on U for $g \in K_{X_\Sigma}$. By Lemma 1.21,

$$\begin{aligned}\mathcal{L}_D(U) &= \{f \in K_{X_\Sigma} \mid (f\cdot g) \geq 0 \text{ on } U\} = \{f \in K_{X_\Sigma} \mid f\cdot g \in \mathcal{O}(U)\}\\ &= g^{-1}\mathcal{O}(U).\end{aligned}$$

Multiplication by g provides the module isomorphism

$$\mathcal{L}_D(U) = g^{-1}\mathcal{O}(U) \quad \xrightarrow{\cdot g} \quad \mathcal{O}(U).$$

□

Let us recall the definition of a tensor product.

2.5 Definition. Let A, B be modules over a ring R. Consider all formal linear combinations of elements of $A \times B$ with coefficients in R. Then, an R-module M_0 is obtained. In M_0, we define the submodule a generated by all elements

$$(a + a', b) - (a, b) - (a', b),$$

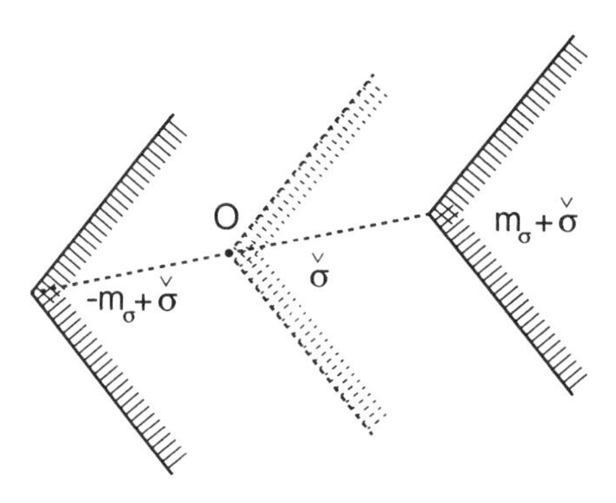

FIGURE 2.

$$(a, b + b') - (a, b) - (a, b'),$$
$$(ra, b) - r(a, b),$$

and

$$(a, rb) - r(a, b),$$

where $a, a' \in A, b, b' \in B, r \in R$ are arbitrary. Then $A \otimes B := M_0/\mathfrak{a} =: M$ is an R-module which we call the *tensor product* of A and B.

From the definition we readily see the following lemma:

2.6 Lemma. *The tensor multiplication in the definition of a tensor product is a map*

$$\begin{array}{cccc} \otimes: & A \times B & \longrightarrow & M \\ & (a, b) & \longmapsto & a \otimes b \end{array}$$

satisfying the following properties for all $a, a' \in A, b, b' \in B, r \in R$,

$$(a + a') \otimes b = a \otimes b + a' \otimes b,$$
$$a \otimes (b + b') = a \otimes b + a \otimes b',$$

and

$$(ra) \otimes b = a \otimes (rb) = r(a \otimes b).$$

Example 1. Let V, V' be vector spaces over a field K, $\dim V = n$, $\dim V' = n'$. We write the elements of V as column vectors, with respect to some basis of V, and the elements of V' as row vectors with respect to a basis of V'. The matrix product

$$a \otimes b := \begin{pmatrix} a_1 \\ \vdots \\ a_n \end{pmatrix} (b_1 \cdots b_{n'}) = \begin{pmatrix} a_1 b_1 & \cdots & a_1 b_{n'} \\ \vdots & & \vdots \\ a_n b_1 & \cdots & a_n b_{n'} \end{pmatrix}$$

defines a tensor product in a natural way. As a result, $V \otimes_K V' =: M(n, n')$ is the K-module of all $n \times n'$-matrices with entries taken from K which is an $n \cdot n'$-dimensional vector space over K.

Note that not all $n \times n'$-matrices are of type $a \otimes b$. For example, the unit matrix $\begin{pmatrix} 1 & 0 \\ 0 & 1 \end{pmatrix}$ is readily seen *not* to be a tensor product.

Example 2. Let $R = \mathbb{C}$, and let A, B be rings of Laurent polynomials with coefficients in $\mathbb{C}$. For $f \in A, g \in B$, we set

$$f \otimes g := f \cdot g, \qquad \text{in particular,} \quad z^k \otimes z^r := z^{k+r},$$

and, in this way, obtain $A \otimes_{\mathbb{C}} B$. In particular, if $A := \mathbb{C}[z_1]$, $B := \mathbb{C}[z_2]$, the element $z_1 + z_2$ of $A \otimes_{\mathbb{C}} B \cong \mathbb{C}[z_1, z_2]$ is *not* the tensor product of an element of A and an element of B.

2.7 Definition. Let $\mathcal{F}, \mathcal{G}$ be sheaves of $\mathcal{O}_{X_\Sigma}$-modules. For any open subset U of X_Σ, we set

$$(\mathcal{F} \otimes \mathcal{G})(U) = (\mathcal{F} \otimes_{\mathcal{O}_{X_\Sigma}} \mathcal{G})(U) := \mathcal{F}(U) \otimes_{\mathcal{O}_{X_\Sigma(U)}} \mathcal{G}(U),$$

and call $\mathcal{F} \otimes \mathcal{G}$ the *tensor product* of the sheaves $\mathcal{F}$ and $\mathcal{G}$.

2.8 Lemma. *$\mathcal{F} \otimes \mathcal{G}$ is a sheaf of $\mathcal{O}_{X_\Sigma}$-modules.*

PROOF. By definition of the tensor product of modules over a ring, we know that $(\mathcal{F} \otimes \mathcal{G})(U)$ is an $\mathcal{O}_{X_\Sigma}(U)$-module. We have

$$(f \otimes g)|_U = (f|_U) \otimes (g|_U).$$

All defining properties of a sheaf follow by definition. □

2.9 Lemma. *Let $\mathcal{F}, \mathcal{G}, \mathcal{H}$ be sheaves of $\mathcal{O}_{X_\Sigma}$-modules. Then,*

$$(\mathcal{F} \otimes \mathcal{G}) \otimes \mathcal{H} \cong \mathcal{F} \otimes (\mathcal{G} \otimes \mathcal{H}) \qquad \text{and} \qquad \mathcal{F} \otimes \mathcal{G} \cong \mathcal{G} \otimes \mathcal{F}.$$

PROOF. By

$$(a \otimes b) \otimes c \quad \longmapsto \quad a \otimes (b \otimes c), \qquad \text{for any } a \in \mathcal{F}, b \in \mathcal{G}, c \in \mathcal{H},$$

a bijection is given between the generating elements of $(\mathcal{F} \otimes \mathcal{G}) \otimes \mathcal{H}$, $\mathcal{F} \otimes (\mathcal{G} \otimes \mathcal{H})$, respectively. Its extension to linear combinations provides an isomorphism. Similarly, $\mathcal{L}$ commutativity is shown to hold. □

2.10 Theorem.

(a) *Let $\mathcal{F}$ be an invertible sheaf of rational functions on X_Σ. Then, $\mathcal{F} = \mathcal{L}_{D_\mathcal{F}}$ for some Cartier divisor $D_\mathcal{F}$ on X.*

(b) *Let $\mathcal{F}_1 = \mathcal{L}_{D_{\mathcal{F}_1}}$ and $\mathcal{F}_2 = \mathcal{L}_{D_{\mathcal{F}_2}}$. Then, $\mathcal{F}_1 \otimes \mathcal{F}_2 = \mathcal{L}_{D_{\mathcal{F}_1}+D_{\mathcal{F}_2}}$ or, equivalently, $D_{\mathcal{F}_1 \otimes \mathcal{F}_2} = D_{\mathcal{F}_1} + D_{\mathcal{F}_2}$.*

(c) *Two sheaves $\mathcal{F}_1 = \mathcal{L}_{D_{\mathcal{F}_1}}$ and $\mathcal{F}_2 = \mathcal{L}_{D_{\mathcal{F}_2}}$ are isomorphic if and only if $D_{\mathcal{F}_1}$ and $D_{\mathcal{F}_2}$ are linearly equivalent.*

PROOF.

(a) As we have seen above for each of the sets U of the given covering we have a module isomorphism

$$\varphi_U : \mathcal{F}(U) \longrightarrow \mathcal{O}(U).$$

We set $f_U := \varphi_U^{-1}(1)$. Then, $\mathcal{F}(U) = \mathcal{O}(U) \cdot f_U$. Each f_U defines a principal divisor (f_U) on U. We assert that two such divisors (f_U), $(f_{U'})$ coincide on $U \cap U'$. In fact, $\frac{f_U}{f_{U'}}$ is a section of $(\mathcal{F} \otimes \mathcal{F}^{-1})(U \cap U') = \mathcal{O}(U \cap U')$ and, hence, is regular. The same is true for $\frac{f_{U'}}{f_U}$. So, $\frac{f_U}{f_{U'}}$ has no zeros and poles on $U \cap U'$ and represents a unit f_0 of $R_{U \cap U'}$. Therefore, $f_U = f_{U'} \cdot f_0$, so that f_U and $f_{U'}$ represent the same divisor.

We associate with $\mathcal{F}$ a Cartier divisor as follows. Given a prime divisor D, we choose a Zariski open set U for which $D \cap U \neq \emptyset$, and we set

$$n_{D,\mathcal{F}} := -n_{D,f_U}.$$

As we have seen, the number $n_{D,\mathcal{F}}$ does not depend on the special choice of U. We set

$$D_{\mathcal{F}} := \sum_{D \text{ prime}} n_{D,\mathcal{F}} D$$

which is readily seen to be a Cartier divisor; hence, $\mathcal{F} = \mathcal{L}_{D_{\mathcal{F}}}$. It satisfies

$$\begin{aligned}\mathcal{L}_{D_{\mathcal{F}}}(U) &= \{f \in K_{X_\Sigma} \mid (f) + D_{\mathcal{F}} \geq 0 \text{ on } U\}\\ &= \{f \in K_{X_\Sigma} \mid \frac{f}{f_U} \in \mathcal{O}(U)\}\\ &= f_U \cdot \mathcal{O}(U) = \mathcal{F}(U).\end{aligned}$$

(b) Let $\mathcal{F}_1, \mathcal{F}_2$ be two such sheaves. We assert that $\mathcal{L}_{D_{\mathcal{F}_1}} \otimes \mathcal{L}_{D_{\mathcal{F}_2}} = \mathcal{L}_{D_{\mathcal{F}_1}+D_{\mathcal{F}_2}}$. In fact, $\mathcal{L}_{D_{\mathcal{F}_1}}(U) = f \cdot \mathcal{O}(U)$, $\mathcal{L}_{D_{\mathcal{F}_2}}(U) = g \cdot \mathcal{O}(U)$, hence $\mathcal{L}_{D_{\mathcal{F}_1}+D_{\mathcal{F}_2}}(U) = f \cdot g \cdot \mathcal{O}(U) = (f \cdot \mathcal{O}(U)) \otimes (g \cdot \mathcal{O}(U)) = \mathcal{L}_{D_{\mathcal{F}_1}}(U) \otimes \mathcal{L}_{D_{\mathcal{F}_2}}(U)$. By Lemma 1.5, the two sheaves coincide.

(c) Now, $\mathcal{L}_{D_{\mathcal{F}_1}} \cong \mathcal{L}_{D_{\mathcal{F}_2}}$ implies $\mathcal{L}_{D_{\mathcal{F}_1}-D_{\mathcal{F}_2}} \cong \mathcal{L}_{D_{\mathcal{F}_1}} \otimes \mathcal{L}_{-D_{\mathcal{F}_1}} = \mathcal{O}(X_\Sigma)$. Let $f = \varphi^{-1}(1)$ generate $\mathcal{L}_{D_{\mathcal{F}_1}-D_{\mathcal{F}_2}}$. Then, $(f) = D_{\mathcal{F}_1} - D_{\mathcal{F}_2}$. Hence, $D_{\mathcal{F}_1}$ and $D_{\mathcal{F}_2}$ are linearly equivalent. The converse is also true.

□

2.11 Theorem. *The invertible sheaves of $\mathcal{O}_{X_\Sigma}$-modules define a commutative group with respect to tensor multiplication (and after identifying isomorphic copies).*

PROOF. The group properties follow from Lemmas 2.8, 2.9, the definition of "invertible", and by setting $\mathcal{F}^{-1} = \mathcal{L}_{-D_{\mathcal{F}}}$ for each $\mathcal{F} = \mathcal{L}_{D_{\mathcal{F}}}$. □

2.12 Definition. The group introduced in Theorem 2.11 is called the *Picard group* Pic X_Σ of X_Σ.

2.13 Theorem. *Any invertible sheaf $\mathcal{F}$ on X_Σ is isomorphic to one of the sheaves introduced in Theorem* 1.10.

PROOF. By Theorem 2.10, $\mathcal{F} = \mathcal{L}_{D_{\mathcal{F}}}$ for some Cartier divisor $D_{\mathcal{F}}$. The restriction of $D_{\mathcal{F}}$ to the big torus T is, by Lemma 1.17, a principal divisor (f_0) for some rational function f_0 on X_Σ. Let

$$D_0 := D_{\mathcal{F}} - (f_0),$$

and let $\mathcal{F}_0$ be the invertible sheaf defined by D_0 (see Theorem 1.22). Then, by Theorem 2.10,

$$\mathcal{F} = \mathcal{L}_{D_{\mathcal{F}}} \cong \mathcal{L}_{D_0} = \mathcal{F}_0.$$

Therefore, it is sufficient to prove that $\mathcal{F}_0$ satisfies the conditions of Theorem 1.10.

Note that the sets of zeros and poles define invariant divisors on X_Σ.

We choose some $\sigma \in \Sigma$. To each 1-face of σ, there corresponds an invariant divisor D_i, $i = 1, \ldots, k$. The intersection is a closed orbit $O_\sigma := D_1 \cap \cdots \cap D_k$.

Given a $p \in O_\sigma$, the Cartier divisor $-D_0$ can be represented on some neighborhood U of p by a principal divisor (f), that is, $(f) = -D_0$ on U. We can find a monomial z^m such that $(f \cdot z^m) \geq 0$ on U. By Lemma 1.21, $f \cdot z^m$ is regular on U, and we may represent $f \cdot z^m$ as a quotient $f \cdot z^m = \frac{g}{h}$, where g, h are Laurent polynomials regular on $X_{\check\sigma}$ and $h(p) \neq 0$. Finally, $(f) = (g \cdot z^{-m}) = -D_0$ on some neighborhood of p, and $f' := g \cdot z^{-m}$ is a Laurent polynomial.

Consider the divisor $(f') + D_0$. The prime divisors $D_1, \dots, D_k$ do not occur in $(f') + D_0$. The other divisors occur with positive coefficient since they intersect the torus and f' is regular on the torus. Therefore, $(f') + D_0 \geq 0$. Analogously, $(f'') + D_0 \geq 0$, and we find $f' = \sum_{i=1}^k a_i z^{m_i}$. By Lemmas 1.18 and 1.19 $(z^{m_i}) + D_0 \geq 0$, which means, by Lemma 1.21, that $\frac{z^{m_i}}{f'}$ is regular at p. For at least one such function, $\frac{z^{m_i}}{f'}(p) \neq 0$, since $\sum_i a_i \frac{z^{m_i}}{f'} = \frac{f'}{f'} = 1$. We obtain $(z^{m_i}) = (f) = -D_0$ in some neighborhood of p. But, in the representation of (z^{m_i}) and $-D_0$, only the prime divisors $D_1, \dots, D_k$ occur. Their coefficients are equal since the divisors are the same on some neighborhood which intersects $D_1, \dots, D_k$. Thus, $-D_0 = (z^{m_i})$ on $X_{\check\sigma}$ which implies

$$\begin{aligned}\mathcal{F}_0(X_{\check\sigma}) &= \{f \in K_{X_{\check\sigma}} \mid (f) + D_0 \geq 0\} = \{f \in K_{X_{\check\sigma}} \mid (f \cdot z^{-m}) \geq 0\} \\ &= z^m \mathcal{O}(X_{\check\sigma}).\end{aligned}$$

The other sheaf properties are evident. □

We can also express the Picard group in terms of Cartier divisors and in terms of invariant Cartier divisors.

2.14 Theorem.

(a) *The Cartier divisors of $X = X_\Sigma$ define a group* $\operatorname{Div}_C X$ *under addition.*

(b) *The principal divisors are a subgroup* $\operatorname{Div}_P X$ *of* $\operatorname{Div}_C X$.

(c) *The T-invariant Cartier divisors of $X = X_\Sigma$ define a group* $\operatorname{Div}_C^T X$ *under addition.*

(d) *The principal T-invariant Cartier divisors provide a subgroup* $\operatorname{Div}_P^T X$ *of* $\operatorname{Div}_C X$.

(e) $\operatorname{Pic} X \cong \operatorname{Div}_C X / \operatorname{Div}_P X \cong \operatorname{Div}_C^T X / \operatorname{Div}_P^T X$.

PROOF.

(a) to (d) readily follow from the definitions.

(e) We assign to each D the sheaf $\mathcal{F} = \mathcal{L}_D$ (Theorem 1.22). Then, by Theorem 2.10, we obtain a homomorphism whose kernel is the group of principal divisors. This shows the first isomorphism. The second readily follows from Theorem 2.13.

□

2.15 Theorem. *For any toric variety X_Σ, the combinatorial Picard group* $\operatorname{Pic} \Sigma$ *(V, 5) and the Picard group of X_Σ are isomorphic (as groups),*

$$\operatorname{Pic} \Sigma \cong \operatorname{Pic} X_\Sigma.$$

PROOF. By Theorem 2.13, each sheaf $\mathcal{F}$ of rational functions on X_Σ is isomorphic to a sheaf as introduced in Theorem 1.10. The latter sheaves are determined by the systems m_σ, $\sigma \in \Sigma$. Now, the theorem readily follows from V, Lemma 5.5. □

2.16 Theorem. *Let X_Σ be an arbitrary n-dimensional toric variety where Σ contains at least one n-cone, and let, for any $\sigma \in \Sigma$, the space of linear dependencies of the generators of σ be denoted by L_σ. If $\sigma_1, \ldots, \sigma_p$ are all maximal cones which are not simplex cones, we set*

$$L := L_{\sigma_1} + \cdots + L_{\sigma_p} \quad \textit{and} \quad \lambda := \dim L.$$

For k being the number of 1-cones of Σ, we obtain

$$\operatorname{Pic} X_\Sigma \cong \mathbb{Z}^{k-n-\lambda}.$$

PROOF. See V, Theorem 5.9. □

Example 3. $\operatorname{Pic} X_\Sigma = \{0\}$ if X_Σ is affine (Σ consisting of one cone σ and the faces of σ), as is true, in general, for affine varieties in algebraic geometry.

Example 4. $\operatorname{Pic} X_\Sigma = \{0\}$ is also possible for compact toric varieties, as is seen from V, 5, Example 4.

Example 5. $\operatorname{Pic} X_\Sigma = \mathbb{Z}^{k-n}$ if Σ is simplicial and contains an n-cone.

Example 6. $\operatorname{Pic} \mathbb{P}^n = \mathbb{Z}$, since, as a toric variety, $\mathbb{P}^n = X_\Sigma$ is given by a simplicial fan with $k = n + 1$.

2.17 Definition. We call $\mu(X_\Sigma) := k - n - \lambda$ the *Picard number* of X_Σ.

Exercises

1. Find $\mu(X_\Sigma)$ for Σ being spanned by the faces of a pyramid with basis Q.
2. If the singularities of a toric variety X_Σ are resolved according to VI, section 8, how does $\mu(X_\Sigma)$ change?
3. In any dimension n, find a compact toric variety X_Σ for which $\mu(X_\Sigma) = 0$.
4. Find a centrally symmetric rational realization of a dodecahedron Δ, and determine $\operatorname{Pic} X_\Sigma$ for the fan spanned by the faces of Δ.

3. Projective toric varieties

In this section, we shall study a condition under which a compact toric variety X_Σ may appropriately be "embedded" into a projective space $\mathbb{P}^r$. The respective condition for Σ is that of "strong polytopality" (V, Definition 4.3). First, we recall some facts on projective spaces and introduce some basic notions in a way which fits into the special situation we will discuss.

Each point of $\mathbb{P}^r$ can be described by homogeneous coordinates $[x_0, \ldots, x_r]$ (see VI, section 3), which are defined up to common multiples by nonvanishing complex numbers. We will consider the polynomial ring $\mathbb{C}[x_0, \ldots, x_r]$.

3.1 Definition. By a *form of degree* k we mean a homogeneous polynomial of degree k. The vector space of all forms of degree k will be denoted by $\mathbb{C}[x_0, \ldots, x_r]_k$.

Note that each $f \in \mathbb{C}[x_0, \ldots, x_r]$ can be represented as $f = f_{i_1} + \cdots + f_{i_l}$ where $0 \leq i_1 < \cdots < i_l \leq k$ and $f_{i_j} \in \mathbb{C}[x_0, \ldots, x_r]_{i_j}$, $j = 1, \ldots, l$.

The forms of degree k do not define functions on $\mathbb{P}^r$, since $[\lambda x_0, \ldots, \lambda x_r] = [x_0, \ldots, x_r]$ for $\lambda \neq 0$ but $f(\lambda x_0, \ldots, \lambda x_r) = \lambda^k f(x_0, \ldots, x_r)$ which, in general, does not coincide with $f(x_0, \ldots, x_r)$. However, the zero set of the form is well defined.

3.2 Definition. Let $F_1, \ldots, F_k$ be forms on $\mathbb{P}^r$.

A set $Z = \{x \in \mathbb{P}^r \mid F_1(x) = \cdots = F_k(x) = 0\}$ is called an *algebraic subset* of $\mathbb{P}^r$.

As in the case of algebraic subsets of affine varieties, each algebraic subset Z of $\mathbb{P}^r$ defines the homogeneous ideal

$$\mathfrak{i}_{Z,\mathbb{P}^r} := \{f_{i_1} + \cdots + f_{i_l} \mid f_{i_j}|_Z = 0, \; j = 1, \ldots, l\}.$$

So, the ideals are of the same type as those for affine sets, and, hence, have the same properties. In particular, they are finitely generated.

3.3 Lemma and Definition. *The algebraic subsets of* $\mathbb{P}^r$ *may be considered as the closed sets of a topology on* $\mathbb{P}^r$, *called the Zariski topology on* $\mathbb{P}^r$.

PROOF. This follows from the above remarks about the ideals $\mathfrak{i}_{Z,\mathbb{P}^r}$. □

On $\mathbb{P}^r$, we can define rational functions $f = \frac{F_1}{F_2}$ where $F_1, F_2 \in \mathbb{C}[x_0, \ldots, x_r]_k$ for some $k \geq 0$. We may assume F_1, F_2 to be relatively prime. Then, f is a well defined, regular function on the set $X = \mathbb{P}^r \setminus \{x \mid F_2(x) = 0\}$, and it satisfies

$$f(\lambda x_0, \ldots, \lambda x_r) = \frac{\lambda^k F_1(x_0, \ldots, x_r)}{\lambda^k F_2(x_0, \ldots, x_r)} = f(x_0, \ldots, x_r)$$

for any $\lambda \neq 0$. Let U be a subset of $\mathbb{P}^r$. The set R_U of all rational functions regular on U is readily seen to be a ring. In analogy to VI, Definition 1.8, we call it the *ring of regular functions on* U. One should keep in mind, however, that its elements are not functions of the homogeneous coordinates $x_0, \ldots, x_r$, but of the points they represent. Furthermore, the degrees of the forms F_1, F_2 are supposed to be equal.

Example 1. We consider the rings $R_{U_0}, \ldots, R_{U_r}$ where $U_0, \ldots, U_r$ are the affine charts of $\mathbb{P}^r$. Recall that $U_i = \{[x_0, \ldots, x_r] \mid x_i \neq 0\}, i = 0, \ldots, r$.

We see that $R_{U_i} = \{f = \frac{F}{x_i^k} \mid k \geq 0, F \in \mathbb{C}[x_0, \ldots, x_r]_k\} = \mathbb{C}[\frac{x_0}{x_i}, \ldots, \frac{x_{i-1}}{x_i}, \frac{x_{i+1}}{x_i}, \ldots, \frac{x_r}{x_i}] \cong \mathbb{C}[y_1, \ldots, y_r]$ where $y_j = \frac{x_{j-1}}{x_i}$ for $j = 1, \ldots, i$ and $y_j = \frac{x_j}{x_i}$ for $j = i+1, \ldots, r$. Moreover, we have a bijection $\varphi_i : U_i \longrightarrow \mathbb{C}^r$ defined by $[x_0, \ldots, x_r] \longmapsto (\frac{x_0}{x_i}, \ldots, \frac{x_{i-1}}{x_i}, \frac{x_{i+1}}{x_i}, \ldots, \frac{x_r}{x_i})$. The corresponding ring homomorphism

$$\varphi_i^* : \mathbb{C}[y_1, \ldots, y_r] = R_{\mathbb{C}^r} \longrightarrow R_{U_i}$$

is an isomorphism. Hence, φ_i is an isomorphism in the sense of VI, Definition 1.33.

Thus, $\mathbb{P}^r$ can be covered by open sets U_i that are isomorphic to $\mathbb{C}^r$, a result which we have already found by considering $\mathbb{P}^r$ (for $r = n$) as a toric variety X_Σ, where Σ has $n+1$ cones $\sigma_0, \ldots, \sigma_n$ of dimension n and $U_i = X_{\check{\sigma}_i}, i = 0, \ldots, n$ (compare VI, section 3).

3.4 Lemma. *The Zariski topology on $\mathbb{P}^r$, as defined in Lemma* 3.3, *coincides with the Zariski topology for $\mathbb{P}^r$ considered as a toric variety X_Σ (compare* VI, *Definition* 1.24*).*

PROOF. It is sufficient to prove that, on each U_i (introduced in Example 1), the topology defined by regular functions and the topology induced by the Zariski topology on X_Σ are equal.

First, let Z be the zero set of finitely many forms

$$Z = \{p \in \mathbb{P}^r \mid F_j(p) = 0,\ j = 1, \ldots, s\}.$$

Then, $Z \cap U_i = \{p \in U_i \mid \frac{F_j}{x_i^{k_j}}(p) = 0,\ F_j \in \mathbb{C}[x_0, \ldots, x_r]_{k_j}, j = 1, \ldots, s\}$. Hence, $Z \cap U_i$ is the zero set of regular functions which defines a closed set in the induced topology of X_Σ.

Conversely, let $Z \subset U_i$ be the zero set of regular functions of R_{U_i}. Then,

$$Z = \{p \in U_i \mid f_j(p) = 0,\ f_j \in R_{U_i},\ j = 1, \ldots, s\}.$$

Setting $F_j := x_i^{k_j} f_j$ we associate a form $F_j \in \mathbb{C}[x_0, \ldots, x_r]_{k_j}$ with each $f_j = \frac{F_j}{x_i^{k_j}}$ such that Z is described as

$$\begin{aligned} Z &= \{p \in U_i \mid F_j(p) = 0,\ j = 1, \ldots, s\} \\ &= U_i \cap \{p \in \mathbb{P}^r \mid F_j(p) = 0,\ j = 1, \ldots, s\}. \end{aligned}$$

This completes the proof of the Lemma. □

3.5 Definition. We call a compact toric variety X_Σ *projective* if there exists an injective morphism

$$\Phi : X_\Sigma \hookrightarrow \mathbb{P}^r$$

of X_Σ into some projective space such that $\Phi(X_\Sigma)$ is Zariski closed in $\mathbb{P}^r$. We consider $\mathbb{P}^r$ as a toric variety whose big torus T has dimension r. We say Φ is *equivariant*, if it is equivariant in the sense of VI, Definition 6.3, where the image $\Phi(T_0)$ of the big torus T_0 acts on $\Phi(X_\Sigma)$ as a subgroup of T. If an equivariant Φ exists, we call X_Σ *equivariantly projective*.

Let $\varphi : X_\Sigma \hookrightarrow \mathbb{P}^r$ be an equivariant morphism. We obtain

$$\begin{array}{rcc} \varphi|_{T_0} : T_0 & \longrightarrow & T = \{[x_0, \dots, x_r] \mid x_i \neq 0,\ i = 0, \dots, r\} \\ z = (z_1, \dots, z_n) & \longmapsto & [z^{m_0}, \dots, z^{m_r}]. \end{array}$$

The monomials z^{m_j} are defined up to a common multiple z^m, so that $(z^{m_0+m}, \dots, z^{m_r+m})$ may also be chosen as a representative.

In V, section 5, we introduced the notion of an associated polytope P of a strongly polytopal fan Σ. We obtain P by translating and intersecting the duals of all cones of Σ so that $\Sigma = \Sigma(-P)$ becomes the fan of normal cones of $-P$ (see also I, section 4):

$$P = (a_1 + \operatorname{pos}(P - a_1)) \cap \dots \cap (a_q + \operatorname{pos}(P - a_q)).$$

Figure 3 illustrates σ_i, $\check{\sigma}_i$ and $-\check{\sigma}_i = \operatorname{pos}(a_i - P)$ for some $i \in \{1, \dots, q\}$. Sometimes, we write m_{σ_i} instead of a_i.

If $b_1, \dots, b_q \in \mathbb{Z}^n$ are chosen so that they satisfy the compatibility condition (1) in V, section 5, instead of P, we obtain either $\emptyset$ or a polytope

$$Q = (b_1 + \check{\sigma}_1) \cap \dots \cap (b_q + \check{\sigma}_q),$$

which can be lower dimensional. For sufficiently large $r \in \mathbb{Z}_{\geq 0}$, Q is a summand of rP (see V, Theorem 5.15). If Σ is not strongly polytopal, Q can still be a polytope. However, it does not carry enough information to reconstruct Σ from it.

3.6 Definition. If a sheaf $\mathcal{F}$ of $\mathcal{O}_{X_\Sigma}$-modules is given according to Theorem 1.10, and if $\Sigma = \Sigma(-P)$ is the fan of a polytope $-P$, we call $P = P(\mathcal{F})$ an *$\mathcal{F}$-polytope*.

Clearly:

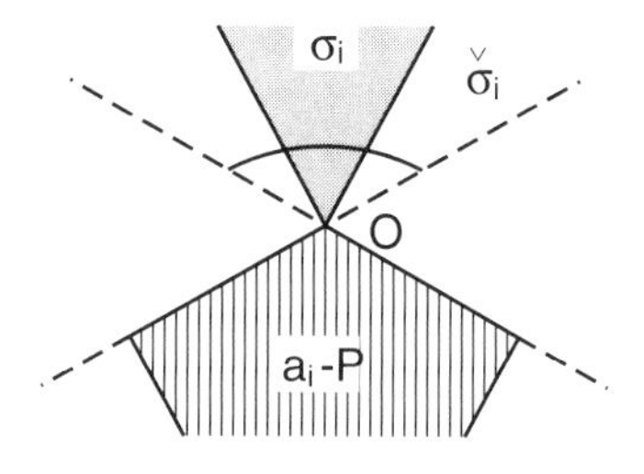

FIGURE 3.

3.7 Lemma. *If $P = P(\mathcal{F})$ is an $\mathcal{F}$-polytope, then any lattice polytope strictly combinatorially isomorphic to P is also an $\mathcal{F}$-polytope.*

3.8 Lemma. *Let $\Sigma = \Sigma(-P)$ be the fan of a polytope $-P$. Then, $\bar{P} := nP$ has the following property. For any lattice point a of $\bar{P}$, the generators of the monoid $\check{\sigma}_a \cap \mathbb{Z}^n := [\mathrm{pos}(\bar{P} - a)] \cap \mathbb{Z}^n$ lie in $\bar{P} - a$.*

PROOF. First, let a be a vertex of $\bar{P}$. We decompose $\check{\sigma}_a$ into simplex cones $\tau_1, \dots, \tau_s$ (see V, Theorem 1.12). For each monoid $\tau_i \cap \mathbb{Z}^n$, we see from the proof of Gordan's lemma (V, Lemma 3.4) that there are generators in the parallelepiped spanned by the simple lattice vectors $a_{i1}, \dots, a_{in}$ for which $\tau_i = \mathrm{pos}\{a_{i1}, \dots, a_{in}\}, i = 1, \dots, s$. This parallelepiped is readily seen to lie in $\bar{P} - a$.

Let $a \in \mathrm{relint}\, F$ where F is a face of P, $1 \le \dim F \le n$. We set $\check{\sigma}_a := \mathrm{pos}(\bar{P} - a) = \{f_1(x) \ge 0\} \cap \dots \cap \{f_k(x) \ge 0\}$ for functionals $f_1, \dots, f_k$. For any vertex v of F, we find $\check{\sigma}_v := \mathrm{pos}(\bar{P} - v) = \{f_1(x) \ge 0\} \cap \dots \cap \{f_l(x) \ge 0\}$ where $l > k$. Let $\tau := \mathrm{pos}(F - v) = \{f_1(x) = \dots = f_k(x) = 0\} \cap \{f_{k+1}(x) \ge 0\} \cap \dots \cap \{f_l(x) \ge 0\}$. Since $a \in \mathrm{relint}\, F$, $\sum_{i=0}^{s} \alpha_i (m_i - a) = 0$ where $\alpha_i \in \mathbb{Z}_{>0}$ and m_i are the vertices of F, $i = 0, \dots, s$. We may assume $v = m_0$. Then, $a - v = \sum_{j=1}^{s} \alpha_i (m_i - a) + (\alpha_0 - 1)(v - a)$ which means that $a - v$ is in the monoid M generated by all $m - a$ for $m \in P$. Hence, all elements $m - v = (m - a) + (a - v)$ are in M so that, by the first part of the proof, $\check{\sigma}_v \cap \mathbb{Z}^n \subset M$. Note that $f_{k+1}(a - v) > 0, \dots, f_l(a - v) > 0$ since $a \in \mathrm{relint}\, F$. Now, for $p \in \check{\sigma}_a \cap \mathbb{Z}^n$, $f_{k+1}(p + \lambda(a - v)) > 0, \dots, f_l(p + \lambda(a - v)) > 0$ for sufficiently large λ. Moreover, $f_1(p + \lambda(a - v)) \ge 0, \dots, f_l(p + \lambda(a - v)) \ge 0$. Hence, $p + \lambda(a - v) \in \check{\sigma}_v$ and $p \in \check{\sigma}_v + \lambda(v - a) \subset M$, so that $M = \check{\sigma}_a \cap \mathbb{Z}^n$ follows. □

3.9 Definition. By a *rational map* $\varphi : X \dashrightarrow Y$, we mean a morphism from a Zariski open set U of X to Y, $\varphi : U \to Y$.

Each sequence $(z^{m_0}, \dots, z^{m_r})$ of monomials defines a rational map to $\mathbb{P}^r$, possibly a morphism, which, then, is readily seen to be uniquely determined.

3.10 Lemma. *A rational map, defined by $(z^{m_0}, \dots, z^{m_r})$, is a morphism if and only if, for each $p \in X_\Sigma$, we can find a monomial z^m such that the monomials $(z^{m+m_0}, \dots, z^{m+m_r})$ are regular at p and do not all vanish there.*

PROOF. Let $X_{\check{\sigma}}$ be a smallest affine piece of X_Σ which contains p. It is sufficient to find an extension of φ on $X_{\check{\sigma}}$. Assume $z^{m+m_i}(p) \ne 0$. Then, also, $z^{m+m_i} \ne 0$ on all of $X_{\check{\sigma}}$, and we obtain a well defined morphism

$$\begin{array}{rcc} \varphi_\sigma : X_{\check{\sigma}} & \longrightarrow & U_i \\ x & \longmapsto & (z^{m_0 - m_i}, \dots, 1, \dots, z^{m_r - m_i}). \end{array}$$

All functions $z^{m_j - m_i} = \frac{z^{m_j + m}}{z^{m_i + m}}$ are regular on $X_{\check{\sigma}}$. □

3.11 Theorem. *A toric variety X_Σ is equivariantly projective if and only if Σ is (complete and) strongly polytopal.*

PROOF. Let Σ be strongly polytopal, and let P be an associated polytope with a set of vertices $\{m_\sigma \mid \sigma \in \Sigma^{(n)}\}$, so that $P = \bigcap_{\sigma\in\Sigma^{(n)}}(m_\sigma + \check{\sigma})$. We also consider the cones $\check{\sigma}_a$ as introduced in Lemma 3.8. For any $a \in P$, we may assume the generators of each $\check{\sigma}_a \cap \mathbb{Z}^n$ to be contained in $P - a$; if this is not so, we replace P by nP (see Lemma 3.8). Let $m_0, \dots, m_r$ be all lattice points in P, and let $\check{\sigma}_i := \operatorname{pos}(P - m_i), i = 0, \dots, r$ (compare Lemma 3.8). We consider the rational map

$$(1) \qquad \begin{array}{rcl} \varphi : X_\Sigma & \dashrightarrow & \mathbb{P}^r \\ p & \longmapsto & [z^{m_0}(p), \dots, z^{m_r}(p)] \end{array}$$

where $z^{m_j}(p) = p^{m_j}$ in case $p = (p_1, \dots, p_n) \in T$ (but the monomials z^{m_j} are not defined in each point of X_Σ). Then,

$$\begin{array}{rcl} \varphi|_{X_{\check{\sigma}_i}} : X_{\check{\sigma}_i} & \longrightarrow & U_i \\ p & \longmapsto & (z^{m_0-m_i}(p), \dots, z^{m_r-m_i}(p)) \end{array}$$

is a morphism, for each $i = 0, \dots, r$. Therefore, φ is a morphism. Moreover, since $R_{U_i} = \mathbb{C}[\frac{x_0}{x_i}, \dots, \frac{x_{i-1}}{x_i}, \frac{x_{i+1}}{x_i}, \dots, \frac{x_r}{x_i}]$, $R_{X_{\check{\sigma}}} = \mathbb{C}[z^{m_0-m_i}, \dots, z^{m_r-m_i}]$ is generated by $z^{m_0-m_i}, \dots, z^{m_r-m_i}$, $(\varphi|_{X_{\check{\sigma}}})^* : R_{U_i} \to R_{X_{\check{\sigma}}}$ is a surjection. Hence, by VI, Lemma 1.34, $\varphi|_{X_{\check{\sigma}}}$ is a closed embedding.

We still have to show that φ is a bijection. Suppose $\varphi(p) = \varphi(q) \in U_i$. Then all $z^{m_j-m_i}$ are regular at p and q, $j = 0, \dots, r$, hence, $p, q \in X_{\check{\sigma}}$. But $\varphi|_{X_\sigma}$ is an embedding, so that $p = q$.

Conversely, let $\varphi : X_\Sigma \to \mathbb{P}^r$ be an equivariant embedding, given by rational functions $z^{m_0}, \dots, z^{m_r}$. For $\sigma \in \Sigma^{(n)}$ and $O_\sigma = \{p\}$, $\varphi(p) \in U_i$ for some i. As φ is equivariant, $\varphi(T_0)$ is a subgroup of the big torus T of $\mathbb{P}^r$. Hence, $\varphi^{-1}(U_i) = \varphi^{-1}(U_i \cap \varphi(X_\Sigma)) \cong U_i \cap \varphi(X_\Sigma)$ is an affine open, T_0-invariant subset of X_Σ and contains $X_{\check{\sigma}}$. Hence, $X := \varphi^{-1}(U_i) = \bigcup_{i\in I_0} X_{\check{\sigma}_i} \supset X_{\check{\sigma}}$ for some index set I_o, and $\sigma_i \in \Sigma$. We consider the cone $\sigma' := \bigcap_{i\in I_0}\check{\sigma}_i$. The intersection $\sigma' \cap \mathbb{Z}^n$ represents all monomials which are regular on X. $\varphi|_{\varphi^{-1}(U_i)} : \varphi^{-1}(U_i) \to U_i$ is a closed embedding. Then, by VI, Lemma 1.34, φ^* is a surjection. The functions $\varphi^*(\frac{x_0}{x_i}) = z^{m_0-m_i}, \dots, \varphi^*(\frac{x_r}{x_i}) = z^{m_r-m_i}$ are regular on U_i and span $R_{\varphi^{-1}(U_i)}$. Since $\varphi^{-1}(U_i)$ is affine, it equals $X_{\sigma'}$, and $O_{\sigma'}$ is its only minimal orbit. Hence, $O_{\sigma'} = \{p\}$ and $X_{\sigma'} = X_{\check{\sigma}}$ is the smallest open invariant subset containing p.

Finally, we see that, for each $\sigma \in \Sigma^{(n)}$, the $m_j - m_i$, $j = 0, \dots, r$, (i as above), generate $\check{\sigma} \cap \mathbb{Z}^n$. We choose $m_i = m_\sigma$ for the cones $\sigma \in \Sigma^{(n)}$. For any $\tau \in \Sigma$, we choose a $\sigma \supset \tau$, $\sigma \in \Sigma^{(n)}$ and set $m_\tau := m_\sigma$. Note that the system $\{m_\sigma + \check{\sigma} \mid \sigma \in \Sigma^{(1)}\}$ is a virtual polytope. The polytope $P = \bigcap_{\sigma\in\Sigma^{(1)}}(m_\sigma + \check{\sigma})$ contains $m_0, \dots, m_r$ and is of maximal dimension. The m_σ, for $\sigma \in \Sigma^{(n)}$, are clearly the vertices of P. The associated fan $\Sigma(-P)$ is seen to equal Σ. Hence, Σ is strongly polytopal. □

3.12 Theorem. *Let X_Σ be equivariantly projective, and let*

$$\Phi : X_\Sigma \hookrightarrow \mathbb{P}^r$$

be the embedding which is induced by the rational map φ in (1). *Then, $\Phi(X_\Sigma)$ is the set of common solutions of finitely many monomial equations*

$$x_{i_0}^{\alpha_0} \cdots x_{i_k}^{\alpha_k} = x_{i_{k+1}}^{\alpha_{k+1}} \cdots x_{i_r}^{\alpha_r} \tag{2}$$

which arise from affine relationships

$$\alpha_0 m_{i_0} + \cdots + \alpha_k m_{i_k} = \alpha_{k+1} m_{i_{k+1}} + \cdots + \alpha_r m_{i_r}$$
$$\alpha_0 + \cdots + \alpha_k = \alpha_{k+1} + \cdots + \alpha_r, \quad \alpha_j \in \mathbb{Z}_{\geq 0}, \ j = 1, \cdots, r.$$

PROOF. We consider the ideal $\mathfrak{i} \subset \mathbb{C}[x_0, \ldots, x_r]$, generated by all forms $F \in \mathbb{C}[x_0, \ldots, x_r]$, for which $F(z^{m_0}, \ldots, z^{m_r})$ is the zero Laurent polynomial (as, for example, $F(x_0, x_1, x_2, x_3) = x_0 x_2 - x_1 x_3$ in Example 3 below which satisfies $F(z^{m_0}, z^{m_1}, z^{m_2}, z^{m_3}) = z^{m_0+m_2} - z^{m_1+m_3} \equiv 0$ since $m_0 + m_1 = m_2 + m_3$). We know, by Hilbert's basis theorem (or as a consequence of V, Lemma 3.10), that $\mathfrak{i}$ is finitely generated. It defines a subvariety

$$Z := \{x \in \mathbb{P}^r \mid F(x) = 0 \text{ for all } F \in \mathfrak{i}\}.$$

We assert that $Z = \Phi(X_\Sigma)$. In fact, for U_i as above,

$$\begin{aligned} \mathfrak{i}_{Z \cap U_i, U_i} &= \{\frac{F}{x^k} \mid F \in \mathbb{C}[x_0, \ldots, x_r]_k \cap \mathfrak{i}, \quad k \geq 0\} \\ &= \{f \in \mathbb{C}[\frac{x_0}{x_i}, \ldots, \frac{x_{i-1}}{x_i}, \frac{x_{i+1}}{x_i}, \ldots, \frac{x_r}{x_i}] \\ &\qquad \mid f(z^{m_0-m_i}, \ldots, z^{m_{i-1}-m_i}, z^{m_{i+1}-m_i}, \ldots, z^{m_r-m_i}) \equiv 0\}. \end{aligned}$$

On the other hand, we deduce from $\Phi|_{X_{\check{\sigma}_i}} : X_{\check{\sigma}_i} \to U_i$ being an embedding that

$$\begin{aligned} \mathfrak{i}_{\Phi(X_\Sigma) \cap U_i, U_i} &= \ker(\varphi^* : \mathbb{C}[\frac{x_0}{x_i}, \ldots, \frac{x_{i-1}}{x_i}, \frac{x_{i+1}}{x_i}, \ldots, \frac{x_r}{x_i}] \\ &\qquad \longrightarrow \mathbb{C}[z^{m_0-m_i}, \ldots, z^{m_{i-1}-m_i}, z^{m_{i+1}-m_i}, \ldots, z^{m_r-m_i}]) \\ &= \mathfrak{i}_{Z \cap U_i, U_i}. \end{aligned}$$

Hence, $\Phi(X_\Sigma) \cap U_i = Z \cap U_i$ for $i = 0, \ldots, r$, and we conclude that $\Phi(X_\Sigma) = Z$.

Let F be any form of degree k in $\mathfrak{i}$, and write $F = \sum_{i \in I} a_i x^i$ where I is a finite set of vectors $i = (i_0, \ldots, i_r) \in (\mathbb{Z}_{\geq 0})^{r+1}$ such that $i_0 + \cdots + i_r = k$. Let $\bar{m} := (m_0, \ldots, m_r)$ and $\bar{m} \cdot i := m_0 i_0 + \cdots + m_r i_r$. Then, $F(z^{m_0}, \ldots, z^{m_r}) = \sum_{i \in I} a_i z^{\bar{m} \cdot i} = 0$. For $I_m := \{i \in I \mid \bar{m} \cdot i = m\}$ and $M := \{\bar{m} \cdot i \mid i \in I\}$, we obtain $F(z^{m_0}, \ldots, z^{m_r}) = \sum_{m \in M} (\sum_{i \in I_m} a_i) z^m \equiv 0$, and, hence, $\sum_{i \in I_m} a_i = 0$. If for $i \in I_m$, $a_i \neq 0$, then, there is an $a_j \neq 0$ with $j \in I_m$. By subtracting $a_i(x^i - x^j)$ from F, we find a form with a number of monomials smaller than F. Proceding by induction we find that $F = \sum b_i(x^i - x^j)$ where $\bar{m} \cdot i = \bar{m} \cdot j$. Hence, $m_0 i_0 + \cdots + m_r i_r = m_0 j_0 + \cdots + m_r j_r$ and $i_0 + \cdots + i_r = j_0 + \cdots + j_r$. This proves the theorem. □

Example 2. If Σ has generators $e_1, \ldots, e_n, -e_1 - \cdots - e_n$ (see VI, section 3, Example 5), we may choose, as an associated polytope, the simplex $P := \operatorname{conv}\{0, -e_1, \ldots, -e_n\}$. Since the vertices of P are the only lattice points of P

and since there are no affine relations between them, each point $[y_0, \ldots, y_n]$ of $\mathbb{P}^n$ represents a point of X_Σ, that is,

$$X_\Sigma \cong \mathbb{P}^n .$$

Example 3. Hirzebruch surfaces $\mathcal{H}_0 = \mathbb{P}^1 \times \mathbb{P}^1$ and $\mathcal{H}_1$ (see VI, 3, Example 3 with Σ reflected in 0).

For $\mathcal{H}_0 = \mathbb{P}^1 \times \mathbb{P}^1$, we can choose a square with side length 1 as $\bar{P}$, so there are four coordinates (y_0, y_1, y_2, y_3), and we obtain an embedding of $\mathbb{P}^1 \times \mathbb{P}^1$ into $\mathbb{P}^3$ by one equation $y_0 y_2 = y_1 y_3$. For $\mathcal{H}_1$, we find an embedding into $\mathbb{P}^4$ which is represented by the following equations in the coordinates $y_i := z^{m_i}$ (Figure 4),

$$y_0 y_2 = y_1 y_4, \qquad y_1 y_3 = y_2^2, \quad \text{and} \quad y_0 y_3 = y_2 y_4.$$

It is readily seen that none of these monomial equations is a consequence of the others.

Remark. The search for equations can be achieved as follows. For each vertex m of $\bar{P}$, consider all $m - m_i$, for $m_i \in (\bar{P} \cap \mathbb{Z}^n) \setminus \{m\}$ as generators of $\check{\sigma}$ and look for a basis of the space of all positive linear relations in the sense of V, 3. Use them to represent $X_{\check{\sigma}}$ according to VI, Theorem 2.7, and make the monomial equations homogeneous (by setting $\xi_i = \frac{y_i}{y_0}$, $i = 1, \ldots, k$, and multiplying the monomial equations by an appropriate power of y_0). Collect all equations obtained in this way, and sort out those which are consequences of others. This search is not at all trivial and leads to questions of linear programming.

For $k > 1$, the number of equations for a Hirzebruch surface increases rapidly (compare Exercise 1). So, the equivariant projective representation of $\mathcal{H}_k$ according to Theorem 2.2 is, in general, rather awkward. The original definition of $\mathcal{H}_k$ by one equation in $\mathbb{P}^2 \times \mathbb{P}^1$ is much more elegant. Furthermore, it can be shown by other methods that $\mathcal{H}_k$ is always embeddable into a $\mathbb{P}^5$.

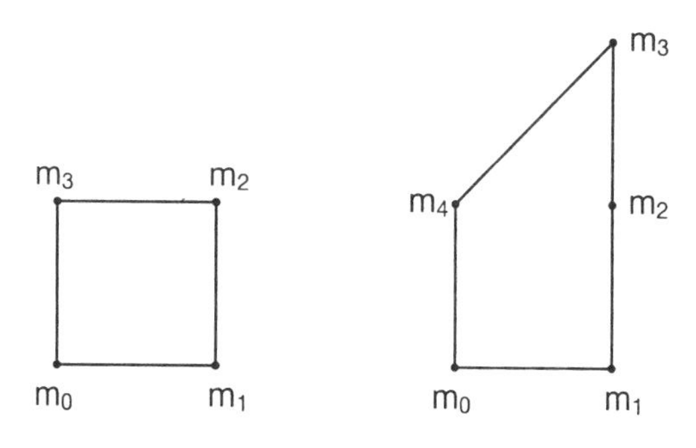

FIGURE 4.

3.13 Theorem. *Given a compact toric variety X_Σ, there exists a projective toric variety $X_{\Sigma'}$ and an equivariant morphism $\psi : X_{\Sigma'} \longrightarrow X_\Sigma$ (induced by the identity map of $\mathbb{R}^n$).*

PROOF. This is a conclusion from V, Theorem 4.5. □

3.14 Definition. An invertible sheaf $\mathcal{F}$ of rational functions is called *very ample* if, for a basis $s_0, \ldots, s_r$ of the vector space $\mathcal{F}(X_\Sigma)$ of global sections (see Definition 1.11), the assignment $x \mapsto [s_0(x), \ldots, s_r(x)]$ provides a closed embedding $\Phi : X_\Sigma \to \mathbb{P}^r$. We say $\mathcal{F}$ is *ample* with *ampleness factor* k if $\mathcal{F}^{(k)} := \mathcal{F} \otimes \cdots \otimes \mathcal{F}$ (k times) is very ample. If $\mathcal{F} = \mathcal{L}_D$ (see Theorem 1.22) is ample, we also say that D is *ample*.

If X_Σ is equivariantly projective, hence, possesses an $\mathcal{F}$-polytope $P = P(\mathcal{F})$, then, by Lemma@-5, $\mathcal{F}$ is ample with n as an ampleness factor. The invariant global sections correspond to the lattice points of P. We have used this fact implicitly in the proof of Theorem 3.12.

Exercises

1. Find an embedding of the Hirzebruch surface $\mathcal{H}_2$ into $\mathbb{P}^5$ by setting up five quadratic monomial equations none of which is a consequence of the others and which imply all equations (2).
2. Find an embedding $\mathbb{P}^1 \times \mathbb{P}^1 \times \mathbb{P}^1$ into $\mathbb{P}^7$ by an appropriate 3-cube as polytope $\bar{P}$. Find nine quadratic monomial equations none of which is a consequence of the others and which imply all equations (2).
3. Find a polytope $\bar{P}$ for the weighted projective space introduced in VI, 3, Example 8.
4. Represent $\mathbb{P}^p \times \mathbb{P}^q$, $p, q \in \mathbb{Z}_{\geq 0}$, as a toric variety, and find a polytope $\bar{P}$ for the embedding $\mathbb{P}^p \times \mathbb{P}^q \hookrightarrow \mathbb{P}^r$ according to Theorem 3.12.

4. Support functions and line bundles

In section 2 we have shown that the Picard group of a toric variety has four isomorphic characterizations

$$\operatorname{Pic} X_\Sigma \cong \operatorname{Div}_C X_\Sigma / \operatorname{Div}_P X_\Sigma \cong \operatorname{Div}_C^T X_\Sigma / \operatorname{Div}_P^T X_\Sigma \cong \operatorname{Pic} \Sigma.$$

We shall add three more descriptions. This illustrates nicely the interaction of different algebraic geometric and combinatorial concepts in the special case of toric varieties. We use seven "languages" to express the same facts.

First, we introduce piecewise linear functions defined on the point set $|\Sigma|$ which, in the case of a complete projective toric variety, are the negative support functions of polytopes $-P$, where $\Sigma = \Sigma(-P)$ (see I, Definition 4.14).

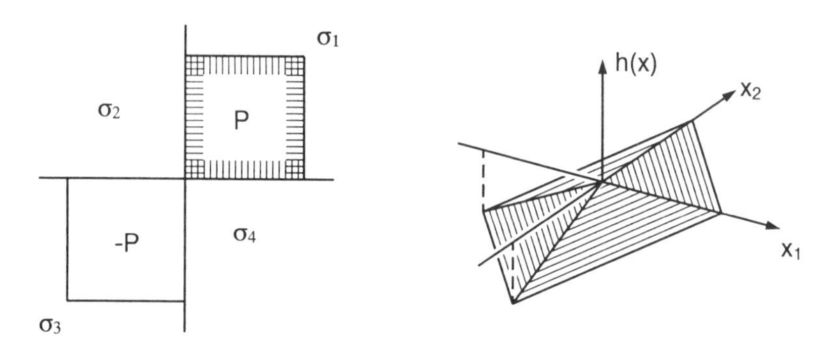

FIGURE 5.

Example 1. Let Σ be the fan of $\mathbb{P}^1 \times \mathbb{P}^1$ consisting of the four quadrants $\sigma_1, \ldots, \sigma_4$ of $\mathbb{R}^2$ and their faces. We choose as P the square with vertices $m_{\sigma_1} = 0$, $m_{\sigma_2} = e_1, m_{\sigma_3} = e_1 + e_2, m_{\sigma_4} = e_2$. We obtain $P = (m_{\sigma_1} + \check{\sigma}_1) \cap \cdots \cap (m_{\sigma_4} + \check{\sigma}_4)$, and, for the corresponding sheaf, we have the defining modules $z^{m_{\sigma_i}} \mathcal{O}(X_{\check{\sigma}_i})$, $i = 1, \ldots, 4$. We set

$$h(x) = -h_{-P}(x) = \begin{cases} \langle 0, x \rangle = 0 & \text{for } x \in \sigma_1, \\ \langle e_1, x \rangle & \text{for } x \in \sigma_2, \\ \langle e_1 + e_2, x \rangle & \text{for } x \in \sigma_3, \\ \langle e_2, x \rangle & \text{for } x \in \sigma_4. \end{cases}$$

Clearly, $h(e_1) = -h_{-P}(e_1) = 0$, $h(e_2) = -h_{-P}(e_2) = 0$, $h(-e_1) = -h_{-P}(-e_1) = -1$, and $h(-e_2) = -h_{-P}(-e_2) = -1$ (Figure 5).

Example 2. We choose the same Σ as in Example 1 but characterize, in Figure 6, a virtual polytope, which is not a polytope, by a function h whose negative is not convex.

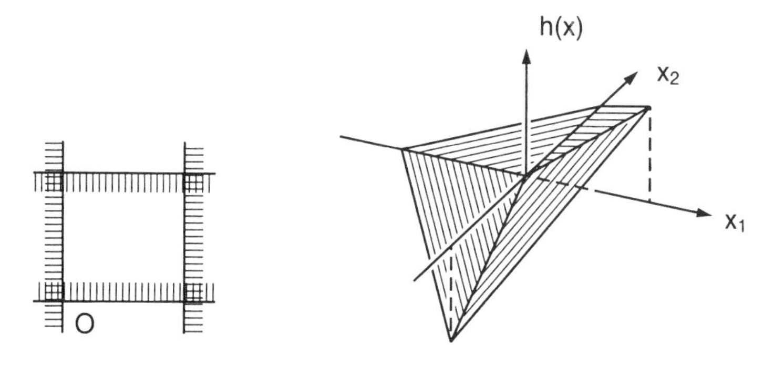

FIGURE 6.

4.1 Definition. Given a complete fan Σ, we call a function

$$h : |\Sigma| \longrightarrow \mathbb{R}$$

a *virtual support function* of Σ if it satisfies the following conditions:

(1) h is continuous.
(2) For each $\sigma \in \Sigma$, there exists an $m_\sigma \in \mathbb{Z}^n$ such that

$$h(x) = \langle m_\sigma, x\rangle \quad \text{for all } x \in \sigma.$$

An immediate consequence of Definition 4.1 and V, Definition 5.3 is the following.

4.2 Lemma. *h is a virtual support function if and only if the set $\{m_\sigma + \check{\sigma}\}_{\sigma\in\Sigma}$ is a virtual polytope, that is, the set $\{m_\sigma\}_{\sigma\in\Sigma}$ satisfies the compatibility condition*

$$m_\sigma - m_{\sigma_0} \in \operatorname{cospan} \check{\sigma}_0 \text{ if } \sigma_0 \text{ is a face of } \sigma. \tag{3}$$

The following theorem is also readily obtained by using the results of V, section 5:

4.3 Theorem.

(a) *The virtual support functions of Σ define a group* $\mathrm{SF}(\Sigma)$ *under addition.*
(b) *The linear functions obtained in case all m_σ equal the same lattice point define a subgroup* $\mathrm{LF}(\Sigma)$ *of* $\mathrm{SF}(\Sigma)$.
(c) *We have*

$$\operatorname{Pic} \Sigma \cong \mathrm{SF}(\Sigma)/\,\mathrm{LF}(\Sigma).$$

Example 3. We consider the fan Σ introduced in V, 5, Example 4. As we remarked there, Σ cannot be spanned by the (planar) faces of a spherical polyhedron. Hence, there is no continuous function which is linear on each $\sigma \in \Sigma$. So, no virtual support function and hence no Cartier divisor exists, which proves again $\operatorname{Pic} X_\Sigma = \{0\}$.

Now we turn to a notion that is closely related to that of sheaves and is used in many parts of algebraic geometry and complex analysis.

4.4 Definition. Let $\{U_\alpha\}_{\alpha\in I}$ be a Zariski open covering of a toric variety X_Σ, and let $g_{\alpha\beta}$ for $\alpha, \beta \in I$ be invertible maps on $U_\alpha \cap U_\beta$ satisfying the following cocycle condition:

$$\text{For any } \alpha, \beta, \gamma \in I,\ g_{\beta\gamma} \cdot g_{\gamma\alpha} = g_{\beta\alpha} \text{ on } U_\alpha \cap U_\beta \cap U_\gamma.$$

We glue together each pair $U_\alpha \times \mathbb{C}$, $U_\beta \times \mathbb{C}$ by the map $g_{\alpha\beta} \times \mathrm{id}$:

$$\begin{array}{cccc} g_{\alpha\beta} \times \mathrm{id} : & (U_\alpha \cap U_\beta) \times \mathbb{C} & \longrightarrow & (U_\alpha \cap U_\beta) \times \mathbb{C} \\ & \cap & & \cap \\ & U_\alpha \times \mathbb{C} & & U_\beta \times \mathbb{C} \end{array}$$

As a result, we obtain a point set $\mathcal{L}$. We say $(\mathcal{L}, \{g_{\alpha\beta}\})$, or briefly $\mathcal{L}$, is a *line bundle* with *transition maps* $g_{\alpha\beta}$. The map $\pi : \mathcal{L} \to X_\Sigma$, defined by $(x, c) \mapsto x$ in any $U_\alpha \times \mathbb{C}$, is called the *projection map* of $\mathcal{L}$.

By definition, each $g_{\alpha\beta}$ preserves all fibers $\pi^{-1}(x)$. Moreover, $\pi^{-1}(x)$ has the structure of a one-dimensional vector space over $\mathbb{C}$.

Example 4. $\mathcal{L} = X_\Sigma \times \mathbb{C}$ with $g_{\alpha\beta} = \mathrm{id}$ for each pair $\alpha, \beta \in I$ (if any covering is given) is called the *trivial* line bundle.

Note that, for any open set U contained in some U_α, we can find an isomorphism

$$f_{U,\alpha} : \pi^{-1}(U) \longrightarrow U \times \mathbb{C} \tag{5}$$

such that, for any $x \in U$,

$$f_{U,\alpha}|_{\pi^{-1}(x)} : \pi^{-1}(x) \longrightarrow \{x\} \times \mathbb{C}$$

is an isomorphism of one-dimensional vector spaces.

4.5 Definition. Let $(\mathcal{L}, g_{\alpha\beta})$, $(\mathcal{L}', g'_{\alpha\beta})$, be two line bundles on X_Σ with respect to open coverings $\{U_\alpha\}_{\alpha\in I}$, $\{U_{\alpha'}\}_{\alpha'\in I'}$ and with projection maps π, π', respectively. We say $\mathcal{L}$ and $\mathcal{L}'$ are *isomorphic* if there exists a bijective map $\psi : \mathcal{L} \to \mathcal{L}'$ which leaves each fiber as a whole fixed, $\pi' \circ \psi = \pi$, such that, for any $U \subset U_\alpha \cap U_\beta$, $\alpha \in I, \beta \in I'$ and $f_{U,\alpha}$ as in (5), the map composition

$$U \times \mathbb{C} \xrightarrow{f_{U,\alpha}^{-1}} \pi^{-1}(U) \xrightarrow{\psi|_{\pi^{-1}(U)}} \pi'^{-1}(U) \xrightarrow{f_{U,\beta}} U \times \mathbb{C}$$

is of the form

$$(x, t) \longmapsto (x, \varphi(x)t)$$

for some regular and invertible function φ on U.

Remark. The definition of a line bundle does not depend on the special covering $\{U_\alpha\}_{\alpha\in I}$. If we proceed to a finer covering $\{U_{\alpha'}\}_{\alpha'\in I'}$, then, for each $\alpha' \in I'$, $U_\alpha = \bigcup_{U_{\alpha'}\subset U_\alpha} U_{\alpha'}$ and we set $g'_{\alpha'\beta'}(t) := g_{\alpha\beta}(t)$ for $t \in U_{\alpha'} \cap U_{\beta'} \subset U_\alpha \cap U_\beta$. It is readily checked that the $g'_{\alpha'\beta'}$ are well defined and $(\mathcal{L}, g_{\alpha\beta})$, $(\mathcal{L}', g'_{\alpha'\beta'})$ are isomorphic.

Lemma 4.6 is an immediate consequence of Definition 4.5.

4.6 Lemma. *Let $\{U_\alpha\}_{\alpha\in I}$ be a Zariski open covering of X_Σ, and let $(\mathcal{L}, g_{\alpha\beta})$, $(\mathcal{L}', g'_{\alpha\beta})$ be two line bundles on X_Σ with respect to this covering. $\mathcal{L}$ and $\mathcal{L}'$ are isomorphic if and only if there exist invertible functions g_α on U_α such that*

$$g_{\alpha\beta} = g_\alpha g'_{\alpha\beta} g_\beta^{-1} \quad \text{on } U_\alpha \cap U_\beta.$$

Next, we introduce tensor products of line bundles.

4.7 Definition. Let $(\mathcal{L}, g_{\alpha\beta})$, $(\mathcal{L}', g'_{\alpha\beta})$ be two line bundles on X_Σ (relative to the same covering $\{U_\alpha\}_{\alpha\in I}\}$). By the *tensor product* $\mathcal{L} \otimes \mathcal{L}'$ of $\mathcal{L}$ and $\mathcal{L}'$ we mean the line bundle defined by the transition maps

$$g''_{\alpha\beta} := g_{\alpha\beta}\, g'_{\alpha\beta}.$$

The line bundle given by

$$\widetilde{g}_{\alpha\beta} := g_{\alpha\beta}^{-1}$$

is said to be the *inverse* line bundle $\mathcal{L}^{-1}$ of $\mathcal{L}$. The group of all line bundles on X_Σ (up to isomorphisms) under $\otimes$ is denoted by Lb X_Σ.

4.8 Theorem. Pic $X_\Sigma \cong$ Lb X_Σ.

PROOF. Given an invertible sheaf $\mathcal{F}$, we introduce, for each Zariski open subset U of X_Σ, the function f_U as in the proof of Theorem 2.10. In particular, for a covering $\{U_\alpha\}_{\alpha\in I}$, we set $f_\alpha := f_{U_\alpha}$ and obtain, on $U_\alpha \cap U_\beta$, an invertible function $g_{\alpha\beta} := f_\alpha \cdot f_\beta^{-1}$. If f'_α defines the same element of the Picard group as f_α does, there exists an invertible function g_α and a rational function f such that $f'_\alpha = f \cdot g_\alpha \cdot f_\alpha$. So, $g_{\alpha\beta}$ satisfies the condition in Lemma 4.6, and we obtain a line bundle $(\mathcal{L}, g_{\alpha\beta})$ which, up to an isomorphism, is unique. Tensor products of line bundles, then, correspond to tensor products of invertible sheaves, and it is readily seen that we obtain a group isomorphism of Pic X_Σ and Lb X_Σ (after identifying isomorphic copies). □

We can construct, explicitly, an invertible sheaf from a line bundle. To do so, we introduce sections (compare the sections of sheaves in Definition 1.11).

4.9 Definition. Let $(\mathcal{L}, g_{\alpha\beta})$ be a line bundle on X_Σ with respect to the covering $\{U_\alpha\}_{\alpha\in I}$, and let U be any Zariski open subset of X_Σ. By a *section* of $\mathcal{L}$ on U, we mean a map $s : U \to \pi^{-1}(U)$ given by

$$\begin{array}{rcl} s_\alpha : U_\alpha \cap U & \longrightarrow & (U_\alpha \cap U) \times \mathbb{C} \\ x & \longmapsto & (x, g_{s_\alpha}(x)) \end{array}$$

where g_{s_α} is regular on U_α and satisfies $g_{s_\alpha}/g_{s_\beta} = g_{\alpha\beta}$ for each $\beta \in I$. The set of sections on U is denoted by $\Gamma(U, \mathcal{L})$.

$\Gamma(U, \mathcal{L})$ has the structure of an $\mathcal{O}(U)$-module. Sums $s + s'$ are defined by the sums $g_{s_\alpha} + g'_{s_\alpha}$. For $f \in R_U = \mathcal{O}(U)$, the product $f \cdot s$ is defined by the functions $f \cdot g_{s_\alpha}$.

For any fixed $\alpha \in I$, we consider the injective map

$$\begin{array}{rcl} i_\alpha : \Gamma(U, \mathcal{L}) & \longrightarrow & K_{X_\Sigma} \\ s & \longmapsto & g_{s_\alpha}. \end{array}$$

We set $\mathcal{F}(U) := i_\alpha(\Gamma(U, \mathcal{L}))$ and obtain Lemma 4.10.

4.10 Lemma. *$\mathcal{F}$ is an invertible sheaf of rational functions.*

Analogously to invariant Cartier divisors, we now introduce the following toric notion for line bundles.

4.11 Definition. Let $(\mathcal{L}, g_{\sigma\tau})$ be a line bundle on X_Σ with respect to the covering $\{X_{\check{\sigma}}\}_{\sigma\in\Sigma}$. If for each element ι of the big torus T the map

$$\begin{array}{cccccc} \psi_t & : & \mathcal{L} & \longrightarrow & \mathcal{L} \\ \psi_t|_{X_{\check{\sigma}}\times\mathbb{C}} & : & X_{\check{\sigma}} \times \mathbb{C} & \longrightarrow & X_{\check{\sigma}} \times \mathbb{C} \\ & & (x, c) & \longmapsto & (tx, t^{m_\sigma}c) \end{array}$$

satisfies

$$t^{m_\sigma - m_\tau} \cdot g_{\sigma\tau}(x) = g_{\sigma\tau}(tx),$$

then, we call $\mathcal{L}$ *equivariant.*

Example 5. Let $\mathcal{L} = X_\Sigma \times \mathbb{C}$. For any $m \in \mathbb{Z}^n$, we define the torus action ψ by $\psi_t(x, c) = (tx, t^m c)$. Denote the set of equivariant ("trivial") line bundles, thus obtained, by $\operatorname{Tr} X_\Sigma$.

We denote the group of all equivariant line bundles (under $\otimes$) by $\operatorname{Elb} X_\Sigma$. Forgetting about torus action, we obtain a morphism $\operatorname{Elb} X_\Sigma \to \operatorname{Lb} X_\Sigma$ whose kernel is $\operatorname{Tr} X_\Sigma$ (as introduced in Example 5). Moreover, we readily obtain Theorem 4.12.

4.12 Theorem. *For the group* $\operatorname{Elb} X_\Sigma$ *of equivariant line bundles,*

$$\operatorname{Lb} X_\Sigma \quad \cong \quad \operatorname{Pic} \Sigma \quad \cong \quad \operatorname{Elb} X_\Sigma / \operatorname{Tr} X_\Sigma.$$

In summarizing, we obtain the following diagram of "languages" and a sevenfold characterization of $\operatorname{Pic} X_\Sigma$:

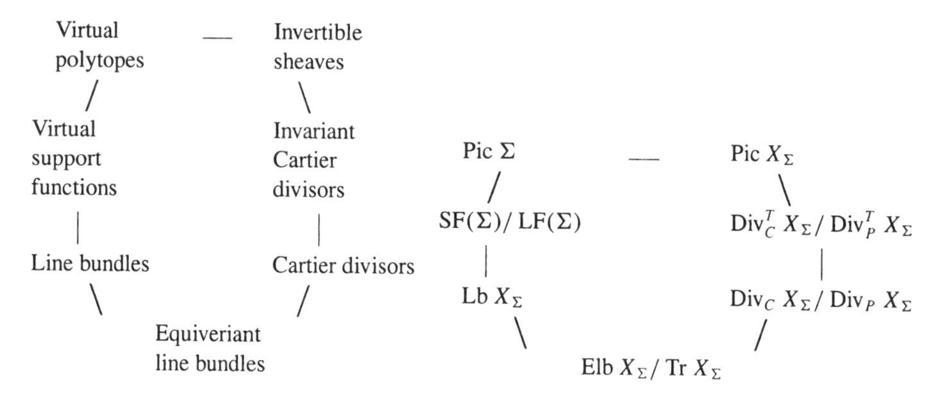

Exercises

1. Find the statements analogous to Example 1 for any Hirzebruch surface.
2. Find the transition functions $g_{\sigma\tau}$ of line bundles on X_Σ where (a) $X_\Sigma = \mathbb{P}^n$, (b) $X_\Sigma = \mathbb{P}^1 \times \mathbb{P}^1 \times \mathbb{P}^1$, and (c) X_Σ is a Hirzebruch surface.
3. In the language of virtual polytopes, verify directly that Pic $X_\Sigma = 0$ in Example 3.
4. Consider $\Sigma := \{\mathbb{R}_{\geq 0}\, e_1, \mathbb{R}_{\geq 0}(-e_1), \{0\}\}$ as a fan in $\mathbb{R}^2$. Describe Pic X_Σ and its analogs in all seven languages, as mentioned in the text.

5. Chow ring

The theory of divisors and virtual support functions, as considered in the preceding sections, was based on combining linearly toric $(n-1)$-subvarieties of a toric variety X_Σ. We wish to extend the theory to linear combinations of toric subvarieties of any dimension k, $0 \leq k \leq n-1$, and embed Pic X_Σ into a ring built up by such varieties. We restrict ourselves to Σ being regular and complete.

To achieve this, we could use a standard procedure in algebraic geometry, that is, defining "k-cycles" and "rational equivalence" as analogs of "divisors" and "linear equivalence". It is, however, more convenient for us to proceed somewhat differently, making use of the various "languages" presented in section 4.

5.1 Definition. Let X_Σ be a complete smooth toric variety, and let $\varrho_1, \ldots, \varrho_k$ be the one-dimensional cones of Σ. To each ϱ_i, we assign an indeterminate variable $U_i, i = 1, \ldots, k$, and consider the polynomial ring $\mathbb{Z}[U_1, \ldots, U_k]$. We introduce the following ideals in this ring.

(1) Let $\mathfrak{a}$ be generated by all monomials $U_{i_1} \cdots U_{i_s}$, $1 \leq i_1 < \cdots < i_s \leq k$ for which $\varrho_{i_1} + \cdots + \varrho_{i_s} \notin \Sigma$.

(2) Let $\mathfrak{b}$ be generated by all linear polynomials $a_{1j}U_1 + \cdots + a_{kj}U_k$, $j = 1, \ldots, n$ where $a_i := (a_{i1}, \ldots, a_{in}) \in \mathbb{Z}^n$ is the simple vector for which $\varrho_i = \mathbb{R}_{\geq 0}\, a_i, i = 1, \ldots, k$.

Then, we call $\mathbb{Z}[U_1, \ldots, U_k]/(\mathfrak{a} + \mathfrak{b})$ the *Chow ring* $\mathrm{Ch}(X_\Sigma)$ of X_Σ. We set u_i for the residue class determined by $U_i, i = 1, \ldots, k$.

Example 1. Let X_Σ be the Hirzebruch surface with generators $a_1 = (1, 0)$, $a_2 = (0, 1), a_3 = (-1, 0), a_4 = (r, -1)$. Then, $\mathfrak{a}$ is generated by U_1U_3 and U_2U_4. The defining linear polynomials of $\mathfrak{b}$ are $U_1 - U_3 + rU_4$ and $U_2 - U_4$. Using Lemma 5.3 below, we see that $\mathrm{Ch}(X_\Sigma) = \{\alpha u_1 + \beta u_2 + \gamma u_1 u_2 + \delta \mid \alpha, \beta, \gamma, \delta \in \mathbb{Z}\}$.

5.2 Lemma. Pic X_Σ *is isomorphic to the additive subgroup of all linear polynomials of* $\mathrm{Ch}(X_\Sigma)$.

PROOF. We represent Pic X_Σ as $\mathrm{SF}(\Sigma)/\mathrm{LF}(\Sigma)$ according to Theorem 4.3. To each linear polynomial $\alpha_1 u_1 + \cdots + \alpha_k u_k$, we assign a piecewise linear function $h : |X_\Sigma| \longrightarrow \mathbb{R}$ by setting $h(a_i) := -\alpha_i$, $i = 1, \ldots, k$, and extending linearly on each cone of Σ. Since Σ is simplicial, the extension is well defined. So (1) and (2) in section 4 are satisfied. Conversely, any h satisfying (1) and (2) in section 4 defines a linear polynomial by setting $\alpha_i := -h(a_i)$, $i = 1, \ldots, k$.

The linear functions of $\mathrm{LF}(\Sigma)$ are integral linear combinations of the functions $\langle e_i, x\rangle$ where $e_1, \ldots, e_n$ is the canonical basis of $\mathbb{R}^n$. From

$$\langle e_i, a_j\rangle = a_{ji}, \qquad i = 1, \ldots, n, \quad j = 1, \ldots, k,$$

we see that $\mathfrak{b} \cong \mathrm{LF}(\Sigma)$. Since $\mathfrak{a}$ does not contain linear elements, the lemma follows. □

5.3 Lemma ("Shifting away" lemma). *Let Σ be regular and complete. If $\sigma = \varrho_{i_1} + \cdots + \varrho_{i_s} \in \Sigma$, we set $P_\sigma := U_{i_1} \cdots U_{i_s}$ and $p_\sigma := [P_\sigma] = P_\sigma + \mathfrak{a} + \mathfrak{b}$. If σ is a face of $\sigma' \in \Sigma$ and $\dim \sigma > 0$, then there exist cones $\sigma_j \in \Sigma$, $\dim \sigma_j = \dim \sigma$, $j = 1, \ldots, q$, and integers c_j such that no σ_j is a face of σ' and*

$$p_\sigma = c_1 p_{\sigma_1} + \cdots + c_q p_{\sigma_q}.$$

PROOF. Let τ be a face of σ such that $\dim \sigma = 1 + \dim \tau$ (for $\dim \sigma = 0$, there is nothing to prove). Since Σ is complete, we may assume $\dim \sigma' = n$. Up to renumbering the $\varrho_i = \mathbb{R}_{\geq 0}\, a_i$, $i = 1, \ldots, k$, let $\sigma' = \varrho_1 + \cdots + \varrho_n$, $\sigma = \varrho_1 + \cdots + \varrho_s$, and $\tau = \varrho_2 + \cdots + \varrho_s$ (or $\{0\}$ if $\dim \sigma = 1$). By definition of $\mathfrak{b}$ and the regularity of Σ,

$$(1) \quad \begin{aligned} a_{11}u_1 + \cdots + a_{k1}u_k &= 0 \\ &\ \vdots \\ a_{1n}u_1 + \cdots + a_{kn}u_k &= 0 \end{aligned} \qquad (2) \quad \det(a_1 \ldots a_n) = \pm 1.$$

Solving (1) for $u_1, \ldots, u_n$ by Cramer's rule and by using (2), we obtain $u_1, \ldots, u_n$ as integral linear combinations of $u_{n+1}, \ldots, u_k$. In particular,

$$u_1 = b_{n+1}u_{n+1} + \cdots + b_k u_k, \qquad b_{n+1}, \ldots, b_k \in \mathbb{Z}.$$

Since Σ is complete, $k > n$, so that at least one $b_i \neq 0$, $n + 1 \leq i \leq k$. Now,

$$\begin{aligned} p_\sigma = u_1 \cdots u_s &= (b_{n+1}u_{n+1} + \cdots + b_k u_k)u_2 \cdots u_s \\ &= b_{n+1}u_{n+1}u_2 \cdots u_s + \cdots + b_k u_k u_2 \cdots u_s, \end{aligned}$$

where all those $u_i u_2 \cdots u_s$ are nonzero for which $\varrho_i + \tau \in \Sigma$, $i = n+1, \ldots, k$. This proves the lemma. □

Example 2. Let $a_1 := e_1$, $a_2 := e_2$, $a_3 := e_3$, and $a_4 := -e_1 - e_2 - e_3$ be the generators of Σ for $X_\Sigma = \mathbb{P}^3$. We find that $\mathfrak{a}$ is generated by $U_1U_2U_3U_4$, and $\mathfrak{b}$ by $U_1 - U_4, U_2 - U_4, U_3 - U_4$. Hence $u_1 = u_2 = u_3 = u_4$ and $u_1u_2u_3u_4 = u_1^4 = 0$. We may represent $\mathrm{Ch}(\Sigma)$ either by $\{\alpha u_1 + \beta u_1^2 + \gamma u_1^3 + \delta \mid \alpha, \beta, \gamma, \delta \in \mathbb{Z}\}$ or by

$\{\alpha u_1 + \beta u_1 u_2 + \gamma u_1 u_2 u_3 + \delta \mid \alpha, \beta, \gamma, \delta \in \mathbb{Z}\}$. We consider $\sigma' := \varrho_1 + \varrho_2 + \varrho_3$, $\sigma := \varrho_1 + \varrho_2$. Then, for $\sigma_1 := \varrho_1 + \varrho_4$, $p_\sigma = p_{\sigma_1}$.

Example 3. In $\mathbb{R}^2$, let $\varrho_1 := \mathbb{R}_{\geq 0}\, e_1, \varrho_2 := \mathbb{R}_{\geq 0}\, e_2, \varrho_3 := \mathbb{R}_{\geq 0}(-e_1 + e_2), \varrho_4 := \mathbb{R}_{\geq 0}(-e_1), \varrho_5 := \mathbb{R}_{\geq 0}(-e_1 - e_2), \varrho_6 := \mathbb{R}_{\geq 0}(-e_2)$, and $\sigma = \varrho_1, \sigma' = \varrho_1 + \varrho_2$. We find that $u_1 - u_3 - u_4 - u_5 = 0$, and, hence, $p_{\varrho_1} = u_1 = u_3 + u_4 + u_5$ (Figure 7).

5.4 Definition. If $\sigma = \varrho_{i_1} + \cdots + \varrho_{i_s} \in \Sigma$, we call the square-free monomial $p_\sigma = u_{i_1} \cdots u_{i_s}$ a *face element* of $\mathrm{Ch}(X_\Sigma)$. For $\sigma = \{0\}$, we set $p_\sigma = 1$. We denote, by $\mathrm{Ch}^{(s)}(X_\Sigma)$, the subgroup of $\mathrm{Ch}(X_\Sigma)$ generated by the face elements for a fixed degree. We say p_σ *represents* the toric subvariety $X_{(\sigma)} := X_{\Sigma/\sigma}$ determined by $\Sigma/\sigma = \pi(\mathrm{st}(\sigma, \Sigma))$.

5.5 Theorem. *The Chow ring can be decomposed as follows:*

$$\mathrm{Ch}(X_\Sigma) = \mathrm{Ch}^{(0)}(X_\Sigma) \oplus \cdots \oplus \mathrm{Ch}^{(n)}(X_\Sigma).$$

PROOF. Suppose $u_{i_1}^{r_1} \cdots u_{i_t}^{r_t}$ is a monomial of $\mathrm{Ch}(X_\Sigma)$, $i_1 < \cdots < i_t$. If $\varrho_{i_1} + \cdots + \varrho_{i_t} \notin \Sigma$, the monomial equals zero. So, let $\sigma = \varrho_{i_1} + \cdots + \varrho_{i_t} \in \Sigma$. In the case $r_1 > 1$ we replace u_{i_1} according to Lemma 5.2, by a linear combination of u_j which do not belong to $u_{i_1}, \ldots, u_{i_t}$. Continuing in this way, we replace $u_{i_1}^{r_1} \cdots u_{i_t}^{r_t}$ by an integral linear combination of square–free monomials. This proves the theorem. □

Example 4. Example 2 can readily be generalized to show that for $\sigma_i := \mathbb{R}_{\geq 0}\, e_1 + \cdots + \mathbb{R}_{\geq 0}\, e_i$, $\sigma_0 := \{0\}$, $\mathrm{Ch}(\mathbb{P}^n) = \mathbb{Z}\, p_{\sigma_0} \oplus \mathbb{Z}\, p_{\sigma_1} \oplus \cdots \oplus \mathbb{Z}\, p_{\sigma_n}$.

Multiplication in the Chow ring is closely related to the intersection of toric subvarieties. From Theorem 5.5 and the preceding definitions we derive Lemma 5.6.

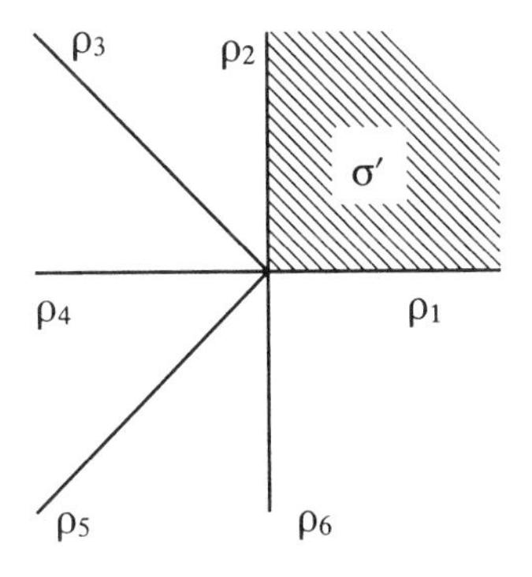

FIGURE 7.

5.6 Lemma. *If $\sigma, \tau \in \Sigma$ and $\sigma \cap \tau = \{0\}$ and p_σ, p_τ are the face elements which represent $X_{(\sigma)}$, $X_{(\tau)}$, then, $p_\sigma \cdot p_\tau$ represents the intersection $X_{(\sigma)} \cap X_{(\tau)}$.*

This relationship will be considered again in chapter VIII.

Exercises

1. Find $\mathrm{Ch}(X_\Sigma)$ of Example 3 according to Theorem 5.5.
2. Find $\mathrm{Ch}(X_\Sigma)$ for the higher dimensional analogs of Hirzebruch surfaces (VI, 3, Example 3).
3. Show that the assumption "Σ is complete" can be weakened such that Theorem 5.5 still holds.
4. Find examples of nonregular toric varieties for which the shifting away lemma can be proved.

6. Intersection numbers. Hodge inequality

If n toric hypersurfaces of an n-dimensional toric variety intersect in finitely many points, we are interested in counting the numbers of points of intersection, counted with "multiplicity". This leads to what is known as the intersection numbers of divisors. First, we discuss some of the basic ideas. Throughout this section, we assume X_Σ to be a smooth and compact toric variety of dimension n.

Let $\varrho_1, \ldots, \varrho_n$ be 1-cones of Σ and $D_{\varrho_1}, \ldots, D_{\varrho_n}$ the corresponding invariant divisors (toric $(n-1)$-subvarieties defined by $\mathrm{st}(\varrho_i, \Sigma)$, $i = 1, \ldots, n$). If $\sigma := \varrho_1 + \cdots + \varrho_n$ is an n-cone of Σ, the D_{ϱ_i} are the closures of the coordinate hyperplanes of $X_{\check{\sigma}}$ and, hence, intersect "transversally" in one point. So, we associate intersection number 1 with $D_{\varrho_1}, \ldots, D_{\varrho_n}$. If $\varrho_1 + \cdots + \varrho_n$ is not contained in a cell of Σ, the subvarieties D_{ϱ_i} have empty intersection, hence, intersection number 0. This is a natural starting point for intersection numbers.

Not equally trivial, but still "natural", is the requirement that an intersection number does not change if one divisor is replaced by a linearly equivalent one. In particular, self-intersection numbers can be defined by n linearly equivalent, different copies of the same divisor. We shall see that such intersection numbers are possibly negative.

Finally, we want intersection numbers to be linear in each component. For example, if a D_{ϱ_i} is replaced by kD_{ϱ_i}, each point of intersection with other divisors is to be counted k times. Also, the additivity is geometrically plausible.

We use these intuitive requirements for a definition.

6.1 Definition. Let $\mathcal{D} := \mathrm{TCDiv}(X_\Sigma)$ be the group of T-invariant Cartier divisors on a smooth compact n-dimensional toric variety X_Σ. Then a mapping of the n-fold cartesian product

$$\mathcal{D} \times \cdots \times \mathcal{D} \underset{(\cdot,\dots,\cdot)}{\longrightarrow} \mathbb{Z}$$

is called an *intersection number* for toric divisors if it satisfies the following conditions for all $D_1, D_1', D_2, \ldots, D_n \in \mathcal{D}$.
(1) $(D_1.\cdots.D_n) = (D_{\pi(1)}.\cdots.D_{\pi(n)})$ for any permutation π of $1, \ldots, n$.
(2) $(D_1 + D_1'.D_2.\cdots.D_n) = (D_1.D_2.\cdots.D_n) + (D_1'.D_2.\cdots.D_n)$.
(3) $(kD_1.D_2.\cdots.D_n) = k(D_1.D_2.\cdots.D_n)$, for $k \in \mathbb{Z}$.
(4) $(D_1.D_2.\cdots.D_n) = (D_1'.D_2.\cdots.D_n)$, if D_1, D_1' are linearly equivalent.
(5) For 1-cones $\varrho_1, \ldots, \varrho_n$ of Σ,

$$(D_{\varrho_1}.\cdots.D_{\varrho_n}) = \begin{cases} 1 & \text{if } \varrho_1 + \cdots + \varrho_n \in \Sigma^{(n)}, \\ 0 & \text{if } \varrho_1 + \cdots + \varrho_n \notin \Sigma. \end{cases}$$

We shall prove existence and uniqueness of intersection numbers only for compact smooth projective X_Σ, since, in this case, there is a natural relationship to mixed volumes. First will we look at some examples.

Example 1. Let $n = 1$ and $\Sigma = \{\varrho_1 := \mathbb{R}_{\geq 0}\, e_1, \varrho_2 := \mathbb{R}_{\geq 0}(-e_1), \{0\}\}$. Any T-invariant divisor $D = -\langle m_1, e_1\rangle D_{\varrho_1} - \langle m_2, -e_1\rangle D_{\varrho_2} = -m_1 D_{\varrho_1} + m_2 D_{\varrho_2}$ $(m_1, m_2 \in \mathbb{Z})$ can be characterized by a function h which may be convex or concave.

The corresponding invertible sheaf has rings $z^{m_1}\mathcal{O}_{X_\Sigma}(X_{\check{\varrho}_1})$, $z^{m_2}\mathcal{O}_{X_\Sigma}(X_{\check{\varrho}_2})$, $\mathbb{C}[z, z^{-1}]$. It has nonconstant global sections if $m_1 < m_2$ (h concave). Up to linear equivalence (adding a linear function to h or applying a translation to $P = (m_1 + \check{\varrho}_1) \cap (m_2 + \check{\varrho}_2) = [m_1, m_2]$), we can assume the global sections of the invertible sheaf to be the polynomials $\alpha_0 + \alpha_1 z + \cdots + \alpha_{m_2-m_1} z^{m_2-m_1}$. The maximal degree which occurs is $m_2 - m_1$, also called the *degree* of the invertible sheaf. The corresponding divisor D is an $(m_2 - m_1)$-fold point so that $m_2 - m_1$ provides the self–intersection number of D. Also, in the case $m_2 - m_1 \leq 0$ (h linear or convex), we call $m_2 - m_1$ the self-intersection number of D.

6.2 Lemma. *For $n = 2$, let τ be a 1-cone of a regular complete fan Σ, $\tau = \mathbb{R}_{\geq 0}\, a$, and let $\sigma_1 := \mathbb{R}_{\geq 0}\, a + \mathbb{R}_{\geq 0}\, b$, $\sigma_2 := \mathbb{R}_{\geq 0}\, a + \mathbb{R}_{\geq 0}\, c$ be the adjacent 2-cones of τ in Σ, a, b, c simple. If $\alpha \in \mathbb{Z}$ is the integer for which*

$$\alpha a + b + c = 0,$$

then, D_τ has self-intersection number

$$(D_\tau.D_\tau) = \alpha.$$

PROOF. We set $\varrho_1 := \mathbb{R}_{\geq 0}\, b$, $\varrho_2 := \mathbb{R}_{\geq 0}\, c$. We may assume $a = e_1$. As a Cartier divisor, D_τ is given by the piecewise linear function h, defined by $h(b) = h(c) = 0$, $h(a) = -1$, $h(x) = 0$ for $\sigma \in \mathbb{R}^2 \setminus (\sigma_1 \cup \sigma_2)$ (Figure 8) so that h defines an invertible sheaf on X_Σ. We can add the linear function $\langle a, \cdot\rangle$ to it and obtain an equivalent function $\bar{h}$,

$$\bar{h}(x) := h(x) + \langle a, x\rangle, \qquad \text{for } x \in \mathbb{R}^2.$$

Clearly, $\bar{h}(x) = 0$ for $x \in \tau$. The restriction of $\bar{h}$ to $\tau^\perp$ defines an invertible sheaf on D_τ. By Example 1, $(D_\tau.D_\tau) = -(\bar{h}(b) + \bar{h}(c))$. From $\bar{h}(b) + \bar{h}(c) = h(b) + \langle a, b\rangle + h(c) + \langle a, c\rangle = \langle a, b+c\rangle = -\alpha\langle a, a\rangle = -\alpha$, we find $(D_\tau.D_\tau) = \alpha$. □

Example 2. The projective plane $\mathbb{P}^2$ is given by a fan with generators $a = e_1$, $b = e_2$, $c = -e_1 - e_2$, and, hence, $a + b + c = 0$. Therefore $(D_\tau.D_\tau) = 1$ for each coordinate line of $\mathbb{P}^2$. This is intuitively obvious. Any two different lines of $\mathbb{P}^2$ are projectively equivalent and intersect in one point.

Example 3. If $a = b + c$, we may consider the fan Σ' obtained from Σ by deleting τ, σ_1, σ_2 and introducing $\sigma_1 \cup \sigma_2$ as a new cone. Then, X_Σ is obtained from $X_{\Sigma'}$ by a blowup, and the exceptional projective line has self-intersection number -1.

6.3 Theorem. *Let X_Σ be a smooth, compact, projective toric variety, $\Sigma = \Sigma(-P)$. Then, intersection numbers are well defined and uniquely determined. For Cartier divisors $D_1, \ldots, D_n$ whose piecewise linear functions are convex and, hence, are represented by polytopes $P_1, \ldots, P_n$, respectively,*

$$(D_1.\cdots.D_n) = n!V(P_1, \ldots, P_n).$$

For Cartier divisors represented by virtual polytopes $\mathcal{P}_1 = \mathcal{P}(P_1) - \mathcal{P}(r_1 P), \ldots, \mathcal{P}_n = \mathcal{P}(P_n) - \mathcal{P}(r_n P)$ (see V, Theorem 5.15*), we can calculate $(D_1.\cdots.D_n)$ successively from the intersection numbers of the Cartier divisors $\tilde{D}_1, \ldots, \tilde{D}_n$ given by $\mathcal{P}(P_1), \ldots, \mathcal{P}(P_n)$, respectively ($D$ given by P),*

$$(\tilde{D}_1.\cdots.\tilde{D}_n) = n!V(P_1, \ldots, P_n) = (D_1 + r_1 D.\cdots.D_n + r_n D).$$

PROOF. By IV, Lemma 3.4 and Lemma 3.6, the mixed volume function $V(\cdot, \ldots, \cdot)$ satisfies (1), (2), and, in the case $k \geq 0$, also (3).

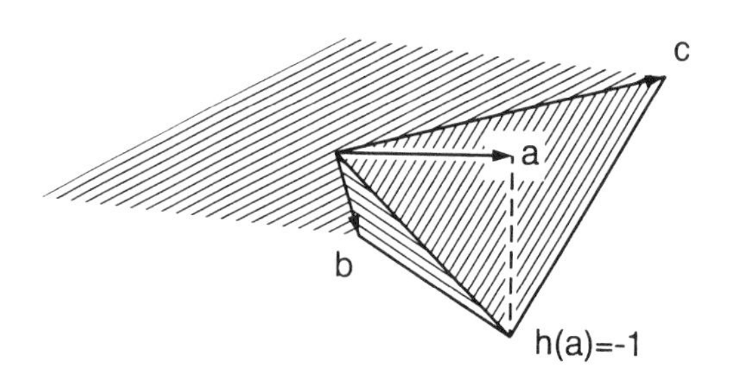

FIGURE 8.

Let $k = -1$. We associate with $-D_1$ the virtual lattice polytope (V, Definition 53) $\mathcal{P}$ for which

$$\mathcal{P} + \mathcal{P}(P_1) = \mathcal{P}(\{0\}).$$

Since $V(\{0\}, P_2, \ldots, P_n) = 0$ (V, Lemma 3.6), we obtain

$$(-D_1.D_2.\cdots.D_n) + (D_1.D_2.\cdots.D_n) = 0.$$

Now, (3) follows for all $k \in \mathbb{Z}$.

To prove (4), we apply the following characterization of linear equivalence (compare Theorem 4.6). The piecewise linear functions h, h' of D_1, D_1', respectively, differ only by a linear function. Hence, the virtual polytopes differ only by the addition of a point, that is, by a translation. $V(\cdot, \ldots, \cdot)$ is, however, invariant under translations (IV, Lemma 3.6).

Concerning (5), we remind ourselves that, for a 1-cone $\varrho \in \Sigma$, the Cartier divisor D_ϱ can be represented by the virtual polytope $\mathcal{P}_\varrho := \{\bar{m}_\varrho + \check{\varrho}\} \cup (\check{\Sigma} \setminus \{\check{\varrho}\})$.

The sum $\mathcal{P}_\varrho + \mathcal{P}(P) =: \mathcal{P}(P_\varrho)$ is always a polytope element. Let $\varrho_1, \ldots, \varrho_n$ be n different 1-cones of Σ. If D is again given by P,

$$(D_{\varrho_1} + D.D_{\varrho_2} + D.\cdots.D_{\varrho_n} + D) = n!V(P_{\varrho_1}, \ldots, P_{\varrho_n}).$$

Each P_{ϱ_i} is obtained from

$$P = (m_{\varrho_i} + \check{\varrho}_i) \cap \Big(\bigcap_{\substack{\varrho \neq \varrho_i \\ \varrho \in \Sigma^{(1)}}} (m_\varrho + \check{\varrho}) \Big)$$

by "moving out" the supporting hyperplane $m_{\varrho_i} + \varrho_i^\perp$ of P such that, between $m_{\varrho_i} + \varrho_i^\perp$ and $m_{\varrho_i} + \bar{m}_{\varrho_i} + \varrho_i^\perp$, there are no lattice points (Figure 9).

First, we illustrate the idea of the proof for $n = 2$. If $\varrho_1 + \varrho_2 \notin \Sigma$, that is if ϱ_1, ϱ_2 are not adjacent in Σ, we obtain

$$(1) \qquad 2V(P_{\varrho_1}, P_{\varrho_2}) = V(P_{\varrho_1} + P_{\varrho_2}) - V(P_{\varrho_1}) - V(P_{\varrho_2}).$$

Let $A_1 := P_{\varrho_1} \setminus P$, $A_2 := P_{\varrho_2} \setminus P$. Then, we obtain $V(P_{\varrho_1} + P_{\varrho_2}) = 4V(P) + 2V(A_1) + 2V(A_2)$ and $V(P_{\varrho_1}) = V(P) + V(A_1)$, $V(P_{\varrho_2}) = V(P) + V(A_2)$. Hence, $(D_{\varrho_1} + D.D_{\varrho_2} + D) = 2V(P_{\varrho_1}, P_{\varrho_2}) = V(P_{\varrho_1} + P_{\varrho_2}) - V(P_{\varrho_1}) - V(P_{\varrho_2}) = 2V(P) + V(A_1) + V(A_2)$.

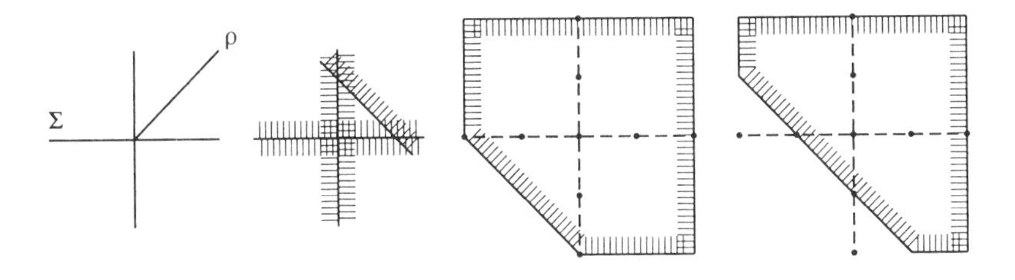

FIGURE 9.

Furthermore,

$$\begin{aligned}(D_{\varrho_1}.D) &= (D_{\varrho_1} + D.D) - (D.D) = 2V(P_{\varrho_1}, P) - 2V(P)\\ &= V(P_{\varrho_1} + P) - V(P_{\varrho_1}) - 3V(P)\\ &= 4V(P) + 2V(A_1) - V(A_1) - 4V(P) = V(A_1).\end{aligned}$$

Similarly, $(D_{\varrho_2}.D) = V(A_2)$. Therefore,

$$(D_{\varrho_1}.D_{\varrho_2}) = (D_{\varrho_1} + D.D_{\varrho_2} + D) - (D_{\varrho_1}.D) - (D_{\varrho_2}.D) - (D.D) = 0.$$

In the case $\varrho_1 + \varrho_3 \in \Sigma$, the calculation of $(D_{\varrho_1}.D_{\varrho_3})$ works similarly, except that an additional area B occurs in $P_{\varrho_1} + P_{\varrho_3}$ (hatched in the lower right part of Figure 10). Clearly, $V(B) = 1$. Therefore

$$(D_{\varrho_1}.D_{\varrho_3}) = 1.$$

The proof for arbitrary n proceeds analogously. If $k \le n$ of the cones $\varrho_1, \dots, \varrho_n$ are different faces of a $\tau \in \Sigma$, say $\varrho_1, \dots, \varrho_k$, a k-dimensional parallelotope B occurs in the sum $P_{\varrho_1} + \cdots + P_{\varrho_n}$, and $V(B) = 1$ or 0 according to $k = n$ or $k < n$, respectively. Expressing mixed volumes by volumes according to IV, Theorem 3.7, a calculation, as in the case $n = 2$, is readily carried out so as to prove (5).

It still remains to be shown that intersection numbers are uniquely determined by (1)–(5). Applying (1)–(4) implies that we need only prove $(D_{\varrho_1}.\cdots.D_{\varrho_n})$ to be uniquely determined for any $\varrho_1, \dots, \varrho_n \in \Sigma^{(1)}$. The case where they are all different is clear from (5). So, we assume inductively that $(D_{\varrho_1}.\cdots.D_{\varrho_n})$ is determined by (5) if, at most, $n - k < n$ of $\varrho_1, \dots, \varrho_n$ coincide. Suppose uniqueness is true if at least k of the D_{ϱ_i} are different, and let only $k-1$ of the D_{ϱ_i} in $(D_{\varrho_1}.\cdots.D_{\varrho_n})$ differ, say $\varrho_1, \dots, \varrho_{k-1}$. If $\varrho_1 + \cdots + \varrho_n \notin \Sigma$, then, $(D_{\varrho_1}.\cdots.D_{\varrho_n}) = 0$. So, let $\varrho_1 + \cdots + \varrho_n \in \Sigma$, hence, $\sigma := \varrho_1 + \cdots + \varrho_n = \varrho_1 + \cdots + \varrho_{k-1} \in \Sigma$ (and σ face of $\sigma' \in \Sigma^{(n)}$). By the shifting away lemma (Lemma 5.3), we can set

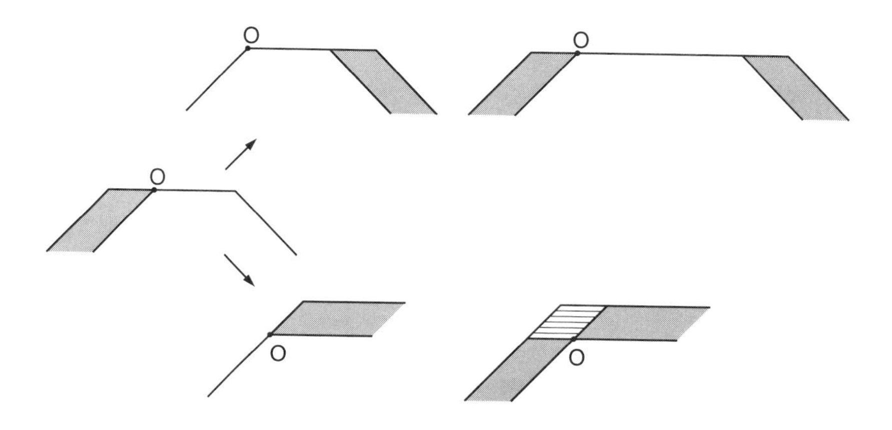

FIGURE 10.

$D_{\varrho_k} = \beta_1 D_{\varrho_{i_1}} + \cdots + \beta_s D_{\varrho_{i_s}}$ where $\varrho_{i_j} \notin \{\varrho_1, \ldots, \varrho_{k-1}\}$, $j = 1, \ldots, s$. Now, in

$$(D_{\varrho_1}.\cdots.D_{\varrho_n}) = \sum \beta_i (D_{\varrho_1}.\cdots.D_{\varrho_{k-1}}.D_{\varrho_{j_i}}.D_{\varrho_{k+1}}.\cdots.D_{\varrho_n}),$$

each term has k different factors and hence is uniquely determined. Since, by (4), the intersections numbers do not change under linear equivalence of the D_{ϱ_i}, the intersection number $(D_{\varrho_1}.\cdots.D_{\varrho_n})$ is uniquely determined. □

Example 4. Let X_Σ be a del Pezzo surface whose fan Σ has generators $a_1 = e_1 + e_2$, $a_2 = e_2$, $a_3 = -e_1$, $a_4 = -e_1 - e_2$, $a_5 = -e_2$, and $a_6 = e_1$. The associated polytope P is shown in Figure 11.

We denote the Cartier divisor of P by D and set $D_i := D_{\mathbb{R}_{\geq 0} a_i}$, $i = 1, \ldots, 6$. We wish to calculate $(D_1.D_1)$. For this purpose, we observe that

$$(D_1.D_1) = (D_1 + D.D_1 + D) - 2(D_1 + D.D) + (D.D).$$

The divisor $D_1 + D$ is represented by the polytope P_1 obtained from P by joining a triangle (dotted area in Figure 11) to it. From $(D_1 + D.D_1 + D) = 2V(P_1)$, $(D_1 + D.D) = 2V(P_1, P) = V(P_1 + P) - V(P_1) - V(P)$, and $V(P_1) = 3\frac{1}{2}$, $V(P) = 3$, $V(P_1 + P) = 13\frac{1}{2}$, we obtain $(D_1.D_1) = -1$, in accordance with Example 3.

6.4 Theorem (Toric Hodge inequality). *Given a smooth, compact, projective toric variety, let D be an arbitrary T-invariant Cartier divisor, and let $D', D_1, \ldots, D_{n-2}$ be T-invariant Cartier divisors which are represented by polytopes. Then,*

$$(D.D'.D_1.\cdots.D_{n-2})^2 \geq (D.D.D_1.\cdots.D_{n-2})(D'.D'.D_1.\cdots.D_{n-2}).$$

PROOF. If D is also represented by a polytope, this is a direct consequence of Theorem 6.3 and the Alexandrov-Fenchel inequality (see IV, 5).

So, let D be arbitrary. According to V, Theorem@515, we replace D by $D+kD'$ for sufficiently large $k > 0$ such that $D + kD'$ is represented by a polytope. Then, by using (2), (3), we obtain the theorem from

$$(D + kD'.D'.D_1.\cdots.D_{n-2})^2$$

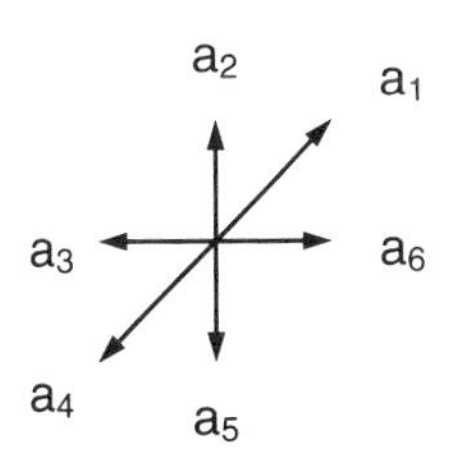

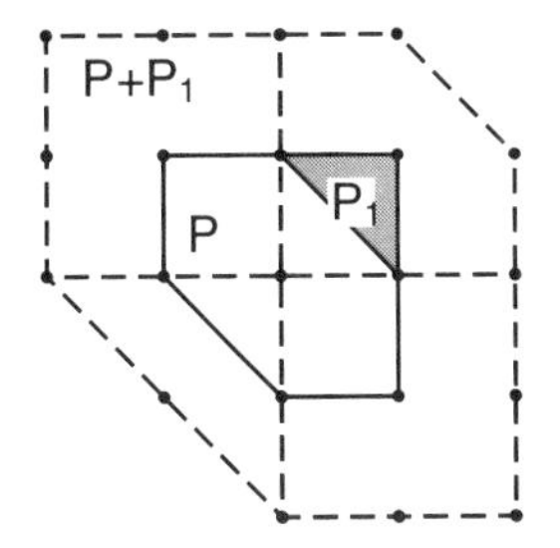

FIGURE 11.

$$\geq (D + kD'.D + kD'.D_1.\cdots.D_{n-2})(D'.D'.D_1.\cdots.D_{n-2}).$$

□

Remark. If D', $D_1, \ldots, D_{n-2}$ are not represented by polytopes, the Hodge inequality need not be true. For example, let a del Pezzo surface be given as in Example 4. Then, $(D_1.D_1) = (D_3.D_3) = -1$, but $(D_1.D_3) = 0$, so that the Hodge inequality is violated.

6.5 Theorem (A special toric Hodge index theorem). *Given a smooth, compact, projective toric variety, let D be an arbitrary T-invariant Cartier divisor, and let D', $D_1, \ldots, D_{n-2}$ be T-invariant Cartier divisors which are ample represented by polytopes. If $(D.D'.D_1.\cdots.D_{n-2}) = 0$ and $(D.D.D_1.\cdots.D_{n-2}) \neq 0$, $(D'.D'.D_1.\cdots.D_{n-2}) \neq 0$. Then*

$$(D.D.D_1.\cdots.D_{n-2}) < 0.$$

PROOF. This is an immediate consequence of Theorem 6.4. □

Exercises

1. Extend Lemma 6.2 to the n-dimensional case and prove the toric Nakai–Moishezon criterion for ampleness. Let X_Σ be a smooth compact projective toric variety. A T-invariant Cartier divisor D is ample if and only if, for any $(n-1)$-face $\varrho_1 + \cdots + \varrho_{n-1} \in \Sigma$ ($\varrho_i \in \Sigma^{(1)}$, $i = 1, \ldots, n-1$), $(D.D_{\varrho_1}.\cdots.D_{\varrho_{n-1}}) < 0$.
2. In VI, 3, Example 4 find all types of intersection numbers $(D_{\varrho_1}.D_{\varrho_2}.D_{\varrho_3})$ which occur ($\varrho_1, \varrho_2, \varrho_3$ arbitrary 1-cones of the fan).
3. Generalize condition (5) in the definition of intersection numbers if Σ is allowed to be an arbitrary, complete simplicial fan.
4. Discuss equality cases of Theorem 6.5; use the results of IV, 5, in particular, Example 2.

7. Moment map and Morse function

There are direct relationships between an affine or a projective toric variety and the defining cone or polytope, respectively, which we shall study now. They provide new insight into the structure of toric varieties, and they produce tools for giving the Chow ring a topological meaning (in Chapter VIII).

7.1 Theorem. *Let $\check{\sigma} = \mathrm{pos}\{m_1, \ldots, m_n\}$ be an n-dimensional simplex cone, and suppose $m_1, \ldots, m_n, \ldots, m_k$ are simple lattice vectors which generate the monoid $\check{\sigma} \cap \mathbb{Z}^n$. For $u = (u_1, \ldots, u_k) = (z^{m_1}, \ldots, z^{m_n}, \ldots,$*

z^{m_k}), *consider the map*

$$\begin{array}{rcl} \varphi : X_{\check{\sigma}} & \longrightarrow & \mathbb{C}^n \\ u & \longmapsto & u_1 m_1 + \cdots + u_n m_n \end{array}.$$

Then,

(1) φ *maps the big torus* $T = \{u \mid u_i \neq 0, i = 1, \ldots, k\}$ *onto the torus* $T' := \{u_1 m_1 + \cdots + u_n m_n \mid u_i \neq 0, i = 1, \ldots, n\}$.

(2) $\varphi|_{X^{\geq 0}_{\check{\sigma}}}$ *maps* $X^{\geq 0}_{\check{\sigma}}$ *bijectively onto* $\check{\sigma}$, *in particular each orbit of* $X^{\geq 0}_{\check{\sigma}}$ *onto the relative interior of a face of* $\check{\sigma}$ *(compare* VI, *Definition 3.8).*

PROOF.

(1) is true by definition of φ.

(2) Among the $|\lambda_i|$ roots of $x_i^{\lambda_i}$, there is only one which is real and positive. Therefore, $T^{>0}$ is mapped bijectively onto $T'^{>0} = \{y \in T' \mid y_i > 0, i = 1, \ldots, n\}$.

Since $X_{\check{\sigma}}$ is the disjoint union of tori in dimensions $0, \ldots, n$, we may apply the same arguments as for $X_{\check{\sigma}}$ to the subvarieties X_τ of $X_{\check{\sigma}}$, where τ is a face of $\check{\sigma}$.

□

Example 1. Let $\sigma := \text{pos}\{e_1, e_1 + 2e_2\}$, hence, $\check{\sigma} = \text{pos}\{2e_1 - e_2, e_2\} = \text{pos}\{m_1, m_2\}$ (see Example 3 in VI, 2).

The degree of φ is 2. The point $1 \cdot m_1 + 1 \cdot m_2 = 2e_1$ in $\check{\sigma}$ has two inverse images $u^{(1)} = (1, 1, 1)$, $u^{(2)} = (1, 1, -1)$. However, m_1, m_2 (as generators of 1-faces of $\check{\sigma}$) have unique inverse images $v = (1, 0, 0)$, $w = (0, 1, 0)$, respectively. $X^{\geq 0}_{\check{\sigma}}$ consists of the points $(u_1, u_2, +\sqrt{u_1 u_2})$ for $u_1 \geq 0, u_2 \geq 0$ (shaded area in Figure 12).

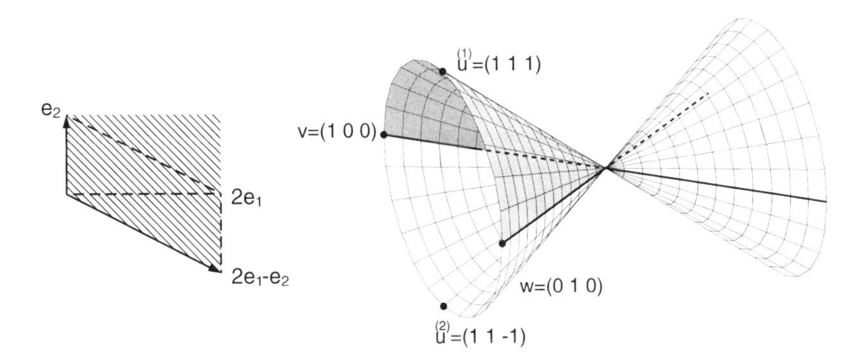

FIGURE 12.

Example 2. If $X_{\check{\sigma}}$ is regular, φ provides an isomorphism for $X_{\check{\sigma}} \cong \mathbb{C}^{n-d} \times \mathbb{C}^{*d}$ according to VI, Theorem 2.12.

The map φ can also be defined for nonsimplicial σ. We will not discuss this but will proceed to maps of compact projective toric varieties onto polytopes.

As in section 2, we assign a complex coordinate (parameter function) $x_i := z^{m_i}$, $i = 0, \ldots, r$ to each lattice point m_i of an n-dimensional lattice polytope P. We assume P to be "large enough", that is, multiply it, possibly by a positive integer, such that we obtain an embedding

$$X_{\Sigma(P)} \hookrightarrow \mathbb{P}^r$$

according to Definition 3.5. Let $\{m_0, \ldots, m_r\} = P \cap \mathbb{Z}^n$.

7.2 Definition. Let $f : X_\Sigma \longrightarrow P$ be given by

$$f(x) := \frac{|x_0|m_0 + \cdots + |x_r|m_r}{|x_0| + \cdots + |x_r|}.$$

We call f the *moment map* of X_Σ.

Furthermore, let $g : X_\Sigma \longrightarrow P$ be defined by

$$g(x) := \frac{|x_0|^2 m_0 + \cdots + |x_r|^2 m_r}{|x_0|^2 + \cdots + |x_r|^2}.$$

We say g is the *modified moment map* of X_Σ.

Clearly, $f(x)$ and $g(x)$ are convex combinations of $m_0, \ldots, m_r$, hence, in P.

7.3 Theorem. *f and g are continuous surjective maps with the following properties.*

(1) *f, g map the big torus T of X_Σ onto the interior of P, and, more generally, each orbit onto the relative interior of a face of P.*

(2) *$f_0 := f|_{X_\Sigma^{\geq 0}}$ and $g_0 := g|_{X_\Sigma^{\geq 0}}$ are bijective. In particular, they map each closure of an orbit bijectively onto a face of P.*

PROOF. It is sufficient to undertake the proof for f since the transformation $(|x_0|, \ldots, |x_r|) \longmapsto (|x_0|^2, \ldots, |x_r|^2)$ preserves (1) and (2).

If $(x_0, \ldots, x_r)$ is replaced by $(\lambda x_0, \ldots, \lambda x_r)$, for any $\lambda \neq 0$, then, f does not change. So, we can assume $|x_0| + \cdots + |x_r| = 1$. Clearly, f is continuous on X_Σ.

Both statements (1) and (2) refer to the tori T_F of which X_Σ is composed. Each of them corresponds to a face F of P where $T = T_P$. But any face F defines a subvariety of X_Σ, given by those $x_i = z^{m_i}$ for which $m_i \in F \cap \mathbb{Z}^n$, and by the affine relations between such m_i (see section 3). Therefore, it is sufficient to prove (1) and (2) for the big torus T.

First, we show that f is a surjective map of X_Σ onto P. Let $y \in P$,

(a) $$y = y_0 m_0 + \cdots + y_r m_r,$$

(b) $$y_0 + \cdots + y_r = 1,$$

(c) $$y_i \geq 0, \qquad i = 0, \ldots, r.$$

We consider the set C_y of all vectors $(y_0, \ldots, y_r)$ which represent the same y according to the conditions (a), (b), and (c). C_y is a point only if the face F of P for which $y \in \operatorname{relint} F$ is a simplex. From one representation of y we obtain others by adding representations of 0,

(a′) $$0 = \alpha_0 m_0 + \cdots + \alpha_r m_r,$$

(b′) $$\alpha_0 + \cdots + \alpha_r = 0.$$

The set of vectors $a := (\alpha_0, \ldots, \alpha_r)$ that satisfy (a′), (b′) is an affine subspace C' of $\mathbb{R}^{r+1}$, and $C_y \subset y + C'$.

We wish to select one among the possible representations of y which is, in some way, "canonical". This can be achieved by choosing y such that it minimizes the function

$$\Phi(y_0, \ldots, y_r) := y_0(-1 + \log y_0) + \cdots + y_r(-1 + \log y_r)$$

on C_y. The function Φ can be defined for all points $y \in \mathbb{R}^{r+1}$ with $y_i \geq 0$, $i = 0, \ldots, r$ (using $y \log y \to 0$ as $y \to 0^+$). Restricted to any real torus $\tilde{T} \subset \mathbb{R}^{r+1}$, $0 < \dim \tilde{T} \leq r + 1$, Φ has a Hessian matrix

$$\begin{pmatrix} y_{i_1}^{-1} & & O \\ & \ddots & \\ O & & y_{i_v}^{-1} \end{pmatrix}$$

for appropriate indices $i_1, \ldots, i_v$. Suppose Φ had two minima on C_y, attained in y^*, y^{**} say. Since the Hessian is positively definite on the line segment joining y^*, y^{**}, we obtain a contradiction. Therefore $y^* = y^{**}$.

Let $a \in C'$ be chosen arbitrarily (so that (a′), (b′) are satisfied). Up to renumbering, we can set $a = (\alpha_0, \ldots, \alpha_p, \alpha_{p+1}, \ldots, \alpha_s, 0, \ldots, 0)$, where $\alpha_i > 0$ for $i \in \{0, \ldots, p\} =: A_+$ and $\alpha_j < 0$ for $j \in \{p+1, \ldots, s\} =: A_-$.

Let $y(t) := y^* + ta$, $t \in \mathbb{R}$. We claim that y^* satisfies

(3) $$y_0^{*\alpha_0} \cdots y_p^{*\alpha_p} = y_{p+1}^{*\alpha_{p+1}} \cdots y_s^{*\alpha_s}.$$

There are four different cases.

Case 1

For sufficiently small $\varepsilon > 0$, for all $i \in A_+ \cup A_-$ and all $-\varepsilon < t < \varepsilon$, $0 < y_i(t) < 1$. Then, $y(t) \in C_y$, and

(4) $$\frac{d}{dt}\Phi(y(t))|_{t=0} = \alpha_0 \log y_0^* + \cdots + \alpha_s \log y_s^* = 0$$

(because of the minimality of Φ in y^*). From (4) we readily obtain (3).

Case 2

If case 1 fails to be true for at least one i, we see that $y_i^* = 0$ (unless $y = m_i$, a trivial case). First, let $y_j^* = 0$ also be true such that $i \in A_+$, $j \in A_-$. Then, (3) is trivially satisfied.

Case 3

$y_i^* = 0$ for an $i \in A_+$ but $y_j^* > 0$ for $j \in A_-$. Then, for suitably small $\varepsilon > 0$, $0 < y_i(t) < 1$ is true for all $i \in A_+ \cup A_-$ and for all $0 < t < \varepsilon$. Therefore,

$$\frac{d}{dt}\Phi(y(t)) = \alpha_0 \log y_0(t) + \cdots + \alpha_s \log y_s(t)$$

exists in y^*. Since $\alpha_i > 0$ and $\log y_i(t) < 0$,

$$\lim_{t \to 0^+} \frac{d}{dt}\Phi(y(t)) = -\infty.$$

But, then, in any neighborhood of y^*, there is a point of C_y in which Φ is smaller than in y^*, a contradiction.

Case 4

$y_i^* = 0$ for an $i \in A_-$ but $y_j^* > 0$ for $j \in A_+$. Here, we proceed analogously to Case 3.

This completes the proof of the surjectivity of f.

In order to show injectivity for f restricted to $X_\Sigma^{\geq 0}$, let $y \in P$ be given according to (a), and let $(y_0, \ldots, y_r) = (z^{m_0}, \ldots, z^{m_r})$. Then

$$(5) \qquad \begin{aligned} m_{01} \log z_1 + \cdots + m_{0n} \log z_n &= \log y_0 \\ &\vdots \\ m_{r1} \log z_1 + \cdots + m_{rn} \log z_n &= \log y_r. \end{aligned}$$

We see from (5) that the vector $(\log y_0, \ldots, \log y_r)$ is a linear combination of the vectors $\tilde{m}_j := (m_{0j}, \ldots, m_{rj})$, $j = 1, \ldots, n$. Each $\tilde{m}_j$, however, is orthogonal to the affine space aff $C_y = \{(y_0, \ldots, y_r) \mid y_0 m_0 + \cdots + y_r m_r = y\}$, and, hence, the same is true for $(\log y_0, \ldots, \log y_r)$. On the other hand, $(\log y_0, \ldots, \log y_r)$ is the gradient of $\Phi(y_0 \ldots, y_r)$. Therefore Φ attains its minimal value in $(y_0, \ldots, y_r)$, and, hence, $y_0 = y_0^*, \ldots, y_r = y_r^*$, so that the monomials $z^{m_0} = y_0^*, \ldots, z^{m_r} = y_r^*$ of $z = (z_1, \ldots, z_n)$ are uniquely determined.

We assert that $z_1, \ldots, z_n$ are also uniquely determined: In (5), we replace $y_0, \ldots, y_r$ by $y_0^*, \ldots, y_r^*$, respectively. Since, among $m_0, \ldots, m_r$ there are n linearly independent vectors, the solution $(\log z_1, \ldots, \log z_n) \in \mathbb{R}^n$ is uniquely determined. Hence $z_1 > 0, \ldots, z_n > 0$ are also uniquely determined. □

Example 3. We consider $m_0 := 0$, $m_1 := 2e_1 - e_2$, and $m_2 := e_2$ so that P is a triangle in $\mathbb{R}^2$. X_Σ is obtained as the projective extension of $X_{\check{\sigma}}$ in Example 1. The cone $\text{pos}\{2e_1 - e_2, e_2\}$ is replaced by the triangle P. f and g project the hatched area in Figure 12 extended by the "points at infinity" bijectively onto P.

Now we combine the modified moment map g with a linear functional $\langle m, \cdot\rangle$, $m \in \mathbb{Z}^n$.

7.4 Definition. Let P be a lattice polytope, $\{m_0, \ldots, m_r\} = P \cap \mathbb{Z}^n$, such that P is associated with the fan Σ and gives rise to an embedding $X_\Sigma \hookrightarrow \mathbb{P}^r$. We choose an $m \in \mathbb{Z}^n$, such that $\langle m, m_0\rangle, \ldots, \langle m, m_r\rangle$ are different and nonzero, and define ($x = \mathbb{C}(x_0, \ldots, x_r) \in \mathbb{P}^r$)

$$f_{(m)} : X_\Sigma \longrightarrow \mathbb{R},$$

$$f_{(m)}(x) := \frac{|x_0|^2\langle m, m_0\rangle + \cdots + |x_r|^2\langle m, m_r\rangle}{|x_0|^2 + \cdots + |x_r|^2}.$$

Then, $f_{(m)}$ is called the *Morse function* of X_Σ with respect to m.

7.5 Theorem. *Let P and X_Σ be given as in the definition of the Morse function, and suppose, in addition, that X_Σ is regular. Up to renumbering the m_j, let $m_1, \ldots, m_n$ be the lattice points adjacent to a vertex m_0 of P on the edges emanating from m_0. The one-point orbit $\overset{\circ}{x} := g^{-1}(m_0)$ is the zero point of an affine piece $X_{\check{\sigma}}$ of X_Σ. If we choose m_0 to be the zero of $\mathbb{R}^n$, then $f_{(m)}(x)$ is expressed in the local coordinates $u_j = \frac{x_j}{x_0} =: v_j + \sqrt{-1}w_j$ ($v_j, w_j \in \mathbb{R}$) of $X_{\check{\sigma}}$ as*

$$f_{(m)}(x) = (v_1^2 + w_1^2)\langle m, m_1\rangle + \cdots + (v_n^2 + w_n^2)\langle m, m_n\rangle + \mathcal{F}, \tag{6}$$

where $\mathcal{F}$ is a power series in $v_1, \ldots, v_n, w_1, \ldots, w_n$ whose terms are of degree greater than 2.

By the choice of m, $\langle m, m_i\rangle \neq 0$, $i = 1, \ldots, n$.

PROOF. We divide numerator and denominator of $f_{(m)}(x)$ by $|x_0|^2$ and, from $|u_j|^2 = u_j\bar{u}_j = v_j^2 + w_j^2$, obtain

$$f_{(m)}(x) = \frac{(v_1^2 + w_1^2)\langle m, m_1\rangle + \cdots + (v_r^2 + w_r^2)\langle m, m_r\rangle}{1 + (v_1^2 + w_1^2) + \cdots + (v_r^2 + w_r^2)}.$$

Since $m_1, \ldots, m_n$ generate $\mathbb{Z}^n$ integrally, any $m_j \notin \{m_1, \ldots, m_n\}$ is a nonnegative, integral, linear combination of $m_1, \ldots, m_n$ and gives rise to a monomial $u_1^{k_1} \cdots u_n^{k_n} = u_j$, $k_j \in \mathbb{Z}_{\geq 0}$, $k_1 + \cdots + k_n > 1$. Then, $v_j^2 + w_j^2$ is a polynomial in $v_1, \ldots, v_n, w_1, \ldots, w_n$ of degree greater than 2. By expansion of $f_{(m)}(x)$ into a Taylor series we obtain (6). □

7.6 Definition. A point x of X_Σ is called *critical* with respect to $f_{(m)}$ if, in real local coordinates according to Theorem 7.5, $f_{(m)}$ has vanishing partial derivatives (in v_j, w_j, $j = 1, \ldots, n$) at x. The *index* of $f_{(m)}$ in $\overset{\circ}{x}$ is twice the number of positive coefficients $\langle m, m_j\rangle$ in (6).

The following two theorems are immediate consequences of Theorem 7.5:

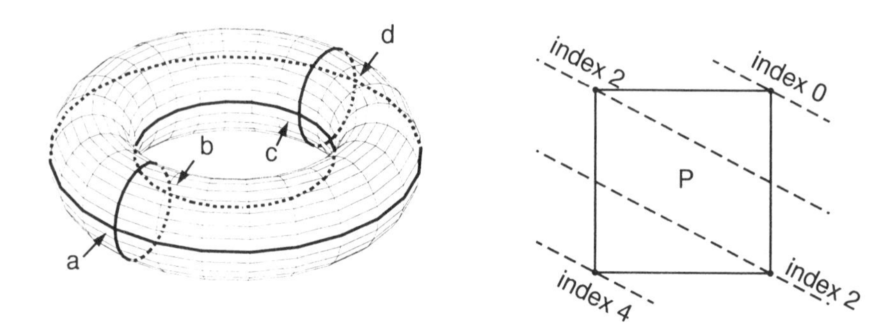

FIGURE 13.

7.7 Theorem. *The critical points of X_Σ with respect to any Morse function $f_{(m)}$ of X_Σ are the zero-dimensional orbits and only these.*

7.8 Theorem. *For $m \in \mathbb{Z}^n \setminus \{0\}$, let the open half-space H^+ be defined by $\{y \in \mathbb{R}^n \mid \langle m, y\rangle > 0\}$. Then the index of $f_{(m)}$ in the one-point orbit $\overset{\circ}{x}$ of X_Σ (see Theorem 7.5) equals twice the number of edges of P emanating from m_0 and lying in H^+.*

Example 4. Let $X_\Sigma = \mathbb{P}^1 \times \mathbb{P}^1$, $m = e_1 + 2e_2$, and $P = \operatorname{conv}\{0, e_1, e_2, e_1 + e_2\}$. Denote the cones $\sigma_0, \ldots, \sigma_3$ as in VI, 3, Example 2, and let a, b, c, d be the zero point of $X_{\check{\sigma}_0}, X_{\check{\sigma}_1}, X_{\check{\sigma}_2}, X_{\check{\sigma}_3}$, respectively. The index of $f_{(m)}$ is then 4 in a, 2 in b, 2 in c and 0 in d, respectively (Figure 13).

Remark. If X_Σ is restricted to its real points (not only positive ones!) or defined over $\mathbb{R}$ instead of $\mathbb{C}$, then the Morse function in (6) attains the form ($v = (v_1, \ldots, v_n)$)

$$f_{(m)}(v) = v_1^2\langle m, m_1\rangle + \cdots + v_n^2\langle m, m_n\rangle + F_0$$

for a series F_0 in $v_1, \ldots, v_n$ whose terms are of degree greater than 2. The index is, then, defined as the number of positive $\langle m, m_i\rangle$, $i \in \{1, \ldots, n\}$, and equals the number of edges emanating from m_0 and lying in H^+.

Exercises

1. Let $\sigma := \operatorname{pos}\{(k+1, k, k), (k, k+1, k,), (k, k, k+1)\}$ for some natural number $k > 0$. Find the map φ according to Theorem 7.1 and determine its degree.
2. Determine explicitly the moment map of $X_\Sigma = \mathbb{P}^p \times \mathbb{P}^q$, p, q natural numbers or 0.

3. Calculate explicitly the Morse function of $X_\Sigma = \mathbb{P}^1 \times \mathbb{P}^1 \times \mathbb{P}^1$ with respect to $m = (1, 2, 3)$ and each one-point orbit.
4. Discuss the restriction of the Morse function $f_{(m)}$ to the compact torus T_{cp} of X_Σ (analogously to the above Remark).

8. Classification theorems. Toric Fano varieties

In this section, we will collect a few classification theorems for smooth, compact, toric varieties, which are translations of the combinatorial results in V, sections 7 and 8, into the algebraic geometric language. Concerning the general case, we restrict ourselves to dimensions two and three. (Numerous results for arbitrary dimensions are known; see the Appendix to this section). We, then, concentrate on an interesting special class of n-dimensional toric varieties defined by Fano polytopes.

For $n = 1$ we have seen that there is only one compact toric variety, namely $\mathbb{P}^1$. For $n = 2$, we derive Theorem 8.1.

8.1 Theorem.
(a) *Any smooth, compact two-dimensional toric variety can be blown down to $\mathbb{P}^2$ or to a Hirzebruch surface $\mathcal{H}_k$, $k \in \mathbb{Z}\setminus\{1, -1\}$.*
(b) *Each Hirzebruch surface $\mathcal{H}_k$ is a $\mathbb{P}^1$-fiber bundle (see* VI, 6*) over $\mathbb{P}^1$.*

PROOF. (a) This follows from V, Theorem 7.5. (b) has already been stated in VI, 6, Example 3. □

In the case $n = 3$, no general theorem exists as for $n = 2$. We impose conditions on the Picard number of X_Σ (see section 3) which, for a smooth compact X_Σ, is $f_1(\Sigma) - n$ ($f_1(\cdot)$ the number of 1-cones).

8.2 Theorem. *Any smooth, compact three-dimensional toric variety, with Picard number at most* 3*, can be blown down to one of the following types of toric varieties:*
(1) $\mathbb{P}^3$,
(2) *a $\mathbb{P}^1$-fiber bundle over $\mathbb{P}^2$,*
(3) *a $\mathbb{P}^2$-fiber bundle over $\mathbb{P}^1$, or*
(4) *a $\mathbb{P}^1$-fiber bundle over a Hirzebruch surface $\mathcal{H}_k$.*

PROOF. This follows from V, Theorem 7.7. □

8.3 Definition. Let X_Σ be a smooth and compact toric variety, and let $\varrho_1, \ldots, \varrho_r$ be the 1-cones of Σ. The divisor

$$-K := D_{\varrho_1} + \cdots + D_{\varrho_r}$$

is called the *anticanonical divisor* of X_Σ.

The name is seen from the following background. Consider the ("logarithmic") differential form $\frac{dz_1}{z_1} \wedge \cdots \wedge \frac{dz_n}{z_n}$ on the big torus T. It has the property of being invariant (possibly up to the sign) if we replace the torus coordinates $(z_1, \ldots, z_n) = (z^{e_1}, \ldots, z^{e_n})$ by the coordinates $(u_1, \ldots, u_n) = (z^{m_1}, \ldots, z^{m_n})$ of an affine piece $X_{\check{\sigma}}$ of X_Σ, where $\check{\sigma} = \operatorname{pos}\{m_1, \ldots, m_n\}$ and $\det(m_1, \ldots, m_n) = \pm 1$. This can be seen by direct calculation. For example, in the case $n = 2$,

$$
\begin{aligned}
\frac{du_1}{u_1} \wedge \frac{du_2}{u_2} &= z^{-m_1-m_2}(m_{11}z_1^{-1}z^{m_1}dz_1 + m_{12}z_2^{-1}z^{m_2}dz_2) \\
&\quad \wedge (m_{21}z_1^{-1}z^{m_1}dz_1 + m_{22}z_2^{-1}z^{m_2}dz_2) \\
&= z_1^{-1}z_2^{-1}(dz_1 \wedge dz_2)\det(m_1, m_2) = \pm\frac{dz_1}{z_1} \wedge \frac{dz_2}{z_2}.
\end{aligned}
$$

So $z_1^{-1} \cdots z_n^{-1}$ is a regular function on T which extends to an invertible sheaf $\mathcal{K}$ of rational functions on X_Σ where $\mathcal{K}(\mathcal{O}_{X_{\check{\sigma}}}) = z^{-(m_1+\cdots+m_n)}\mathcal{O}(X_{\check{\sigma}})$. It corresponds to a toric Cartier divisor $K = \alpha_1 D_1 + \cdots + \alpha_r D_r$ where the coefficients are readily seen to be $\alpha_1 = \cdots = \alpha_r = -1$. K is called the *canonical divisor* of X_Σ. Then, $-K$ is the anticanonical divisor. The piecewise linear function $-h$, by which it is described, has value 1 on each generator a_i, where $\varrho_i = \mathbb{R}_{\geq 0}\, a_i$, $i = 1, \ldots, r$.

If $-h$ happens to be strictly convex, it is the support function of a polytope P^* whose vertices are the generators $a_1, \ldots, a_r$ of $\varrho_1, \ldots, \varrho_r$, respectively. So, P^* is a Fano polytope.

8.4 Definition. If the anticanonical divisor $-K = D_{\varrho_1} + \cdots + D_{\varrho_r}$ of a smooth, compact, projective toric variety is ample, we call it a *toric Fano variety.*.

Clearly,

8.5 Lemma. *A smooth compact projective toric variety X_Σ is a toric Fano variety if and only if Σ is spanned by a Fano polytope P^* (polar to an associated polytope P of Σ).*

8.6 Definition. We call a k-dimensional, toric Fano variety a *del Pezzo variety* $Y_{(k)}$ if it is obtained from $\mathbb{P}^1 \times \cdots \times \mathbb{P}^1$ (k-fold) by blowing up in two T-invariant points. $Y_{(2)}$ is said to be a *del Pezzo surface.*

From V, Theorem 8.2, we derive Theorem 8.7.

8.7 Theorem. *There exist five types of two-dimensional toric Fano varieties,*

(1) $\mathbb{P}^2$,

(2) $\mathbb{P}^1 \times \mathbb{P}^1$,

(3) $\mathcal{H}_1$ *(Hirzebruch surface),*

(4) $\mathbb{P}^1 \times \mathbb{P}^1$ *blown up in one invariant point,*

(5) *del Pezzo surface.*

8.8 Lemma. *Any del Pezzo variety is characterized by a del Pezzo polytope* $P_{(k)} = \operatorname{conv}\{e_1, \ldots, e_k, -e_1, \ldots, -e_k, e_1 + \cdots + e_k, -e_1 - \cdots - e_k\}$ *(see* V, 8*).*

PROOF. This follows from V, Theorem 8.4. □

8.9 Theorem. *Any n-dimensional toric Fano variety* X_Σ *with a centrally symmetric, Fano polytope is the product of projective lines and del Pezzo varieties,*

$$X_\Sigma = \underbrace{\mathbb{P}^1 \times \cdots \times \mathbb{P}^1}_{r} \times Y_{(k_1)} \times \cdots \times Y_{(k_s)}, \qquad r + k_1 + \cdots + k_s = n.$$

PROOF. This is a conclusion from V, Theorem 8.4. □

For $n = 2$ only (2) and (5) in Theorem 8.7 have centrally symmetric Fano polytopes.

The following three-dimensional Fano variety does not belong to those characterized by Theorem 8.9.

Examples. (see V, 8, Figure 31). $\mathbb{P}^1 \times \mathcal{H}_1$ blown up in an appropriate T-invariant projective line.

As in V, 8, we remark that there are precisely 18 types of three-dimensional toric Fano varieties and 121 four-dimensional ones (Batyrev's theorem).

Exercises

1. Characterize the toric Fano varieties with Fano polytopes as in V, 8, Exercises 1 and 2.
2. Find a four-dimensional toric Fano variety whose Fano polytope is not centrally symmetric.
3. Calculate $(-K, D_\varrho)$ for $\varrho \in \Sigma^{(1)}$ where Σ is the fan of any two-dimensional toric Fano variety and does not split into lower-dimensional polytopes.
4. A compact, smooth, toric variety X_Σ is a toric Fano variety if and only if the following is true. Let $\varrho_1, \ldots, \varrho_{n-1}$ be 1-cones of Σ such that $\varrho_1 + \cdots + \varrho_{n-1}$ is an $(n-1)$-face of Σ, and let $-K$ be the anticanonical divisor of X_Σ. Then, $(-K.D_{\varrho_1}.\ldots.D_{\varrho_{n-1}}) > 0$.

VIII

Cohomology of toric varieties

1. Basic concepts

In this last chapter, we wish to study topological properties of toric varieties which are hidden in the structure of the Chow ring.

We shall need some knowledge of algebraic topology. To fix the notation and to give those readers who are not yet familiar with homology theory some introductory information, we will present a short survey of the homological terminology and describe the origins of some of the facts used. For more details, we recommend Fulton [1995].

One of the original problems of topology is that of classifying manifolds or more general topological spaces up to bijective, bicontinuous mappings, called homeomorphisms. Toric isomorphisms, for example, are homeomorphisms, though they possess algebraic (and differential geometric) properties which are neglected if we concentrate on topological features. Another example is given by the boundaries of convex bodies of the same dimension. Any two of them are homeomorphic, where the homeomorphism can easily be obtained by translation and central projection.

Working on classification problems usually means looking for invariants under homeomorphisms (or more general continuous maps). The Euler characteristic, for example, as considered in chapter III, is the same for the boundary of any polytope of n-dimensions, and also for other cell-decomposed spheres.

Here "invariance" has an even stronger meaning than remaining the same under homeomorphisms. It refers also to the cell-decomposition which is chosen. We have shown this invariance for polytopes (in III, 3); the general proof is, however, rather complicated. (We note that our interest in cell complexes was not motivated by topology but by combinatorial data used in the algebraic discussion of toric varieties).

Often the invariants are of an algebraic nature, mainly groups or modules. Consider, for example, closed Jordan curves on a 2-manifold $\mathcal{M}$. If $\mathcal{M}$ is a compact two-dimensional torus, we may consider each such curve continuously "deformed" in $\mathcal{M}$ so as to obtain the join of a k-fold "meridian" a and an m-fold "equator" b through a definite point 0. The new curve is said to be "homotopic" and, for specific deformations to be introduced below, "homologous" to the original one.

The abelian group with generators a, b, that is, $\mathbb{Z} \times \mathbb{Z}$, is, then, called the first homology group of $\mathcal{M}$. If $\mathcal{M}$ is a sphere, however, any closed curve can be "deformed" into a point, and the first homology group assigned to $\mathcal{M}$ is the trivial group.

If we want to make the definition of homology groups precise, one of the main problems is to fix the "equivalence under deformation" of the curves and, more generally, of k-cycles (k-spheres) in $\mathcal{M}$. Linear equivalence of divisors in toric varieties is, in particular, such an equivalence, but, of course, not the most general one. A classical way to handle the problem is that of "simplicial homology".

We look for a decomposition of the given space into (or an approximation of it by) a simplicial complex $\mathcal{C}$, consider k-cycles consisting of cells of $\mathcal{C}$, and define homology groups. Then, we show the invariance of such groups if one decomposition is replaced by another. The main steps of defining the homology groups are as follows.

Let $S_k := \operatorname{conv}\{x_0, \ldots, x_k\}$ be a k-simplex. If an even permutation is applied to the vertices of S_k, it represents an orientation-preserving map of S_k onto itself. An orientation-reversing map is assigned to an odd permutation. So, we assign two orientations to S_k. By $+\langle x_0, \ldots, x_k\rangle$ we denote the class of simplices obtained from S_k by the even permutations, and, by $-\langle x_0, \ldots, x_k\rangle$, those obtained by odd permutations.

Now, formal integral multiples of simplices attain a geometric meaning. $m\langle x_0, \ldots, x_k\rangle$ is, for $m > 0$, an m-fold copy of the oriented simplex S_k, whereas $(-m)\langle x_0, \ldots, x_k\rangle = -m\langle x_0, \ldots, x_k\rangle$ is an m-fold copy of S_k with the opposite orientation. $0 \cdot \langle x_0, \ldots, x_k\rangle$ is a formal zero element. The abelian group, obtained from all k-dimensional simplices of $\mathcal{C}$ as generators, is called the *kth chain group* $C_k(\mathcal{C})$ of $\mathcal{C}$, and its elements are said to be *k-chains*. $C_k(\mathcal{C})$ is a free abelian group with $f_k(\mathcal{C})$ generators. If $\mathcal{C}$ has no k-cell (for example, if $k < 0$), we set $C_k(\mathcal{C}) = 0$ (zero group).

The $(k-1)$-face of S_k obtained by deleting x_i is analogously assigned an orientation and denoted by $\langle x_0, \ldots, \hat{x}_i, \ldots, x_k\rangle$. We define the *boundary* of $\langle x_0, \ldots, x_k\rangle$ by

$$\partial_k\langle x_0, \ldots, x_k\rangle = \begin{cases} \sum_{i=0}^{k}(-1)^i\langle x_0, \ldots, \hat{x}_i, \ldots, x_k\rangle & \text{for } k > 0, \\ 0 & \text{for } k = 0. \end{cases}$$

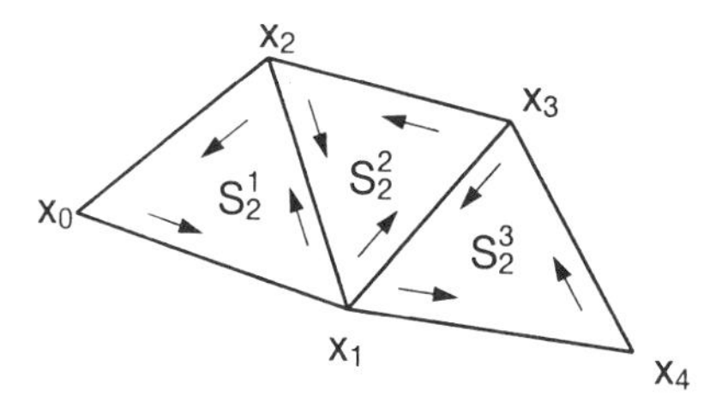

FIGURE 1.

Let $S_k^1, \ldots, S_k^q$ be the generating simplices of $C_k(\mathcal{C})$. Then, we extend ∂_k to an arbitrary element $c := \alpha_1 S_k^1 + \cdots + \alpha_q S_k^q$ ($\alpha_i \in \mathbb{Z}; i = 1, \ldots, q$) by

$$\partial_k c := \alpha_1 \partial_k S_k^1 + \cdots + \alpha_q \partial_k S_k^q \in C_{k-1}(\mathcal{C}),$$

and obtain a group homomorphism ($k > 0$)

$$\partial_k : C_k(\mathcal{C}) \longrightarrow C_{k-1}(\mathcal{C}),$$

called a *boundary operation.*

Example 1. In the complex $\mathcal{C}$ of Figure 1, $C_0(\mathcal{C}) \cong \mathbb{Z}^5$, $C_1(\mathcal{C}) \cong \mathbb{Z}^7$, $C_2(\mathcal{C}) \cong \mathbb{Z}^3$, and $\partial_2 S_2^1 = \langle x_0, x_1\rangle + \langle x_1, x_2\rangle + \langle x_2, x_0\rangle$, $\partial_2 S_2^2 = \langle x_1, x_3\rangle + \langle x_3, x_2\rangle + \langle x_2, x_1\rangle$, $\partial_2 S_2^3 = \langle x_1, x_4\rangle + \langle x_4, x_3\rangle + \langle x_3, x_1\rangle$. Since $\langle x_1, x_2\rangle = -\langle x_2, x_1\rangle$ (odd permutation), $\partial_2(S_2^1 + S_2^2) = \langle x_0, x_1\rangle + \langle x_1, x_3\rangle + \langle x_3, x_2\rangle + \langle x_2, x_0\rangle$.

A little direct calculation shows that

$$\partial_k \partial_{k+1} C_{k+1}(\mathcal{C}) = \{0\}. \tag{1}$$

We call a chain $c \in C_k(\mathcal{C})$ *closed*, if $\partial_k c = 0$, and a *bounding chain*, if there exists a chain $c' \in C_{k+1}(\mathcal{C})$ such that $\partial_{k+1} c' = c$. We say two chains $c_1, c_2 \in C_k$ are *homologous* $c_1 \sim c_2$, if their difference $c_1 - c_2$ is a bounding chain.

The notion "homologous" now gives a precise definition of the sort of local "deformations" that are allowed. We can replace a $(k-1)$-face $c := \langle x_0, \ldots, \hat{x}_i, \ldots, x_k\rangle$ of $\langle x_0, \ldots, x_k\rangle$ by

$$\begin{aligned}\bar{c} := &- \langle \hat{x}_0, \ldots, x_k\rangle - \cdots - (-1)^{i-1}\langle x_0, \ldots, \hat{x}_{i-1}, \cdots x_k\rangle \\ &- (-1)^{i+1}\langle x_0, \ldots, \hat{x}_{i+1}, \ldots, x_k\rangle - \cdots - (-1)^k\langle x_0, \ldots, \hat{x}_k\rangle,\end{aligned}$$

since $(-1)^i c - \bar{c}$ is the boundary of $\langle x_0, \ldots, x_k\rangle$ and, hence, homologous to 0. The opposite operation is also allowed. Intuitively speaking, we may "pull" a chain $\in C_{k-1}(\mathcal{C})$ over simplices of $C_k(\mathcal{C})$, and, hence, apply a special "homotopic" deformation.

So, in Example 1, $\langle x_0, x_1\rangle + \langle x_1, x_2\rangle + \langle x_2, x_0\rangle$ can be "deformed" into $\langle x_0, x_1\rangle + \langle x_1, x_3\rangle + \langle x_3, x_2\rangle + \langle x_2, x_0\rangle$ and, then, into $\langle x_0, x_1\rangle + \langle x_1, x_4\rangle + \langle x_4, x_3\rangle + \langle x_3, x_2\rangle + \langle x_2, x_0\rangle$, and, on the other side, $\langle x_0, x_1\rangle + \langle x_1, x_2\rangle + \langle x_2, x_0\rangle$ is homologous to 0 since it bounds S_2^1. So, all three chains are homologous to 0. This is no longer true if one of the 2-cells is deleted from the simplicial complex $\mathcal{C}$.

However, "homologous" includes much more than "deformable into each other". For example, let an ordinary torus be simplicially decomposed, and let T, T' be two triangles (2-cells) of the simplicial complex. Denote by $\mathcal{C}$ the complex obtained by deleting T, T' but leaving $\partial_2 T, \partial_2 T'$ in the complex. Then, $\partial_2 T - \partial_2 T'$ bounds the 2-chain consisting of all 2-cells of $\mathcal{C}$, hence, is homologous to 0. But $\partial_2 T$ cannot be "deformed" into $\partial_2 T'$ inside $\mathcal{C}$, and $\partial_2 T - \partial_2 T'$ is not deformable into a point. (Deformations are treated explicitly in so-called homotopy theory).

The closed chains in $C_k(\mathcal{C})$ form a subgroup $Z_k(\mathcal{C})$ of $C_k(\mathcal{C})$, as do the bounding chains $B_k(\mathcal{C})$, and, because of (1), $B_k(\mathcal{C})$ is also a subgroup of $Z_k(\mathcal{C})$. Moreover,

$Z_k(\mathcal{C}) = \ker \partial_k$ (kernel of the homomorphism ∂_k), and $B_k(\mathcal{C}) = \operatorname{im} \partial_{k+1}$ (image of ∂_{k+1}). We call

(2) $$H_k(\mathcal{C}) := Z_k(\mathcal{C})/B_k(\mathcal{C}) = \ker \partial_k / \operatorname{im} \partial_{k+1}$$

the kth *homology group* of $\mathcal{C}$. For $k \neq 0, \ldots, n$ (n the maximal dimension of a cell of $\mathcal{C}$), we set $H_k(\mathcal{C}) := \{0\}$.

If $\mathcal{C}$ consists of a simplex and its faces, clearly, $H_k(\mathcal{C}) = \{0\}$ for $k \neq 0$. Also, in the example of Figure 1, each 1-chain is homologous to 0, and, hence, $H_1(\mathcal{C}) = \{0\}$. If, however, S_2^1 is deleted from $\mathcal{C}$, we find for the new simplicial complex $\mathcal{C}'$, that $H_1(\mathcal{C}') = \mathbb{Z}$. In the case where $\mathcal{C}$ is connected (any two points can be joined by a polygon in $\mathcal{C}$), $H_0(\mathcal{C}') = \mathbb{Z}$, whereas, in the case where $\mathcal{C}$ consists of p connected, mutually disjoint subcomplexes (components), $H_0(\mathcal{C}) = \mathbb{Z}^p$.

The relationship "homologous to 0" can be generalized for $c, c' \in C_k(\mathcal{C})$ by saying that c is *homologous* to c' *relative to a subcomplex* $\mathcal{C}_0$ of $\mathcal{C}$ if $c - c'$ is homologous to a chain $c_0 \in C_k(\mathcal{C}_0)$ in the ordinary sense ($c - c'$ can be "pulled into $\mathcal{C}_0$"). Then, all the above definitions carry over, so that, by adding the word "relative" everywhere, the kth relative chain group $C_k(\mathcal{C}, \mathcal{C}_0)$, the group $Z_k(\mathcal{C}, \mathcal{C}_0)$ of relative cycles, and the group $B_k(\mathcal{C}, \mathcal{C}_0)$ of relative boundary chains are defined. Then,

$$H_k(\mathcal{C}, \mathcal{C}_0) := Z_k(\mathcal{C}, \mathcal{C}_0)/B_k(\mathcal{C}, \mathcal{C}_0)$$

is said to be the kth *homology group* of $\mathcal{C}$ with respect to $\mathcal{C}_0$.

If a topological space X can be decomposed into or approximated by simplicial complexes $\mathcal{C}$ so that $H_k(\mathcal{C})$ does not depend on the special choice of $\mathcal{C}$, we set $Z_k(X) := Z_k(\mathcal{C})$, $B_k(X) := B_k(\mathcal{C})$, $H_k(X) := H_k(\mathcal{C})$ and call $H_k(X)$ the kth *homology group* of X. Analogously, the kth *relative homology group* $H_k(X, Y)$ of X, with respect to a subspace Y of X, is defined. In the next sections, we shall assume $H_k(X)$ and $H_k(X, Y)$ to be defined for a toric variety X and a toric subvariety Y of X, both considered as real topological spaces. So we shall not care any more about simplicial decompositions or approximation. (For references see the Appendix to this section).

To homology groups, dual groups are assigned in the same way as dual vector spaces are assigned to vector spaces. To any pair A, G of commutative groups we assign

$$\operatorname{Hom}(A, G) := \{\varphi \mid \varphi : A \longrightarrow G \text{ a homomorphism}\},$$

which is a group with operation defined by

$$(\varphi \cdot \psi)(x) := \varphi(x) \circ \psi(x) \qquad \text{for all } x \in A$$

("$\cdot$" the group operation in $\operatorname{Hom}(A, G)$, "$\circ$" that in G).

The dual maps can also be defined as in the case of vector spaces. If $f : A \longrightarrow B$ is a homomorphism, f^* is defined by $f^*(\psi) = \psi \circ f$ for all $\psi \in \operatorname{Hom}(B, G)$,

$$\begin{array}{ccc} A & \xrightarrow{f} & B \\ \operatorname{Hom}(A, G) & \xleftarrow{f^*} & \operatorname{Hom}(B, G). \end{array}$$

In this way we assign to the homomorphism ∂_k its dual homomorphism $\delta_{k-1} := \partial_k^*$ and call it *coboundary operation*. Furthermore, we set $C^k(X; G) := \mathrm{Hom}(C_k(X), G)$, and obtain

$$\text{(3)} \qquad \cdots \xrightarrow{\partial_{k+1}} C_k(X) \xrightarrow{\partial_k} C_{k-1}(X) \xrightarrow{\partial_{k-1}} \cdots,$$

$$\text{(4)} \qquad \cdots \xleftarrow{\delta_k} C^k(X; G) \xleftarrow{\delta_{k-1}} C^{k-1}(X; G) \xleftarrow{\delta_{k-2}} \cdots.$$

It is readily seen that

$$\text{(5)} \qquad \delta_{k+1}\delta_k C^k(X; G) = 0, \qquad \text{for all } k \in \mathbb{Z},$$

analogously to (1).

Now we define the *kth cohomology group* of X with coefficients in G by

$$H^k(X; G) := \ker \delta_k / \operatorname{im} \delta_{k-1},$$

and $H^k(X; G) = 0$ in the case $k \notin \{0, \dots, n\}$.

In fact, the ideas of homology and cohomology groups turn out to be fruitful for purely algebraic considerations. We replace $C_k(X)$ and $C^k(X; G)$ by abelian groups and ∂_k, δ_k by homomorphisms satisfying (1) or (5), respectively. Then homology and cohomology groups can be defined, as above, by $\ker \partial_k / \operatorname{im} \partial_{k+1}$ and $\ker \delta_k / \operatorname{im} \delta_{k-1}$, respectively.

A system (3) or (4) or its generalization to groups is called a *chain complex* if it satisfies (1) or (5). (One should be aware that "complex" in Chapter III has a different meaning).

If subspaces of topological spaces are investigated, sometimes, several chain complexes are to be compared. So let two chain complexes $\mathcal{A}$, $\mathcal{A}'$ be given together with homomorphisms $f_k : A'_k \longrightarrow A_k$,

$$\begin{array}{llccccccc}
\mathcal{A}' & \cdots \longrightarrow & A'_{k+1} & \xrightarrow{\partial'_{k+1}} & A'_k & \xrightarrow{\partial'_k} & A'_{k-1} & \xrightarrow{\partial'_{k-1}} & \cdots \\
 & & \downarrow f_{k+1} & & \downarrow f_k & & \downarrow f_{k-1} & & \\
\mathcal{A} & \cdots \longrightarrow & A_{k+1} & \xrightarrow{\partial_{k+1}} & A_k & \xrightarrow{\partial_k} & A_{k-1} & \xrightarrow{\partial_{k-1}} & \cdots
\end{array}$$

such that all "square diagrams" are commutative, that is, $f_k \circ \partial'_{k+1} = \partial_{k+1} \circ f_{k+1}$ for each $k \in \mathbb{Z}$. Then, we write, briefly,

$$\mathcal{A}' \xrightarrow{f} \mathcal{A},$$

and call $f := (f_k)_{k \in \mathbb{Z}}$ a *chain mapping*. Its main application can be found in a combination of exact sequences $0 \longrightarrow A'_k \xrightarrow{f} A_k \xrightarrow{g} A''_k \longrightarrow 0$ in chain complexes $\mathcal{A}'$, $\mathcal{A}$, $\mathcal{A}''$,

$$\begin{array}{ccccccccc}
 & & 0 & & 0 & & 0 & & \\
 & & \downarrow & & \downarrow & & \downarrow & & \\
\cdots & \longrightarrow & A'_{k+1} & \xrightarrow{\partial'_{k+1}} & A'_k & \xrightarrow{\partial'_k} & A'_{k-1} & \longrightarrow & \cdots \\
 & & \downarrow f_{k+1} & & \downarrow f_k & & \downarrow f_{k-1} & & \\
\cdots & \longrightarrow & A_{k+1} & \xrightarrow{\partial_{k+1}} & A_k & \xrightarrow{\partial_k} & A_{k-1} & \longrightarrow & \cdots \\
 & & \downarrow g_{k+1} & & \downarrow g_k & & \downarrow g_{k-1} & & \\
\cdots & \longrightarrow & A''_{k+1} & \xrightarrow{\partial''_{k+1}} & A''_k & \xrightarrow{\partial''_k} & A''_{k-1} & \longrightarrow & \cdots . \\
 & & \downarrow & & \downarrow & & \downarrow & & \\
 & & 0 & & 0 & & 0 & &
\end{array}$$

More simply, we say that the sequence

$$(6) \qquad 0 \longrightarrow \mathcal{A}' \xrightarrow{f} \mathcal{A} \xrightarrow{g} \mathcal{A}'' \longrightarrow 0$$

of chain complexes is *exact*.

Let $H'_k := \ker \partial'_k / \operatorname{im} \partial'_{k+1}$, $H_k := \ker \partial_k / \operatorname{im} \partial_{k+1}$, and $H''_k := \ker \partial''_k / \operatorname{im} \partial''_{k+1}$ be the homology groups defined by $\mathcal{A}'$, $\mathcal{A}$, $\mathcal{A}''$, respectively. We can order them as follows in an exact sequence, called the *long, exact sequence* of the (short) exact sequence (6).

$$(7) \qquad \begin{array}{l} \cdots \longrightarrow H''_{k+1} \xrightarrow{(d_*)_{k+1}} H'_k \xrightarrow{(f_*)_k} H_k \\ \qquad\qquad \xrightarrow{(g_*)_k} H''_k \xrightarrow{(d_*)_k} H'_{k-1} \longrightarrow \cdots . \end{array}$$

Here, $(f_*)_k$, $(g_*)_k$ are the homomorphisms which are naturally induced by f_k, g_k, respectively. We wish to define $(d_*)_{k+1}$. To any element $x''_{k+1} \in \ker \partial''_{k+1} \subset A''_{k+1}$ we assign an element $x'_k \in A'_k$ as follows (illustrated by the diagram below).

$$\begin{array}{ccc}
 & & x'_k \\
 & & \downarrow f_k \\
x_{k+1} & \xrightarrow{\partial_{k+1}} & x_k \\
\downarrow g_{k+1} & & \downarrow g_k \\
x''_{k+1} & \xrightarrow{\partial''_{k+1}} & 0.
\end{array}$$

From the exactness of $A_{k+1} \xrightarrow{g_{k+1}} A''_{k+1} \longrightarrow 0$, we see that g_{k+1} is surjective, hence, $x''_{k+1} = g_{k+1}(x_{k+1})$ for some $x_{k+1} \in A_{k+1}$. We set $x_k := \partial_{k+1}(x_{k+1})$ and obtain from $0 = \partial''_{k+1}(x''_{k+1}) = (\partial''_{k+1} \circ g_{k+1})(x_{k+1}) = (g_k \circ \partial_{k+1})(x_{k+1}) = g_k(x_k)$ that $x_k \in \ker g_k = \operatorname{im} f_k$. Therefore, $x_k = f_k(x'_k)$ for some $x'_k \in A'_k$. We assign x'_k to x''_{k+1}. This assignment is, in general, not unique. But it can be shown, by brief calculation, that we obtain uniqueness if we consider the homology classes $x''_{k+1} + \operatorname{im} \partial''_{k+2}$ and $x'_k + \operatorname{im} \partial'_{k+1}$. We obtain a homomorphism $(d_*)_{k+1} : H''_{k+1} \longrightarrow H'_k$, called a *connecting homomorphism.* Again, some direct calculations are needed to prove the exactness of (7).

Example 2. A standard example is the following. Let $\mathcal{C}$ be a simplicial complex, and let $\mathcal{C}_0$ be a subcomplex of $\mathcal{C}$. For each k, there is an exact sequence

$$0 \longrightarrow C_k(\mathcal{C}_0) \xrightarrow{i_k} C_k(\mathcal{C}) \xrightarrow{j_k} C_k(\mathcal{C}, \mathcal{C}_0) \longrightarrow 0,$$

i_k being the injection ($i_k(c_0) = c_0$ for each $c_0 \in \mathcal{C}_0$), and j_k being the projection ($j_k(c) = c + C_k(\mathcal{C}_0)$ for each $c \in C_k(\mathcal{C})$). We set $\mathcal{A}' := (C_k(\mathcal{C}_0))_{k \in \mathbb{Z}}$, $\mathcal{A} := (C_k(\mathcal{C}))_{k \in \mathbb{Z}}$, $\mathcal{A}'' := (C_k(\mathcal{C}, \mathcal{C}_0))_{k \in \mathbb{Z}}$. The long exact homology sequence (7) provides, in many cases, useful information about $\mathcal{C}$.

In closing this short survey, let us return, once again, to simplicial homology. First, we note that we can replace simplicial complexes by more general cell complexes. The cells are topological balls which are convex polytopes or, more generally, can be decomposed into such polytopes.

If $\mathcal{C}$ is a cell complex whose cells are polytopes, we construct the dual cell complex $\mathcal{C}^*$ as follows. We apply the barycentric subdivision β to $\mathcal{C}$ and consider, for each vertex v of $\mathcal{C}$, the sets $|\operatorname{st}(v, \beta(\mathcal{C}))|$ as new maximal cells, that is, cells of maximal dimension. All other cells are defined by successive intersection of maximal cells (Figure 2).

In the case where $\mathcal{C}$ is the boundary complex $\mathcal{B}(P)$ of a polytope P, we can directly set $\mathcal{C}^* := \mathcal{B}(P^*)$, where P^* is a polar polytope of P (compare II, 2).

We can show that $H^{n-k}(\mathcal{C}, \mathbb{Z}) \cong H_{n-k}(\mathcal{C}^*)$, and deduce from it what is known as *Poincaré duality*, that is,

$$H_k(\mathcal{C}) \cong H^{n-k}(\mathcal{C}; \mathbb{Z}) \tag{8}$$

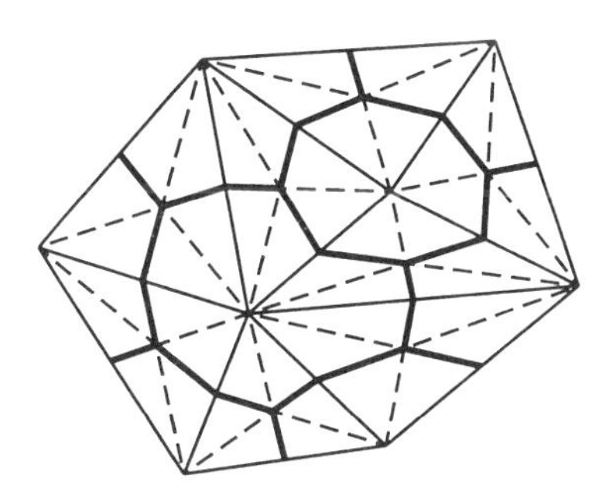

FIGURE 2.

if $|\mathcal{C}|$ is a compact, orientable manifold. Again, Poincaré duality (in generalized versions Alexander duality or Serre duality), then, applies to manifolds which are approximated by $\mathcal{C}$ or have $\mathcal{C}$ as a decomposition.

Exercises

1. Find a simplicial complex $\mathcal{C}$ whose point set is homeomorphic to a two-dimensional compact torus (built up, for example, by 18 triangles in $\mathbb{R}^3$). Verify $H_0(\mathcal{C}) = \mathbb{Z}$, $H_1(\mathcal{C}) = \mathbb{Z}^2$, and $H_2(\mathcal{C}) = \mathbb{Z}$.
2. Work out explicitly the long, exact, homology sequence in the case where $\mathcal{C}$ consists of the boundary complex of a 3-simplex from which one triangle F is deleted, and $\mathcal{C}_0$ consists of the vertices and edges of F.
3. Consider $\mathbb{P}^1$ as a real two-dimensional manifold, find a simplicial complex $\mathcal{C}$ whose point set is homeomorphic to $\mathbb{P}^1$, and calculate all homology and cohomology groups $H_k(\mathcal{C})$, $H^k(\mathcal{C})$, and $k \in \mathbb{Z}$.
4. Rewrite the introduction of chain complexes, chain mappings, long exact sequences, etc. in terms of δ instead of ∂, that is, for cohomological groups.

2. Cohomology ring of a toric variety

In this section, we will consider smooth, projective varieties X_Σ. Let $H_k(X_\Sigma)$ be the kth homological group of X_Σ (as $2n$-dimensional topological space) introduced either by simplicial decomposition or in some other standard way of topology. Also, let $H^k(X_\Sigma; \mathbb{Z})$ be the kth cohomology group of X_Σ with coefficients in $G = \mathbb{Z}$ (we restrict ourselves to $G = \mathbb{Z}$ throughout). $H^\bullet(X_\Sigma; \mathbb{Z}) := \bigoplus_k H^k(X_\Sigma; \mathbb{Z})$ has a ring structure, which will be described precisely, and is called the *cohomology ring* of X_Σ. Our aim in this section is to prove the following:

2.1 Theorem. *Let X_Σ be a smooth, projective toric variety.*

(a) *The combinatorial Chow ring and the cohomology ring of X_Σ are isomorphic as rings,*

$$\mathrm{Ch}(X_\Sigma) \cong H^\bullet(X_\Sigma; \mathbb{Z}).$$

(b) *The odd cohomology groups vanish,*

$$H^{2l+1}(X_\Sigma; \mathbb{Z}) = 0, \qquad \text{for } l \in \mathbb{Z}.$$

(c) *The Betti numbers of $H^{2l}(X_\Sigma; \mathbb{Z})$ can be calculated from the face numbers (f-vector) of Σ,*

$$\beta_{2l} := \operatorname{rank} H^{2l}(X_\Sigma; \mathbb{Z}) = \sum_{i=l}^{n} (-1)^{i+l} \binom{i}{l} f_{n-i},$$

where f_i is the number of i-dimensional cones of Σ.

Remark. The theorem is also true for compact, nonprojective toric varieties.

The proof of Theorem 2.1 proceeds in several steps. The starting point for combinatorial Chow rings had been the $(n-k)$-cones σ of Σ assigned to the k-dimensional faces F of a polytope P associated with Σ (negative polar polytope of a polytope which spans Σ). In X_Σ, a toric k-subvariety is defined by F, which can also be considered the closure $\bar{O}_\sigma$ of a k-dimensional orbit O_σ of X_Σ. Since the cycles $\bar{O}_\sigma$ can also be chosen as representatives of homology classes in $H_k(X_\Sigma)$, we obtain a natural assignment

$$p_0 : U_i \quad \longmapsto \quad [|D_{\varrho_i}|],$$

where D_{ϱ_i} denotes the invariant divisor defined by the 1-cone ϱ_i, $i = 1, \ldots, r$, $|D_{\varrho_i}|$ the point set it carries, and

$$p_0 : U_{i_1} \cdots U_{i_k} \quad \longmapsto \quad [|D_{\varrho_{i_1}}| \cap \cdots \cap |D_{\varrho_{i_k}}|] \in H_k(X_\Sigma).$$

By Poincaré duality,

$$d : H_k(X_\Sigma) \quad \longrightarrow \quad H^{n-k}(X_\Sigma; \mathbb{Z}),$$

and $d \circ p_0$ induces a map

$$p : \mathrm{Ch}(X_\Sigma) \quad \longrightarrow \quad H^\bullet(X_\Sigma; \mathbb{Z})$$

which we must show to be an isomorphism.

(1) $\qquad p_0(\mathfrak{a} + \mathfrak{b}) = 0, \qquad$ so p is well defined.

In fact, let $U_{i_1} \cdots U_{i_k}$ be a generating element of $\mathfrak{a}$, that is $\varrho_{i_1} + \cdots + \varrho_{i_k} \notin \Sigma$. Any two $D_i, D_j, i, j \in \{i_1, \ldots, i_k\}, i \neq j$, intersect transversally (in a toric subvariety) or not at all. At least two of them do not intersect (otherwise $\varrho_{i_1} + \cdots + \varrho_{i_k} \in \Sigma$), hence, $p_0(U_{i_1} \cdots U_{i_k}) = 0$ (zero element of $H_k(X_\Sigma)$). Therefore, $p_0(\mathfrak{a}) = 0$.

$\mathfrak{b}$ is spanned by linear polynomials

$$\langle a_1, m\rangle U_1 + \cdots + \langle a_r, m\rangle U_r$$

which represent principal divisors $(z^m) =: D_0$.

As a topological cycle, D_0 is homologous to 0 as is true for any principal divisor (f): One considers the real interval $[0, \infty]$ and the space $E := \{f^{-1}(t) \mid t \in [0, \infty]\}$. The boundary of E consists of the point sets $|D_+| := \{f^{-1}(0)\}$ and $|D_-| := \{f^{-1}(\infty)\}$. If we consider each point of $|D_+| \cup |D_-|$ with its multiplicity, we obtain cycles D_+, D_- also in the sense of homology theory (for example, by simplicial approximation).

Now, $D_0 - D_+ - D_-$ is homologous to 0. This implies $p_0(\mathfrak{b}) = 0$. From $p_0(\mathfrak{b}) = p_0(\mathfrak{a}) = 0$, we obtain $p_0(\mathfrak{a} + \mathfrak{b}) = 0$.

Now, we construct a basis for $H_k(X_\Sigma)$ consisting of classes $[\bar{O}_\sigma]$, $\dim \sigma = n - k$. We use a Morse function as defined in VII, section 7.

Let $m \in \mathbb{Z}^n$ be chosen such that (up to renumbering), for the zero orbits $\overset{(1)}{x}, \ldots, \overset{(s)}{x}$,

(2) $$f_{(m)}(\overset{(1)}{x}) < \cdots < f_{(m)}(\overset{(s)}{x}).$$

We define sets

$$X_i := \{x \in X_\Sigma \mid f_{(m)}(x) < f_{(m)}(\overset{(i)}{x})\}.$$

There is a largest face F_i of P which culminates in the vertex m_i of P which corresponds to $\overset{(i)}{x}$, that is, F_i belongs to the dual shelling of P determined by m (see III, 6). Let $\bar{O}_i$ be the closure of the orbit (toric subvariety) defined by F_i. Then, by (2),

$$\bar{O}_i \subset X_{i+1}.$$

We express $f_{(m)}(x)$ in a neighborhood of $\overset{(i)}{x} =: \overset{(0)}{x}$, according to VII, Theorem 7.5, as

$$f_{(m)}(x) = (v_1^2 + w_1^2)\langle m, m_1\rangle + \cdots + (v_n^2 + w_n^2)\langle m, m_n\rangle + \mathcal{F}$$

where, in local coordinates $x_j = z^{m_j - m_0}$, $x_j = v_j + \sqrt{-1}w_j$, $j = 1, \ldots, n$, and $\mathcal{F}$ as in VII, Theorem 7.5. Up to renumbering, we may assume $\langle m, m_j\rangle > 0$, for $j = 1, \ldots, \lambda_i$, and $\langle m, m_j\rangle \leq 0$ for $j = \lambda_i + 1, \ldots, n$, where $2\lambda_i$ is the index of $f_{(m)}$ in $\overset{(i)}{x}$.

The coordinates of $\bar{O}_i$ are given by the vertices of F_i. Therefore, the points of $\bar{O}_i$ in the affine piece of X_Σ which contains $\overset{(i)}{x}$ are characterized by

$$x_{\lambda_i+1} = \cdots = x_n = 0. \tag{3}$$

Now, we apply some results of Morse theory (see, for example, Milnor [1969]). The affine space (3) ("handle") has dimension $2n - 2\lambda_i$ and generates an element of the relative homology group $H_{2n-2\lambda_i}(X_{i+1}, X_i)$, its boundary lying in X_i (compare Example 2 of section 1). Since it is part of $\bar{O}_i$, we obtain, in a natural way, a homomorphism

$$h : H_{2n-2\lambda_i}(X_{i+1}) \longrightarrow H_{2n-2\lambda_i}(X_{i+1}, X_i),$$

which is surjective.

Furthermore (Milnor [1969], p. 29),

$$H_k(X_{i+1}, X_i) = \begin{cases} \mathbb{Z} & \text{for } k = 2n - 2\lambda_i, \\ 0 & \text{otherwise.} \end{cases} \tag{4}$$

We obtain the long, exact, homology sequence (based on chain groups as in Example 2 of 1)

$$\begin{aligned} \cdots \longrightarrow\ & H_{2n-2\lambda_i+1}(X_{i+1}, X_i) \longrightarrow H_{2n-2\lambda_i}(X_i) \\ \longrightarrow\ & H_{2n-2\lambda_i}(X_{i+1}) \overset{h}{\longrightarrow} H_{2n-2\lambda_i}(X_{i+1}, X_i) \longrightarrow \cdots. \end{aligned}$$

By (4), the first of these groups vanishes and the last one is isomorphic to $\mathbb{Z}$. So we have the short exact sequence

$$0 \longrightarrow H_{2n-2\lambda_i}(X_i) \longrightarrow H_{2n-2\lambda_i}(X_{i+1}) \overset{h}{\longrightarrow} \mathbb{Z} \longrightarrow 0.$$

This sequence splits. Therefore,

$$(5)\qquad \begin{aligned} H_{2n-2\lambda_i}(X_{i+1}) &\cong H_{2n-2\lambda_i}(X_i) \oplus H_{2n-2\lambda_i}(X_{i+1}, X_i) \\ &\cong H_{2n-2\lambda_i}(X_i) \oplus \mathbb{Z}\,. \end{aligned}$$

By repeated application of (5), we find (since $X_r = X_\Sigma$)

$$(6)\qquad H_k(X_\Sigma) = H_{2n-2\lambda_i}(X_\Sigma) = \bigoplus_i \mathbb{Z}_{(i)}\,,$$

where $\mathbb{Z}_{(i)} \cong \mathbb{Z}$ and the summation is carried out for all i such that $2\lambda_i = 2n - k$.

Proof of part (a) of Theorem 2.1. From (6), it follows that the map $p : \mathrm{Ch}(X_\Sigma) \longrightarrow H^\bullet(X_\Sigma; \mathbb{Z})$ is surjective. Since it maps formal linear combinations onto formal linear combinations, it is a homomorphism of groups. Moreover, it preserves the multiplicative structure, since multiplication in $H^\bullet(X_\Sigma; \mathbb{Z})$ is defined by intersection of cycles and p_0 has been defined by $p_0(U_{i_1} \cdots U_{i_k}) = [|D_{\varrho_{i_1}}| \cap \cdots \cap |D_{\varrho_{i_k}}|]$. So it remains to be shown that p is injective or, equivalently, possesses an inverse map. In fact, we can assign, to each orbit closure $\bar{O}_i$, uniquely, a face F of P which is an intersection $F = \overset{(1)}{F} \cap \cdots \cap \overset{(k)}{F}$ of facets of P. To each $\overset{(i)}{F}$, there corresponds a divisor D_{ϱ_i}. We assign to F the square-free monomial $U_1 \cdots U_k =: q(\bar{O}_i)$ (up to renumbering). By the "shifting away" Lemma 5.3 in Chapter VII, these monomials generate the Chow ring. Now, q induces the inverse map of p.

Part (b) of Theorem 2.1 follows from (4).

Part (c) of the theorem is a consequence of III, Theorem 6.8. □

Exercises

1. Find, explicitly, the Chow rings of the Hirzebruch varieties $\mathcal{H}_r$, $r \in \mathbb{Z}$.
2. Find all Betti numbers β_{2l} for any projective space $\mathbb{P}^n$.
3. How do the Betti numbers change if a smooth, compact toric variety is blown up along a T-invariant subvariety?
4. Calculate the Betti numbers of a smooth, compact toric variety X_Σ which is an $X_{\Sigma'}$-fiber bundle over X_{Σ_0} ($X_{\Sigma'}$, X_{Σ_0} also smooth; see VI, 6) from the Betti numbers of $X_{\Sigma'}$ and X_{Σ_0}.

3. Čech cohomology

Having calculated homology and cohomology groups for smooth, compact projective toric varieties X_Σ, we now turn to the study of invertible sheaves on X_Σ by cohomological means. We restrict ourselves to calculating the Čech cohomology of invertible sheaves as introduced in VII, 1 and base on it, in the next section, a version of the Riemann–Roch–Hirzebruch theorem. To understand the relationship between simplicial cohomology, general cohomology of sheaves on a toric variety, and Čech cohomology, some more work must be done. Simplicial cohomology coincides with the general cohomology of a constant sheaf (a sheaf of continuous functions to a group Z considered with discrete topology). On the other hand, the general cohomology of any sheaf (in particular, simplicial cohomology) can be calculated by the Čech complex of a so-called acyclic covering (Leray's theorem).

We only mention that the affine covering $\{X_{\check{\sigma}}\}_{\sigma\in\Sigma}$ is acyclic for any invertible sheaf, so that we can use Čech cohomology of this covering for defining the cohomology of invertible sheaves on a toric variety.

Let Σ be any fan. We denote by 1_σ the function which has constant value 1 on σ and is 0 elsewhere. Then, each cochain (in the sense of section 1) of the cell complex Σ with coefficients in $\mathbb{C}$ can be interpreted as a linear combination of such functions. In short,

$$C^k(\Sigma, \mathbb{C}) \cong \mathbb{C}\cdot 1_{\sigma_1} \oplus \cdots \oplus \mathbb{C}\cdot 1_{\sigma_s} \cong \mathbb{C}^s,$$

where $\{\sigma_1, \ldots, \sigma_s\} = \Sigma^{(n-k)}$.

We may look at these cochain groups in terms of sheaf theory. Σ can be considered as a covering of the set $|\Sigma|$. If the sheaf axioms are extended to nonopen coverings of a topological space, we readily see that a sheaf $\mathbb{C}_{|\Sigma|}$ of piecewise constant, complex-valued functions on $|\Sigma|$ is obtained if we set

$$\mathbb{C}_{|\Sigma|}(\sigma) := \mathbb{C}\cdot 1_\sigma$$

and extend it to unions of cones of Σ. (We could also consider open ε-neighborhoods of the cones σ and apply the original sheaf axioms of open coverings).

The covering Σ of $|\Sigma|$ corresponds to the open covering $\{X_{\check{\sigma}}\}_{\sigma\in\Sigma}$ of X_Σ, and $\sigma \cap \tau$ is the counterpart of $X_{\check{\sigma}} \cap X_{\check{\tau}}$.

To each group $\mathbb{C}_{|\Sigma|}(\sigma)$, we assign the $\mathcal{O}_{X_\Sigma}$-module $\mathcal{O}_{X_\Sigma}(X_{\check{\sigma}})$, that is, to $\mathbb{C}_{|\Sigma|}$, there corresponds the structure sheaf $\mathcal{O}_{X_\Sigma}$ of X_Σ.

In general, let $U := \{U_j\}$ be a covering of a set X. For our purposes, it suffices to assume that U has only finitely many elements $U_1, \ldots, U_r$. Furthermore, we suppose that an n-dimensional cell complex $\mathcal{C}$ exists such that

(a) there is a bijective map φ of U into the set of maximal cells of $\mathcal{C}$, $\varphi(U_j) =: \sigma_j$, $j = 1, \ldots, r$,

(b) each $(n-k)$-cell of $\mathcal{C}$ is the intersection of at most $k+1$ maximal cells, and

(c) $\varphi(U_{i_1}) \cap \cdots \cap \varphi(U_{i_m})$ is a cell of $\mathcal{C}$ for any $U_{i_1}, \ldots, U_{i_m} \in U$.

Let $\mathcal{F}$ be a sheaf of functions defined in some way on the U_j.

As examples, we have

$$X = |\Sigma|, \quad U = \Sigma, \quad \mathcal{F} = \mathbb{C}_{|\Sigma|},$$

$$X = X_\Sigma, \text{ and } \quad U = \{X_{\check{\sigma}}\}_{\sigma\in\Sigma}, \quad \mathcal{F} = \mathcal{O} \text{ the structure sheaf},$$

in both cases $\mathcal{C} := \Sigma$.

If τ is an $(n-k)$-dimensional cell of $\mathcal{C}$ and if $\tau = \sigma_{i_0} \cap \cdots \cap \sigma_{i_k}$ ($\sigma_{i_0}, \ldots, \sigma_{i_k}$ maximal), then, we set

$$U_\tau = U_{i_0\cdots i_k} := U_{i_0} \cap \cdots \cap U_{i_k}. \tag{1}$$

Let $\tau_1, \ldots, \tau_s$ be all $(n-k)$-cells of $\mathcal{C}$. We call

$$\check{C}^k(U, \mathcal{F}) := \mathcal{F}(U_{\tau_1}) \oplus \cdots \oplus \mathcal{F}(U_{\tau_s})$$

the k *cochain group* of U with coefficients in $\mathcal{F}$.

We wish to define a coboundary operator for these cochain groups. For any $f \in \check{C}^k(U, \mathcal{F})$, we set

$$f_{i_0 \cdots i_k} := f|_{U_{i_0 \cdots i_k}}$$

and

$$(d^k f)_{i_0 \cdots i_{k+1}} := \sum_{j=0}^{k+1} (-1)^{j+1} f_{i_0 \cdots \hat{i}_j \cdots i_{k+1}}|_{U_{i_0 \cdots i_{k+1}}},$$

$f_{i_0 \cdots \hat{i}_j \cdots i_{k+1}}$ meaning $f_{i_0 \cdots i_{j-1} i_{j+1} \cdots i_{k+1}}$. For $k \in \mathbb{Z} \setminus \{0, \ldots, n\}$, we define

$$d^k f = 0 \qquad \text{(zero map)}.$$

It is readily seen that we obtain a homomorphism

$$d^k : \check{C}^k(U, \mathcal{F}) \longrightarrow \check{C}^{k+1}(U, \mathcal{F}).$$

In particular, for $k = 0$,

$$(d^0 f)_{ij} = (f_i - f_j)|_{U_i \cap U_j}, \qquad \text{if } \varphi(U_i) \cap \varphi(U_j) \text{ is an } (n-1)\text{-cell of } \mathcal{C}.$$

As in simplicial homology it is readily shown that $d^k d^{k-1} = 0$ so that $\operatorname{im} d^{k-1} \subset \ker d^k$.

3.1 Definition. The group

$$\check{H}(U, \mathcal{F}) := \ker d^k / \operatorname{im} d^{k-1}$$

is called the *kth Čech cohomology group* of X with respect to the covering $U = \{X_{\check{\sigma}}\}_{\sigma \in \Sigma}$ and with coefficients in the sheaf $\mathcal{F}$.

Example 1. Let Σ be a complete fan in $\mathbb{R}^2$ with two-dimensional (successive) cones $\sigma_1, \ldots, \sigma_r$, and let $\mathcal{F}$ be an invertible sheaf as defined in VII, 1. For $f = (f_1, \ldots, f_r) \in \mathcal{F}(X_{\check{\sigma}_1}) \oplus \cdots \oplus \mathcal{F}(X_{\check{\sigma}_r})$, $d^0 f = (f_1 - f_2, f_2 - f_3, \ldots, f_{r-1} - f_r, f_r - f_1)$, and $\check{H}^0(\{X_{\check{\sigma}}\}_{\sigma \in \Sigma}, \mathcal{F}) = \ker d^0 = \{(f_1, \ldots, f_1)\}$.

Exercises

1. Consider the (noncomplete) fan $\Sigma := \{\mathbb{R}_{\geq 0}\, e_1, \mathbb{R}_{\geq 0}\, e_2, \mathbb{R}_{\geq 0}(-e_1 - e_2), \{0\}\}$ in $\mathbb{R}^2$, and let $\mathcal{F}$ be an invertible sheaf as in VII, 1. Find $\check{H}^0(\{X_{\check{\sigma}}\}_{\sigma \in \Sigma}, \mathcal{F})$.
2. Find the Čech cohomology groups for Σ, consisting of a single n-cone and its faces.
3. Prove explicitly that $d^{k+1} d^k = 0$.
4. Find all Čech cohomology groups for one-dimensional toric varieties.

4. Cohomology of invertible sheaves

We wish to calculate $\check{H}^k(\{X_{\check{\sigma}}\}_{\sigma\in\Sigma}, \mathcal{F})$ for an invertible sheaf $\mathcal{F}$ on X_Σ as considered in VII, 1. First, we show that

4.1 Lemma. Each element of a torus action on X_Σ leaves the Čech cohomology groups $\check{H}^k(\{X_{\check{\sigma}}\}_{\sigma\in\Sigma}, \mathcal{F})$ invariant.

PROOF. φ leaves each U_τ fixed, hence, induces an isomorphism for each $\mathcal{F}(U_\tau)$ and, therefore, an isomorphism of $\check{C}^k(U, \mathcal{F}) = \mathcal{F}(U_{\tau_1}) \oplus \cdots \oplus \mathcal{F}(U_{\tau_s})$. This readily implies the lemma. □

If z^m occurs in each $f|_{X_{\check{\tau}}}$, $\dim\tau = n-k$, with nonzero coefficient, then, z^m generates a subgroup $\check{H}^k_m$ of $\check{H}^k(\{X_{\check{\sigma}}\}_{\sigma\in\Sigma}, \mathcal{F})$. If not, we set $\check{H}^k_m = \{0\}$.

Clearly, $\check{H}^k_m \cap \check{H}^k_{m'} = \{0\}$ if $m \neq m'$. Therefore, we have a natural decomposition:

4.2 Lemma. $\check{H}^k(\{X_{\check{\sigma}}\}_{\sigma\in\Sigma}, \mathcal{F}) = \bigoplus_{m\in\mathbb{Z}^n} \check{H}^k_m$.

So, the calculation of Čech cohomology is reduced to the question of finding the $\check{H}^k_m$ for all $m \in \mathbb{Z}^n$. This question will be answered for invertible sheaves which correspond to polytopes. For that purpose, we use generalizations of the sheaf $\mathbb{C}_{|\Sigma|}$, as introduced in section 3.

To begin with, we consider an arbitrary invertible sheaf $\mathcal{F}$ and the (real-valued) piecewise linear function h (on $|\Sigma|$) by which it is described. For any $m \in \mathbb{Z}^n$, we set

$$\Sigma_{(m)} := \{\sigma \in \Sigma \mid \langle m, x\rangle \geq h(x) \quad \text{for all } x \in \sigma\}.$$

$\Sigma_{(m)}$ is a subfan of Σ and defines a toric variety $X_{\Sigma_{(m)}} \subset X_\Sigma$. The piecewise constant functions $g : |\Sigma_{(m)}| \longrightarrow \mathbb{C}$ satisfying

$$g|_\sigma = \begin{cases} c_\sigma \cdot 1_\sigma & \text{if } \sigma \in \Sigma_{(m)} \\ 0 & \text{if } \sigma \notin \Sigma_{(m)} \end{cases}$$

define a subsheaf of $\mathbb{C}_{|\Sigma|}$ which we denote by

$$\mathbb{C}_{(m)}\,.$$

There is a natural isomorphism of $\mathbb{C}_{(m)}$ to $\mathbb{C}_{|\Sigma_{(m)}|}$ (defined as $\mathbb{C}_{|\Sigma|}$ in section 3). $\mathbb{C}_{(0)} = \mathbb{C}_{|\Sigma|}$ if $h(x) \leq 0$ for all $x \in |\Sigma|$.

Furthermore, we define the following sheaf. Let

$$\Sigma^{(m)} := \{\sigma \setminus |\Sigma_{(m)}| \mid \sigma \in \Sigma\}$$

be a set of convex cones. They are, in general, neither open nor closed. Nevertheless we can define piecewise constant, complex-valued functions on such cones and obtain a sheaf which we denote by $\mathbb{C}^{(m)}$ because of

$$(\sigma \setminus |\Sigma_{(m)}|) \cap (\tau \setminus |\Sigma_{(m)}|) = (\sigma \cap \tau) \setminus |\Sigma_{(m)}|$$

for any pair $\sigma, \tau \in \Sigma$.

Example 1. Let Σ be the fan of $\mathbb{P}^1 \times \mathbb{P}^1$ in $\mathbb{R}^2$ (see VI, 3, Example 2) and define a piecewise linear function h on Σ (Figure 3). Thus,

$$h(x) = \begin{cases} -x_1 + x_2 & \text{for } x \in \sigma_0, \\ x_1 + x_2 & \text{for } x \in \sigma_1, \\ -x_1 - x_2 & \text{for } x \in \sigma_2, \\ x_1 - x_2 & \text{for } x \in \sigma_3. \end{cases}$$

We find for $m = 0$, $\Sigma_{(m)} = \{\mathbb{R}_{\geq 0}\, e_1, \mathbb{R}_{\geq 0}(-e_1), \{0\}\}$, $\Sigma^{\langle m \rangle} = \{\sigma_i \setminus \mathbb{R}\, e_1$ for $i = 0, 1, 2, 3$, $\mathbb{R}_{>0}\, e_2$, $\mathbb{R}_{>0}(-e_2)\}$. The toric variety $X_{\Sigma_{(m)}}$ is the complement of two "meridians" in X_Σ.

We define a homomorphism

$$\mu : \check{H}^0(\Sigma, \mathbb{C}_{|\Sigma|}) \longrightarrow \check{H}^0(\Sigma^{\langle m \rangle}, \mathbb{C}^{(m)})$$

as follows. Let $f \in \check{C}^0(\Sigma, \mathbb{C}_{|\Sigma|})$. Then,

$$\mu(f|_\sigma) = \begin{cases} f|_{\sigma \setminus |\Sigma_{(m)}|} & \text{if } \sigma \notin \Sigma_{(m)} \\ 0 & \text{if } \sigma \in \Sigma_{(m)}. \end{cases}$$

4.3 Lemma. The sequence

$$0 \longrightarrow \ker \mu \overset{\iota}{\longrightarrow} \check{H}^0(\Sigma, \mathbb{C}_{|\Sigma|}) \overset{\mu}{\longrightarrow} \check{H}^0(\Sigma^{\langle m \rangle}, \mathbb{C}^{(m)}) \longrightarrow 0$$

is exact (ι the inclusion map).

PROOF. If $\sigma \notin \Sigma_{(m)}$, then, $\sigma \setminus |\Sigma_{(m)}|$ contains all relative interior points of σ so that the restriction map μ is surjective. If $\sigma \in \Sigma_{(m)}$, then, $\sigma \setminus |\Sigma_{(m)}| = \emptyset$, and, by setting $\mu(f) = 0$ for all $f \in \mathbb{C}_{|\Sigma|}(\sigma)$, we also obtain a surjective map. ι is, by definition, injective. □

We wish to calculate the groups $\check{H}^k(\Sigma, \mathbb{C}_{(m)})$ from the groups $\check{H}^k(\Sigma, \mathbb{C}_{|\Sigma|})$ and $\check{H}^k(\Sigma^{\langle m \rangle}, \mathbb{C}^{(m)})$. For this purpose, we set up the following

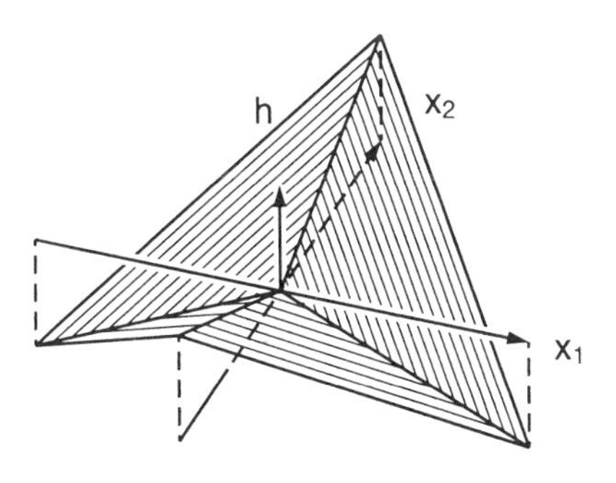

FIGURE 3.

diagram of chain complexes.

$$\begin{array}{ccccccccc}
 & & 0 & & 0 & & 0 & & \\
 & & \downarrow & & \downarrow & & \downarrow & & \\
0 & \longrightarrow & \check{C}^0(\Sigma, \mathbb{C}_{(m)}) & \stackrel{\iota}{\longrightarrow} & \check{C}^0(\Sigma, \mathbb{C}_{|\Sigma|}) & \stackrel{\mu}{\longrightarrow} & \check{C}^0(\Sigma^{\langle m\rangle}, \mathbb{C}^{\langle m\rangle}) & \longrightarrow & 0 \\
 & & \downarrow d^0 & & \downarrow d^0 & & \downarrow d^0 & & \\
0 & \longrightarrow & \check{C}^1(\Sigma, \mathbb{C}_{(m)}) & \stackrel{\iota}{\longrightarrow} & \check{C}^1(\Sigma, \mathbb{C}_{|\Sigma|}) & \stackrel{\mu}{\longrightarrow} & \check{C}^1(\Sigma^{\langle m\rangle}, \mathbb{C}^{\langle m\rangle}) & \longrightarrow & 0 \\
 & & \downarrow d^1 & & \downarrow d^1 & & \downarrow d^1 & & \\
 & & \vdots & & \vdots & & \vdots & &
\end{array}$$

By Lemma 4.3, we find $\check{C}^0(\Sigma, \mathbb{C}_{(m)}) = \ker \mu$ (μ applied to $\check{C}^0(\Sigma, \mathbb{C}_{|\Sigma|})$). From the definition of the coboundary operators d^i, we readily see that the exactness of the first "horizontal" sequence (Lemma 4.3) carries over to the exactness of all "horizontal" sequences. We obtain a long, exact, cohomology sequence

$$\begin{array}{l}
0 \longrightarrow \check{H}^0(\Sigma, \mathbb{C}_{(m)}) \stackrel{\iota^*}{\longrightarrow} \check{H}^0(\Sigma, \mathbb{C}_{|\Sigma|}) \stackrel{\mu^*}{\longrightarrow} \check{H}^0(\Sigma^{\langle m\rangle}, \mathbb{C}^{\langle m\rangle}) \longrightarrow \cdots \\
\cdots \longrightarrow \check{H}^k(\Sigma, \mathbb{C}_{(m)}) \stackrel{\iota^*}{\longrightarrow} \check{H}^k(\Sigma, \mathbb{C}_{|\Sigma|}) \stackrel{\mu^*}{\longrightarrow} \check{H}^k(\Sigma^{\langle m\rangle}, \mathbb{C}^{\langle m\rangle}) \\
\qquad \stackrel{\delta^*}{\longrightarrow} \check{H}^{k+1}(\Sigma, \mathbb{C}_{(m)}) \longrightarrow \cdots .
\end{array}$$

From now on, we need additional assumptions for Σ and $\Sigma_{(m)}$.

4.4 Lemma. Suppose $\operatorname{int}(|\Sigma| \setminus |\Sigma_{(m)}|)$ is an open ball. Then,

(a) $\check{H}^k(\Sigma, \mathbb{C}_{|\Sigma|}) = 0 \quad$ for $k > 0$,

(b) $\check{H}^k(\Sigma^{\langle m\rangle}, \mathbb{C}^{\langle m\rangle}) = 0 \quad$ for $k > 0$.

PROOF.

(a) We remind ourselves that $\check{H}^k(\Sigma, \mathbb{C}_{|\Sigma|})$ is an interpretation of ordinary cohomology of Σ with coefficients in $\mathbb{C}$. Using the fact proved in algebraic topology that $H^k(|\Sigma|, \mathbb{C})$ does not depend on the decomposition of $|\Sigma|$, we obtain

$$\check{H}^k(\Sigma, \mathbb{C}_{|\Sigma|}) = H^k(\Sigma, \mathbb{C}) = H^k(|\Sigma|, \mathbb{C}).$$

The latter group is known to be 0 for $k > 0$, so (a) follows.

(b) Here, the arguments are essentially the same as in (a), it does not matter if a cone is closed or not.

□

4.5 Lemma. Let Σ be complete, and let $\mathcal{F}$ be an invertible sheaf on X_Σ which can be represented by a convex (not necessarily strictly convex) piecewise linear function h (so that X_Σ is not necessarily projective; see VII, Theorem 2.2).

(a) If for $m \in \mathbb{Z}^n$

(1) $\quad \langle m, x\rangle \geq h(x) \quad$ for all $x \in \mathbb{R}^n$

then, $\check{H}^k_m(\{X_{\check{\sigma}}\}_{\sigma\in\Sigma}, \mathcal{F}) \cong \check{H}^k(\Sigma, \mathbb{C}_{(m)}) = 0$ for all $k > 0$.

(b)

$$\check{H}^0_m(\{X_{\check{\sigma}}\}_{\sigma\in\Sigma}, \mathcal{F}) = \begin{cases} \mathbb{C} & \text{if } m \in P, \\ 0 & \text{otherwise.} \end{cases}$$

Proof. We consider the above long, exact, cohomology sequence. If we apply Lemma 4.4 to it, we see that $\check{H}^k(\Sigma, \mathbb{C}_{(m)}) = 0$ for all $k > 0$. So it remains to be shown that $\check{H}^k_m \cong \check{H}^k(\Sigma, \mathbb{C}_{(m)})$.

A rational function f lies in ker d^0 if and only if it is the restriction of a regular function on each $X_{\check{\sigma}}$, $\sigma \in \Sigma$, that is if it is given by a global section of $\mathcal{F}$, $f \in \mathcal{F}(X_\Sigma)$. From the definition of $\check{H}^0_m$, $f \in \check{H}^0_m$ if $f = cz^m$, $c \in \mathbb{C}$.

Now, from VII, Theorem 1.10 and Definition 3.6, we know that $f \in \mathcal{F}(X_\Sigma)$ if and only if m lies in the $\mathcal{F}$-polytope P. This is, in turn, equivalent to (1). Therefore, for $m \in P$, each $\sigma \in \Sigma$ lies in $\Sigma_{(m)}$ so that $\check{H}^0_m(\{X_{\check{\sigma}}\}_{\sigma\in\Sigma}, \mathcal{F}) = \mathbb{C}$. If $m \notin P$, then, $cz^m \notin \ker d^0$, and, hence, $\check{H}^0_m(\{X_{\check{\sigma}}\}_{\sigma\in\Sigma}, \mathcal{F}) = 0$. But, we also have $\check{H}^0(\Sigma, \mathbb{C}_{(m)}) = \mathbb{C}$ or 0, $\Sigma = \Sigma_{(m)}$ or $\Sigma \setminus \Sigma_{(m)} \neq \emptyset$, respectively. Therefore the starting points of the boundary operations, by which the higher Čech cohomology sequences are defined, are the same, so that $\check{H}^k_m \cong \check{H}^k(\Sigma, \mathbb{C}_{(m)})$ for all $k \in \mathbb{Z}$. □

Lemma 4.5 and Lemma 4.2 imply Theorem 4.6.

4.6 Theorem. *Let Σ be complete, and let $\mathcal{F}$ be an invertible sheaf of $\mathbb{C}$-valued rational functions on X_Σ which can be represented by a piecewise linear convex function h. Then*

$$\check{H}^k(\{X_{\check{\sigma}}\}_{\sigma\in\Sigma}, \mathcal{F}) = 0 \qquad \textit{for } k > 0,$$

and

$$\operatorname{rank} \check{H}^0(\{X_{\check{\sigma}}\}_{\sigma\in\Sigma}, \mathcal{F}) = G(P),$$

the number of lattice points in the polytope P with support function h (see IV, 6*).*

Exercises

1. Show that Lemma 4.2 does not hold for Example 1.
2. Find $\check{H}^k(\{X_{\check{\sigma}}\}_{\sigma\in\Sigma}, \mathcal{F})$ for Σ consisting of one n-cone and its faces and $\mathcal{F}$ an arbitrary T-invariant invertible sheaf on X_Σ.
3. Work out explicitly the long, exact, cohomological sequence in the case of $\mathbb{P}^1 \times \mathbb{P}^1$ (a fan, as in Example 1) and $\mathcal{F}$ defined by $P = \operatorname{conv}\{0, e_1\}$.
4. Let $\mathcal{F}_1, \mathcal{F}_2$ be two invertible sheaves on X_Σ. Find $\check{H}^k(\{X_{\check{\sigma}}\}_{\sigma\in\Sigma}, \mathcal{F}_1 \otimes \mathcal{F}_2)$ from $\check{H}^k(\{X_{\check{\sigma}}\}_{\sigma\in\Sigma}, \mathcal{F}_i)$, $i = 1, 2$.

5 The Riemann–Roch–Hirzebruch theorem

Let X_Σ be a smooth, projective toric variety. We will apply the results of section 4 to a study of the Euler characteristic of a toric Cartier divisor assigned to a polytope P and discuss its calculation by combinatorial data of P. For a definition of the Euler characteristic (following Hirzebruch), we use the language of invertible sheaves.

5.1 Definition. If $\mathcal{F}$ is an invertible sheaf of rational functions on a toric variety X_Σ, we call

$$\chi(X_\Sigma, \mathcal{F}) := \sum_{j=0}^{n} (-1)^j \operatorname{rank} \check{H}^j(\{X_{\check{\sigma}}\}_{\sigma \in \Sigma}, \mathcal{F})$$

the *Euler characteristic* of $\mathcal{F}$.

Theorem 4.6 implies Theorem 5.2.

5.2 Theorem. *If $\mathcal{F}$ is given by an upper convex piecewise linear function h such that $-h$ is the support function of a polytope P, then,*

$$\chi(X_\Sigma, \mathcal{F}) = G(P),$$

the number of lattice points in P.

It is possible to calculate $\chi(X_\Sigma, \mathcal{F})$ via the Chow ring of X_Σ by using the so-called Riemann–Roch–Hirzebruch theorem. At the moment, the formula obtained is not known to have a "toric" proof. We assume it to be true without proof. Because of Theorem 5.2, it provides a method for counting the number of lattice points in P. As we mentioned in the introduction to this book, the Riemann–Roch–Hirzebruch theorem is the only example we present of combinatorial information obtained from algebraic-geometric information. In all other cases we have deduced algebraic-geometric facts from combinatorial-geometric ones.

5.3 Definition. Let $U_1, \ldots, U_k$ be the variables in the Chow ring, as introduced in VII, 5. We set

$$c_p := \sum_{i_1 < \cdots < i_p} U_{i_1} \cdots U_{i_p}, \qquad \{i_1, \ldots, i_p\} \subset \{1, \ldots, k\}$$

and call c_p the *pth Chern class* of X_Σ, $p = 1, \ldots, n$. We, also we set $c_0 = 1$.

Remark. The U_i can be identified with the divisors D_{ϱ_i} which are also bijectively assigned to the 1-cones of Σ. Then, c_1 is identified with the anticanonical divisor which we have introduced in VII, 8. We do not discuss, here, the general topological meaning of Chern classes. Our definition is pragmatic and concentrates on the toric situation.

5.4 Definition. Let the symbols $\gamma_1, \ldots, \gamma_n$ (in an algebraic extension of the Chow ring) be indirectly defined from

$$(1+\gamma_1)\cdots(1+\gamma_n) = c_0 + c_1 + \cdots + c_n$$

as solutions of the equations

$$\sum_{j_1<\cdots<j_p} \gamma_{j_1}\cdots\gamma_{j_p} = c_p, \qquad \{j_1, \ldots, j_p\} \subset \{1, \ldots, n\},$$

which show on the left sides the elementary symmetric functions in $\gamma_1, \ldots, \gamma_n$. Then, we call the term of order p in the formal Taylor expansion of

$$\frac{\gamma_1}{1-e^{-\gamma_1}} \cdots \frac{\gamma_n}{1-e^{-\gamma_n}} \tag{1}$$

the *pth Todd class* Td_p of X_Σ, $p = 0, \ldots, n$, where $Td_0 := 1$.

Remark. It can be shown from the multiplication rules in the Chow ring that $Td_p = 0$ for $p > n$. The first terms are

$$Td_0 = 1, \quad Td_1 = \frac{1}{2}c_1, \quad Td_2 = \frac{1}{12}(c_1^2 + c_2), \quad \text{and } Td_3 = \frac{1}{24}c_1c_2.$$

Considering the bijective assignments for $\varrho_i \in \Sigma^{(1)}$,

$$U_i \longleftrightarrow \varrho_i \longleftrightarrow D_{\varrho_i},$$

we also have the biunique correspondence

$$U_{i_1}\cdots U_{i_n} \longleftrightarrow (D_{\varrho_{i_1}}.\cdots.D_{\varrho_{i_n}})$$

between the monomials of greatest order in Ch X_Σ and the intersection form for the respective divisors. So, we assign to each $U_{i_1}\cdots U_{i_n}$ a number in $\mathbb{Z}$. For the sake of abbreviation, we write $U_i = D_{\varrho_i}$ and $U_{i_1}\cdots U_{i_n} = (D_{\varrho_{i_1}}.\cdots.D_{\varrho_{i_n}}) = D_{\varrho_{i_1}}\cdots D_{\varrho_{i_n}}$.

Now, we present the main result in a toric version ($\varrho_i = \mathbb{R}_{\geq 0}\, a_i, i = 1, \ldots, k$).

5.5 Theorem (Riemann–Roch–Hirzebruch). *Let X_Σ be a smooth, projective toric variety and let $D = -\sum_{i=1}^k h(a_i)D_{\varrho_i}$ be a Cartier divisor assigned to a polytope P with support function h. For the invertible sheaf $\mathcal{F}$ defined by P,*

$$\chi(X_\Sigma, \mathcal{F}) = G(P) = \sum_{j=0}^{n} \frac{1}{j!}\Big(-\sum_{i=1}^{k} h(a_i)D_{\varrho_i}\Big)^j \cdot Td_{n-j}. \tag{2}$$

Example 1. Hirzebruch surface $\mathcal{H}_r$ (compare VII, 5, Example 1): We consider the polytope P as in Figure 4.

The support function h of P has values

$$h(a_1) = -1, \quad h(a_2) = -1, \quad h(a_3) = 0, \quad \text{and } h(a_4) = 0.$$

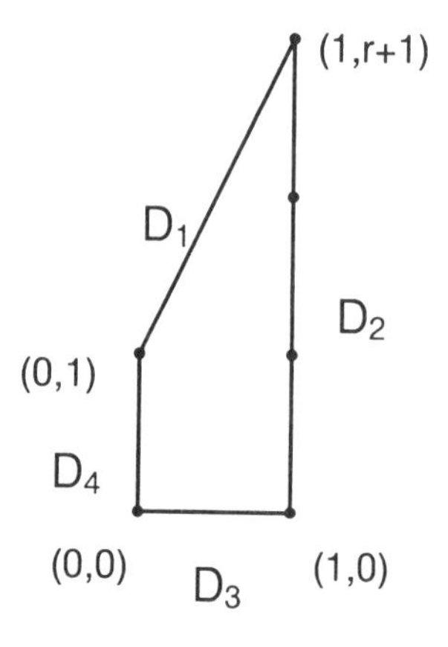

FIGURE 4.

Therefore,

$$\begin{aligned}
\chi(X_\Sigma, \mathcal{F}) &= G(p) \\
&= \frac{1}{0!}\cdot 1\cdot\frac{1}{12}\Big((D_{\varrho_1}+D_{\varrho_2}+D_{\varrho_3}+D_{\varrho_4})^2+4\Big) \\
&\quad + \frac{1}{1!}(D_{\varrho_1}+D_{\varrho_2})\cdot\frac{1}{2}(D_{\varrho_1}+D_{\varrho_2}+D_{\varrho_3}+D_{\varrho_4}) \\
&\quad + \frac{1}{2!}(D_{\varrho_1}+D_{\varrho_2})^2\cdot 1 \\
&= 1+\frac{1}{2}(r+4)+\frac{1}{2}(r+2) = r+4
\end{aligned}$$

which is, in fact, the number of lattice points of P, as directly obtained from Figure 4.

It is remarkable that Theorem 5.5 provides a formula for $G(P)$ in which only data of Σ and the values of the support function h of P in the direction of facet normals occur. So, in particular, if P is replaced by a strictly combinatorial equivalent copy P', only the values of $h(a_i)$ in (2) change. If P' becomes very large, a direct counting of the lattice points of P might fail, but (2) remains calculable.

Remark. Further results about counting lattice points can be obtained by combining the Riemann–Roch–Hirzebruch theorem and Ehrhart's theorem (IV, 6.); see Fulton's book [1993] (further references in the Appendix to this section).

Exercises

1. By Theorem 5.2, verify that $G(P) = 7$ in the hexagon associated with a del Pezzo surface (see VII, 8).
2. Given any cube associated with the fan of $\mathbb{P}^1 \times \mathbb{P}^1 \times \mathbb{P}^1$, calculate $G(P)$ by Theorem 5.2.
3. Find an alternative proof of Ehrhart's theorem (IV, 6) by using Theorem 5.5.
4.
   a. Prove a theorem analogous to Minkowski's theorem (IV, Theorem 3.2), and introduce mixed lattice point numbers.
   b. Find properties of mixed lattice point numbers analogous to those of mixed volumes (IV, 3).

Summary: A Dictionary

Combinatorial Convexity	**Algebraic Geometry**
Lattice cone σ	Affine toric variety X_σ
–regular	–smooth
–simplex cone	–quasi-smooth
Face τ of lattice cone σ	Subvariety X_τ of X_σ
Lattice point $m = (m_1, \ldots, m_n)$	Monomial $z^m = z_1^{m_1} \cdots z_n^{m_n}$
Sum $m + m'$	Product $z^m \cdot z^{m'} = z^{m+m'}$
Monoid $\sigma \cap \mathbb{Z}^n$	Spectrum spec $\mathbb{C}[\sigma \cap \mathbb{Z}^n]$
Lattice $\mathbb{Z}^n$	Torus $T = \mathbb{C}^{*n} = \operatorname{spec} \mathbb{C}[\mathbb{Z}^n]$
Fan Σ	Toric variety X_Σ
–complete	–compact
–regular	–smooth
–simplicial	–quasi–smooth
–strongly polytopal	–projective
Join $\Sigma' \cdot \Sigma''$ of fans Σ', Σ''	Fiber bundle $X_{\Sigma' \cdot \Sigma''}$ with typical fiber $X_{\Sigma'}$
Regular stellar subdivision	Blowup
Inverse regular stellar subdivision	Blowdown
Star st(σ, Σ) of $\sigma \in \Sigma$	Invariant toric subvariety of X_Σ
Unimodular transformation of $\mathbb{R}^n$	Algebraic isomorphism of X_Σ
Map of fans L	Toric morphism $\bar{L}$
Lattice polytope P associated with $\Sigma = \Sigma(P)$	Projective toric variety $X_\Sigma = X_{\Sigma(P)}$
Fano polytope	Toric Fano variety
Simplex	Weighted projective space
Regular simplex	Projective space
Direct sum $P_1 \oplus P_2$ of summands P_1, P_2 of P	Product space $X_{\Sigma(P_1)} \times X_{\Sigma(P_2)}$
Lattice polytope strictly combinatorially isomorphic to P	Ample T-invariant Cartier divisor of $X_{\Sigma(P)}$
Virtual polytope $\mathcal{P}$	Invertible sheaf $\mathcal{F}$
Combinatorial Picard group Pic Σ	Picard group Pic X_Σ
Polytope element $\mathcal{P}(P')$	Ample invertible sheaf $\mathcal{F}(P')$
Minkowski sum $P' + P''$	Tensor product $\mathcal{F}(P') \otimes \mathcal{F}(P'')$

Support function of P' strictly combinatorially isomorphic to P	Ample T-invariant Cartier divisor $D_{P'}$ of $X_{\Sigma(P)}$
Virtual support function	T-invariant Cartier divisor
1-cone ϱ of Σ	Special Cartier divisor D_ϱ
System $\Sigma^{(1)} = \{\varrho_1, \dots, \varrho_k\}$ of all one-dimensional cones of Σ	Canonical divisor (first Chern class) $D_{\varrho_1} + \cdots + D_{\varrho_k}$
Volume $V(P')$ of P' multiplied by $n!$	Self-intersection number $(D_{P'}.\cdots.D_{P'})$ of $D_{P'}$
Mixed volume $V(P_1, \dots, P_n)$ of summands $P_1, \dots, P_n$ of P	Intersection number $(D_{P_1}.\cdots.D_{P_n})$ of the Cartier divisors $D_{P_1}, \dots, D_{P_n}$
Alexandrov-Fenchel inequality	Hodge inequality
Number of edges $[m_0, m']$ of P, m_0 a vertex of P for which $\langle m, m' - m_0\rangle > 0$ relative to given $m \in \mathbb{Z}^n$	Half the index of the Morse function $f_{(m)}$ of $X_{\Sigma(P)}$ in the zero-dimensional orbit defined by m_0
$\sum_{i=l}^{n}(-1)^{i+l}\binom{i}{l} f_i$, where f_i is the number of i-dimensional faces of a simplicial fan $\Sigma = \Sigma(P)$	Betti number $\beta_{2l} = \operatorname{rank} H^{2l}(X_\Sigma; \mathbb{Z})$
Number $G(P_0)$ of lattice points in the summand P_0 of P	Euler characteristic $\chi(X_{\Sigma(P)}, \mathcal{F}(P_0))$

Appendix

Historical notes, commentaries, additional exercises, research problems, suggestions for further reading

HN	Historical notes
C	Commentaries
AE	Additional exercises
RP	Research problems
SFR	Suggestions for further reading.

Introduction

HN For more about Minding's contributions (around 1841) and the recent work of Russian authors (Bernstein, Kouchnirenko), see the historical notes in Khovanskii [1988]. Concerning the development of toric varieties see Demazure [1970], Kempf, Knudson, Mumford, and Saint Donat [1973], Danilov [1978], Oda [1988].

PART 1 Combinatorial Convexity

Chapter I Convex bodies

HN The general theory of convex bodies was developed by Minkowski around 1900, relating it to the isoperimetric problem and other parts of analysis. A first survey of results is Bonnesen and Fenchel [1933]; see also Blaschke [1916]. A.D. Alexandrov introduced convex body theory into differential geometry; see Busemann [1958], Burago and Zalgaller [1988]. In the latter book, one finds more on inequalities in convex body theory; see also Leichtweiß [1987]. We do not quote references concerning the influence of convex body theory in functional analysis. About measure theoretic aspects, geometric probability, and integral geometry, see Schneider [1993].

1. Convex sets

AE 5. If $M \subset \mathbb{R}^n$ is convex, then the set of interior points of M and the set of relative interior points of M are also convex.
6. Affine transformations $L : \mathbb{R}^n \longrightarrow \mathbb{R}^m$ preserve convexity.

7. Given a set $X \subset \mathbb{R}^n$, define $j(X) := \bigcup_{x,y \in X}[x, y]$. Find the least k for which $j^k(X)$ $(= j \circ \cdots \circ j(X))$ equals conv X.
8. Suppose a ray $\{x + ta \mid t \in \mathbb{R}_{\geq 0}\}$ emanating from a point x of a convex set $C \subset \mathbb{R}^n$ is contained in C. Then for each $y \in C$ also $\{y + ta \mid t \in \mathbb{R}_{\geq 0}\} \subset C$.

2. Theorems of Radon and Carathéodory

HN See references about Radon and Carathéodory in Grünbaum [1967].
AE, SFR About Helly type theorems, see Grünbaum [1967], p. 22 and 126. About Radon's theorem, see Eckhoff [1979], [1993].

3. Nearest point map and supporting hyperplanes

HN,C The use of the nearest point map for developing the basics of convex body theory in P. McMullen, G.C. Shephard [1971] had been proposed by this author. We shall also use it in what follows. About Busemann and Feller's lemma, see the literature quoted in Busemann [1958].

4. Faces and normal cones

AE 5. A closed convex set K is bounded if and only if $\Sigma(K)$ covers $\mathbb{R}^n$.

5. Support function and distance function

AE 5. Define $w_K(u) := h_K(u) + h_K(-u)$ for unit vector u to be the *width* of K *in direction* u.
(a) $w_K(u) = w_{K+x}(u)$ for any $x \in \mathbb{R}^n$.
(b) Find an example of a planar convex body of constant width ($w_K(u)$ independent on u) which is not a circular disc. (Use three circular arcs having midpoints on the other arcs.)
(c) Find analogs of (b) in $\mathbb{R}^3$.
(d) The spherical balls are the only centrally symmetric convex bodies of constant width.
A series of nice problems about convex bodies of constant width can be found in Jaglom and Boltjanskij [1956].

6. Polar bodies

AE 5. Find K^* for K in Figure 6.
6. Find an example of a convex body K which has a finite positive number of one-dimensional faces and no two-dimensional faces. What does K^* look like?

Chapter II Combinatorial theory of polytopes and polyhedral sets

HN The first systematic treatment of this subject is in the book of Grünbaum [1967]. We largely follow McMullen–Shephard [1970] (compare our remarks in **HN,C** of I, 3).
SFR More material on the combinatorial theory of polytopes can be found in Ziegler [1994]. The reader interested in connections to optimization may also consult Stoer and Witzgall [1970].

1. The boundary complex of a polyhedral set

HN Theorem 1.2: see Krein–Milman [1940], which refers, more generally, to convex sets in vector spaces of arbitrary dimensions.

2. Polar polytopes and quotient polytopes

C Our definition of P/F is more concrete than the definition given in McMullen and Shephard [1971], p. 69–71.
SFR For quotient polytopes a "global" counterpart to the "local" notion used here is that of fiber polytopes; see Billera and Sturmfels [1991], Reiner and Ziegler [1994], and Ziegler [1995].

3. Special types of polytopes

SFR A class of polytopes generalizing cyclic polytopes is that of so-called neighborly polytopes; see McMullen and Shephard [1971], pp. 90–93. About regular polytopes, see Coxeter [1973].

4. Linear transforms and Gale transforms

HN We follow the treatment in McMullen [1979] which, in turn, is partly based on Ewald and Voß [1973]. Application of Gale transforms to polytope theory was introduced by Perles (see Grünbaum [1967], p. 108).
SFR About Gale transforms and toric varieties, see also Oda and Park [1991].

5. Matrix representation of transforms

AE 5. Let X again be a finite set of points, and let G be a group of affine transformations mapping X onto itself. Then, we can assign to X a Gale transform $\bar{X}$ with an affine symmetry group $\bar{G}$ such that the same assignment applied to $\bar{X}$, $\bar{G}$ leads back to X, G. (See Ewald and Voß [1973] and McMullen [1979]).

6. Classification of polytopes

HN, SFR Theorem 6.4 can be generalized to polytopes with $n + 3$ vertices, Lloyd [1970]. Also, the classification problem has been solved for a number of other classes of polytopes; c.f. Bokowski and Shemer [1987].
AE 5. Find a 3-polytope P other than a pyramid which is combinatorially equivalent to P^*.
RP The combinatorial classification of polytopes under reasonable restrictions is a worthwhile field of research. Besides restricting the number of vertices, assumptions on combinatorial symmetries can be made (Ewald and Voß [1972], Ewald [1979], Ewald, Kleinschmidt, and Schulz [1976]).

Chapter III Polyhedral spheres

1. Cell complexes

HN Cell complexes have played an essential role in the development of algebraic topology, originally called combinatorial topology. In text books on topology, they are often restricted to simplicial complexes. Cell complexes are of particular importance in the so-called p.l. topology (piecewise linear). For references, see Rourke and Sanderson [1972].

2. Stellar operations

HN See Rourke and Sanderson [1972] about stellar subdivisions in topology. Theorem 2.1: see Grünbaum [1967], p. 80.
C According to an unpublished paper by Morelli, the problem mentioned in the text is solved in the affirmative.
AE 5. Call the maximal distance between vertices of a cell F the *diameter* of F. Given any cell complex and any $\varepsilon > 0$, there exists a $k \in \mathbb{Z}_{\geq 0}$, such that each cell of $\beta^k(\mathcal{C})$ ($\beta^k := \beta \circ \cdots \circ \beta$, k times) has diameter less than ε.
6. Subdivide a tetrahedron and a cube (together with their faces considered as cell complexes) by a finite succession of stellar subdivisions, such that the resulting cell complexes are isomorphic.
7. Let Σ be a fan in $\mathbb{R}^3$ for which pos $|\Sigma| = \mathbb{R}^3$. Find a complete fan $\Sigma' \supset \Sigma$ with $\Sigma'^{(1)} = \Sigma^{(1)}$.
8. Find a fan Σ in $\mathbb{R}^4$ for which pos $|\Sigma| = \mathbb{R}^4$ but each complete $\Sigma' \supset \Sigma$ satifies $\Sigma'^{(1)} \neq \Sigma^{(1)}$.
SFR Ewald and Shephard [1974], Ewald [1978], Pachner [1987], [1991].

3. Euler and Dehn–Sommerville equations

HN The proof of Theorem 3.1 generalizes an idea of Grünbaum [1967], Chapter 8, applied to polytopes. The same is true for Theorem 3.3.
SFR See an extended Euler theorem in Björner and Kalai [1988]; also compare Barvinok [1992], Björner [1986]; Björner and Kalai [1989]; Chan [1994]. About generalized Dehn–Sommerville equations, see Bayer and Klapper [1991]. One of the greatest achievements in polytope theory is the characterization of all lattice vectors which are f-vectors of polytopes. McMullen had stated a conjecture (McMullen and Shephard [1971], p. 179). The necessity of his conditions were shown by Stanley [1980] (using commutative algebra and toric varieties). McMullen [1993b], [1995] has replaced the algebraic geometric arguments in Stanley's proof by those of polytope algebra. Billera and Lee [1980] proved the sufficiency. See also Oda [1991].

4. Schlegel diagrams, n-diagrams, and polytopality of spheres

HN Schlegel diagrams: see Schlegel [1883], Grünbaum [1967]. About diagrams not isomorphic to Schlegel diagrams, see Grünbaum [1967], 11.5. Barnette sphere: see Barnette [1970]. The Steinitz problem has been solved (in the affirmative) for polytopes with, at most, $n + 3$ vertices by Mani [1972] and Kleinschmidt [1976].
SFR For a survey article on polytope complexes, see Klee and Kleinschmidt [1992]. See also Bokowski and Sturmfels [1989].
RP It is one of the big unsolved problems in combinatorial geometry to find necessary and sufficient conditions for the polytopality of polyhedral spheres (Steinitz problem). Under suitable restrictions, the problem can be attacked. About a new development, see Richter-Gebert [1995], Richter-Gebert and Ziegler [1995].

5. Embedding problems

HN The example in Theorem 5.2 is due to C. Schulz [1979]. Theorem 5.3: see Ewald and Schulz [1992].
SFR About related embedding problems, see Sarkaria [1988], [1991].

6. Shellings

HN Theorem 6.1: Bruggesser and Mani [1971].
SFR See a detailed analysis about shellings in Danaraj and Klee, [1974]; see also Ziegler [1994], Lickorish [1991].
About shellability and Cohen–Macaulay complexes, see Kind and Kleinschmidt [1979].

7. Upper bound theorem

HN Stanley [1975] has extended the upper bound theorem to simplicial spheres by using Cohen and Macaulay rings. See a further paper on "polytope algebra": McMullen [1991]. See Barnette [1973], Kleinschmidt and Lee [1984], Brøndtstedt [1982], McMullen and Walkup [1971].
SFR As an analog to the upper bound theorem, there is a lower bound theorem. Detailed information about this can be found in Brøndstedt [1983]. Also compare Lee [1992].

Chapter IV Minkowski sum and mixed volume

HN The theory of mixed volumes is a vital part of Minkowski's original work on convex bodies. It provides the basis for a solution of the isoperimetric problem. For a survey of recent results, see Schneider [1985], [1993]. After Bernstein [1976] and Teissier [1979] discovered the relationship between mixed volumes and intersection numbers in algebraic geometry, the theory has attracted new interest.

1. Minkowski sum

AE 5. Find convex bodies K, L of any given dimension $n > 1$ which are not balls, such that $K + L$ is a ball (compare additional Exercise 5 of I, 5).
6. The Minkowski sum of affine subspaces L, M of $\mathbb{R}^n$ is an affine subspace. How does the dimension of $L + M$ depend on the dimensions of L and M?
7. Find three sets A, B, C in $\mathbb{R}^2$, such that $A + B$, $A + C$, $B + C$ are not convex but $A + B + C$ is convex.
SFR About zonotopes, see McMullen [1971], Shephard [1974], Stanley [1989]. About decomposability of polytopes (with respect to Minkowski sum), see Smilansky [1986], and Meyer [1974].

2. Hausdorff metric

HN Theorem 2.7: Our proof is different from the original one by Blaschke [1916]; see Hadwiger [1955], p. 20.
AE 5. Extend the Hausdorff metric to arbitrary compact subsets of $\mathbb{R}^n$.
6. In the definition of Hausdorff distance replace the unit ball by a fixed n-dimensional convex body. Show that convergence in the new distance is equivalent to that in Hausdorff metric.

3. Volume and mixed volume

AE 5. Any centrally symmetric convex body in $\mathbb{R}^2$ is the limit of a sequence of zonotopes (see Exercise 4) in the Hausdorff metric.
SFR A survey of recent results about mixed volumes is in Sangwine-Yager [1993].

4. Further properties of mixed volumes

SFR More on shadow boundaries is in Shephard [1972]. About a characterization of mixed volumes, see Firey [1976].

5. Alexandrov and Fenchel's inequality

HN A detailed version of the classical proof by Alexandrov and Fenchel can be found in Leichtweiß [1980] and in Schneider [1993]. Concerning equality, see Ewald [1988c]; Ewald and Tondorf [1994].
SFR Schneider [1985], and Stanley [1981].
RP There is a good chance to find new results on the equality case by using the method presented here. Compare Ewald and Tondorf [1994], Gärtner [1996].

6. Ehrhart's theorem

HN Original proof in Ehrhart [1977].
SFR An alternative proof can be found in Brion [1988], simplified by Ishida [1990]. See also Stanley [1986], MacDonald [1971], Hibi [1991], Wills [1989], a survey article: Gritzmann and Wills [1993], 4.2. About a related question, see Brion [1992].

7. Zonotopes and arrangements of hyperplanes

HN Theorem 7.5 is due to Coxeter [1962].

Chapter V Lattice polytopes and fans

1. Lattice cones

C We use the terminology "simplex cone" for what is usually called a "simplicial cone". As polytopes are called "simplicial" if all their proper faces are simplices, it seems more reasonable to reserve the notion "simplicial" to cones whose proper faces are simplex cones.

2. Dual cones and quotient cones

AE 5. Generalize the notion of a cone to $\sigma = \text{pos}\, M$ for any set $M \subset \mathbb{R}^n$. Then, the dual cone $\check{\sigma}$ is also well-defined.
(a) Does "M closed" always imply "$\sigma = \text{pos}\, M$ closed"?
(b) Find a cone $\sigma = \text{pos}\, M$ with apex 0, M a set of lattice points that is not a polyhedral cone but topologically closed.
(c) If $\sigma = \text{pos}\, M$ for a set of lattice points, then, also, $\check{\sigma} = \text{pos}\, M'$ for a set M' of lattice points.

3. Monoids

AE 5. Let σ be a two-dimensional cone in $\mathbb{R}^2$ with apex 0.

(a) conv$\{(\sigma \cap \mathbb{Z}^2) \setminus \{0\}\}$ is a polyhedral (unbounded) set with at most five vertices.
(b) Find generators for the monoid $\sigma \cap \mathbb{Z}^2$.
6. Find a (nonsaturated) monoid S, such that $S \setminus \{a\}$ is no monoid for any $a \in S$.
SFR About the algebraic theory of monoids, see Trung and Hoa [1986]. More about monoids and toric varieties in Rentzsch [1991].

4. Fans

HN Shephard's criterion (Theorem 4.3) in Shephard [1971]. Theorem 4.4 is a new result. See Oda's criterion (Theorem 4.6) in Oda [1978], [1988], Miyake and Oda [1975]. About the relationship between zonotopes and systems of hyperplanes (arrangements), see IV, 7.

5. Combinatorial Picard group.

HN Theorem 5.9 for complete Σ is due to Eikelberg [1992a], partly based on results by Smilansky [1986a]. Here we present a simplified version of the proof. The most general version of Theorem 5.2 can be found in Rentzsch [1991] (formulated in algebro-geometric language; compare VII. 2 below).

6. Regular stellar operations

HN Theorem 6.1 in Miyake and Oda [1975]. Farey's lemma (Lemma 6.2): see Rademacher [1964]. Proof of Oda's weak conjecture for $n = 3$ proved in Danilov [1983]; a gap in the proof filled in by Ewald [1987]. For arbitrary n, Oda's weak conjecture has been shown by Włodarczyk [1992] and by Morelli [1994]. Oda's strong conjecture for arbitrary n is supposed to have been solved by R. Morelli.

7. Classification problems

HN, SFR A large number of classification results can be found in the books of Oda [1978], [1988]. In Kleinschmidt [1988], all regular complete n-dimensional fans with combinatorial Picard number, at most, 2 have been characterized. Batyrev [1991] has extended the result to Picard number 3. In Gretenkort, Kleinschmidt, and Sturmfels [1990] it is shown that a strongly polytopal fan need not have a combinatorially isomorphic copy which is regular.
RP Admit Picard number larger than 3 but restrict the fans by the existence of symmetries (central, axial, or others), and find further classification results. (For the combinatorial aspects, compare **RP** in II,6).

8. Fano Polytopes

HN Theorem 8.2: Our proof is a simplified version of the original proof in Klyachko and Voskresenskij [1985].
RP 1. The maximal possible number of vertices a Fano polytope can have is limited. The exact bounds for $n > 3$ are not known. Batyrev's conjecture: If P is an n-dimensional Fano polytope with $f_0(P)$ vertices, then, $f_0(P) \leq 3n$ for n even, $f_0(P) \leq 3n - 1$ for n odd. Some results in Klyachko and Voskresenskij [1985]; Batyrev [1982b], [1986], Evertz [1988].
2. Investigate generalized Fano polytopes defined as polytopes P with only 0 as interior point and facet hyperplanes of the form $\langle a, x\rangle = 1$, a a simple lattice vector. Compare Batyrev [1992b], Koelman [1992].
SFR Batyrev [1982a], [1992]; Demin [1981]; Ewald [1988a]

PART 2 Algebraic Geometry

Chapter VI Toric varieties

HN, SFR As standard texts about the introduction of toric varieties into the theory, we mention Oda [1988], Kempf, Knudsen, Mumford and Saint Donat [1973]; Danilov [1978]; and the paper Demazure [1970]. Short introductions can also be found in Teissier [1982]; Brylinski [1977]; Audin [1991]; and Fulton [1993]. About another relationship between combinatorial and algebraic geometry, see Barthel, Hirzebruch and Höfer [1987].

1. Ideals and affine algebraic sets

HN, C Hilbert's Nullstellensatz: We follow the proof in Brodmann [1989], p. 59, which assumes the underlying field to have uncountably many elements. There exist many more general versions and proofs for the theorem.
SFR As standard books on algebraic geometry, we mention Hartshorne [1977] and Shafarevic [1974]. For the reader who understands German, we recommend Brodmann [1989], a book with detailed and well-illustrated proofs for all theorems.

2. Affine toric varieties

C In all available texts on the introduction of toric varieties, some knowledge of algebraic groups is preassumed, and a coordinate-free definition of toric varieties is presented. Our less elegant approach is accessible without knowledge of algebraic geometry or algebraic groups.
AE 5. Let σ, τ be cones such that $\sigma - \sigma$ and $\tau - \tau$ are complementary linear spaces. How does $X_{\sigma+\tau}$ depend on X_σ and X_τ?
RP In Theorem 2.7 if the number of binomials can be limited to $q - n$, we call X_σ a *complete intersection* of its *relation hypersurfaces* defined by the binomials. Find all such complete intersections, possibly under suitable restrictions for σ. (About affine complete intersections in the ordinary sense, see Nakayima [1986] and Hamm [1990]).

3. Toric varieties

AE 5. Describe the toric variety obtained from $X_\Sigma = \mathbb{P}^2 \times \mathbb{P}^1$ (Example 4) if all three-dimensional cones are deleted in Σ.
6. Find $Mc(X_\Sigma)$ for Examples 6 and 7.
7. If the fans Σ', Σ'' are contained in complementary subspaces of $\mathbb{R}^n$, describe $X_{\Sigma'} \times X_{\Sigma''}$ as a toric variety.
SFR About higher-dimensional analogs of Hirzebruch varieties (Hirzebruch [1951]), see Kleinschmidt [1988]. Concerning manifolds with corners (which will indirectly occur again in VII, 7), compare Jurkiewicz [1985]; Oda [1988]. On quotients of toric varieties (related to fiber polytopes; compare II, 2 SFR), see Kapranov, Sturmfels, and Zelevinsky [1991]. An interesting property of toric varities is in Włodarczyk [1993]. Compare also Batyrev [1993], [1994], Batyrev and Meln'nikov [1986], Fine [1989].

4. Invariant toric subvarieties

AE 5. For any $n > 2$, find a complete toric variety X_Σ whose invariant toric subvarieties (except X_Σ) are regular, but each $X_{\tilde{\sigma}}$, $\dim \sigma = n$, is not regular.

6. If k is the maximal dimension of faces of a fan Σ in $\mathbb{R}^n$, what is the minimal dimension of an invariant toric subvariety of X_Σ?

5. Torus action

SFR For a general background in algebraic groups, we recommend Borel [1969] and Springer [1981]. About a relation between toric varieties and complex-symmetric spaces, see Lehmann [1989]. Compare also Bialynicki-Birula [1973], Bialynicki-Birula and Sommese [1973]; Fine [1993].

6. Toric morphisms and fibrations

C Example 3: The real part of $\mathcal{H}_1$ is a typical "nonorientable" surface, as considered in topology. One should note, however, that $\mathcal{H}_1$, as a complex surface, that is, a topological four-dimensional manifold, is orientable: All smooth complex manifolds have this property, as is shown in complex analysis.
SFR About toric morphisms, Reid [1983], Crauder [1984].

7. Blowups and blowdowns

HN The idea of "blow–ups" is due to H. Hopf (first version in 1931) and is also called Hopf's σ-process. Its characterization for toric varieties by a regular stellar subdivision has been discovered by Miyake and Oda [1973].
SFR The Oda conjectures reduce to combinatorial problems; see Appendix of V, 6.

8. Resolution of singularities

AE 5. Let σ consist of a cone σ with apex 0 and all faces of σ, so that X_Σ can be embedded in a $\mathbb{C}^q$ according to section 2. Suppose $0 \in \mathbb{C}^q$, is an isolated singularity of X_Σ. Consider $\mathbb{C}^q$, also as an X_{Σ_0} with Σ_0 consisting of a regular q-dimensional simplex cone and its faces. How many blowups of X_{Σ_0} in 0 are needed to resolve the singularity of X_Σ?
HN, SFR Toric varieties of dimension two admit at most so-called cyclic quotient singularities. The minimal resolutions of these singularities were first studied by F. Hirzebruch [1953] by means of continued fractions. A combinatorial version in Oda [1988], pp. 24–37. About quotient singularities in higher dimensions, see Ehlers [1975]. A contribution also in Ewald and Spazier [1994]. — For canonical combinatorial algorithms of resolving three-dimensional toric singularities, see Aguzzoli and Mundici [1994], Bouvier and Gonzales and Sprinberg [1992], [1995]. — Important classes of singularities are canonical and terminal singularities were introduced by Reid [1980]. They occur in Mori theory (a survey can be found in Reid [1987], Kollar [1987]). About the toric version of Mori theory, see Reid [1983a]; Oda [1988]; Oda and Park [1991]. Reid [1983a] and Ishida and Iwashita [1986] gave a combinatorial classification of canonical three-dimensional toric singularities. Similar classification results for canonical three-dimensional and terminal four-dimensional quotient singularities can be found in Morrison and Stevens [1984] and Morrison [1985]; see also Pouyanne [1992]. — So-called Gorenstein singularities are special canonical singularities. About their combinatorial characterization, see Stanley [1978]; Ishida [1980], Miranda [1985], Reid [1989]. On "crepant" resolutions, see Markushevic, Olshanetzky, and Perelonor [1987], Roan and Yau [1987], Roan [1989] for the case of abelian quotient singularities, Batyrev [1994] for the general case. A constructive crepant desingularization algorithm following Reid's general reduction strategy (Reid [1980], [1983b]) in

Dais, Henk, and Ziegler [1996]. For various topological investigations and desingularization methods of hypersurfaces or complete intersections being embedded in affine toric varieties (or appropriate compactifications thereof), see Khovanskij [1977], [1978], Kouchnirenko [1976], Danilov and Khovanskij [1987], Oka [1979] to [1993b]; Hamm [1990], Morales [1984]; Tsuchichashi [1991a], [1991b]; Ishida [1991]; Batyrev [1993a]; Batyrev and Cox [1994]; Cattani, Cox and Dickenstein [1995]; Cox [1995c]; Ishii [1995]. — About an ideal-theoretic approach to the general desingularization process of toric varieties (Theorem 8.5) see Kempf, Knudsen, Mumford, and Saint and Donat [1973], pp. 31–40, Brylinski [1980]. — The so-called deformation theory of toric varieties and toric singularities has mainly been developed by K. Altmann [1991] to [1995].
RP As one of the numerous open problems about singularities of toric varieties, we mention the classification of canonical and terminal singularities in higher dimensions following Morrison and Stevens [1984] and Morrison [1985]. A combinatorial problem, hereby, is that of finding a substitute for White's lemma [1964] which holds for $n = 3$ but not for $n > 3$ (for counterexamples see Wessels [1989]).

9. Completeness and compactness

SFR About alternative proofs for Theorem 9.1, see Oda [1988], p. 16. Concerning Theorem 9.3, see Sumihiro [1974], Oda [1988], p. 17. Related questions of compactification in Ash, Mumford, Rapoport, and Tai [1975].

Chapter VII Sheaves and projective toric varieties

1. Sheaves and divisors

C We choose a restricted definition of sheaf. About the general notion, see, for example, Hartshorne [1977] or Shafarevic [1974], where sheaves also occur in the definition of generalized algebraic varieties (schemes).
SFR A simplified definition of sheaves can also be found in Springer [1981].

2. Invertible sheaves and Picard group

HN About the equivalent group Pic Σ, see V, 5 and the references there. In the case X_Σ is compact, the results on Pic X_Σ can also be obtained by topological means; see Fieseler [1991].
AE 5. If X_Σ is an $X_{\Sigma'}$-fiber bundle over X_{Σ_0}, how is Pic X_Σ related to Pic X_{Σ_0} and Pic $X_{\Sigma'}$?
6. Let X_Σ be smooth. How does Pic X_Σ change if an equivariant blowup is applied?

3. Projective toric varieties

AE 5. Let $P := \operatorname{conv}\{0, e_1, e_2, e_1 + e_2, e_1 + e_2 + re_3\}$ be a polytope in $\mathbb{R}^3$, $r > 1$, $r \in \mathbb{Z}$. Consider the projective toric variety X_Σ for $\Sigma = \Sigma(P)$, and let the invertible sheaf $\mathcal{F} = \mathcal{F}(P)$ be defined by $\mathcal{F}$. Show explicitly that $\mathcal{F} \otimes \mathcal{F}$ is very ample.
RP 1. Call a hypersurface defined by a monomial equation a Σ-*hyperface* of $\mathbb{P}^r$. We say X_Σ is a *complete intersection* of Σ-hypersurfaces if its embedding in $\mathbb{P}^r$ is the intersection of $r - n$ Σ-hyperfaces. We conjecture that the only smooth examples of such intersections are $\mathbb{P}^n$ and $\mathbb{P}^1 \times \mathbb{P}^1$. (Compare **RP** in VI, 2).
2. For which regular X_Σ in Theorem 3.13 is $X_{\Sigma'}$ also regular? The respective fans have combinatorially been classified by Grünbaum [1972].

SFR About hyperplane sections of polytopes and toric varieties, see Khovankij [1986]. Call the least k for which $\mathcal{F}^{\otimes k}$ is very ample, $\mathcal{F}$ an ample invertible sheaf, the least *ampleness factor* k_0 of $\mathcal{F}$. In Ewald and Wessels [1991], it is shown that $k_0 \leq n - 1$ and that the bound is sharp. (An alternative proof in Liu, Trotter, and Ziegler [1992]). Smooth toric varieties with Picard number, at most, 3 are necessarily projective; this is shown in Kleinschmidt and Sturmfels [1989]. See also Ebihara [1992].

4. Support functions and line bundles

C By and large, our notation follows that of Oda [1988].

5. Chow ring

HN Our definition of the Chow ring is based on the work of Jurkiewicz [1980], [1985].
SFR For more details on the Chow ring, see Jurkiewicz [1980], [1985], and Danilov [1978].

6. Intersection numbers. Hodge inequality

HN, C The relationship between Alexandrov and Fenchel's inequality (AF) and the Hodge index theorem has been discovered by Teissier [1981], [1982]. He deduces AF from the Hodge index theorem, as do Danilov [1978] and Oda [1988]. Here, we proceed in the opposite direction after the intersection numbers have been introduced appropriately.
RP Teissier's proof of AF does not contribute more to characterizing the equality case than Alexandrov's original paper does. Recent progress in the combinatorial characterization of equality in AF still has to be transferred into algebraic-geometric conclusions. Also, what equality generally means in Theorems 6.2 and 6.3 has not yet been clarified.
SFR About general intersection theory, see Fulton [1984]; Goresky and MacPherson [1980], [1983]; about intersection in toric varieties, see Fulton [1993], Fulton and Sturmfels [1993]; Wessels [1993].

7. Moment map and Morse function

HN The moment map for toric varieties seems to have been introduced independently by Atiyah [1983] and Jurkiewicz [1980]. The proof of Theorem 7.3 presented here is due to Carl Lee [1988].
SFR In Audin [1991], torus actions on symplectic manifolds are studied with an extensive use of moment maps and Morse functions. A special chapter is dedicated to the case of toric varieties.

8. Classification theorems. Toric Fano varieties

HN See the notes about Fano polytopes in V, 8.
RP Translate classification results for polytopes and fans as discussed in **RP** of V, 7 into algebraic-geometric language, and deduce structure theorems for the varieties.
SFR Batyrev from [1982a] to [1994], Dais [1994], Ito and Reid [1994]. For details on classification theorems, see Oda [1978], [1988], Batyrev [1991], [1992], Demin [1981], Ewald [1988a]. See also Fischli [1991].
HN, SFR Recently, toric geometry has also been used for the construction of Calabi and Yau manifolds which are of particular significance for mathematical physics (string theory). For the constructive techniques, invariants (Hodge numbers and others), and the so-called mirror symmetries (which, in many cases, correspond to the polarity of polytopes), we refer

to Roan [1991]; Morrison [1993]; Batyrev [1993a], [1993b], [1994]; Batyrev and Borisov [1994a], [1994b], [1995]; Batyrev and Dais [1994]; Batyrev and van Straten [1995]; Borisov [1993]; Dais [1995]; Dolgachev [1994]; Kobayashi [1994]; Wagner [1995].

Chapter VIII Cohomology of toric varieties

1. Basic concepts

SFR As text books on algebraic topology, in particular homology and cohomology, we recommend Fulton [1995], Massey [1991], Spanier [1966]; for those who can read German, also the geometrically oriented book Stöcker and Zieschang [1988].

2. Cohomology ring of a toric variety

HN This section is mainly based on the work of Jurkiewicz [1980], [1985]; see also Ehlers [1975].
SFR About the singular case, see McConnell [1989]. See also Bifet [1993], de Meyer, Ford, and Miranda [1993]

3. Čech cohomology

SFR A recent introduction into Čech cohomology can be found in Gunning [1990].

4. Cohomology of invertible sheaves

HN The proof of Theorem 4.6 presented here gets along without so-called spectral sequences and is due to R. Lehmann (personal communication).
SFR Goresky and McPherson [1980], [1983] have presented a theory of intersection numbers ("intersection homology") which is applicable to algebraic varieties, with singularities. About the case of toric varieties, see Stanley [1987], Fischli and Yavin [1991], Yavin [1991], and Fieseler [1991]. See also section 3 of Ishida [1990].

5. Riemann–Roch–Hirzebruch theorem

HN, SFR The original proof of Theorem 5.5 in Hirzebruch [1962]. The general theory is thoroughly treated in Fulton [1984]; c.f. also Fulton [1992] for the toric case. A related calculation of $G(P)$ also in Brion [1988]; c.f. Cappell and Shaneson [1994]. About Chern classes in singular toric varieties, see Barthel, Brasselet, and Fieseler [1992]. A generalization of Theorem 5.5 to the quasi-smooth case in Pommersheim [1993], [1995a], [1995b]. About the further development, see Fulton [1993], Kantor and Khovanskij [1993]. A purely combinatorial version of the Riemann–Roch–Hirzebruch theorem can be found in Morelli [1993a].

References

Aguzzoli, S. and Mundici, D.

[1994] An algorithmic desingularization of three-dimensional toric varieties. Tohoku Math. J. 46, 557–572.

Alon, N. and Kalai, G.

[1985] A simple proof of the upper bound theorem. Europ. J. Combinatorics 6, 211–214.

Altmann, K.

[1991] Equisingular deformations below the Newton boundary. Compositio Math. 80, 257–283.

[1993] Deformations of affine toric varieties. Contrib. Alg. and Geom. 34, 119–150.

[1994] Computation of the vector space T^1 for affine toric varieties. J. Pure Appl. Alg. 95, 239–259.

[1995] Minkowski sums and homogeneous deformations of toric varieties. Tohoku Math. J. 47, 151–184.

Altshuler, A., Bokowski, J., and Steinberg, L.

[1980] The classification of simplicial 3-spheres with nine vertices into polytopes and nonpolytopes. Discrete Math. 31, 115–124.

Altshuler, A. and Steinberg, L.

[1984] Enumeration of the quasisimplicial 3-spheres and 4-polytopes with eight vertices. Pac. J. Math. Vol. 113, No. 2. 269–288.

Atiyah, M.F.

[1983] Angular momentum, convex polyhedra and algebraic geometry. Proc. Edinburgh Math. Soc. 26, 121–138 .

Ash, A., Mumford, D., Rapoport, M., and Tai, Y.

[1975] Smooth compactification of locally symmetric varieties. Brookline, Mass: Math. Sci. Press.

Audin, M.

[1991] The topology of torus actions on symplectic manifolds. Basel, Boston, Berlin: Birkhäuser.

Barnette, D.

[1970] Diagrams and Schlegel diagrams, combinatorial structures and their applications. New York: Gordon and Breach. pp. 1–4.

Barthel, G., Brasselet, J.-P., and Fieseler, K.H.

[1992] Classes de Chern des variétés toriques singulières. C. R. Acad. Sci. Paris Sér. I Math. 315, 187–192.

Barthel, G., Brasselet, J.-P., Fieseler, K.H., and Kaup, L.
[1996a] Diviseurs invariants et homomorphisme de Poincaré de variétés toriques complexes. To appear in Tohoku Math. J.
[1996b] Invariante divisoren und schnitthomologie von torischen varietäten. To appear in Proc. Workshop "Parameter Spaces" in the Banach Center Publications, Warsaw.

Barthel, G., Hirzebruch, F., and Höfer, T.
[1987] Geradenkonfigurationen und algebraische flächen. Aspekte der Math. D4 Vieweg.

Barvinok, A. I.
[1992] On equivariant generalization of Dehn-Sommerville equations. Europ. J. Combinatorics 13, 419–428.

Batyrev, V.V.
[1982a] Toroidal Fano 3-folds. Math. USSR Izv. 19, No. 1, 13–25.
[1982b] Boundedness of the degree of multidimensional toric Fano varieties. Bull. Moskow Univ. Math., Vol. 37, No. 1, pp. 22–27 .
[1991] On the classification of smooth projective toric varieties. Tohoku Math. J. 43, 569–585.
[1992] On the classification of toric Fano 4-folds. Preprint.

Batyrev, V.V.
[1993a] Variations of the mixed Hodge structure of affine hypersurfaces in algebraic tori. Duke Math. J. 69, 349–409.
[1993b] Quantum cohomology rings of toric manifolds. "Journées de géométrie algébrique d'Orsay", Astérisque 218, 9–34.
[1994] Dual polyhedra and mirror symmetry for Calabi–Yau hypersurfaces in toric varieties. J. Alg. Geom. 3, 493–535.

Batyrev, V.V. and Borisov, L.A.
[1994a] Dual cones and mirror symmetry for generalized Calabi–Yau manifolds. Preprint. To appear in "Mirror Symmetry II", (S.–T. Yau ed.,), 65–80.
[1994b] On Calabi–Yau complete intersections in toric varieties. To appear in Proc. of the Int. Conf. "Higher Dimensional Complex Variables" Trento.
[1995] Mirror duality and string–theoretic Hodge numbers. Preprint.

Batyrev, V.V. and Cox, D.
[1994] On the Hodge structure of projective hypersurfaces in toric varieties. Duke Math. J. 75, 293–338.

Batyrev, V.V. and Dais, D.I.
[1994] Strong McKay correspondence, string–theoretic Hodge numbers and mirror symmetry. MPI-preprint 94–115, to appear in Topology.

Batyrev, V.V. and Meln'nikov, D.A.
[1986] A theorem on nonextensibility of toric varieties. Bull. Moskow Univ. Math., Vol. 41, No. 3, pp. 20–24.

Batyrev, V.V. and van Straten, D.
[1995] Generalized hypergeometric functions and rational curves on Calabi–Yau complete intersections in toric varieties. Comm. Math. Phys. 168, 493–533.

Bayer, M. and Klapper, A.
[1991] A new index for polytopes. Discrete Comp. Geom. 6, 33–47.

Bayer, M. and Billera, L.
[1985] Generalized Dehn-Sommerville relation for polytopes, spheres and Eulerian partially ordered sets. Invent. Math. 79, 143–157.

Bernstein, D.N.

[1976] The number of roots of a system of equations. Funct. Anal. Appl. 10, 223–224.

Bertin, J. and Markushevich, D.

[1994] Singularités quotient non abéliennes de dimension three et variétés de Calabi–Yau. Math. Ann. 299, 105–116.

Betke, U.

[1992] Mixed volumes of polytopes. Archiv Math. 58, 388–391.

Betke, U. and McMullen, P.

[1985] Lattice Points in Lattice Polytopes. Mh. Math. 99, 253–265.

Bialynicki–Birula, A.

[1973] Some theorems on actions of algebraic groups. Ann. Math. 98, 480–497.

Bialynicki–Birula, A. and Sommese, A.J.

[1973] A conjecture about compact quotients by tori. Adv. Stud. Pure Math. 8, 59–68.

Bifet, E.

[1993] Cohomology, symmetry, and perfection. Publ. Mat. 36, 2A, 407–420.

Billera L.J. and Lee, C.W.

[1980] A proof of the sufficiency of McMullen's conditions for f-vectors of simplicial convex polytopes. J. Comb. Theory, Vol. 31, Series A, No. 3, 237–255.

Billera, L.J. and Sturmfels, B.

[1992] Fiber polytopes. Ann. Math. 135, 527–549.

[1994] Iterated fiber polytopes. Mathematika 41, 348–363.

Björner, A.

[1986] Face numbers of complexes and polytopes. Proc. Int. Cong. Math., Berkeley USA , 1408–1418.

Björner, A. and Kalai, G.

[1988] An extended Euler–Poincaré theorem. Acta Math. 161, 279–303.

[1989] On f-vectors and homology. Ann. NY Acad. Sci., 555, 63–80.

Björner, A., Las Vergnas, M., Sturmfels, B., White, N., and Ziegler, G.M.

[1993] Oriented matroids, Cambridge (GB): Cambridge U. Press.

Blaschke, W.

[1916] Kreis und Kugel. Leipzig: Veit.

Bögvad, R.

[1994] On the homogeneous ideal of a projective nonsingular toric variety. Preprint to appear in Tohoku Math. J.

Bokowski, J., Ewald, G., and Kleinschmidt, P.

[1984] On combinatorial and affine automorphisms of polytopes. Israel J. Math. 47, 123–130.

Bokowski, J. and Shemer, I.

[1987] Neighborly 6-polytopes with 10 vertices. Israel J. Math. 58, 103–124.

Bokowski, J. and Sturmfels, B.

[1989] Computational synthetic geometry. Lecture Notes in Math. 1355. Berlin: Springer.

Bonnesen, T., Fenchel, W.

[1934] Theorie der konvexen Körper. Ergebnisse der Math. Berlin, Heidelberg, New York, Tokyo: Springer.

Borel, A.
[1969] Linear algebraic groups. Math. Lecture Notes Series. New York, Amsterdam: W.A. Benjamin.

Borisov, A. A. and Borisov, L. A.
[1993] Singular toric Fano varieties. Russ. Acad. Sc. Sb. 75, 277–283.

Borisov, L.A.
[1993] Toward the mirror symmetry for Calabi–Yau complete intersections in Gorenstein toric Fano varieties. Preprint.

Bouvier, C. and Gonzalez–Sprinberg, G.
[1992] G-désingularisations de variétés toriques. C. R. Acad. Sci. Paris Sér. I Math. 315, 817–820.
[1995] Système générateur minimal, diviseurs essentiels et G–désingularisation de variétés toriques. Tohoku Math. J. 47, 125–149.

Brion, M.
[1988] Points entiers dans les polyèdres convexes. Ann. Sci. Ecole Norm. Sup., Serie 21, 653–663.
[1992] Polyèdres et réseaux. Enseig. Math. 38, 71–88.

Brodmann, M.
[1989] Algebraische Geometrie. Basel, Boston, Berlin: Birkhäuser.

Brøndsted, A.
[1983] An introduction to convex polytopes. Graduate Texts in Math. New York, Heidelberg, Berlin: Springer.

Bruggesser, H. and Mani, P.
[1971] Shellable decompositions of cells and spheres. Math. Scand. 29, 197–205.

Brylinski, J.L.
[1979] Décomposition simpliciale d'une réseau, invariante par un groupe fini d'automorphismes. C. R. Acad. Sci. Paris, 288, 137–139.
[1980] Eventails et variétés toriques. Lecture Notes in Math. 777. Springer, Berlin, Heidelberg, New York: 247–288.

Burago, Y. D. and Zalgaller, V.A.
[1988] Geometric Inequalities. Grundlehren der math. Wiss. 285. Berlin, Heidelberg, New York: Springer.

Busemann, H.
[1958] Convex Surfaces. New York: Interscience.

Cappell, S.E. and Shaneson, J.
[1994] Genera of algebraic varieties and counting of lattice points. Bull. AMS 30, 62–69.

Cattani, E., Cox, D., and Dickenstein, A.
[1995] Residues in toric varieties. Preprint.

Chan, C.
[1994] On Subdivisions of Simplicial Complexes: Characterizing local h-vectors. Discrete Comp. Geom. 11, 465–476.

Cox, D.
[1995a] The functor of a smooth toric variety. Tohoku Math. J. 47, 251–262.
[1995b] The homogeneous coordinate ring of a toric variety. J. Alg. Geom. 4, 17–50.
[1995c] Toric residues. To appear in Ark. Mat.

Coxeter, H.S.M.
[1962] The classification of zonohedra by means of projective diagrams. J. Math. Pure Appl. (9) 41, 137–156.
[1973] Regular polytopes. Corrected reprint (Dover, New York) of second edition, New York: Macmillam, 1963.

Crauder, B.
[1984] Two reduction theorems for threefold birational morphisms. Math. Ann. 269, 13–26.

Dais, D.I.
[1995] Enumerative combinatorics of invariants of certain complex threefolds with trivial canonical bundle. Bonner Math. Schriften, Bd. 279.

Dais, D.I., Henk, M., and Ziegler, G.M.
[1996] Constructive methods of crepant resolutions of three-dimensional canonical toric singularities. Algorithms Appl. Preprint.

Danaraj, G. and Klee, V.
[1974] Shellings of spheres and polytopes. Duke Math. J. 41, 443–451.
[1978] Which spheres are shellable? Ann. Discrete Math. 2, 33–52.

Danilov, V.I.
[1978] The geometry of toric varieties. Russian Math. Surv. 33, 97–154.
[1983] The birational geometry of toric three-folds. Math. USSR Izv. 21, 269–280.
[1991] de Rham complex on a toroidal variety. In: Algebraic geometry, Lecture Notes in Math. 1479. pp. 26–38. Berlin, Heidelberg, New York: Springer.

Danilov, V.I. and Khovanskii, A.G.
[1987] Newton polyhedra and an algorithm for computing Hodge–Deligne numbers. Math. USSR Izv. 29, 279–298.

Dantzig, G.B.
[1963] Linear programming and extensions. Princeton University Press, Princeton, NJ.

Demazure, M.
[1970] Sous-groupes algébriques de rang maximum du groupe de Cremona. Ann. Sci. Ecole Norm. Sup. (4) 3, 507–588.

De Meyer, F.R. and Ford, T.J.
[1993] On the Brauer group of toric varieties. Trans. Am. Math. Soc. 335, 559–577.

De Meyer, F.R., Ford, T.J., and Miranda, R.
[1993] The cohomological Brauer group of a toric variety. J. Alg. Geom. 2, 137–154.

Demin, I.V.
[1981] Fano threefolds representable in the form of line bundles. Math. USSR Izv. 17, No. 1, 219–226.

Dolgachev, I.V.
[1994] Mirror symmetry for lattice polarized $K3$-surfaces. Preprint.

Ebihara, M.
[1992] On unirationality of threefolds which contain toric surfaces with ample normal bundles. J. Fac. Sci. Univ. Tokyo Sec. IA Math. 39, 87–139.

Eckhoff, J.
[1979] Radon's theorem revisited. In "Contributions to Geometry", Proc. Geom. Sympos. Siegen (I. Tölke and J. Wills, eds.) Basel, Boston, Berlin, Birkhäuser, 164–185.

[1993] Helly, Radon, and Carathéodory type theorems. Handbook of convex geometry (P. Gruber and J. Wills, ed.) Elsevier Science Publishers B.V. Amsterdam, pp. 389–448.

Ehlers, F.
[1975] Eine Klasse komplexer Mannigfaltigkeiten und die Auflösung einiger isolierter Singularitäten. Math. Ann. 218, 127–156.

Ehrhart, E.
[1977] Polynomes arithmétiques et méthode des polyèdres en combinatoire. Basel, Boston, Berlin: Birkhäuser.

Eikelberg, M.
[1992] The picard group of a compact toric variety. Results Math. 22, 509–527.
[1993] Picard groups of compact toric varieties and combinatorial classes of Fans. Results in Math. 23, 251–293.

Evertz, S.
[1988] Zur klassifikation four-dimensionaler Fano varietäten. Diplomarbeit, Bochum.

Ewald, G.
[1977] Über stellare äquivalenz konvexer polytope. Resultate Math. (1), 54–60.
[1979] Über polytopale sphären mit symmetriegruppe. Archiv Math. 33, 492–495.
[1986a] Über stellare unterteilung von simplizialkomplexen. Arch. Math. 48, 153–158.
[1986b] Spherical complexes and nonprojective toric varieties. Discrete Comp. Geom. 1, 115–122.
[1987] Blowups of smooth toric 3–varieties. Abh. Math. Sem. Univ. Hamburg 57, 193–201.
[1988a] On the classification of Toric Fano varieties. Discrete Comp. Geom. 3, 49–54.
[1988b] An embedding theorem for smooth projective toric varieties. Aequationes Math. 35, 267–271.
[1988c] On the equality case in Alexandrov–Fenchel's inequality for convex bodies. Geometriae Dedicata 28, 213–220.
[1993] Algebraic geometry and convexity. Handbook of convex geometry (P. Gruber and J. Wills, eds.) Elsevier Science Publishers B.V. Amsterdam, 603–626.

Ewald, G., Kleinschmidt, P., and Schulz, C.
[1976] Kombinatorische klassifikation symmetrischer polytope. Abh. Math., Sem. U. Amsterdam Elsevier Science Publishers B.V. Hamburg 45, 191–206.

Ewald, G. and Schmeinck, A.
[1993] A representation of the Hirzebruch-Kleinschmidt varieties by Quadrics. Contrib. Alg. Geom. 34, 151–156.

Ewald, G. and Schulz, C.
[1988] 3–Sphären mit kleinen eckenvalenzen. Aequationes Math. 35, 277–280.
[1992] Non-Starshaped spheres. Archiv Math. 59, 412–416.

Ewald, G. and Shephard, G.C.
[1974] Stellar subdivisions of boundary complexes of convex polytopes. Math. Ann. 210, 7–16.

Ewald, G. and Spazier, B.
[1994] On the resolution of singularities in affine toric 3–varieties. Results in Math. 25, 234–241.

Ewald, G. and Teissier, B. (eds.)
[1996] Combinatorial convex geometry and toric varieties. To appear in Birkhäuser Verlag.

Ewald, G. and Tondorf, E.
[1994] A contribution to equality in Alexandrov–Fenchel's inequality. Geometriae Dedicata 50, 217–233.

Ewald, G. and Voss, K.
[1973] Konvexe Polytope mit Symmetriegruppe. Comm. Math. Helvetici, Vol 48. Fasc. 2, 137–150.

Ewald G. and Wessels, U.
[1991] On the ampleness of invertible sheaves in complete projective toric varieties. Results Math. 19, 275–278.

Fieseler, K.H.
[1991] Rational intersection cohomology of projective toric varieties. J. Reine Angew. Math. 413, 88–98.

Fine, J.
[1989] On varieties isomorphic in codimension one to torus embeddings. Duke Math. J. 58, 79–88.
[1993] Another proof of Sumihiro's theorem on torus embeddings. Duke Math. J. 69, 243–245.

Firey, W.J.
[1976] A functional characterization of certain mixed volumes. Israel J. Math., Vol. 24, No. 3–3.

Fischli, S. and Yavin, D.
[1994] Which 4–manifolds are toric varieties? Math. Z. 215, 179–185.

Ford, T.J.
[1995] Topological invariants of a fan associated to a toric variety. Preprint.

Fulton, W.
[1984] Intersection theory. Ergebnisse der Math. Berlin, Heidelberg, New York, Tokyo: Springer.
[1993] Introduction to toric varieties. Ann. Math. Studies 131, Princeton University Press. Princeton, NJ.
[1995] Algebraic Topology. A First Course. Graduate Texts in Math. New York, Berlin, Heidelberg: Springer.

Fulton, W., McPherson, R., Sottile, F., and Sturmfels, B.
[1995] Intersection theory on spherical varieties. J. Alg. Geom. 4, 181–193.

Fulton, W. and Sturmfels, B.
[1995] Intersection theory on toric varieties. To appear in Topology.

Gärtner, J.
[1996] Untersuchungen zum Gleichheitsfall der torischen Hodge–Ungleichung. Dissertation, Bochum.

Gelfand, I.M., Goresky R.M., McPherson, R.D., and Serganova, V.
[1987] Combinatorial geometries, convex polyhedra and Schubert cells. Adv. in Math. 63, 301–316.

Gelfand, I.M. and Serganova, V.V.
[1987] Combinatorial geometries and torus strata on homogeneous compact manifolds. Russian Math. Surveys 42, 133–168.

Gelfand, I.M., Kapranov, M.M., and Zelevinskij, A.V.
[1990] Newton polytopes of principal A-determinants. Soviet Math. Doklady 40, 278–281.
[1994] Discriminants, Resultants and Multidimensional Determinants. Theory & Applications (V. Kadison and I.M. Singer, ed.). Boston: Birkhäuser.

Gonzalez–Sprinberg, G. and Lejeune-Jalabert, M.
[1991] Modèles canoniques plongés I. Kodai Math. J. 14, 194–209.

Goresky, M. and MacPherson, R.
[1980] Intersection homology theory. Topology Vol. 19, 135–162.

[1983] Intersection homology II. Invent. Math. 71, 77–129.

Gretenkort, J., Kleinschmidt, P., and Sturmfels, B.
[1990] On the existence of certain smooth toric varieties. Discrete Comp. Geom. 5, 255–262.

Gritzmann, P. and Wills., J.
[1993] Lattice points. Handbook of convex geometry, Section 3.7 (P. Gruber and J. Wills eds.), Elsevier Science Publishers B.V. Amsterdam.

Gruber, P. and Wills., J.
[1993] Handbook of convex geometry. Vol. A and B, Elsevier Science Publishers B.V. Amsterdam.

Grünbaum, B.
[1967] Convex polytopes. London, New York, Sydney: Interscience Publ.
[1970] Polytopes, graphs, and complexes. Bull. Amer. Math. Soc. 76, 1131–1201.
[1972] Arrangements and spreads. Amer. Math. Soc. Regional Conf. Series in Math. 10, Rhode Island.

Guest, M.A.
[1995] The topology of the space of rational curves on a toric variety. Acta Math. 174, 119–145.

Gunning, R.C.
[1990] Introduction to holomorphic functions of several complex variables III: Homological theory. Belmont (Cal.): Wadsworth and Brooks/Cole.

Hadwiger, H.
[1955] Altes und neues über konvexe körper. Basel, Boston, Berlin, Birkhäuser.

Halmos, P.R.
[1958] Finite-dimensional vector spaces. London, Toronto: Van Nostrand.

Hamm, H.
[1990] Affine-complete intersections and Newton polyhedra. Münster, Manuscript.

Hartshorne, R.
[1977] Algebraic geometry. Graduate Texts in Math. New York, Heidelberg, Berlin: Springer.

Henk, M.
[1995] A Note on the height of Hilbert bases. TU Berlin. Preprint.

Hibi, T.
[1991a] Ehrhart polynomials of convex polytopes, h-vectors of simplicial complexes, and nonsingular projective toric varieties. DIMACS Series in Discrete Mathematics and Theoretical Computer Science, Vol. 6., A.M.S., 165–177, Princeton, NJ.
[1991b] A lower bound theorem for Erhart polynomials of convex polytopes. Hokkaido University Preprint Series in Math. 117.

Hirzebruch, F.
[1951] Über eine Klasse von einfach–zusammenhängenden komplexen Mannigfaltigkeiten. Math. Ann. 124, 77–86.
[1953] Über vierdimensionale Riemannsche Flächen mehrdeutiger analytischer Funktionen von zwei komplexen Veränderlichen. Math. Ann. 126, 1–22.
[1962] Neue topologische Methoden in der algebraischen Geometrie. Ergebnisse der Math. Berlin, Heidelberg, New York, Tokyo: Springer.

Ishida, M.N.
[1980] Torus embeddings and dualizing complexes. Tohoku Math. J. 32, 11–146.
[1990] Polyhedral Laurent series and Brion's equalities. Int. J. Math. 1, No. 3, 251–265.

[1989] Cusp singularities given by reflections of stellable cones. Manuscript.
[1991] The algebraic surface defined by a sum of four monomials. Comment. Math. University St. Paul 40, 39–51.

Ishida, M.N. and Iwashita, N.
[1986] Canonical cyclic quotient singularities of dimension three. Adv. Stud. Pure Math. 8, 135–151.

Ishii, S.
[1995] The canonical modifications by weighted blowups. Preprint.

Ito, Y.
[1994] Crepant resolution of trihedral singularities. Proc. Japan Acad., Ser. A, 70, 131–136.
[1995a] Crepant resolution of trihedral singularities and the orbifold Euler characteristic. Intern. J. Math. 6, 33–43.
[1995b] Gorenstein quotient singularities of monomial type in dimension three. To appear in J. Math. Sci., University Tokyo.

Ito, Y., Reid, M.
[1994] The McKay correspondence for finite subgroups of $SL(3, \mathbb{C})$. University of Tokyo, preprint. 94–66. To appear in Proc. Int. Conf. "Higher dimensional complex variables" Trento.

Iversen, B.
[1972] A fixed point formula for action or tori on algebraic varieties. Invent. Math. 16, 229–236.

Jaczewski, K.
[1994] Generalized Euler sequence and toric varieties. Contemp. Math. 162, 227–247.

Jaglom, I.M. and Boltjanskij, W.G.
[1956] Konvexe Figuren. Berlin: VEB Deutscher Verl. d. Wiss.

Jurkiewicz, J.
[1980] Chow ring of projective non-singular torus embeddings. Colloquium Mathematicum XLIII. Fasc. 2, 261–270.

Jurkiewicz, J.
[1985] Torus embeddings, polyhedra, k^*-actions and homology. Disertationes mathematicae CCXXXVI, Warsaw.

Kaneyama, T.
[1975] On equivariant vector bundles on an almost homogeneous variety. Nagoya Math. J. 57, 65–86.
[1988] Torus-equivalent vector bundles on projective spaces. Nagoya Math. J. 111, 25–40.

Kantor, J.-M. and Khovanskii, A.G.
[1992] Integral points in convex polyhedra, combinatorial Riemann-Roch theorem and generalized Euler–MacLaurin formula. Inst. des Hautes Etudes Sci. 92/37.

Kapranow, M.M, Sturmfels, B., and Zelevinsky, A.V.
[1991] Quotients of toric varieties. Math. Ann. 290, 643–655.
[1992] Chow polytopes and general resultants. Duke Math. J. 67, 189–218.

Kaup, B. and Kaup, L.
[1983] Holomorphic functions of several variables. An introduction to the fundamental theory. Berlin: de Gruyter.

Kempf, G., Knudsen, F., Mumford, D., and Saint-Donat, B.
[1973] Toroidal embeddings I. Lecture notes in math. 339. Berlin, Heidelberg, New York: Springer.

Khovanskii, A.G.

[1977] Newton polyhedra and toroidal varieties. Funct. Anal. Appl. 11, 289–296.

[1978] Newton polyhedra and the genus of complete intersections. Funct. Anal. Appl. 12, 51–61.

[1986] Hyperplane sections of polyhedra, toroidal manifolds, and discrete groups in Lobachevskii space. Ego Pribzheniya 20, 50–61. New York: Plenum Publishing.

[1988] Algebra and mixed volumes. Addendum in Burago—Zalgaller [1988], 182–320.

Kind, B. and Kleinschmidt, P.

[1979] Schälbare Cohen–Macauley–Komplexe und ihre Parametrisierung. Math. Z. 167, 173–179.

Klee, V. and Kleinschmidt, P.

[1987] The d-step conjecture and its relatives. Math. Operations Res. 12, 718–755.

[1988] A classification of toric varieties with few generators. Aequationes Math. 35, 254–266.

[1995] Polytopal complexes and their relatives. In Handbook of Combinatorics (R. Graham, M. Grötschel, and L. Lovász, ed.), Amsterdam: Holland/Elsevier.

Kleinschmidt, P.

[1976a] Sphären mit wenig ecken. Geometriae Dedicata 5, 307–320.

[1976b] On facets with non-arbitrary shapes. Pac. J. Math., 65, 97–101.

[1978] Stellare Abänderungen und Schälbarkeit von Komplexen und Polytopen. J. Geom 11, 161–176.

[1988] A classification of toric varieties with few generators. Aequationes Math. 35, 254–266.

Kleinschmidt, P. and Kind, B.

[1979] Schälbare Cohen–Macauley–Komplexe und ihre Parametrisierung. Math. Z. 167, 173–179.

Kleinschmidt, P. and Lee, C.W.

[1984] On k-stacked polytopes. Discrete Math. 48, 125–127.

Kleinschmidt, P. and Sturmfels, B.

[1991] Smooth toric varieties with small Picard number are projective. Topology 30, 289–299.

Klyachko, A.A.

[1989] Toric bundles and problems of linear algebra. Funct. Anal. Appl. 23, 135–137.

[1990] Equivariant bundles on toral varieties. Math. USSR Izvestiya 35, No. 2, 337–375.

Klyachko, A.A. and Voskresenskii, V.E.

[1985] Toroidal Fano varieties and root systems. Math. USSR Izv. 24, No. 2, 221–244.

Kobayashi, M.

[1994] Duality of weights, mirror symmetry, and Arnold's strange duality. Preprint.

Koelman, R.J.

[1993a] A criterion for the ideal of a projectively embedded toric surface to be generated by quadrics. Contributions Alg. Geom. 34, 57–62.

[1993b] Generators for the ideal of a projectively embedded toric surface. Tohoku Math. J. 45, 385–392.

Kollar, J.

[1987] The structure of algebraic threefolds–an introduction to Mori's program. Bull. AMS 17, 211–273.

Kouchnirenko, A.G.

[1976] Polyèdres de Newton et nombres de Milnor. Invent. Math. 32, 1–31.

Laterveer, R.

[1994] Linear systems on toric varieties. Preprint to appear in Tohoku Math. J.

Lehmann, R.
[1989] Complex–symmetric spaces. Ann. Inst. Fourier, Grenoble, 39, 373–416.
[1991] Singular complex–symmetric torus embeddings. Archiv d. Math. 56, 68–80.

Lee, C.W.
[1990] Some results on convex polytopes. Contemp. Math. 114, 3–19.
[1992] Winding numbers and the generalized lower-bound conjecture. Center for Discrete Mathematics and Theoretical Computer Science (DIMACS) Series in Discrete Math. 6, 209–219, Princeton, NJ.

Leichtweiß, K.
[1980] Konvexe Mengen. Berlin: Springer und VEB Verlag d. Wiss.

Lickorish, W.B.R.
[1991] Unshellable triangulations of spheres. Europ. J. Combinatorics 12, 527–530.

Liu, J., Trotter, L.E., and Ziegler, G.M.
[1993] On the Height of the Minimal Hilbert Basis. Results in Math. 23, 374–376.

Lloyd, E.K.
[1970] The number of d-polytopes with $d + 3$ vertices. Mathematica 17, 120–132.

MacDonald, I.G.
[1971] Polynomials associated with finite cell-complexes. J. London Math. Soc. 2, Vol. 4, 181–192.

Mani, P.
[1972] Spheres with few vertices. J. Comb. Theory (A) 13, 346–352.

Markushevich, D.G., Olshanetsky, M.A., and Perelomov, A.M.
[1987] Description of a class of superstring compactifications related to semi-simple Lie algebras. Comm. in Math. Phys. 11, 247–274.

Massey, W.S.
[1967] Algebraic Topology. An Introduction. Graduate Texts in Math. New York, Heidelberg, Berlin: Springer.

McConnell, M.
[1989] The rational homology of toric varieties is not a combinatorial invariant. Proc. AMS 4, Vol. 105, 986–991.

McMullen, P.
[1977] Valuations and Euler–type relations on certain classes of convex polytopes. Proc. London Math. Soc. 35, 113–135.
[1971] On zonotopes. Trans. Ann. Math. Soc. 159, 91–109.
[1973] Representations of polytopes and polyhedral sets. Geometriae Dedicata 2, 83–99.
[1979] Transforms, Diagrams, and Representations. In: Proc. Geometry Symposium in Siegen 1978; (J. Tölke and J. Wills, ed.) Basel, Boston, Berlin: Birkhäuser, pp. 92–130.
[1989] The polytope algebra. Adv. in Math. 87, 76–130.
[1993a] On simple polytopes. Invent. Math. 113, 419–444.
[1993b] Separation in the polytope algebra. Contributions Alg. Geom. 34, 15–30.
[1994] Weights on polytopes. To appear in Discrete Comp. Geom.
[1995] Polytope algebras, tensor weights and piecewise polynomials. Intern. Conf. Intuitive Geom. Budapest. To appear.

McMullen, P. and Shephard, G.C.
[1971] Convex polytopes and the upper bound conjecture. Cambridge (GB): Cambridge University Press.

Meyer, W.
[1974] Indecomposable polytopes. Trans. Am. Math. Soc. 190, 77–86.

Milnor, J.W.
[1969] Morse theory. Princeton University Press, Princeton, NJ.

Miranda, R.
[1985] Gorenstein toric threefolds with isolated singularities and cyclic divisor class group. Rocky Mountain J. Math. 39–45.

Miyake, K. and Oda, T.
[1975] Almost homogeneous algebraic varieties under algebraic torus action. Manifolds: Tokyo 1973 (A. Hattori, ed.) University of Tokyo Press 373–381.

Mnëv, N.E.
[1988] The universality theorems on the classification problem of configuration varieties and convex polytope varieties. In Topology and Geometry, Lecture Notes in Math. 1346. Viro, O.Ya., ed. Berlin, Heidelberg, New York: Springer.

Morales, M.
[1984] Polyèdre de Newton et genre géométrique d'une singularité intersection complète. Bull. Soc. Math. France 112, 325–341.

Morelli, R.
[1993a] A theory of polyhedra. Adv. Math. 97, 1–73.
[1993b] Translation scissors congruence. Adv. Math. 100, 1–27.
[1993c] The K-theory of a toric variety. Adv. Math. 100, 154–182.
[1993d] Pick's theorem and the Todd class of a toric variety. Adv. Math. 100, 183–231.
[1994] The birational geometry of toric varieties. Preprint.

Morrison, D.R.
[1985] Canonical quotient singularities in dimension three. Proc. AMS 93, 393–396.
[1993] Mirror symmetry and rational curves on quintic 3–folds: A guide for mathematicians. J. AMS 6, 223–247.

Morrison, D.R. and Stevens, G.
[1984] Terminal quotient singularities in dimension three and four. Proc. AMS 90, 15–20.

Musson, I.M.
[1987] Rings of differential operators on invariant rings of tori. Trans. Am. Math. Soc. 303, 805–827.
[1994] Differential operators on toric varieties. J. Pure Appl. Algebra 95, 303–315.

Nakajima, H.
[1986] Affine torus embeddings which are complete intersections. Tohoku Math. J. 38, 85–98 .

Oda, T.
[1978] Torus embeddings and applications. Bombay: Tata Institute.
[1988] Convex bodies and algebraic geometry. An introduction to the theory of toric varieties. Ergeb. d. Math. Berlin, Heidelberg, New York, London, Paris, Tokyo: Springer.
[1991a] Geometry of toric varieties. In: Proc. Hyderabad conf. algebraic groups. Ed. by S: Ramanan, C. Musili, and N.M. Kumar, Madras: Manoj Prakashan, pp. 407–440.
[1991b] Simple convex polytopes and the strong Lefschetz theorem. J. Pure Appl. Algebra 71, 265–286.
[1993] The algebraic De Rham theorem for toric varieties. Tohoku Math. J. 45, 231–247.

Oda, T. and Park, H.S.
[1991] Linear Gale transforms and Gelfand–Kapranov–Zelevinskij decompositions. Tohoku Math. J. 43, No. 3, 375–399 .

Oka, M.

[1979] On the bifurcation for the multiplicity and topology of the Newton boundary. J. Math. Soc. Japan 31, 435–450.

[1980] On the topology of the Newton boundary II. J. Math. Soc. Japan 32, 65–92.

[1982] On the topology of the Newton boundary III. J. Math. Soc. Japan 34, 541–549.

[1983] On the stability of the Newton boundary. Proc. Symp. Pure Math. 40, Part 2, 259–268.

[1984] On the resolution of two-dimensional singularities. Proc. Japan Acad., Ser. A, 60, 174–177.

[1986a] On the resolution of hypersurface singularities. Adv. Stud. Pure Math. 8, 405–436.

[1986b] On the resolution of three dimensional Brieskorn singularities. Adv. Stud. Pure Math. 8, 437–460.

[1988] Examples of algebraic surfaces, A. fete in topology, Boston, San Diego, New York: Academdic Press, 355–363.

[1989] On the simultaneous resolution of a negligble truncation of the Newton boundary. Contemp. Math. AMS 90, 199–210.

[1993a] Finiteness of fundamental group of compact convex integral polyhedra. Kodai Math. J. 16, 181–195.

[1993b] Nondegenerate complete intersection singularities. Book in preparation.

Pachner, U.

[1987] Diagonalen in Simplizialkomplexen. Geometriae Dedicata 24, 1–28.

[1991] P.L. homeomorphic manifolds are equivalent by elementary shellings. Europ. J. Combinatorics 12, 129–145.

Park, H.S.

[1992] Algebraic cycles on toric varieties. Ph.D. Thesis, Tokyo University.

[1993] The Chow rings and GKZ-decompositions for Q-factorial toric varieties. Tohoku Math. J. 45, 109–140.

Povyanne, N.

[1992] Une résolution en singularités toriques simpliciales des singularités quotient de dimension trois. Ann. Fac. Sci. Toulouse Math. (6), 1, 363–398.

Pommersheim, J.E.

[1993] Toric varieties, lattice points and Dedekind sums. Math. Ann. 295, 1–24.

[1995a] Products of cycles and the Todd class of a toric variety. Preprint.

[1995b] On a formula of R. Morelli. Preprint.

Rademacher, H.

[1964] Lectures on elementary number theory. New York: Blaisdell Publishing Company.

Rataijski, J.

[1993] On the Betti numbers of the real part of a three-dimensional torus embedding. Colloq. Math. 64, 59–64.

Reid, M.

[1980] Canonical threefolds. Journée de Géométrie Algébrique d'Angers (A. Beauville, ed.), Alphen aan den Rijn; Sijthoff and Noordhoff, 273–310.

[1983a] Decomposition of toric morphisms, Arithmetic and geometry, II. Progress in Math. 36, 395–418.

[1983b] Minimal models of canonical 3-folds. Adv. Stud. in Pure Math. 1, 131–180.

[1987] Young person's guide to canonical singularities. In "Algebraic Geometry, Bowdoin 1985", (S.J. Bloch, ed.), Proc. Symp. in Pure Math. 46, Part 1, AMS, 345–416.

Reiner, V. and Ziegler, G.

[1994] Coxeter–associahedra. Mathematika 41, 364–393.

Rentzsch, H.G.
[1991] Divisoren und differentialformen auf torischen varietäten, dissertation, Bochum.

Richter–Gebert, J.
[1995] Realization spaces of 4–polytopes are universal. Berlin: Habilitationsschrift TU.

Richter–Gebert, J., Ziegler, G.
[1995] Realization spaces of 4–polytopes are universal. Bull. Am. Math. Soc. 32, 403–412.

Roan, S.–S.
[1989] The generalization of Kummer surfaces. J. Diff. Geometry 30, 523–537.
[1991] The mirror of Calabi-Yau orbifold. Int. J. Math. 2, 439–455.
[1994a] On $c_1=0$ resolutions of quotient singularities. Int. J. Math. 5, 523–536.
[1994b] Minimal resolutions of Gorenstein threefolds in dimension three. Inst. Math. Acad. Sinica, Preprint R940606-1, to appear in Topology.

Roan, S.–S. and Yau, S.–T.
[1987] On Ricci-flat 3-folds. Acta Math. Sinica, New Ser. 3, 256–288.

Rourke, C.P. and Sanderson, B.J.
[1972] Introduction to piecewise linear topology. Ergebnisse der Math. 69, Berlin, Heidelberg, New York, Tokyo: Springer.

Sangwine–Yager, J.R.
[1993] Mixed volumes. Handbook of convex geometry Section 1.2. (P. Gruber and J. Wills, ed.) Elsevier Science Publishers B.V. Amsterdam.

Sarkaria, K.S.
[1987] Heawood inequalities. J. Comb. Theory 46, No. 1, 50–78.
[1991] A generalized van Kampen–Flores theorem. Proc. AMS 111, No. 2, 559–565.

Schlegel, V.
[1983] Theorie der homogen zusammengesetzten Raumgebilde. Nova Acta Leopold. Carol. 44, 343–459.

Schneider, R.
[1985] On the Aleksandrov–Fenchel inequality. Ann. of the NJ Acad. of Sci. 440. Discrete geom. and convexity, 132–141 .
[1988] On the Alexandrov–Fenchel inequality involving zonoids. Geometriae Dedicata 27, 113–126.
[1990] On the Aleksandrov–Fenchel inequality for convex bodies, I. Results in Math. 17, 287–295.
[1993] Convex Bodies. The Brunn-Minkowski theory. Encycl. Math. 44, Cambridge (GB): Cambridge University Press.

Schulz, Ch.
[1979] An invertible 3–Diagram with 8 vertices. Discrete Math. 28, 201–205 .
[1985] Dual pairs of non-polytopal diagrams and spheres. Discrete Math. 55, 65–72.

Shafarevich, I.R.
[1994] Basic algebraic geometry, 2nd ed. Berlin, Heidelberg, New York: Springer.

Shephard, G.C.
[1960] Inequalities between mixed volumes of convex sets. Mathematika 7, 125–138.
[1971a] Diagrams for positive bases. J. London Math. Soc. 2, No. 4, 165–175.
[1971b] Spherical complexes and radial projections of polytopes. Israel J. Math. 9, No. 2 257–262.
[1972] Sections and projections of convex polytopes. Mathematika 19, 144–162.

Smilansky, Z.
[1986a] Decomposability of polytopes and polyhedra. Dissertation. Hebrew University Jerusalem.

[1986b] An indecomposable polytope all of whose facets are decomposable. Mathematica 33, 192–196.

Spanier, E.
[1966] Algebraic topology. New York: McGraw–Hill.

Springer, T.A.
[1981] Linear algebraic groups. Progress in Math. Vol. 9. Basel, Boston, Berlin, Birkhäuser.

Stanley, R.P.
[1975] The upper bound conjecture and Cohen-Macaulay rings. Studies in Applied Math. Vol. 54, 135–142.
[1978] Hilbert functions of graded algebras. Adv. in Math. 28, 57–81.
[1980] The number of faces of a simplicial convex polytope. Adv. in Math. 35, 236–238.
[1981] Two combinatorial applications of the Aleksandrov–Fenchel inequalities. J. Comb. Theory 31, 56–65.
[1983] Combinatorics and commutative algebra. Progress in math. 41. Basel, Boston, Berlin: Birkhäuser.
[1986] Enumerative combinatorics. Monterey: Wadsworth and Brooks, Cole Math. Series, Vol. I.
[1987] Generalized H-vectors, intersection cohomology of toric varieties, and related results. In Comm. Algebra and Combinatorics, Adv. Stud. Pure Math. 11, 187–213.

Stembridge, J.R.
[1994] Some permutation representations of Weyl groups associated with the cohomology of toric varieties. Adv. Math. 106, 244–301.

Stoer, I. and Witzgall, C.
[1970] Convexity and optimization in finite dimension I. Grundlehren der math. wiss. 163. Berlin, Heidelberg, New York: Springer.

Stöcker, R. and Zieschang, H.
[1988] Algebraische Topologie. Stuttgart: Teubner.

Sturmfels, B.
[1988] Some applications of affine Gale diagrams to polytopes with few vertices. SIAM J. Discr. Math. 1, 121–133.
[1991] Fiber polytopes: a brief overview. In: "Special Differential Equations" (M. Yoshida, ed.). Kyushu University Fukuoka. pp. 117–124.

Sturmfels, B.
[1992] Asymptotic analysis of toric ideals. Memoirs of the Faculty of Science, Kyushu University, Ser. A Mathematics 46, 217–228.
[1994] On the Newton polytope of the resultant. J. of Alg. Combinatorics 3, 207–236.

Sumihiro
[1974] Equivariant completion I. J. Math. Kyoto University. 14, 1–28.

Teicher, M.
[1987] On toroidal embeddings of 3-folds. Israel J. Math. 57, 49–67.

Teissier, B.
[1979] Géométrie algébrique - Du theorème de l'index de Hodge aux inégalités isopérimétriques. C. R. Acad. Sci. Paris 288, 287–289.
[1980] Variétés toriques et polytopes. Séminaire Bourbakie vol. 1980/81, in: Lecture Notes in Math., Berlin, Heidelberg, New York: Springer 71–84.
[1982] Bonnessen-type inequalities in algebraic geometry I: Introduction to the problem. Seminar on Differential Geometry. Princeton, NJ. 85–105.

[1984] Monômes, volumes et multiplicités. Introduction à la théorie des singularités II. Travaux en Cours 37, 127–141.

Trung, N.V. and Hoa, L.T.
[1986] Affine semigroups and Cohen Macaulay rings generated by monomials. Trans. Am. Math. Soc. 289, No. 1, 145–167.

Tsuchihashi H.
[1991a] Hypersurface sections of toric singularities. Kodai Math. J. 14, 210–221.
[1991b] Simple $K3$ singularities which are hypersurface sections of toric singularities. Publ. RIMS Kyoto University. 27, 783–799.

van der Waerden, B.L.
[1967] Algebra II. Heidelberg: Springer.

Wagner, H.
[1995] Gewichtete projektive Räume und reflexive Polytope. Diplomarbeit. Bochum.

Wessels, U.
[1989] Die sätze von White und Mordell über kritische Gitter von Polytopen in den Dimensionen 4 und 5. Diplomarbeit, Bochum.
[1993] Kombinatorisch-geometrische Kennzeichnung und Berechnung der Schnittzahlen von Cartier-Divisoren in kompakten torischen Varietäten. Dissertation, Ruhr-University Bochum.

White, G.K.
[1964] Lattice tetrahedra. Can. J. Math. 16, 389–396.

Wills, M.
[1989] Nullstellenverteilung zweier konvexgeometrischer Polynome. University Gesamthochschule Siegen. Preprint.

Winter, D.J.
[1974] The structure of fields. Graduate Texts in Math. New York: Springer.

Włodarczyk, J.
[1992] Oda's conjecture for n-dimensional toric varieties. Manuscript.
[1993] Embeddings into toric varieties and prevarieties. J. Alg. Geom. 2, 705–726.

Yavin, D.
[1991] On the intersection homology with twisted coefficients of toric varieties and the homology with twisted coefficients of the torus. Manuscript.

Ziegler, G.
[1992] Combinatorial models for subspace arrangements. Berlin: Habilitationsschrift TU.
[1995] Lectures on polytopes. Graduate Texts in Math. New York: Springer.

List of Symbols

$\mathrm{Div}_C^T\, X_\Sigma$	group of T-invariant Cartier divisors 272
$\mathrm{Div}_C\, X_\Sigma$	group of Cartier divisors 272
$\mathrm{Div}_P^T\, X_\Sigma$	group of principal T-invariant Cartier divisors 272
$\mathrm{Div}_P\, X_\Sigma$	group of principal Cartier divisors 272
∂_k	boundary operation 308
d_K	distance function 21
δ_k	coboundary operation 311
E_F	orthogonal complement of aff F 124
$\mathrm{Elb}\, X_\Sigma$	group of equivariant line bundles 286
F^*	polar face 37
f	moment map 298
$\mathcal{F}$	sheaf of rational functions 260
$f_{(m)}$	Morse function 301
$f(P)$	f-vector 35
$\mathcal{F}(U)$	ring of rational functions in U 260
$[F, P]$	set of faces $F' \supset F$ of P 40
$\mathcal{F} \otimes \mathcal{G}$	tensor product of sheaves 270
$F \cdot \mathcal{C}'$	join of cell and cell complex 68
$F \cdot F'$	join of cells 68
g	modified moment map 298
$\tilde{\mathcal{G}}$	polytope group 168
Γ_M^Q	M-shadow boundary 122
$\Gamma^+(f)$	set $\{(x, \xi) \mid x \in \mathbb{R}^n,\ f(x) \le \xi\} \subset \mathbb{R}^{n+1}$ 20
$G(P)$	number of lattice points in P 135
$\mathcal{H}_k$	Hirzebruch surface 228
$H_k(X)$	kth homology group 310
$H^k(X; G)$	kth cohomology group 311
$H_k(X, Y)$	kth relative homology group 310
$\check{H}(U, \mathcal{F})$	Čech cohomology group 319
$H^\bullet(X_\Sigma; \mathbb{Z})$	Cohomology ring 314
h_K	support function 18
$H_K(u)$	supporting hyperplane 19
int M	interior 6
$\mathfrak{i}_Z$	vanishing ideal 201
K^*	polar body 24
$K + L$	(Minkowski) sum 103
$K \oplus L$	direct sum 105
$-K$	anticanonical divisor 303
$K \circ L$	conv$(K \cup L)$ 105
$\bar{L}$	toric morphism 243
$\mathcal{L}$	line bundle 284
λK	multiple 104
$\mathrm{Lb}\, X_\Sigma$	group of line bundles 285
$\mathcal{L}_D$	sheaf determined by D 266
$\mathrm{LF}(\Sigma)$	group of linear functions 283

Index

Graduate Texts in Mathematics

continued from page ii

119 ROTMAN. An Introduction to Algebraic Topology.
120 ZIEMER. Weakly Differentiable Functions: Sobolev Spaces and Functions of Bounded Variation.
121 LANG. Cyclotomic Fields I and II. Combined 2nd ed.
122 REMMERT. Theory of Complex Functions. *Readings in Mathematics*
123 EBBINGHAUS/HERMES et al. Numbers. *Readings in Mathematics*
124 DUBROVIN/FOMENKO/NOVIKOV. Modern Geometry—Methods and Applications. Part III.
125 BERENSTEIN/GAY. Complex Variables: An Introduction.
126 BOREL. Linear Algebraic Groups.
127 MASSEY. A Basic Course in Algebraic Topology.
128 RAUCH. Partial Differential Equations.
129 FULTON/HARRIS. Representation Theory: A First Course. *Readings in Mathematics*
130 DODSON/POSTON. Tensor Geometry.
131 LAM. A First Course in Noncommutative Rings.
132 BEARDON. Iteration of Rational Functions.
133 HARRIS. Algebraic Geometry: A First Course.
134 ROMAN. Coding and Information Theory.
135 ROMAN. Advanced Linear Algebra.
136 ADKINS/WEINTRAUB. Algebra: An Approach via Module Theory.
137 AXLER/BOURDON/RAMEY. Harmonic Function Theory.
138 COHEN. A Course in Computational Algebraic Number Theory.
139 BREDON. Topology and Geometry.
140 AUBIN. Optima and Equilibria. An Introduction to Nonlinear Analysis.
141 BECKER/WEISPFENNING/KREDEL. Gröbner Bases. A Computational Approach to Commutative Algebra.
142 LANG. Real and Functional Analysis. 3rd ed.
143 DOOB. Measure Theory.
144 DENNIS/FARB. Noncommutative Algebra.
145 VICK. Homology Theory. An Introduction to Algebraic Topology. 2nd ed.
146 BRIDGES. Computability: A Mathematical Sketchbook.
147 ROSENBERG. Algebraic K-Theory and Its Applications.
148 ROTMAN. An Introduction to the Theory of Groups. 4th ed.
149 RATCLIFFE. Foundations of Hyperbolic Manifolds.
150 EISENBUD. Commutative Algebra with a View Toward Algebraic Geometry.
151 SILVERMAN. Advanced Topics in the Arithmetic of Elliptic Curves.
152 ZIEGLER. Lectures on Polytopes.
153 FULTON. Algebraic Topology: A First Course.
154 BROWN/PEARCY. An Introduction to Analysis.
155 KASSEL. Quantum Groups.
156 KECHRIS. Classical Descriptive Set Theory.
157 MALLIAVIN. Integration and Probability.
158 ROMAN. Field Theory.
159 CONWAY. Functions of One Complex Variable II.
160 LANG. Differential and Riemannian Manifolds.
161 BORWEIN/ERDÉLYI. Polynomials and Polynomial Inequalities.
162 ALPERIN/BELL. Groups and Representations.
163 DIXON/MORTIMER. Permutation Groups.
164 NATHANSON. Additive Number Theory: The Classical Bases.
165 NATHANSON. Additive Number Theory: Inverse Problems and the Geometry of Sumsets.
166 SHARPE. Differential Geometry: Cartan's Generalization of Klein's Erlangen Program.
167 MORANDI. Field and Galois Theory.
168 EWALD. Combinatorial Convexity and Algebraic Geometry.